Lecture Notes in Computer S

T0230212

Commenced Publication in 1973
Founding and Former Series Editors:
Gerhard Goos, Juris Hartmanis, and Jan van Leeuwen

Tiziana Calamoneri   Irene Finocchi
Giuseppe F. Italiano (Eds.)

Algorithms
and Complexity

6th Italian Conference, CIAC 2006
Rome, Italy, May 29-31, 2006
Proceedings

 Springer

Volume Editors

Tiziana Calamoneri
Irene Finocchi
Dipartimento di Informatica
Università degli Studi di Roma "La Sapienza"
Via Salaria 113, 00198 Roma, Italy
E-mail: {calamo,finocchi}@di.uniroma1.it

Giuseppe F. Italiano
Dipartimento di Informatica, Sistemi e Produzione
Università di Roma "Tor Vergata"
Via del Politecnico 1, 00133 Roma, Italy
E-mail: italiano@disp.uniroma2.it

Library of Congress Control Number: 2006925893

CR Subject Classification (1998): F.2, F.1, E.1, I.3.5, G.2

LNCS Sublibrary: SL 1 – Theoretical Computer Science and General Issues

ISSN   0302-9743
ISBN-10   3-540-34375-X Springer Berlin Heidelberg New York
ISBN-13   978-3-540-34375-2 Springer Berlin Heidelberg New York

Springer is a part of Springer Science+Business Media

springer.com

© Springer-Verlag Berlin Heidelberg 2006
Printed in Germany

Typesetting: Camera-ready by author, data conversion by Scientific Publishing Services, Chennai, India
Printed on acid-free paper   SPIN: 11758471   06/3142   5 4 3 2 1 0

Preface

The 6th International Conference on Algorithms and Complexity (CIAC 2006) was held in Rome, Italy during May 29–31, 2006. These proceedings contain all contributed papers presented at CIAC 2006, together with the invited lectures delivered at the conference. The Program Committee consisted of:

- Nicola Galesi, Univ. of Rome "La Sapienza", Italy
- John Iacono, Brooklyn Polytechnic, USA
- Giuseppe F. Italiano (Chair), Univ. of Rome "Tor Vergata", Italy
- Pascal Koiran, ENS Lyon, France
- Jaroslav Nešetřil, Charles University, Czech Republic
- Sotiris Nikoletseas, CTI and Univ. of Patras, Greece
- Stephan Olariu, Old Dominion Univ., USA
- Anna Ostlin Pagh, ITU, Denmark
- Andrzej Pelc, Université du Québec en Outaouais, Canada
- Peter Sanders, Universitaet Karlsruhe, Germany
- Bruno Simeone, Univ. of Rome "La Sapienza", Italy
- Uri Zwick, Tel-Aviv Univ., Israel

In response to a call for papers, the Program Committee received 80 submissions, and selected 33 papers for inclusion in the scientific program. In addition to the contributed papers, Kurt Mehlhorn (MPI, Germany), Franco P. Preparata (Brown Univ., USA) and Pavel Pudlák (Academy of Sciences, Czech Republic) were invited to give plenary lectures at the conference. All the work of the Program Committee was done electronically. The selection was based on originality, quality and relevance to theoretical computer science. The submissions were refereed as carefully as time permitted; it is expected that many of them will appear in a more polished form in scientific journals in the future.

We wish to thank all authors who submitted papers for consideration, the Program Committee for its hard work, as well as those external reviewers who assisted the Program Committee in the evaluation process. A special thanks to the Organizing Committee for a very dedicated work.

May 2006

Tiziana Calamoneri
Irene Finocchi
Giuseppe F. Italiano

Organization

External Reviewers

Zoe Abrams
Alexander Ageev
Amitai Armon
Pablo Arrighi
Adi Avidor
Amotz Bar-Noy
Luca Becchetti
Stéphane Bessy
Philip Bille
Somenath Biswas
Maria Blesa
Avrim Blum
Jeremy Buhler
John Byers
Ioannis Caragiannis
Massimiliano Caramia
Nicolò Cesa-Bianchi
Marco Cesati
Bogdan Chlebus
Marek Chrobak
Andrea Clementi
Pierluigi Crescenzi
Gianluca De Marco
Christoph Dürr
Fritz Eisenbrand
Lene Favrholdt
Henning Fernau
Fedor Fomin
Dimitris Fotakis
Leszek Gąsieniec
Ricardo Gavalda
Inge Li Gørtz
Fabrizio Grandoni
Joachim Gudmundsson
Leonid Gurvits
Esben Rune Hansen
Michael Hoffmann
Jan Hubička
Costas Iliopoulos

Spiros Kontogiannis
Guy Kortsarz
Dimitris Koukopoulos
Dariusz Kowalski
Dan Král
Evangelos Kranakis
Jan Kratochvíl
Fabian Kuhn
Oliver Kullmann
Moshe Lewenstein
Chaim Linhart
Zvi Lotker
Rune Bang Lyngsø
Christos Makris
David Manlowe
Euripides Markou
Elvira Mayordomo
Xavier Messeguer
Pat Morin
Dhruv Mubayi
Marcin Mucha
Maurizio Naldi
Giri Narasimhan
Alantha Newman
Sara Nicoloso
Rolf Niedermeier
Bengt J. Nilsson
Gianpaolo Oriolo
Andrea Pacifici
Rasmus Pagh
Viki Papadopoulou
Evi Papaioannou
Kunsoo Park
Christian N. S. Pedersen
David Peleg
Paolo Penna
Ugo Pietropaoli
David Pisinger
Tomasz Radzik

Peter Jonsson
Alex Kaporis
Jyrki Katajainen
Claire Kenyon
Nicolas Schabanel
Elad Schiller
Uwe Schöning
Maria José Serna
Asaf Shapira
Micha Sharir
Riccardo Silvestri
Maurizio Strangio
Stéphan Thomassé

R. Ravi
Oded Regev
Milan Ruzic
Miklos Santha
Peter Tiedemann
Jacobo Toran
Ugo Vaccaro
Gabriel Valiente
Tasos Viglas
Paola Vocca
Magnus Wahlström
Michele Zito

Table of Contents

Session 4

Session 5

Session 6

Session 7

Session 8

Session 9

Session 10

Session 11

Reliable and Efficient Geometric Computing*

Kurt Mehlhorn

Max-Planck-Institut für Informatik, Stuhlsatzenhausweg 85,
66123 Saarbrücken, Germany

Reliable implementation of geometric algorithms is a notoriously difficult task. Algorithms are usually designed for the Real-RAM, capable of computing with real numbers in the sense of mathematics, and for non-degenerate inputs. But, real computers are not Real-RAMs and inputs are frequently degenerate.

In the first part of the talk we illustrate the pitfalls of geometric computing by way of examples [KMP+04]. The examples demonstrate in a lucid way that standard and frequently taught algorithms can go completely astray when naively implemented with floating point arithmetic.

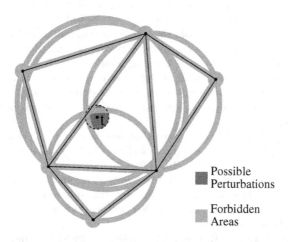

Possible
Perturbations

Forbidden
Areas

Fig. 1. The figure illustrates the concept of controlled perturbation for an incremental Delaunay diagram algorithm. A diagram of six points is already constructed and a seventh point t is to be inserted. The point is replaced by a random point t' in a δ disk centered at t. When t' is inserted, it is subject to sidedness tests with respect to edges of the current diagram and incircle tests with respect to faces of the current diagram. Each edge and each face defines a forbidden region for t'. The forbidden region is either a strip around the edge or an annulus around a circle. If t' lies outside the forbidden regions, the floating point evaluation of the geometric predicates gives the correct results. It is also necessary to guarantee a certain minimal distance between any pair of perturbed points.

In the second part of the talk, we discuss approaches to reliable and efficient geometric computing, in particular the controlled or active perturbation

---

* Partially supported by the IST Programme of the EU under Contract No IST-2005-TODO, Algorithms for Complex Shapes (ACS).

T. Calamoneri, I. Finocchi, G.F. Italiano (Eds.): CIAC 2006, LNCS 3998, pp. 1–2, 2006.

approach introduced by D. Halperin and co-workers [HS98, HR, HL03]. It proposes to slightly perturb the given input in a carefully chosen way so as to avoid degeneracies and so as to reduce the arithmetic demand. The exact solution on the perturbed input (not the original input!) is then computed. The scheme only applies when an approximate result suffices. This is the case whenever inputs are only approximately known.

We build on the work of Halperin et. al. and show that controlled perturbation is a general and simple technique for making a large class of geometric algorithms reliable. We also quantify the relation between the amount of perturbation and the precision of the floating point system. We exemplify the method on examples [FKMS05, MO]. Figure 1 illustrates the technique for the case of a Delaunay diagram computation.

References

[FKMS05] S. Funke, Ch. Klein, K. Mehlhorn, and S. Schmitt. Controlled perturbation for Delaunay triangulations. SODA, pages 1047–1056, 2005. www.mpi-sb.mpg.de/~mehlhorn/ftp/ControlledPerturbation.pdf.

[HL03]   D. Halperin and E. Leiserowitz. Controlled perturbation for arrangements of circles. In *SoCG*, pages 264–273, 2003.

[HR]   D. Halperin and S. Raab. Controlled perturbation for arrangements of polyhedral surfaces with application to swept volumes. available from Halperin's home page; a preliminary version appeared in SoCG 1999, pages 163–172.

[HS98]   Halperin and Shelton. A perturbation scheme for spherical arrangements with application to molecular modeling. *CGTA: Computational Geometry: Theory and Applications*, 10, 1998.

[KMP⁺04] L. Kettner, K. Mehlhorn, S. Pion, S. Schirra, and C. Yap. Classroom examples of robustness problems in geometric computations. In *ESA*, volume 3221 of *LNCS*, pages 702–713, 2004. www.mpi-sb.mpg.de/~mehlhorn/ftp/ClassRoomExample.ps

[MO]   K. Mehlhorn and R. Osbild. Reliable and efficient computational geometry via controlled perturbation (extended abstract). www.mpi-sb.mpg.de/~mehlhorn/ftp/ControlledPerturbationGeneralStrategy.pdf

Beware of the Model: Reflections on Algorithmic Research

Franco P. Preparata

Department of Computer Science, Brown University
`franco@cs.brown.edu`

Over the past four decades the design and analysis of algorithms has been a vibrant area of computer science research, since it was early realized that adoption of a superior algorithm could achieve accelerations unattainable by conceivable technological improvements.

Evaluation of the performance of algorithms must dispense with the details of different platforms and refer to a sort of abstract machine that effectively captures the important features of concrete computers. This abstraction is the computation model, which is intended to be simple to ease formal analysis but at the same time reflective of reality to afford reliable predictions. Indeed, the dialectics of simplicity and reflectivity is the essence of model development.

The Random-Access-Machine (RAM) is the standard model of the sequential processor, and its simplicity has unleashed vigorous algorithmic research. However, simplification means selection of features to be represented in the model, so that details originally judged secondary or irrelevant are likely to reassert their significance when, under the pressure of technological innovations, the model reaches beyond its intended confines.

The first danger is that a model may take a life of its own, thereby becoming itself the reality and defining the "rules of the game". An obvious illustration of this potential danger is the occasional misuse of the "asymptotic-analysis viewpoint", whereby some algorithms declared "optimal" are unlikely to be ever translated into programs. However, there are more subtle shortcomings. Indeed, being remiss in critically scrutinizing the applicability of the model to specific situations may be the source of very serious disappointments. There are several such incidents in the history of algorithmic research. A sample is described below:

1. Computational Geometry adopted (with not much scrutiny) the model of the *real-RAM*, obtained by endowing the RAM with real-number (exact) arithmetic. Inaccurate results may result fatal in the evaluation of the sign of predicates. For example, the efficient BentleyOttmann algorithm for reporting the intersections of a set of segments in the plane, involves a predicate represented by the sign of a thirddegree polynomial in the coordinates. Inaccuracies may invalidate the result. The shortcoming may be avoided, however, by adding integer arithmetic capabilities of specified degree to the original RAM model, i.e., by adopting a sort of the bounded-degree-RAM.

T. Calamoneri, I. Finocchi, G.F. Italiano (Eds.): CIAC 2006, LNCS 3998, pp. 3–4, 2006.

2. The feasibility of parallel computation posed the question of the corresponding model. The discussion centered on the interconnection of modules of the RAM type and an important performance goal was the achievementt of polylog-time computations (NC-class). As usual, processing elements were assumed to have unit-time arithmetic capabilities. In this context, Csanky's algorithm, achieving $O(\log^2 n)$ time for matrix inversion, was an exhilarating surprise. A closer look reveals that, since approximate arithmetic is not known or likely to be applicable to Csanky's method, integer arithmetic requires operand length of $O(n)$ bits for inverting $n \times n$ matrices. Here again, overlooking the arithmetic details of the model, leads to the invalidation of this result.

3. Very-Large-Scale-Integration opened up the possibility of massive parallellism, whose typical model was an interconnection of RAM-type processors with unit-time interprocessor communication. Therefore the emphasis was directed towards small-diameter networks, i.e., trees and hypercubes. However, the area-time theory of layouts reveals that such networks have links of length linear in the problem size. Since in future technologies transmission time is bound to grow with wirelength, hypercubic connections are manifestly nonscalable.

4. Finally, a case study from Computational Biology is not directly concerned with a computation model, but rather with the modeling of the process to which algorithmic research is applied. Sequencingby-Hybridization was presented as a potential alternative for DNA-sequencing. A microarray containing a complete library of oligonucleotides of length k is the platform of a biochemical experiment intended to yield *all* substrings of length k of a target sequence. The algorithmic task is the reconstruction of the target from its substrings. A number of very interesting results were obtained based on the hypothesis of ideal "noiseless" hybridization: a substring is reported if and only if present in the target. A closer look at the biochemical behavior reveals an enormously more complex noisy reality, which casts a negative shadow on the future of the technology.

On Search Problems in Complexity Theory and in Logic (Abstract)

Pavel Pudlák

Mathematical Institute of the Academy of Sciences,
Prague, Czech Republic
pudlak@math.cas.cz

Abstract. A search problem is given by a binary relation $B(x, y)$ in **P**, such that $\forall x \exists y, |y| \leq poly(|x|) B(x, y)$. The computational task is for given x find such a y. We believe that in general this is not possible in polynomial time and oracles are known for which this is the case.

Many-to-one and Turing reductions between search problems are defined in a natural way. We conjecture that there is no complete search problem.

Our aim is to classify search problems and show relations between the computational complexities of them and the proof complexities of the sentences $\forall x \exists y, |y| \leq poly(|x|) B(x, y)$. A typical example of a class of search problems is the class Polynomial Local Search defined as follows.

A **PLS** problem is given by a **P**-time relation $R(p, x)$ and a **P**-time function $F(p, y)$ such that $R(n, n)$ holds for all n. The search problem is for every p and $x \leq p$ to find a $y \leq p$ such that

$$R(p, y) \wedge (\neg F(p, y) < y \vee \neg R(p, F(p, y))).$$

A typical result relating proof complexity and computational complexity of search problems is the following theorem of Buss and Krajíček.

Theorem 1. *A search problem $B(x, y)$ is reducible to a* **PLS** *problem iff*

$$T_2^1 \vdash \forall x \exists y \beta(x, y),$$

for a Σ_0^b formula $\beta(x, y)$ defining the relation $B(x, y)$, (where T_2^1 is a theory that formalizes induction for **NP** *sets.)*

In this lecture we shall present some recent results in this field.

T. Calamoneri, I. Finocchi, G.F. Italiano (Eds.): CIAC 2006, LNCS 3998, p. 5, 2006.
© Springer-Verlag Berlin Heidelberg 2006

Covering a Set of Points with a Minimum Number of Lines

Magdalene Grantson and Christos Levcopoulos

Department of Computer Science, Lund University,
Box 118, 221 Lund, Sweden
{magdalene, christos}@cs.lth.se

Abstract. We consider the minimum line covering problem: given a set S of n points in the plane, we want to find the smallest number l of straight lines needed to cover all n points in S. We show that this problem can be solved in $O(n \log l)$ time if $l \in O(\log^{1-\epsilon} n)$, and that this is optimal in the algebraic computation tree model (we show that the $\Omega(n \log l)$ lower bound holds for all values of l up to $O(\sqrt{n})$). Furthermore, a $O(\log l)$-factor approximation can be found within the same $O(n \log l)$ time bound if $l \in O(\sqrt[4]{n})$. For the case when $l \in \Omega(\log n)$ we suggest how to improve the time complexity of the exact algorithm by a factor exponential in l.

1 Introduction

We consider the minimum line covering problem: given a set S of n points in the plane, we want to find the smallest number l of straight lines needed to cover all n points in S. The corresponding decision problem is: given a set S of n points in the plane and an integer k, we want to know whether it is possible to find k (or fewer) straight lines that cover all n points in S.

Langerman and Morin [7] showed that the decision problem can be solved in $O(nk + k^{2(k+1)})$ time. In this paper we show that the decision problem can be solved in $O(n \log k + (k/2.2)^{2k})$ time.

Kumar *et al.* [6] showed that the minimum line covering problem is APX-hard. That is, unless $P = NP$, there does not exit a $(1 + \epsilon)$-approximation algorithm. In their paper they pointed out that the greedy algorithm proposed by Johnson [5], which approximates the set covering problem within a factor of $O(\log n)$, is the best known approximation for the minimum line covering problem. In this paper we show that a $O(\log l)$-factor approximation for the minimum line covering problem can be obtained in time $O(n \log l + l^4 \log l)$.

We also present an algorithm that solves the line covering problem exactly in $O(n \log l + (l/2.2)^{2l})$ time. This simplifies to $O(n \log l)$ if $l \in O(\log^{1-\epsilon} n)$, and we show that this is optimal in the algebraic computation tree model. That is, we show that the $\Omega(n \log l)$ lower bound holds for all values of l up to $O(\sqrt{n})$. We also suggest more asymptotic improvements for our exact algorithms when $l \in \Omega(\log n)$.

T. Calamoneri, I. Finocchi, G.F. Italiano (Eds.): CIAC 2006, LNCS 3998, pp. 6–17, 2006.

2   Preliminaries

Lemma 1. *Any set S of n points in the plane can be covered with at most $\lceil \frac{n}{2} \rceil$ straight lines.*

Proof. A simple way to show this upper bound is to pick two points at a time, to construct a line through the pair, and then to remove the pair from the set. For the special case when n is odd, we can draw an arbitrary line through the last point. The time complexity of this algorithm is obviously $O(n)$.

Lemma 2. *If a set S of n points can be covered with k lines (k minimal or not), then: for any subset $R \subseteq S$ of at least $k+1$ collinear points (i.e., $|R| \geq k+1$ and $\forall r_1, r_2, r_3 \in R : r_1 \neq r_2 \Rightarrow \exists \alpha \in \mathbb{R} : r_3 = \alpha \cdot (r_2 - r_1) + r_1$), the line through them is in the set of k covering lines.*

Proof. Suppose the line through the points in R was not among the k lines covering S. Then the points in R must be covered with at least $k+1$ lines, since no two points in R can be covered with the same line. (The only line covering more than one point in R is the one through all of them, which is ruled out.) Hence we need at least $k+1$ lines to cover the points in R. This contradicts the assumption that S can be covered with k lines.

Lemma 3. *If a set S of n points can be covered with k lines (k minimal or not), then: any subset of S containing at least $k^2 + 1$ points must contain at least $k+1$ collinear points.*

Proof. Suppose there is a subset $R \subseteq S$ containing at least $k^2 + 1$ points, but not containing $k+1$ collinear points. Then each of the k covering lines must contain at most k points in R. Hence with these at most k covering lines, each containing at most k points, we can cover at most k^2 points. Thus we cannot cover R (nor any superset of R, like S) with the k lines. This contradicts the assumption that S can be covered with k lines.

Corollary 1. *If in any subset of S containing at least $k^2 + 1$ points we do not find $k+1$ collinear points, we can conclude that S cannot be covered with k lines.*

Lemma 4. *If a set S of n points can be covered with l lines, but not with $l-1$ lines (i.e., if l is the minimum number of lines needed to cover S) and $k \geq l$, then: if we generate all lines containing more than k points, the total number of uncovered points will be at most $l \cdot k$.*

Proof. Let R be the set of uncovered points in S after all lines containing more than k points have been generated. Since S can be covered with l lines and $R \subseteq S$, R can be covered with l (or fewer) lines. None of the lines covering points in R can cover more than k points in R (as all such lines have already been generated). Hence there can be at most $l \cdot k$ points in R.

3   General Procedure

Given a set S of n points in the plane, we already know (because of Lemma 1) that the minimum number l of lines needed tocover S is in $\{1, \ldots, \lceil \frac{n}{2} \rceil\}$. In our algorithm, we first check whether $l = 1$, which can obviously be decided in time linear in n. If the check fails (i.e., if the points in S are not all collinear and thus $l \geq 2$), we try to increase the lower bound for l by exploiting Lemmas 2 and 3, which (sometimes) provide us with means of proving that the set S cannot be covered with a certain number k of lines. In the first place, if for a given value of k we find a subset $R \subseteq S$ containing $k^2 + 1$ points, but not containing $k + 1$ collinear points, we can conclude that more than k lines are needed to cover S (because of Corollary 1). On the other hand, if we find $k + 1$ collinear points (details of how this is done are given below), we record the line through them (as it must be among the covering lines due to Lemma 2) and remove from S all points covered by this line. This leads to second possible argument: If by repeatedly identifying lines in this way (always choosing the next subset R from the remaining points), we record k lines while there are points left in S, we can also conclude that more than k lines are needed to cover S.

We check different values of k in increasing order (the exact scheme is discussed below), until we reach a value k_1, for which we fail to prove (with the means mentioned above) that S cannot be covered with k_1 lines. On the other hand, we can demonstrate (in one of the two ways outlined above) that S cannot be covered with k_0 lines, where k_0 is the largest value smaller than k_1 that was tested in the procedure. At this point we know that $l > k_0$.

Suppose that when processing S with $k = k_1$, we identified m_1 lines, $m_1 \leq k_1$. We use a simple greedy algorithm to find m_2 lines covering the remaining points. (Note that m_2 may or may not be the minimum number of lines needed to cover the remaining points. Note also that $m_2 = 0$ if there are no points left to be covered.) As a consequence we know that S can be covered with $m_1 + m_2$ lines (since we have found such lines) and thus that $k_0 < l \leq m_1 + m_2$. We show below that $m_1 + m_2 \in O(l \log l)$ and thus that with the $m_1 + m_2$ lines we selected we obtained an $O(\log l)$ approximation of the optimum.

In a second step we may then go on and determine the exact value of l by drawing on the found approximation (see below for details).

Although in this paper we concentrate on the two-dimensional case, we note that even for higher dimensions one might be able to obtain some improvements by selecting a sample of size $O(k^d)$, using in a similar way the set systems (and applications of those) discussed by Langerman and Morin, see Lemma 4 in [7].

4   Algorithms

In this section we propose approximate and exact algorithms to solve the minimum line covering problem. We use two already known algorithms as subroutines in our algorithms:

1. An algorithm proposed by Guibas *et al.* [4], which finds all lines containing at least $k + 1$ points in a set S of n points in time $O\left(\frac{n^2}{k+1} \log \frac{n}{k+1}\right)$.

2. An algorithm proposed by Langerman and Morin [7], which takes as input a set S of n points and an integer k, and outputs whether S can be covered with k lines in $O(nk + k^{2(k+1)})$ time.

4.1 Finding at Most k Lines Covering More Than k Points

(In all pseudo-code **var** in the declaration of a function's arguments is used to indicate that an argument is passed by reference, not just by value.)

```
function Lines (var S: set of points, k: int) : int;
var m: int;                        (* number of lines found *)
    L₁,..., Lₖ: set of points;     (* points on straight lines found *)
    R: set of points;             (* pool of points for line finding *)
begin
    m := 0; R := ∅;               (* init. line counter and point pool *)
    while m < k do begin          (* try to find at most k lines *)
        while |R| < k² + 1 and S ≠ ∅ do begin
            choose p ∈ S;         (* collect at most k² + 1 points *)
            S := S − {p};         (* into the point pool R *)
            if    ∃i; 1 ≤ i ≤ m : p is collinear with the points in Lᵢ
            then Lᵢ := Lᵢ ∪ {p};  (* point is on already found line *)
            else  R := R ∪ {p};   (* point is on no found line *)
        end
        Lₘ₊₁ := FindLine(R, k);   (* Guibas et al.'s algorithm [4] *)
        if Lₘ₊₁ = ∅ then begin    (* if no line with k + 1 points found *)
            if |R| ≤ k²           (* return the number of lines found *)
            then begin S := S ∪ R; return m; end;
            else return −1;       (* if there are more than k² points, *)
        end                       (* there should be such a line *)
        R := R − Lₘ₊₁; m := m + 1; (* remove the covered points and *)
    end                           (* increment the line counter *)
    if |R| + |S| = 0 return m;    (* return the number of lines found *)
    return −1;                    (* if there are uncovered points, *)
end                               (* more than k lines are needed *)
```

4.2 Approximation for Minimum Line Covering

To greedily cover the at most $k \cdot l$ points (see Lemma 4) that may be left after we removed lines with more than k points, we use the following function:

```
function GreedyCover (S: set of points, k: int) : int;
var m: int;                       (* number of lines found *)
    L: set of points;            (* buffer for points on found line *)
begin
    m := 0;                       (* initialize the line counter *)
    while S ≠ ∅ do begin          (* while not all points are covered *)
        L := FindLine(S, k);      (* Guibas et al.'s algorithm [4] *)
        if L = ∅ then k := k − 1; (* reduce the number of points *)
        else begin m := m + 1; S := S − L; end;
    end;                          (* count the line found and *)
    return m;                     (* remove the covered points *)
end                               (* return the number of lines found *)
```

The approximation algorithm for the line covering problem then is:

```
function ApxLineCover (S: set of points) : int;
var k, m, n : int;                          (* numbers of lines/points *)
    R : set of points;                      (* remaining uncovered points *)
begin
    if all points in S are collinear then return 1;
    k := 2; n := |S|;                       (* initialize variables *)
    while k ≤ ⁸√n do begin
        R := S; m := Lines(R, k);           (* see Subsection 4.1 *)
        if m ≥ 0 then return m + GreedyCover(R, k − 1);
        k := k²;
    end;
    R := S; m := Lines(R, ⁴√n);             (* see Subsection 4.1 *)
    if m ≥ 0 then return m + GreedyCover(R, ⁴√n − 1);
    end;
    k := 2⁴√n;
    while k < ⌈n/2⌉ do begin
        R := S; m := Lines(R, k);           (* see Subsection 4.1 *)
        if m ≥ 0 then return m + GreedyCover(R, k − 1);
        k := 2k;
    end;
    return ⌈n/2⌉;                           (* see Lemma 1 *)
end
```

4.3 Analysis

Approximation ratio. Let k_1 be the value of k in function ApxLineCover, for which the call to function Lines succeeds, i.e., for which it returns a value ≥ 0. Let m_1 be the number of lines found by the function Lines in this case, $n_R = |R|$ the total number of points that are left uncovered, and m_2 the number of lines found by the algorithm GreedyCover applied to these n_R remaining points.

From Lemma 4 we know $n_R \leq k_1 \cdot l$. We also know that the maximum value of k_1 is $(l − 1)^2$, namely when we tried $k_0 = l − 1$, could prove that we cannot cover S with k_0 lines, and then computed $k_1 = k_0^2$. Running the greedy algorithm on the at most $n_R \leq k_1 \cdot l \leq (l − 1)^2 \cdot l \in O(l^3)$ remaining points gives an $O(\log l^3) = O(\log l)$ approximation of the optimum number l_R of lines, with which the n_R remaining points can be covered (as shown by Johnson [5]). Since certainly $l_R \leq l$, we have $m_2 \in O(l \log l)$. The maximum value of m_1 is clearly l, namely when $m_2 = 0$. Thus $m_1 + m_2 \in O(l + l \log l) = O(l \log l)$.

Time complexity of function Lines. For each point p chosen from S, we need to check whether it lies on an already found line or not. This is a fundamental problem in computational geometry called the *point location problem*. A planar subdivision with m lines is called an *arrangement*. Such an arrangement can be preprocessed in time $O(m^2 \log m)$ into a linear size data structure so that a query whether a given point p lies on any of the lines can be answered in $O(\log m)$ time [2, 9]. In principle, we have to construct at most k such data structures, namely each time we find a new line. Thus in the worst case it takes $O(k^3 \log k)$ time to construct all such data structures. (This could be improved, but we do not elaborate it, because it does not affect the overall time complexity.) Since we perform at most n queries—checking whether a chosen point lies on an already constructed line—it takes $O(n \log k)$ to perform all queries. The function FindLine, which is the algorithm by Guibas et al. [4] for identifying at least $k + 1$ collinear points, is called at most k times. Each call takes time $O\left(\frac{(k^2+1)^2}{k+1} \log \frac{k^2+1}{k+1} \right) = O(k^3 \log k)$. Thus the total time complexity of the function Lines is $O\left(k^3 \log k + n \log k + k \cdot k^3 \log k\right) = O(n \log k + k^4 \log k)$.

Time complexity of function GreedyCover. Note that in the function GreedyCover it always holds that all lines covering more than $k + 1$ points have already been found and the covered points removed. Hence we know, because of Lemma 4, that there are always no more than $l \cdot (k + 1)$ points remaining in S. Let $k_1 - 1$ be the value of k with which GreedyCover is called. We group the calls to the function FindLine (which is the algorithm by Guibas et al. [4] for identifying at least $k + 1$ collinear points) according to the values of k: group j, $1 \leq j \leq \lceil \log_2 k_1 \rceil$, contains the values of k with $(k_1 - 1)2^{-j} < k \leq (k_1 - 1)2^{1-j}$. Since all points covered by lines covering more than $k + 1$ points have been removed, we know that for any value of k in the j-th group there are at most $l \cdot ((k_1 - 1)2^{1-j} + 1) \leq l\, k_1 2^{1-j}$ points left to cover. Thus the time bound for a call to the function FindLine with k in the j-th group is

$$O\left(\frac{(l\, k_1 2^{1-j})^2}{k_1 2^{1-j}} \log \frac{l\, k_1 2^{1-j}}{k_1 2^{1-j}} \right) = O\left(2^{-j} l^2 k_1 \log l\right).$$

To estimate the number of calls to function FindLine in the j-th group we have to distinguish two cases: either FindLine returns a line (productive call) or not (unproductive call). Since for each unproductive call we decrement k, there can be at most one unproductive call for each k in the group and thus there are $(k_1 - 1)2^{1-j} - (k_1 - 1)2^{-j} = (k_1 - 1)2^{-j}$ unproductive calls in the j-th group. Each productive call removes at least $(k_1 - 1)2^{-j} + 1 \geq k_1 2^{-j}$ points from S. Thus, since there are at most $l\, k_1 2^{1-j}$ points left for any k in the j-th group, there can be at most $\frac{l \cdot k_1 2^{1-j}}{(k_1-1)2^{-j+1}} \leq 2l$ productive calls. Therefore function FindLine is called at most $2l + (k_1 - 1)2^{-j}$ times in the j-th group and hence a worst case upper bound on the time required for all calls in the j-th group is

$$O\left((2l + (k_1 - 1)2^{-j})2^{-j} l^2 k_1 \log l\right) = O\left((l^3 k_1 2^{-j} + l^2 k_1^2 2^{-2j}) \log l\right)$$

We observe that this upper bound gets smaller by a constant factor, which is smaller than $\frac{1}{2}$, for successive groups (as j increases for successive groups). Hence the total time complexity (sum over all groups) is bounded by twice the above time complexity for $j = 1$. Therefore a worst case bound on the time complexity of the function GreedyCover is $O\left((l^3 k_1 + l^2 k_1^2) \log l\right)$.

Time complexity of function ApxLineCover. ApxLineCover calls function Lines each time the value of k is incremented. There are three cases we have to consider, in each of which k is incremented differently. In any case, let k_1 be the value of k, for which the call to function Lines succeeds, i.e., for which Lines returns a value ≥ 0 (just as above). In addition, let m_1 be the number of lines found by function Lines in this case.

Case 1: $k \leq \sqrt[8]{n}$. In this case we always square k. Immediately before squaring the value of k, we know that $k < l$. Each call to function Lines takes at most $O(n \log k + k^4 \log k)$ time (see above). We observe that this time bound increases by a factor of at least 2 for successive values of k (since we start with $k = 2$ and keep squaring k), leading to a geometric progression of the time complexity. Thus the total time is asymptotically dominated by the last term, which is $O(n \log k_1 + k_1^4 \log k_1)$. Thus the overall time complexity for ApxLineCover is $O(n \log k_1 + k_1^4 \log k_1 + (l^3 k_1 + l^2 k_1^2) \log k_1) = O((n + k_1^4 + l^3 k_1 + l^2 k_1^2) \log k_1)$.

Since we squared the value of k at each step, we know $k_1 < l^2$ (actually $k_1 \leq (l-1)^2$). Thus the total time complexity simplifies to $O(n \log l + l^8 \log l)$. This leads to the following lemma:

Lemma 5. *We can approximate the minimum line covering problem within a factor of $O(\log l)$ in $O(n \log l)$ time if $l \leq \sqrt[8]{n}$.*

However, if we did not succeed in finding a value of $k < \sqrt[8]{n}$, such that function Lines succeeded, we proceed to the second case.

Case 2: $k = \sqrt[4]{n}$. Let k_0 be smallest value of k before it became greater than $\sqrt[8]{n}$. We know that $k_0 \leq \sqrt[8]{n} < l$, trivially implying $k = \sqrt[4]{n} < l^2$. Suppose the function Lines succeeds for $k = \sqrt[4]{n}$, yielding $m_1 \geq 0$ lines. Then the total time invested is the time spent in Case 1, the time for the successful call to function Lines, plus the time for calling function GreedyCover. For the first term (Case 1), we computed above that the worst case time bound is $O(n \log k + k^4 \log k)$, which we have to apply for $k = k_0 \leq l$, which yields $O(n \log l + l^4 \log l)$. For the second term (function Lines) we have (see above) $O(n \log k + k^4 \log k) = O(n \log l + l^8 \log l) = O(n \log l)$. Finally, calling function GreedyCover takes $O((l^3 k + l^2 k^2) \log l) = O((l^5 + l^6) \log l) = O(l^6 \log l)$ time. Thus the overall time complexity is $O(n \log l + l^4 \log l + n \log l + l^6 \log l) = O(n \log l + l^6 \log l)$.

Lemma 6. *We can approximate the minimum line covering problem within a factor of $O(\log l)$ in $O(n \log l)$ time if $l \leq \sqrt[4]{n}$.*

However, if we do not succeed with $k = \sqrt[4]{n}$, we proceed to the third case:

Case 3: $\sqrt[4]{n} < k < \lceil \frac{n}{2} \rceil$. In this case $l > \sqrt[4]{n}$. The time we spent up to this point consists of the time spent in Case 1, that is, $O(k_0^4 \log k_0 + n \log k_0) = O(n \log l + l^4 \log l)$ (see above) plus the time for the (unsuccessful) call to function Lines for $k = \sqrt[4]{n} < l$, which took at most $O(n \log k + k^4 \log k) = O(n \log l + l^4 \log l)$ time. Together this yields $O(n \log l + l^4 \log l)$ time. Afterwards we proceed by doubling the value of k. For each k we call function Lines, spending $O(k^4 \log k + n \log k)$ time, which simplifies to $O(k^4 \log k)$ for $k > \sqrt[4]{n}$. We observe that this calculated worst case upper time bound increases by a factor of at least $2^4 = 16$ for successive calls, leading to a geometric progression of the time complexity. Thus the total time is asymptotically dominated by the last term, which is $O(k_1^4 \log k_1 + n \log k_1 + k_1^2 l^2 \log k_1 + k_1 l^3 \log k_1)$, which simplifies to $O(l^4 \log l)$, since $k \leq 2l$. Summarising all the three cases we obtain the following theorem:

Theorem 1. *We can approximate the minimum line covering problem within a factor of $O(\log l)$ in time $O(n \log l + l^4 \log l)$.*

Corollary 2. *We can approximate the minimum line covering problem within a factor of $O(\log l)$ in $O(n \log l)$ time if $l \in O(\sqrt[4]{n})$.*

4.4 Exact Minimum Line Covering

Drawing on the total number $m_1 + m_2$ of lines obtained by the approximation algorithm, we can determine the exact value of the minimum number of lines l needed to cover a given set S of n points. Let $m = m_1 + m_2$. If we know m, we use function Lines to repeatedly identify lines containing at least $m + 1$ collinear points from subsets $R \subseteq S$ containing $m^2 + 1$ points. Since $m \geq l$, all lines found in this way must be in the optimal solution (see Lemma 2). Hence we can find the optimal solution by running any optimal line covering algorithm on the remaining points, of which there can be no more then $l \cdot m$ (see Lemma 4). An example of such an optimal algorithm is the algorithm proposed by Langerman and Morin [7]. More formally, our algorithm works as follows: To find the minimum number of lines needed to cover the remaining $n_R = m \cdot l$ points we use the following function, which calls the algorithm proposed by Langerman and Morin [7].

```
function ExactCover (S : set of points) : int;
var n, k: int;                (* current number of lines *)
begin
    n = |S|; k := 1;          (* traverse possible numbers of lines *)
    while k < ⌈n/2⌉ do begin
        if IsCoverable(S, k)   (* Langerman and Morin's algorithm [7] *)
        then return k;         (* if S can be covered with k lines *)
        k := k + 1;            (* return the current k, *)
    end;                       (* otherwise go to the next k *)
    return ⌈n/2⌉;             (* S can always be covered with ⌈n/2⌉ lines *)
end
```

The algorithm itself works as follows:

```
function ExactLineCover (S: set of points) : int;
var m: int;                    (* approximate solution *)
begin
    m := ApxLineCover(S);      (* find approximate solution *)
    m := Lines(S, m);          (* find lines that must be in the solution *)
    return m + ExactCover(S);
end                            (* cover the rest with an exact algorithm *)
```

Theorem 2. *The minimum line covering problem can be solved exactly in $O(n \log l + l^{2l+2})$ time. In particular, if $l \in O(\log^{1-\epsilon} n)$, the minimum line covering problem can be solved in $O(n \log l)$ time.*

Proof. We already argued above that the approximation algorithm takes $O(n \log l + l^8 \log l)$ time if $l < \sqrt[8]{n}$. It finds m lines that cover S. Finding all lines containing at least $m + 1$ points takes $O(n \log m + m^4 \log m)$ time as analyzed above. Recall that the approximation algorithm yields at most $m = O(l \log l)$ lines. Therefore the time complexity of finding the lines can also be written as

$$O(n \log(l \log l) + (l \log l)^4 \log(l \log l))$$
$$= O(n \log l + n \log \log l + l^4 \log^5 l + l^4 \log^4 l \cdot \log \log l)$$
$$= O(n \log l + l^4 \log^5 l).$$

Finally, we use the algorithm proposed by Langerman and Morin [7] to find the minimum number of lines covering the most $O(m \cdot l) = O(l^2 \log l)$ points. Their algorithm takes at most $O(nk + k^{2k+2})$ time to decide whether n points can be covered with k lines. It has to be applied for all values k, $1 \leq k \leq l - m_3$, where m_3 is the number of lines identified by applying function Lines with $k = m$. In the worst case, it is $m_3 = 0$. The sum over the time complexities for the different values of k is clearly dominated by the last term, that is, for $k = l$. Thus the time complexity for this step is $O(l^3 \log l + l^{2l+2})$ time. As a consequence, the overall time complexity is $O(n \log l + l^4 \log^5 l + l^3 \log l + l^{2l+2}) = O(n \log l + l^{2l+2})$. This simplifies to $O(n \log l)$ if $l \leq \frac{\log n}{2 \log \log n}$.

Theorem 3. *Given a set S of n points in the plane and an integer k, we can answer whether it is possible to find k lines that cover all the points in the set in $O(n \log k + k^{2k+2})$ time.*

Proof. We apply the approximation algorithm (function ApxLineCover) as described above, but endowing it with the integer k as an additional argument. The algorithm proceeds basically as before, but it terminates and returns -1 (meaning that the set cannot be covered with k lines) if it tries to build the point location structure with more than k^2 lines. Otherwise, it proceeds to the end, to find m lines as described above. Next we call function Lines to find m_3 lines, each covering at least $m + 1$ points. We know that these m_3 lines are in the optimal solution. So our next step will be to determine, whether we can cover the remaining points with $k - m_3$ lines. This can be done by calling the algorithm by Langerman and Morin [7], with $k - m_3$ as the input. (Their algorithm answers whether the remaining points can be covered with $k - m_3$ lines.)

4.5   Producing the Optimal Set of Lines

We remark that after computing the optimal number of lines l, we can also produce the actual lines covering the input point set within the same time bounds. One way is to first use the algorithm proposed by Guibas *et al.* [4] to produce lines covering at least $l + 1$ points. Let l' denote the number of lines covering at least $l + 1$ points and n' the number points left to be covered. We observe that at least one of the remaining $l - l'$ lines cover at least $\frac{n'}{l-l'}$ points. A line is called a *candidate* line if it covers at least $\frac{n'}{l-l'}$ points. Next, we repeat the following step to produce the remaining $l - l'$ lines: If $n' \leq 2(l - l')$ then any line covering at least two points can be included in the optimal solution. Otherwise, for each candidate line we tentatively (temporarily) remove the points covered by it and call Langerman and Morin's algorithm [7] to see whether the remaining points can be covered by $l - l' - 1$ lines. Clearly, the candidate can be included in the optimal solution if and only if the answer is yes.

To calculate the time bound we show that there are at most $\frac{3}{2} \cdot (l - l')^2$ candidate lines. Any point can be covered by no more than $\frac{3}{2} \cdot (l - l')$ candidate lines. (The factor $\frac{3}{2}$ comes from the extreme case when $l - l' = \frac{n'}{3}$, so that each candidate line covers only three points and the point is covered by $\frac{n'-1}{2}$ candidates.) Hence, if we sum for each point the number of candidates it is covered by, we thus get an upper bound of $n' \cdot \frac{3}{2}(l - l')$. But we observe that this sum equals the sum we obtain by adding for each candidate line the number of points it covers. Since each candidate line covers at least $\frac{n'}{l-l'}$ points, the number of candidate lines cannot be larger than $(n' \cdot \frac{3}{2}(l - l'))/\frac{n'}{l-l'}$ and hence not larger than $\frac{3}{2}(l - l')^2$. Therefore we call Langerman and Morin's algorithm [7] at most $\frac{3}{2}(l - l')^2$ times before we produce one more optimal line. In subsequent calls to their algorithm, the number of optimal lines to be produced gets smaller and hence the time complexity gets smaller each time by at least a constant factor, since it is exponential in the number of optimal lines. This results in a geometric progression of the time complexity. Therefore the worst-case bound for the first call asymptotically dominates all subsequent calls.

4.6 Improving the Time Bound When $l \in \Omega(\log n)$

The following theorem is shown.

Theorem 4. *For any input set of n points, it can be decided whether there is a set of lines of cardinality at most k covering the n points in time $O(n \log k + (\frac{k}{2.2194...})^{2k})$. Moreover, an optimal set of covering lines with minimum cardinality l can be produced in time $O(n \log l + (\frac{l}{2.2194...})^{2l})$*

Remaining details of this section have been removed in this conference version because of page limits. The interested reader can download the full technical report, see [3]. There we also suggest how to further improve the asymptotic bounds stated in the above theorem.

5 Lower Bound

In this section we give a lower bound on the time complexity for solving the minimum line cover problem. We make the assumption that the minimum number of lines l needed to cover a set S of n points is at most $O(\sqrt{n})$. (For larger values of l our lower bound may not be very interesting, since the best known upper bounds on the time complexity of the minimum line cover problem are exponential anyway.)

The main result we prove in this section is as follows:

Theorem 5. *The time complexity of the minimum line cover problem is $\Omega(n \log l)$ in the algebraic decision tree model of computation.*

We prove Theorem 5 with a reduction from a special variant of the general set inclusion problem [1], which we call the k-Array Inclusion problem. Set inclusion is the problem of checking whether a set of m items is a subset of the second set of n items with $n \geq m$. Ben-Or [1] showed a lower bound of $\Omega(n \log n)$ for this problem using the following important statement.

Statement 1. *If YES instances of some problem Π have N distinct connected components in $\Re^n$, then the depth of the real computation tree for this problem is $\Omega(\log N - n)$.*

Applying Statement 1 to the complement of Π, we get the same statement for NO instances. We define the k-array inclusion problem as follows:

Definition 1. k**-Array Inclusion Problem:** *Given two arrays $A[1 \ldots k]$ of distinct real numbers and $B[1 \ldots m]$ of (not necessarily distinct) real numbers, $k \leq m$, $m + k = n$, determine whether or not each element in $B[1 \ldots m]$ belongs to $A[1 \ldots k]$.*

Corollary 3. *Any algebraic computation tree solving k-array inclusion problem must have a depth of $\Omega(n \log k)$.*

Proof. This lower bound can be shown in a corresponding way as the lower bound for the set inclusion problem [1]. As already pointed out by Ben-Or, any computational tree will correctly decide the case when $A[1 \ldots k] = (1 \ldots k)$. The number of disjoint connected components, N, for YES instances of the k-array inclusion problem is k^m. This is because, in order to create a YES instance, for each element m_i in $B[1 \ldots m]$, there are k choices concerning which of the k fixed elements in $A[1 \ldots k]$, m_i could be equal to. Since these choices are independent for each m_i, the total number of YES-instances becomes k^m. Applying Statement 1, we get a lower bound of $\Omega(m \log k)$, which is also $\Omega(n \log k)$, since $m > \frac{n}{2}$.

To establish the lower bound in Theorem 5 for the minimum line cover problem, we convert (in linear time) the input of the k-array inclusion problem into a suitable input to the minimum line cover problem as follows: Each real number a_i, $1 \leq i \leq k$, in the array $A[1 \ldots k]$ becomes k points with coordinates (a_i, j) (in total we obtain k^2 points), $1 \leq j \leq k$ and each real number b_j in the array $B[1 \ldots m]$, $1 \leq j \leq m$, becomes a point with coordinates $(b_j, -j)$, all points are in two dimensional space. None of the constructed sets of $n = k + m$ points coincide. If we use any algorithm for the minimum line cover problem to solve the constructed instance, the output will be a set of lines covering these points. To obtain an answer to the k-array inclusion problem, we check whether the total number of lines, denoted by l, obtained for the minimum line cover problem is greater than k. If $l = k$ then each element in $B[1 \ldots m]$ belongs to $A[1 \ldots k]$, otherwise at least one element in $B[1 \ldots m]$ does not belong to $A[1 \ldots k]$. Since the k-array inclusion problem requires $\Omega(n \log k)$ time, it follows that the minimum line cover problem requires $\Omega(n \log k)$ time as well.

According to this construction, if it would be possible to compute the number l in $o(n \log l)$ time, for some $l = O(\sqrt{n})$, then it would also be possible to solve the k-array inclusion problem in time $o(n \log k)$ for the case when $k = l$, which would contradict our lower bound for the k-array inclusion problem.

Acknowledgment. The authors wish to thank Dr. Christian Borgelt for his detailed reading and comments on the paper.

References

1. M. Ben-Or. Lower Bounds for Algebraic Computation Trees. *Proc. 15th Ann. ACM Symp. on Theory of Comp.*, 80–86. ACM Press, New York, NY, USA 1983
2. H. Edelsbrunner, L. Guibas and J.Stolfi. Optimal Point Location in a Monotone Subdivision. *SIAM J. Comput.* 15:317–340.Society for Industrial and Applied Mathematics, Philadelphia, PA, USA 1986
3. M. Grantson and C. Levcopoulos. Covering a Set of Points with a Minimum Number of Lines. *Technical Report LU-CS-TR:2005-236, ISSN 1650-1276 Report 156.* Also at: `http://www.cs.lth.se/home/Magdalene_Grantson/line.pdf`
4. L. Guibas, M. Overmars, J. Robert. The Exact Fitting Problem in Higher Dimensions. *Computational Geometry: Theory and Applications*, 6:215–230. 1996
5. D. Johnson. Approximation Algorithms for Combinatorial Problems. *J. of Comp. Syst. Sci.* 9:256-278. 1974
6. V. Kumar, S. Arya, and H. Ramesh. Hardness of Set Cover With Intersection 1. *Proc. 27th Int. Coll. Automata, Languages and Programming,* LNCS 1853:624-635. Springer-Verlag, Heidelberg, Germany 2000
7. S. Langerman and P. Morin. Covering Things with Things. *Proc. 10th Annual Europ. Symp. on Algorithms (Rome, Italy),* LNCS 2461:662-673. Springer-Verlag, Heidelberg, Germany, 2002
8. N. Megiddo and A. Tamir. On the Complexity of Locating Linear Facilities in the Plane. *Operation Research Letters* 1:194-197. 1982
9. N. Sarnak, and R.E. Tarjan. Planar Point Location Using Persistent Search Tree. *Comm. ACM* 29:669-679. ACM Press, New York, NY, USA 1986

Approximation Algorithms for Capacitated Rectangle Stabbing

Guy Even[1], Dror Rawitz[2], and Shimon (Moni) Shahar[1]

[1] School of Electrical Engineering, Tel-Aviv University, Tel-Aviv 69978, Israel
{guy, moni}@eng.tau.ac.il
[2] Caesarea Rothschild Institute, University of Haifa, Haifa 31905, Israel
rawitz@cri.haifa.ac.il

Abstract. In the rectangle stabbing problem we are given a set of axis parallel rectangles and a set of horizontal and vertical lines, and our goal is to find a minimum size subset of lines that intersect all the rectangles. We study the capacitated version of this problem in which the input includes an integral capacity for each line that bounds the number of rectangles that the line can cover. We consider two versions of this problem. In the first, one is allowed to use only a single copy of each line (*hard capacities*), and in the second, one is allowed to use multiple copies of every line provided that multiplicities are counted in the size of the solution (*soft capacities*).

For the case of d-dimensional rectangle stabbing with soft capacities, we present a $6d$-approximation algorithm and a 2-approximation algorithm when $d = 1$. For the case of hard capacities, we present a bi-criteria algorithm that computes $16d$-approximate solutions that use at most two copies of every line. For the one dimensional case, an 8-approximation algorithm for hard capacities is presented.

1 Introduction

Understanding the combinatorial and algorithmic nature of capacitated covering problems is still an open problem. Only a few capacitated problems were studied including the general case of set-cover [1] and the restricted case of vertex-cover [2, 3]. Capacity constraints appear naturally in many applications, for example, bounded number of clients an antenna can serve. In this paper we consider a capacitated version of a covering problem, called rectangle stabbing. The geometric nature of the problem is used to obtain approximation algorithms.

The problems. The *rectangle stabbing* problem (RS) is a covering problem. The input is a finite set $\mathcal{U}$ of axis parallel rectangles and a finite set $\mathcal{S}$ of horizontal and vertical lines. A *cover* is a subset of $\mathcal{S}$ that intersects every rectangle in $\mathcal{U}$. The goal is to find a cover of minimum size. We denote the set of rectangles that a line S intersects by $\mathcal{U}(S)$. Using this notation, an RS instance is a set-cover instance in which the goal is to find a collection of subsets $\mathcal{U}(S)$, the union of which equals $\mathcal{U}$. W.l.o.g., we assume that the RS instance is discrete in the following sense [4]: rectangle corners have integral coordinates and lines intersect

T. Calamoneri, I. Finocchi, G.F. Italiano (Eds.): CIAC 2006, LNCS 3998, pp. 18–29, 2006.

the axes at integral points. In the one-dimensional version, the set $\mathcal{U}$ consists of horizontal intervals and the set $\mathcal{S}$ consists of points. This is the well known polynomial clique cover problem in interval graphs. RS can be extended to d dimensions (dRS). For $d \geq 3$, $\mathcal{U}$ consists of axis parallel d-dimensional rectangles ("boxes") and the set $\mathcal{S}$ consists of hyper-planes that are orthogonal to one of the d axes ("walls"). In the sequel we stick to the two-dimensional terminology, that is, we refer to $\mathcal{U}$ as a set of rectangles and to $\mathcal{S}$ as a set of lines.

In the *capacitated d-dimensional rectangle stabbing* problem the input includes an integral capacity $c(S)$ for every line $S \in \mathcal{S}$. The capacity $c(S)$ bounds the number of rectangles that S can cover. This means that in the capacitated case one has to specify which line covers each rectangle. The assignment of rectangles to lines may not assign more than $c(S)$ rectangles to a line S. We discuss two variants of capacitated d-dimensional rectangle stabbing, called covering with hard capacities (HARD-dRS) and covering with soft capacities (SOFT-dRS).

A SOFT-dRS cover is formally defined as follows. An *assignment* is a function $A : \mathcal{S} \to 2^{\mathcal{U}}$ where $A(S) \subseteq \mathcal{U}(S)$, for every S. A rectangle u is *covered* by a line S if $u \in A(S)$. An assignment A is a *cover* if every rectangle is covered by some line, i.e., $\bigcup_{S \in \mathcal{S}} A(S) = \mathcal{U}$. The *multiplicity* (or number of copies) of a line $S \in \mathcal{S}$ in an assignment A equals $\lceil |A(S)|/c(S) \rceil$. We denote the multiplicity of S in A by $\alpha(A, S)$. The *size* of a cover A is $\sum_{S \in \mathcal{S}} \alpha(A, S)$. We denote the size of A by $|A|$. The goal is to find a cover of minimum size.

Given the multiplicities of every line in a cover A, one can compute a cover with the same multiplicities by solving a flow problem. We therefore often refer to a cover as a multi-set of lines. The *support* of an assignment A is the set of lines $\{S \in \mathcal{S} : A(S) \neq \emptyset\}$. Note that the support is a set and not a multiset. We denote the support of A by $\sigma(A)$. In HARD-dRS, a line may appear at most once in a cover. Hence, in this case, a cover is an assignment A for which $|A(S)| \leq c(S)$, (or $\alpha(A, S) \leq 1$) for every $S \in \mathcal{S}$. In this setting, we refer to a cover as the set of lines it contains (i.e., its support). Note that SOFT-dRS is a special case of HARD-dRS, since given a SOFT-dRS instance one can always transform it into a HARD-dRS instance by duplicating each line $|\mathcal{U}|$ times.

All the problems mentioned above have weighted versions, in which we are given a weight function w defined on the lines. In this case the cost of a cover A is $w(S) = \sum_{S} \alpha(S) \cdot w(S)$, and the goal is to find a cover of minimum weight.

Previous results. Since 1-RS is equivalent to clique cover in interval graphs, it can be solved in linear time [5]. Hassin and Megiddo [6] showed that RS is NP-hard, for $d \geq 2$. Gaur et al. [4] presented a d-approximation algorithm for dRS that uses linear programming to reduce d dimensions to one dimension.

Capacitated covering problems (even with weights) date back to Wolsey [1] (see also [7, 2]). Wolsey presented a greedy algorithm for weighted set-cover with hard capacities that achieves a logarithmic approximation ratio. Guha et al. [3] presented a 2-approximation primal-dual algorithm for the weighted vertex cover problem with soft capacities. Chuzhoy and Naor [2] presented a 3-approximation algorithm for vertex cover with hard capacities (without weights) which is based on randomized rounding with alterations. They also proved that the weighted version of this problem is as hard to approximate as set cover. Gandhi et al. [8] improved the approximation ratio for capacitated vertex cover to 2.

Our results. We present a 2-approximation algorithm for SOFT-1RS. This algorithm is a dynamic programming algorithm that finds an optimal solution of a certain form. In the full paper we show that this algorithm extends to weighted SOFT-1RS. We present a $6d$-approximation algorithm for SOFT-dRS, where d is arbitrary. This algorithm solves an LP relaxation of the problem, and rounds it using the geometrical structure of the problem. For the case of hard capacities we show that the same technique can be used to obtain a bi-criteria algorithm for HARD-dRS. Our algorithm computes solutions that are $16d$-approximate and use at most two copies of each line. An 8-approximation algorithm for the one dimensional case is also presented. In the full paper, we present two hardness results. The first mimics the hardness result given in [2] to show that weighted HARD-2RS is set-cover-hard, even if all weights are in $\{0, 1\}$. The second result proves that it is NP-hard to approximate dRS with a ratio of $c \cdot \log d$, for some constant c. Note that the dimension d is considered here to be part of the input.

2   Interval Stabbing with Soft Capacities

In this section we present a 2-approximation algorithm for SOFT-1RS. In the one-dimensional case rectangles are simply intervals that we draw as horizontal intervals. To facilitate the task of drawing overlapping intervals, we separate intervals by drawing them at different heights. Hyper-planes in the one dimensional case are simply points. Since intervals are drawn as horizontal intervals with different heights, we refer to the hyper-planes as vertical lines instead of points. To summarize, the input in SOFT-1RS consists of a set $\mathcal{U}$ of horizontal intervals, and a set $\mathcal{S}$ of vertical lines with capacities $c(S)$.

The presentation is divided into two parts. First, we define special covers, called *decisive covers*. We show that restricting the cover to be a decisive cover incurs a penalty that is bounded by a factor of two. Second, we present a dynamic programming algorithm that computes an optimal decisive cover.

Definition 1. *The total order $\prec$ is defined over the set $\mathcal{S}$ of lines as follows: $S \prec S'$ if either (i) $c(S) > c(S')$ or (ii) $c(S) = c(S')$ and S is to the left of S'.*

Consider a cover A. Suppose that the support $\sigma(A)$ of a cover A is $\{S_1, S_2, \ldots, S_k\}$, where $S_1 \prec S_2 \prec \cdots \prec S_k$.

Definition 2. *A cover A is called* decisive *if $A(S_i) = \mathcal{U}(S_i) \setminus \cup_{j<i}\mathcal{U}(S_j)$, for every $1 \le i \le k$.*

In a decisive cover each interval u is covered by the smallest (according to $\prec$) line $S \in \sigma(A)$ that intersects u. Hence, "preference" is given to lines of higher capacity. Given a cover A, the decisive cover A' *induced* by A is the cover obtained by assigning each interval u to the first line $S \in A$ that intersects it. Note that if A' is the decisive cover induced by a cover A, then $\sigma(A') \subseteq \sigma(A)$.

Claim 1. *The decisive cover A' induced by a cover A satisfies $|A'| \leq 2|A|$.*

Proof. We prove the slightly stronger inequality $|A'| \leq |A| + |\sigma(A)|$ using a charging scheme. Suppose that the purchasing power of a coupon is one copy of a vertical line. We say that a fractional distribution of coupons to intervals and lines is *valid* with respect to a cover $\tilde{A}$, if: (i) each line $S \in \sigma(\tilde{A})$ holds at least one coupon, and (ii) each interval $u \in \tilde{A}(S)$ holds at least $1/c(S)$ coupons. Note that if a distribution of coupons is valid with respect to a cover $\tilde{A}$ then the number of coupons distributed to the intervals and lines is not less than the size of $\tilde{A}$. Indeed, if we consider each line $S \in \sigma(\tilde{A})$ separately, then the intervals together with S have at least $1 + |\tilde{A}(S)|/c(S) \geq \alpha(\tilde{A}, S)$ coupons.

We now consider the following distribution of coupons. Every line $S \in \sigma(A)$ gets one coupon and every interval $u \in A(S)$ gets $\alpha(A, S)/|A(S)|$ coupons. Note that (i) the number of coupons distributed to the intervals equals the size of A, (ii) the number of coupons distributed to the vertical lines equals the size of the support $\sigma(A)$. To complete the proof, we show that this distribution of coupons is valid with respect to A'. Consider an interval u. The number of coupons given to u is $\alpha(A, S)/|A(S)| \geq 1/c(S)$. Let S' denote the line assigned to u in A', namely, $u \in A'(S')$. Since $S' \prec S$, it follows that $c(S') \geq c(S)$, and hence the number of coupons assigned to u is at least $1/c(S')$, as required. $\square$

Next, we present a dynamic programming algorithm that finds an optimal decisive cover. According to Claim 1 this cover is 2-approximate.

We use the following notation. Given an interval u, we denote the coordinates of its endpoints by $\ell(u) < r(u)$. We assume, without loss of generality, that the coordinates are integers between 1 and $2|\mathcal{U}|$. Indeed, if two vertical lines intersect the same set of intervals, then we can unite them into one line by deleting the line with the smaller capacity. For every two integers $i < j$, let $\mathcal{U}(i,j)$ denote the set of intervals contained in the range $[i,j]$, namely, $\mathcal{U}(i,j) = \{u \in \mathcal{U} \mid i \leq \ell(u) < r(u) \leq j\}$. Also, let $\mathcal{S}(i,j,k)$ denote the set of vertical lines of capacity at most k whose x-coordinate is in the range $[i,j]$.

The dynamic programming table Π of size $O(n^3)$ is defined as follows. The entry $\Pi(i,j,k)$ equals the size of an optimal decisive cover $A_{i,j,k}$ that covers the intervals in $\mathcal{U}(i,j)$ by lines from $\mathcal{S}(i,j,k)$. We initialize the table as follows $\Pi(i,j,k) = 0$ if $\mathcal{U}(i,j) = \emptyset$, and $\Pi(i,j,k) = \infty$ if there exist an interval $u \in \mathcal{U}(i,j)$ that is not intersected by lines in $\mathcal{S}(i,j,k)$. The remaining table entries $\Pi(i,j,k)$ are calculated in polynomial time as follows. Let x_S denote the x-coordinate of a vertical line $S \in \mathcal{S}$. Let $\alpha(S,i,j)$ denote the number of copies of S required to cover all the intervals it intersects in $\mathcal{U}(i,j)$; namely, $\alpha(S,i,j) = \lceil |\{u \in \mathcal{U}(i,j) \mid \ell(u) \leq x_S \leq r(u)\}|/c(S) \rceil$. The following recurrence is used:

$$\Pi(i,j,k) \leftarrow \min\{\Pi(i,j,k-1),$$
$$\min_{\substack{S \in \mathcal{S}(i,j,k) \\ c(S)=k}} \{\Pi(i, x_S - 1, k-1) + \alpha(S,i,j) + \Pi(x_S + 1, j, k)\}\}$$

Note that, if $i = x_S$ then $\Pi(i, x_S - 1, k - 1) = 0$. Similarly, if $x_S = j$ then $\Pi(x_S + 1, j, k) = 0$.

The justification for the recurrence is as follows. Consider two integers $i < j$. If $\Pi(i, j, k) < \Pi(i, j, k-1)$, then the cover $A_{i,j,k}$ must contain a line of capacity k. Consider the leftmost line S of capacity k in $A_{i,j,k}$. Since $A_{i,j,k}$ is decisive, the line S must cover all the intervals that it intersects. Hence, $\alpha(S, i, j)$ copies of S are required. The remaining intervals are partitioned into intervals to the left of S and intervals to the right of S. The intervals in $\mathcal{U}(i, x_S - 1)$ are covered in $A_{i,j,k}$ by lines of capacity strictly less than k. The recurrence simply considers all possible lines of capacity k between i and j.

3   Fractional Rectangle Stabbing

In this section we present LP relaxations of d-dimensional rectangle stabbing with soft and hard capacities. We then show that the LP relaxations can be seen as network flow problems.

3.1   LP Formulation

Following [2], we consider the linear programming relaxation for HARD-dRS. To simplify notation we write $u \in S$ instead of $u \in \mathcal{U}(S)$.

$$\min \sum_{S \in \mathcal{S}} x(S)$$

$$\text{s.t.} \sum_{S \mid u \in S} y(S, u) \geq 1 \qquad \forall u \in \mathcal{U} \qquad (1)$$

$$\sum_{u \in S} y(S, u) \leq c(S) x(S) \qquad \forall S \in \mathcal{S} \qquad (2)$$

$$y(S, u) \leq x(S) \qquad \forall S, u \qquad (3)$$

$$x(S) \leq 1 \qquad \forall S \in \mathcal{S} \qquad (4)$$

$$x(S), y(S, u) \geq 0 \qquad \forall S, u \qquad (5)$$

We denote this LP by LP-HARD. The variable $x(S)$ indicates the "portion" of S that belongs to the cover. The variable $y(S, u)$ indicates the portion of u that is covered by S. Constraints of type (1) are covering constraints. Capacity constraints are formulated using constraints of types (2) and (3). Constraints of type (4) and (5) are fractional relaxations of $x(S), y(S, u) \in \{0, 1\}$. Note that there is a variable $y(S, u)$ only if $u \in S$. However, to simplify notation, we consider all pairs (S, u). If $u \notin S$, we assign $y(S, u) = 0$.

An LP-relaxation of SOFT-dRS is obtained by omitting constraints of type (4). We denote the LP-relaxation by LP-SOFT.

The integrality gap of both LP-HARD and LP-SOFT is at least $2 - o(1)$ even in the one-dimensional case. Consider an instance that contains $k + 1$ rectangles and two lines of capacity k that intersect all the rectangles. A fractional optimal solution is $x^*(S) = (k + 1)/(2k)$ for each line S and $y^*(S, u) = 1/2$ for every line S and rectangle u. This means that the value of the fractional minimum is $1 + \frac{1}{k}$, while the integral optimum is 2.

The following definitions apply to both LP-HARD and LP-SOFT. We refer to a pair (x, y) as a *partial cover* if it satisfies all the constraints, except (perhaps) constraints of type (1). A rectangle is *covered* if its type (1) constraint is satisfied.

If $\sum_{S\,|\,u\in S} y(S, u) \geq \alpha$, we refer to u as α-*covered*. If $\sum_{S\,|\,u\in S} y(S, u) > 0$ we say that u is *positively covered*.

We denote an optimal solution by (x^*, y^*). The sum $\sum_{S\in\mathcal{S}} x^*(S)$ is denoted by OPT*. W.l.o.g. we assume that the covering constraints are tight, i.e., that $\sum_{S\,|\,u\in S} y^*(S, u) = 1$ for every $u \in \mathcal{U}$.

3.2   A Network Flow Formulation

This section is written in HARD-dRS terms, but similar arguments can be made in the case of SOFT-dRS. It is very useful to view the LP relaxation as a network flow problem [7, 2]. Here we are given a (fractional) set of lines x and wish to find the best possible assignment y.

The network N_x is the standard construction used for bipartite graphs. On one side we have all the lines and on the other side we have all the rectangles. There is an arc (S, u) if $u \in S$. The capacity of an arc (S, u) equals $x(S)$. There is a source s that feeds all the lines. The capacity of each arc (s, S) emanating from the source equals $x(S) \cdot c(S)$. There is a sink t that is fed by all the rectangles. The capacity of every arc (u, t) entering the sink equals 1.

Observation 1. *There is a one-to-one correspondence between vectors y such that (x, y) is a partial cover and flows f in N_x. The correspondence $y \leftrightarrow f_y$ satisfies $f_y(u, t) = \sum_{S\,|\,u\in S} y(S, u)$, for every rectangle $u \in \mathcal{U}$, and $f_y(s, S) = \sum_{u\in S} y(S, u)$, for every line $S \in \mathcal{S}$.*

Proof. Given y simply define f_y as follows.

$$f_y(e) \triangleq \begin{cases} \sum_{u\in S} y(S, u) & \text{if } e = (s, S), \\ y(S, u) & \text{if } e = (S, u), \\ \sum_{S\,|\,u\in S} y(S, u) & \text{if } e = (u, t). \end{cases}$$

The mapping from flows to vectors is defined similarly.   □

We refer to $f_y(s, S)$ as the *flow supplied by S* and to $f_y(u, t)$ as the *flow delivered to u*. For simplicity, we denote $f_y(s, S)$ by $f_y(S)$ and $f_y(u, t)$ by $f_y(u)$. We say that y is *maximum with respect to x* if f_y is a maximum flow in N_x.

Next, we show that we can identify infeasible instances of HARD-dRS.

Observation 2. *Feasibility of a HARD-dRS instance can be verified by computing a maximum integral flow in a network N_x, where $x(S) = 1$, for every $S \in \mathcal{S}$.*

The following observation implies that it suffices to compute a feasible cover (x, y), where x is integral.

Observation 3. *Let (x, y) be a feasible solution of LP-HARD. If x is integral then an integral y' such that (x, y') is a feasible solution can be computed in polynomial time.*

Definition 3. *Let (x, y) and (x, y') be partial covers. We say that y' dominates y if (i) $f_{y'}(u) \geq f_y(u)$, for every $u \in \mathcal{U}$, and (ii) $f_{y'}(S) \geq f_y(S)$, for every $S \in \mathcal{S}$. We write $y' \succeq y$ to denote that y' dominates y.*

Observation 4. *Let (x, y) denote a partial cover. Then one can find in polynomial time a maximum vector y' with respect to x that also dominates y.*

Proof. We use an augmenting path algorithm to compute a maximum flow f' in N_x starting with f_y. The flow f' induces the desired vector $y' \succeq y$ since saturating an augmenting path from s to t never decreases the flow in edges exiting s, or in edges entering t. $\qquad\square$

Let **aug-flow** be an efficient algorithm that given a partial cover (x, y), finds a vector $y' \succeq y$ that is maximum with respect to x. Note that **aug-flow** may change the assignment of lines to rectangles. In terms of the network flow, the flow of certain edges may decrease, but the sum of flows that enters (exits, respectively) every rectangle (line, respectively) does not decrease.

4   Rectangle Stabbing with Soft Capacities

In this section we present a $6d$-approximation algorithm for SOFT-dRS. The algorithm is based on solving LP-SOFT, and then rounding the solution. For the sake of simplicity, the algorithm is presented for the 2-dimensional case ($d = 2$).

Let $\varepsilon = 1/6d$ and let (x^*, y^*) be an optimal solution of LP-SOFT. We define $H \triangleq \{S \mid x^*(S) \geq \varepsilon\}$ and $L \triangleq \{S \mid x^*(S) < \varepsilon\}$. Let $L = L^h \cup L^v$ denote a partition of L into horizontal and vertical lines. We partition the horizontal line in L^h into "contiguous blocks" by accumulating lines in L^h from "left" to "right" until the sum of fractional values $x(S)$ in the block exceeds ε. We denote the blocks by $L_1^h, L_2^h, \ldots, L_{b(h)}^h$ and the (possibly empty) leftover block by $\tilde{L}^h$. By the construction, $\varepsilon \leq \sum_{S \in L_j^h} x^*(S) < 2\varepsilon$ for every $j \leq b(h)$ and $\sum_{S \in \tilde{L}^h} x^*(S) < \varepsilon$. The same type of partitioning is applied to the vertical lines in L^v to obtain the blocks $L_1^v, \ldots, L_{b(v)}^v$ and the leftover block $\tilde{L}^v$.

Observation 5. *The number of blocks (not including the leftover block) in each dimension satisfies $b(h) \leq \frac{1}{\varepsilon} \cdot \sum_{S \in L^h} x^*(S)$ and $b(v) \leq \frac{1}{\varepsilon} \cdot \sum_{S \in L^v} x^*(S)$.*

Let $S_{h,j}^*$ and $S_{v,j}^*$ denote lines of maximum capacity in L_j^h and L_j^v, respectively. Let $L^* \triangleq \{S_{h,j}^* \mid 1 \leq j \leq b(h)\} \cup \{S_{v,j}^* \mid 1 \leq j \leq b(v)\}$.

Definition 4. *We define the partial cover (x, y) as follows. The support of the cover is $H \cup L^*$. For every $S \in H$ and $u \in \mathcal{U}(S)$, we keep $x(S) = x^*(S)$ and $y(S, u) = y^*(S, u)$. For every $S \in L^*$ and $u \in \mathcal{U}(S)$, let $B(S)$ denote the block that contains S. Then, $x(S) = \sum_{S' \in B(S)} x^*(S')$ and $y(S, u) = \sum_{S' \in B(S)} y^*(S', u)$. The remaining components of the solution (x, y) are set to zero.*

Note that if $S = S_{h,j}^*$ and $u \in S_{h,j}^*$, then $y(S_{h,j}^*, u)$ covers u to the same extent that u is covered by lines in L_j^h according to y^*. Hence, rectangles that are intersected by $S_{h,j}^*$ are "locally satisfied". Also notice that $\sum_S x(S) = \sum_S x^*(S)$. We now prove that (x, y) is a indeed partial cover.

Claim 2. (x, y) *is a partial cover.*

Proof. We first show that constraints of type (3) are satisfied. Clearly, this is true for $S \notin L^*$. Consider a line $S^* \in L^*$, and let B denote the block of lines in L that contains S^*. For every rectangle $u \in S^*$, the following holds: $y(S^*, u) = \sum_{S' \in B} y^*(S', u) \le \sum_{S' \in B} x^*(S') = x(S^*)$. Next, we show that constraints of type (2) are satisfied. This trivially holds for $S \notin H \cup L^*$ since both $x(S) = 0$, and $y(S, u) = 0$. Constraint (2) holds for $S \in H$, since $x(S) = x^*(S)$, and $y(S, u) = y^*(S, u)$. It remains to consider lines in $S^* \in L^*$. Let B denote the block of lines in L that contains S^*. It follows that

$$\sum_{u \in S^*} y(S^*, u) = \sum_{u \in S^*} \sum_{S \in B} y^*(S, u) \le \sum_{S \in B} \sum_{u \in S} y^*(S, u)$$
$$\le \sum_{S \in B} c(S) x^*(S) \le \max_{S \in B} c(S) \sum_{S \in B} x^*(S) = c(S^*) x(S^*).$$

where the first inequality follows from the fact that some rectangles may lose part of their flow, the second inequality is due to the feasibility of (x^*, y^*), and the third inequality follows from Def. 4. □

Claim 3. *The coverage of every rectangle u is greater than $(1 - 4d\varepsilon)$ in the partial cover (x, y).*

Proof. Consider a rectangle u. We show that, in each dimension, the coverage of u decreases by less than 4ε due to the transition from y^* to y. By definition, coverage by lines in H is preserved. In addition, if a rectangle u intersects all the lines in a block L_j^h, then the coverage of u by lines in L_j^h is now covered by $S_{h,j}^*$. Namely, $\sum_{S \in L_j^h} y^*(S, u) = y(S_{h,j}^*, u)$. It follows that u may lose coverage only in the "leftmost" and "rightmost" blocks that u intersects. In each such block, the coverage of u is bounded by 2ε. Since u is covered in (x^*, y^*), it follows that $\sum_S y(S, u) > 1 - d \cdot 4\varepsilon$, and the claim follows. □

Since $\varepsilon = 1/6d$, by Claim 3 we get that each rectangle is $1/3$-covered by (x, y). A cover is obtained by scaling as follows. Let $x'(S) = \lceil 3x(S) \rceil$ for every $S \in \mathcal{S}$, and $y'(u) = 3y(u)$ for every $u \in \mathcal{U}$. Clearly, every rectangle is covered by (x', y'). Moreover, by Obs. 3 an integral y'' such that (x', y'') is a cover can by computed in polynomial time. It remains to show that (x', y'') is a $6d$-approximation. It suffices to show that $x'(S) \le 6d \cdot x(S)$, for every $S \in H \cup L^*$. If $x(S) \ge 1/3$ then, $x'(S) \le 3x(S) + 1 \le 6x(S)$. If $x(S) < 1/3$, then $x'(S) = 1$ and $x(S) \ge \varepsilon$ for every line $S \in H \cup L^*$. Therefore $x'(S) = 1 = 6d\varepsilon \le 6d \cdot x(S)$, as required.

5   Rectangle Stabbing with Hard Capacities

We present a bi-criteria approximation algorithm for HARD-dRS that computes $16d$-approximate cover that uses at most two copies of each line. The algorithm is similar to the $6d$-approximation algorithm for SOFT-dRS. We first computed an optimal solution for LP-HARD. Afterwards, we set $\varepsilon = \frac{1}{8d}$ and compute H and L^* using the same algorithm defined in the previous section. Finally, we take two copies of each line in $H \cup L^*$ and use flow to compute an integral cover.

We first show that this a cover. The rounding of the LP-solution yields a $(1 - 4d\varepsilon)$-cover according to Claim 3. We obtain a $1/2$-cover simply by setting $\varepsilon = \frac{1}{8d}$. Note that $x(S) \leq 1$, for every line S, hence two copies are not less than scaling by two and rounding up. Note that we rely on Obs. 2 to insure that there is an integral cover using these two copies of each line in the support of x.

The approximation ratio of $16d$ is proved as follows. Note that $x(S) > 0$ only if $S \in H \cup L^*$. Since we take two copies of lines in $H \cup L^*$, it suffices to prove that $|H \cup L^*| \leq 8d \cdot \sum_{S \in \mathcal{S}} x^*(S)$. Clearly, $|H| \leq \frac{1}{\varepsilon} \cdot \sum_{S \in H} x^*(S)$. Due to the bound on the number of blocks (Obs. 5) we obtain, $|L^*| \leq \frac{1}{\varepsilon} \cdot \sum_{S \in L} x^*(S)$. It follows that $|H \cup L^*| \leq \frac{1}{\varepsilon} \cdot \sum_{S \in \mathcal{S}} x^*(S)$, as required.

6 Interval Stabbing with Hard Capacities

In this section we present an 8-approximation algorithm for HARD-1RS. The algorithm augments the positive cover obtained by Claim 3 with $\varepsilon = 1/4$. A local greedy rule is used to select the line to be added to the partial cover.

6.1 Thirsty Lines and Dams

Throughout this section we consider a partial cover (x, y) such that x is integral and y is maximum with respect to x.

Definition 5. *Let (x, y) be a partial cover such that x is integral and y is maximum with respect to x. A line $S \in x$ is a dam with respect to (x, y) if y remains maximum with respect to x even if the capacity $c(S)$ is (arbitrarily) increased. Otherwise, S is thirsty with respect to (x, y).*

Note that if S is not saturated (i.e., $f_y(S) < x(S) \cdot c(S)$), then obviously S is not thirsty, so S is a dam. However, S may be saturated (i.e., $f_y(S) = c(S)$) and yet not thirsty. Such a case is easily described using the network flow formalism: the arc (s, S) belongs to a min-cut in N_x but not to every min-cut.

Lemma 1. *Let (x, y) be a partial cover such that x is integral and y is maximum with respect to x. If $S \in x$ and S is a dam, then (1) every interval $u \in S$ is covered (i.e., $f_y(u) = 1$), and (2) if $u \in S$ and $y(S', u) > 0$, then S' is also a dam.*

Proof. Proof of (1). If u is not covered, then an increase in $c(S)$ can be used to increase $y(S, u)$, contradicting the assumption that S is a dam.

Proof of (2). First, $S' \in x$ since x is integral and $y(S', u) > 0$. We show that if S' is thirsty, then S is also thirsty. Loosely speaking, we show that increasing $c(S)$ enables an increase in the flow, since $y(S', u)$ can be decreased and this "released" flow can be used to serve another interval. We show this formally by presenting an augmenting path in the residual graph of N_x after the capacity of S is increased. Let p be an augmenting path in N_x obtained when $c(S')$ is increased (p exists since S' is thirsty). Obviously, the first arc in p is (s, S'). Observe that the three arcs $(s, S), (S, u)$, and (u, S') are in the residual graph of N_x after $c(S)$ is increased. This follows since: (i) $f_y(S)$ is less than the increased capacity of S, (ii) $f_y(S, u) \leq 1 - y(S', u) < 1 = x(S)$, and (iii) $y(S', u) > 0$. Thus the path $s \to S \to u \to S'$ concatenated with $p \setminus (s, S')$ is an augmenting path in the residual of N_x after the capacity of S is increased, as required.   □

The following corollary is directly implied by Lemma 1.

Corollary 1. *Let (x, y) be a partial cover such that x is integral and y is maximum with respect to x. Define: $D \triangleq \{S \in x \mid S \text{ is a dam}\}$ and $\mathcal{U}_D \triangleq \{u \in \mathcal{U} \mid \exists S \in D \text{ such that } u \in S\}$. Then, for every $u \in \mathcal{U}_D$, $\sum_{S \in D} y(S, u) = 1$.*

Next, we show that if no thirsty lines exist in a positive partial cover, then the cover is feasible.

Corollary 2. *Let (x, y) be a partial cover such that x is integral and y is maximum with respect to x. If every interval is positively covered and no line is thirsty, then (x, y) is a feasible cover.*

Proof. Since every interval is positively covered and there are no thirsty lines, it follows that $\mathcal{U}_D = \mathcal{U}$, and by Coro. 1, every rectangle is covered.   □

6.2   Decomposition into Strips

Let (x, y) be a partial cover, where x is integral and y is maximum with respect to x. Consider two consecutive dams S_1 and S_2 (i.e., there is no dam between S_1 and S_2). The subproblem induced by S_1 and S_2 consists of the following lines and intervals: (i) the vertical lines that are strictly between S_1 and S_2 and (ii) the intervals that are contained in the open strip, the boundaries of which are S_1 and S_2. We refer to the subproblem induced by two consecutive dams as a *strip* and denote it by $B = (\mathcal{S}^B, \mathcal{U}^B)$. Note that extreme dams induce marginal strips that are bounded just from one side.

Definition 6. *The residual capacity of a line $S \in \mathcal{S}^B$ in a strip $B = (\mathcal{S}^B, \mathcal{U}^B)$ is defined by $c^B(S) = \min\{c(S), |S \cap \mathcal{U}^B|\}$.*

Definition 7. *Let (x, y) be a partial cover, where x is integral and y is maximum with respect to x. Let $B = (\mathcal{S}^B, \mathcal{U}^B)$ denote a strip with respect to (x, y). The flow supplied by f_y to strip B is defined by $f_y(B) \triangleq \sum_{S \in \mathcal{S}^B} \sum_{u \in \mathcal{U}^B} y(S, u)$. The deficit in strip B of a partial cover (x, y) is defined by $\Delta_y(B) \triangleq |\mathcal{U}^B| - f_y(B)$. A strip B is called active if $\Delta_y(B) > 0$.*

Let (x, y) denote a partial cover with an integral x and y that is maximum with respect to x. Let $\{B_i\}_{i \in I}$ denote the set of strips induced by the dams corresponding to (x, y). The following observation uses a "flooding" argument to show that feasibility follows from lack of active strips.

Observation 6. *$\Delta_y(B_i) \leq 0$, for every $i \in I$, if and only if (x, y) is feasible.*

6.3   The Approximation Algorithm

The approximation algorithm for HARD-1RS begins like the bi-criteria approximation algorithm and then applies a new augmentation procedure, called **make-feasible**. The algorithm proceeds as follows: (i) Solve LP-HARD. (ii) Set $\varepsilon = 1/4$. Fix x_0 to be the indicator function of the set $H \cup L^*$. Fix y_0 to be the rounding of the LP solution as described in Def. 4. (iii) Apply **aug-flow**(x_0, y_0) to compute a maximum flow y_0' with respect to x_0 that dominates y_0. (iv) Run **make-feasible**(x_0, y_0') to obtain a cover (x^I, y^F) in which x^I is integral but y^F is fractional. (v) Obtain an integral cover (x^I, y^I) using a maximum flow algorithm (Obs. 4).

Algorithm **make-feasible** iteratively augments the partial cover until a cover is obtained. Since a new line is added to the cover in each iteration, the output component x is integral. By Obs. 6, Algorithm **make-feasible** stops when there are no active strips. Otherwise, a new line is added to the cover as follows: (i) pick an active strip B and a line $S_{\max}$ with the largest residual capacity in B, (ii) add $S_{\max}$ to the partial cover x to obtain x', and (iii) find a maximum flow $y' \succeq y$ with respect to x; by calling **aug-flow**(x', y). The algorithm then recurses with (x', y'). Throughout this section, the x-component of every partial cover is integral. To simplify notation, we treat the x-component as the subset itself. So $x' \leftarrow x \cup \{S\}$ means that x' is the indicator function of the subset corresponding to x together with $\{S\}$.

Algorithm 1. make-feasible$(\mathcal{S}, \mathcal{U}, x, y)$

1: Termination condition: If (x, y) is feasible then **Return**(x, y).
2: Let $B = (\mathcal{S}^B, \mathcal{U}^B)$ denote an active strip with respect to (x, y).
3: Find a max-residual-capacity line $S_{\max} \leftarrow \mathrm{argmax}\{c^B(S) : S \in \mathcal{S}^B \setminus x\}$.
4: Add $S_{\max}$ to x: $x' \leftarrow x \cup \{S_{\max}\}$.
5: Augment flow: $y' \leftarrow$ **aug-flow**(x', y).
6: Recurse: **Return make-feasible**$(\mathcal{S}, \mathcal{U}, x', y')$.

First, we show that Algorithm **make-feasible** finds a feasible cover if one exists. Observe that as long as there is an active strip, we add a line $S_{\max}$ to x. As soon as every strip is not active, the cover is feasible by Obs. 6. Hence, it remains to prove that $S_{\max}$ is well defined.

Claim 4. *If $B = (\mathcal{S}^B, \mathcal{U}^B)$ is an active strip, then $\mathcal{S}^B \setminus x \neq \emptyset$.*

Proof. We assume that the problem is feasible. Hence $(\mathcal{S}, y^*)$ is a feasible cover and $(\mathcal{S}^B, y_B^*)$ is a feasible cover of B, where y_B^* is the restriction of y^* to $\mathcal{S}^B \times \mathcal{U}^B$. Assume for the sake of contradiction that $\mathcal{S}^B \subseteq x$. Since y is maximum with respect to x, it follows that (x, y) is feasible in B, which means by Obs. 6 that B is not active, a contradiction.   □

The algorithm runs in polynomial time, since there are at most $|\mathcal{S}|$ recursive calls, and the running time of each recursive call is polynomial by Obs. 4.

Theorem 1. *The approximation ratio of the algorithm for* HARD-1RS *is* $\frac{2}{\varepsilon} = 8$.

The proof is omitted for lack of space. The main idea in the proof is that each line added by Algorithm **make-feasible** to the partial cover becomes a dam together with at least one of the original thirsty lines. Since there are no more than $\frac{1}{\varepsilon}$OPT* lines in $H \cup L^*$, we reach a total of at most 8OPT* lines.

7   Open Problems

We list a few open problems. The hardness of the one-dimensional rectangle stabbing (soft or hard) is open. An $O(d)$ approximation algorithm for HARD-dRS is also open, as well as an $O(d)$ approximation algorithm for weighted SOFT-dRS. In the full version, we show that weighted HARD-dRS is set-cover hard.

Gaur et al. [4] presented a d-approximation algorithm for weighted dRS that uses linear programming to reduce the problem to d one-dimensional instances. Their analysis relies on the integrality of the LP relaxation in the one dimensional case. Our 2-approximation for weighted SOFT-1RS does not prove a bound on the integrality gap. Our 6-approximation algorithm for unweighted SOFT-1RS proves that integrality gap of the one-dimensional case is bounded by 6. Hence another $6d$-approximation ratio follows by combining a reduction similar to the Gaur et al. [4] and our 6-approximation algorithm for SOFT-1RS.

Acknowledgment. We thank Alexander Ageev for pointing out an error in an earlier version of the paper.

References

1. Wolsey, L.A.: An analysis of the greedy algorithm for the submodular set covering problem. Combinatorica **2** (1982) 385–393
2. Chuzhoy, J., Naor, J.: Covering problems with hard capacities. In: 43nd IEEE Symposium on Foundations of Computer Science. (2002) 481–489
3. Guha, S., Hassin, R., Khuller, S., Or, E.: Capacitated vertex covering. Journal of Algorithms **48**(1) (2003) 257–270
4. Gaur, D.R., Ibaraki, T., Krishnamurti, R.: Constant ratio approximation algorithms for the rectangle stabbing problem and the rectilinear partitioning problem. Journal of Algorithms **43** (2002) 138–152
5. Golumbic, M.C.: Algorithmic Graph Theory and Perfect Graphs. Academic Press, New York (1980)
6. Hassin, R., Megiddo, N.: Approximation algorithms for hitting objects with straight lines. Discrete Applied Mathematics **30**(1) (1991) 29–42
7. Bar-Ilan, J., Kortsarz, G., Peleg, D.: Generalized submodular cover problems and applications. Theoretical Computer Science **250** (2001) 179–200
8. Gandhi, R., Halperin, E., Khuller, S., Kortsarz, G., Srinivasan, A.: An improved approximation algorithm for vertex cover with hard capacities. In: 30th Annual International Colloquium on Automata, Languages and Programming. Volume 2719 of LNCS. (2003) 164–175

In-Place Randomized Slope Selection

Henrik Blunck and Jan Vahrenhold

Westfälische Wilhelms-Universität Münster, Institut für Informatik,
48149 Münster, Germany
{blunck, jan}@math.uni-muenster.de

Abstract. *Slope selection* is a well-known algorithmic tool used in the context of computing robust estimators for fitting a line to a collection $\mathcal{P}$ of n points in the plane. We demonstrate that it is possible to perform slope selection in expected $\mathcal{O}(n \log n)$ time using only constant extra space in addition to the space needed for representing the input. Our solution is based upon a space-efficient variant of Matoušek's *randomized interpolation search*, and we believe that the techniques developed in this paper will prove helpful in the design of space-efficient randomized algorithms using samples. To underline this, we also sketch how to compute the repeated median line estimator in an in-place setting.

1 Introduction

Computing a *line estimator*, i.e., fitting a line to a collection $\mathcal{P}$ of n data points $\{p_1, \ldots, p_n\}$ in the plane is a frequent task in statistical analysis. Some estimators (such as the *least squares* estimator) can be determined with little computational effort, they suffer, however, from corruption of the estimate by data outliers. Therefore, the robustness of an estimator is considered essential, and the additional computational cost needed to compute robust estimators is widely accepted [10]. A frequently used, robust line estimator is the so-called *Theil-Sen* estimator (see [14] and the references therein) which considers all $\binom{n}{2}$ lines induced by the pairs of points in $\mathcal{P}$ and selects the line with median slope. In the Computational Geometry community, this problem is known as the *slope selection* problem for which an $\Omega(n \log n)$ lower bound has been established [6].

To obtain efficient algorithms, the slope selection problem is studied in the dual setting where each point (x, y) is identified with the line $\{(\xi, \upsilon) \mid \upsilon = x \cdot \xi - y\}$ and vice versa. Using properties of this duality transform [8], it is easy to show that selecting the k-th smallest slope is dual to the following problem: Given a set of n lines in the plane, find the k-th leftmost intersection point in the arrangement induced by these lines.

While the slope selection problem has been solved optimally both using deterministic and randomized algorithms, we revisit it in the space-efficient model of computation, where the main focus is to use as little extra space as possible over and above the space needed for representing the input. Besides the fact that investigating a possible dependency between time complexity and space requirement is of its own theoretical interest, there are also practical considerations that motivate the design and analysis of space-efficient algorithms. First

T. Calamoneri, I. Finocchi, G.F. Italiano (Eds.): CIAC 2006, LNCS 3998, pp. 30–41, 2006.

and foremost, algorithms with little space overhead have the potential of using the different stages of hierarchical memory, e.g., caches, to a much higher degree of efficiency. Another motivation, especially for designing algorithms for statistical data analysis, comes from the recently increased interest in sensor networks where small-scale computing devices are used to collect large amounts of data. Since the memory of such sensor devices usually is limited, and since transmitting data is more costly than local computation, it is desirable to process as much data as possible locally before transmitting (intermediate) results.

Related Work. A variety of deterministic algorithms for solving the slope selection problem in optimal $\mathcal{O}(n \log n)$ running time have been presented [4, 6, 11], however, as noted by Matoušek *et al.* [14], all of them are based on relatively complicated concepts such as parametric search, sorting network, expander graphs, or cuttings. To obtain more practical results, *randomized* approaches have been taken which resulted in several algorithms with expected $\mathcal{O}(n \log n)$ running time [7, 13, 17]. Due to space constraints, we refer the reader to the article by Matoušek *et al.* [14] (and to the references therein) for a discussion of how the slope selection problem relates to the computation of robust estimators.

The Model. The goal of investigating space-efficient algorithms is to design algorithms that use very little extra space in addition to the space used for representing the input. The input is assumed to be stored in an array A of size n, thereby allowing random access. We assume that a constant size memory can hold a constant number of words and that each word can hold one pointer, or an $\mathcal{O}(\log n)$ bit integer, and a constant number of words can hold one element of the input array. An *in-place* algorithm uses $\mathcal{O}(1)$ extra words of memory. Some fundamental geometric problems such as 2D convex hulls and closest pairs can be solved *in-place* and in optimal time [1, 2, 5]. More involved problems such as range searching and line-segment intersection (currently) can be solved in-place only in near-optimal running time [3, 18], or, as for the case of 3D convex hulls and related problems, using both (poly-)logarithmic extra space and time [3].

Our Results. In this paper we show how to solve the slope selection problem in expected $\mathcal{O}(n \log n)$ running time while at the same time using only constant extra space. Our algorithm follows the approach of Matoušek [13], and thus we also devise an in-place variant of his *randomized interpolation search* technique. This variant, together with a algorithmic subroutine for efficiently constructing and storing a set of randomly sampled intersections, is of independent interest, since it was introduced by Matoušek to substitute Meggiddo's *parametric search* technique [15]. Furthermore, we sketch how to use these results to obtain an in-place algorithm for computing the *repeated median line estimator*.

2 Space-Efficient Algorithmic Tools

Space-efficient algorithms have been investigated since the 1960's, and there exists a variety of algorithmic building blocks that require little extra space in addition to the space needed for representing the input. *Heapsort* [19] sorts n elements in-place performing $\mathcal{O}(n \log n)$ operations. Also, we can do linear-time median-finding in-place—even while maintaining the data sorted according to some other total order [1]. Finally, we will use the linear-time merging algorithm by Geffert *et al.* [9] that merges two sorted subarrays using constant extra space.

A standard technique in the design of space-efficient algorithms is to encode a single bit by a permutation of two objects q and r [16]. In our case, for lines q, r with $q <_\xi r$ (i.e., with q preceding r in the vertical order at $x = \xi$) the permutation qr encodes a binary zero, and the permutation rq encodes a binary one (by the nature of the problem setting, the input does not contain duplicates). Given this representation, we can encode any integer in the range $[0, \ldots, n - 1]$, e.g., an index referencing an array entry, using $2 \cdot \log n$ lines, and the time needed for decoding or modifying any such pointer is $\mathcal{O}(\log n)$.

As a subroutine of our algorithm, we will need to sort a set of encoded pointers according to some order imposed on the elements referenced by these pointers. If we use *heapsort* to sort a set of r elements (pointers), we perform $\mathcal{O}(r \log r)$ operations. Since we may need to decode and dereference $\mathcal{O}(1)$ pointers to perform one such operation, the overall running time is $\mathcal{O}(r \log r \log n)$. Note, that some pointers may reference lines that are part of a pointer encoding, and thus, whenever the sorting algorithm tells us to swap two pointers, we must not move around the lines representing these pointers. Instead, we simply swap the encoded *values* by making the first pointer reflect the value of the second and vice versa. As a consequence, no line will be more than one position off its correct location (and we can check this in constant time per access). Each swap resp. update can be done in-place and in $\mathcal{O}(\log n)$ time. This results in the following corollary:

Corollary 1. *We can sort r binary encoded pointers of size $\mathcal{O}(\log n)$ bits each in-place and in $\mathcal{O}(r \log r \log n)$ time while preserving the referenced information.*

3   Randomized Interpolation Search

Due to the duality transform discussed in the introduction, the problem of selecting the line with median slope translates to selecting the intersection with median x-coordinate in the arrangement induced by the lines dual to the points in $\mathcal{P}$. Since the duality transform is done read-only by simply reinterpreting the coordinate-based representation of a point as the slope-intercept representation of a line, we assume that our input is given as a set $\mathcal{P}$ of lines in the plane.

Since the arrangement induced by a set of n lines contains $\Theta(n^2)$ intersections, it is infeasible (both with respect to the desired $\mathcal{O}(n \log n)$ running time and with respect to the in-place setting) to compute all intersections. Instead, the algorithm of Matoušek [13] maintains a vertical strip $\langle b, e \rangle := [b, e] \times \mathbb{R} \subset \mathbb{R}^2$ that is guaranteed to contain the k-th smallest intersection point. This strip is iteratively refined by first constructing a sample $\mathcal{R}$ of r intersection points (r is a parameter to be defined later) and by then constructing a (narrower) candidate strip $\langle b', e' \rangle$ based upon the x-coordinates of two appropriately chosen sampled intersection points. The algorithm then checks whether $\langle b', e' \rangle$ indeed contains the k-th smallest intersection point. If this is not the case, the process is repeated for $\langle b, e \rangle$ but using a new sample $\mathcal{R}$, else the algorithm iterates with the refined strip $\langle b', e' \rangle$. Since the above technique refines the strip $\langle b, e \rangle$ based upon a non-constant number of randomly selected samples, it is referred to as *randomized interpolation search*, and Matoušek observes that it can be used as a randomized substitute for Megiddo's *parametric search* technique [15].

The efficiency of the resulting algorithm for slope selection is based upon the following lemma which (applied iteratively) implies that the number $|\mathcal{I}(b, e)|$ of intersections that lie inside $\langle b, e \rangle$ can be reduced to $\mathcal{O}(r)$ using an expected constant number of iterations:

Lemma 1 (Lemma 2.1 in [14]). *Given a set of numbers $\Theta = \{\theta_1, \theta_2, \ldots, \theta_N\}$, an index k ($1 \leq k \leq N$), and an integer $r > 0$, we can compute in $\mathcal{O}(r)$ time an interval $[\theta_{\text{lo}}, \theta_{\text{hi}}]$, such that, with probability $1 - 1/\Omega(\sqrt{r})$, the k-th smallest element of Θ lies within this interval, and the number of elements in Θ that lie within the interval is at most $N/\Omega(\sqrt{r})$.*

We use the algorithm implied by the above lemma to compute the intersection with the k-th smallest x-coordinate among the $N = \binom{n}{2}$ intersections induced by the lines in $\mathcal{P}$. Since Matoušek *et al.* [14] proved that it is possible to choose $r := \lceil N^\beta \rceil$ for any $0 < \beta < 1$ while maintaining the asymptotic efficiency of the resulting algorithm, we set $r := \lceil \sqrt{n} \rceil$ as opposed to the original choice of $r := n$ [13]. As Matoušek [13] computes $[\theta_{\text{lo}}, \theta_{\text{hi}}]$ using $\mathcal{O}(r)$ sampled intersections, the bound on the running time given in Lemma 1 will be replaced by the $\mathcal{O}(n \log n)$ time bound for the in-place sampling algorithm discussed in the next section. In any case, we note that the probability and the bound on the size of the elements within $[\theta_{\text{lo}}, \theta_{\text{hi}}]$ is independent of whether or not the algorithm is implemented in-place. The main algorithm for slope selection (as presented by Matoušek [13]) is given below as Algorithm 1.

It remains to describe how to construct (and store!) the sampled set $\mathcal{R}$ of $r = \lceil \sqrt{n} \rceil$ intersections in an in-place setting, i.e., using only constant extra space. Furthermore, we need to discuss how to compute $|\mathcal{I}(b, e)|$. To this effect, we describe an algorithm for the former task, which also provides a solution for the latter task. Anticipating the results presented in the next section, we combine them with the above lemma and the original analyses of Matoušek *et al.* [13, 14]:

Theorem 1. *The slope selection problem for a set of n input points in the plane can be solved in-place and in expected $\mathcal{O}(n \log n)$ running time.*

4 Constructing a Random Sample $\mathcal{R}$ of r Intersections

In this section, we describe an algorithmic subroutine that will be invoked from Line 5 of the overall slope selection algorithm (Algorithm 1) with three parameters b, e, and r. Its purpose is to draw a random sample (with replacement) of size $r := \lceil \sqrt{n} \rceil$ from the set of all intersections induced by the lines in $\mathcal{P}$ and falling within $\langle b, e \rangle$. We demonstrate that we can compute and represent these r

---

Algorithm 1. Algorithm SELECT($A[0, \ldots, n-1], k, r$) for selecting the k-th leftmost intersection point in the arrangement induced by the lines in A [13]

1: Let $\kappa := k$. Let $b := -\infty$ and $e := \infty$.   {Initial "guess" for the candidate strip.}
2: **while** not finished **do**
3:    Let $N := |\mathcal{I}(b, e)|$.   {Number of intersections inside $\langle b, e \rangle$.}
4:    **if** $N > r$ **then**
5:        Draw (with replacement) a random sample $\mathcal{R}$ of size r from $\mathcal{I}(b, e)$.
6:        Let $\kappa := (r/N)(k - |\mathcal{I}(-\infty, b)|)$, $\kappa_{b'} := max(1, \lfloor \kappa - 3\sqrt{r}/2 \rfloor)$, and
          $\kappa_{e'} := min(r, \lfloor \kappa + 3\sqrt{r}/2 \rfloor)$.
7:        Compute the intersections with ranks κ, $\kappa_{b'}$, $\kappa_{e'}$ in $\mathcal{R}$ (ranks are with respect
          to the order of the x-coordinates).
8:        Let b' be the x-coordinate of the intersection in $\mathcal{R}$ with rank $\kappa_{b'}$, and let e' be
          the x-coordinate of the intersection with rank $\kappa_{e'}$.
9:        **if** $|\mathcal{I}(-\infty, b')| < k < |\mathcal{I}(-\infty, e')|$ **then**
10:           Let $b := b'$ and $e := e'$.   {The k-th leftmost intersection point is in $\langle b', e' \rangle$;
              update b and e.}
11:       **end if**
12:    **else**
13:       Report the intersection with rank κ in $\mathcal{R}$ as the k-th leftmost intersection
          point and finish.
14:    **end if**
15: **end while**

---

intersections in-place and in time $\mathcal{O}(n \log n + r^2 \log n)$. Due to space constraints, we refer the reader to the full version of this paper for the proofs of the respective running times.

4.1   An Overview of the Algorithm

Our subroutine follows the approach of Matoušek [13, Lemma 1]: we first draw (with replacement) a set of r random integers from $\{0, \ldots, |\mathcal{I}(b, e)| - 1\}$ where $|\mathcal{I}(b, e)|$ is the number of intersections in $\langle b, e \rangle$ and sort these numbers. These numbers give the ranks of the intersections with respect to the order in which they are found. The main ingredient used for efficiently computing intersections in $\langle b, e \rangle$ is the following well-known observation: the number of intersections inside $\langle e, b \rangle$ is exactly the number of inversions between the permutation of $\mathcal{P}$ that arranges the lines in sorted $<_b$-order (the vertical order at $x = b$) and the permutation that arranges the lines in sorted $<_e$-order (the vertical order at $x = e$) [12]. Thus, to efficiently compute $|\mathcal{I}(b, e)|$, we can run the classic divide-and-conquer algorithm for inversion counting (see, e.g., [12]). While doing so, we keep track of the total number of inversions/intersections seen so far and "record" an intersection if its rank matches one of the r given ranks. Since we have sorted the ranks, we can process each of them in constant extra time.

Running Matoušek's algorithm in an in-place setting is complicated by two facts: (1) The recursion stack of an divide-and-conquer algorithm may require $\Omega(\log n)$ extra words, and (2) we need to store the r ranks and the intersections computed so far. The problem of performing a recursive algorithm in-place has been discussed earlier [1], and we choose to process the "recursion tree" implicitly in a bottom-up, level-by-level traversal. We overcome the second problem by running the algorithm in three phases: in the first phase, we process the lines stored in $A[0, \ldots, \lfloor n/2 \rfloor - 1]$ and use $A[\lfloor n/2 \rfloor, \ldots, n-1]$ to encode the ranks and the intersections found so far, and in the second phase, we reverse the roles of both subarrays. We finalize the algorithm with a third phase that processes the intersections induced by lines stored in different halves of the array.

4.2   In-Place Data Structures for Recording Intersections

Employing the encoding technique described in Section 2, we use the subarray of size $n/2$ that does not contain the lines to be processed in the current phase to represent three (implicit) data structures $\mathcal{D}_R$, $\mathcal{D}_L$, and $\mathcal{D}_I$. Each of these data structures is represented in a subarray containing $4 \cdot r \log n$ lines. For reasons that will become evident in Section 4.3, we also reserve a scratch space of the same size as the data structure $\mathcal{D}_L$.[1] The array thus is partitioned as follows:

Lines to be processed		$\mathcal{D}_R$	Scratch	$\mathcal{D}_L$	$\mathcal{D}_I$
0		$\frac{1}{2}n$			$n-1$

Storing ranks. The randomly generated ranks in the range $[0, \ldots, n^2 - 1]$ are encoded in a "sorted-list" data structure $\mathcal{D}_R$. During the initialization of $\mathcal{D}_R$, these ranks are sorted using *heapsort* in $\mathcal{O}(r \log r \log n)$ time (see Section 2). Thus, our algorithm will be able to traverse the list and report each rank to be processed in $\mathcal{O}(\log n)$ time.

Storing lines involved in intersections. We use a "sorted-list" data structure $\mathcal{D}_L$ to record (references to) lines involved in all of the sampled intersections found so far. These (references to) lines are maintained in sorted $<_b$-order. Every reference is inserted into $\mathcal{D}_L$ using insertion sort (ignoring duplicates), leading to a $\mathcal{O}(r^2 \log n)$ global cost for maintaining $\mathcal{D}_L$.

Storing intersections. The "linked-list" data structure $\mathcal{D}_I$ records the intersections found so far by indexing into $\mathcal{D}_L$. To add an intersection induced by two lines ℓ_1 and ℓ_2 to $\mathcal{D}_I$, we first insert (references to) ℓ_1 and ℓ_2 in sorted $<_b$-order into $\mathcal{D}_L$ and then append the pair (i, j) referencing (the references in $\mathcal{D}_L$ to) these two lines at the end of $\mathcal{D}_I$. The cost for performing a single insertion into $\mathcal{D}_I$ is in $\mathcal{O}(\log n)$, and thus the global cost is in $\mathcal{O}(r \log n)$.

As a consequence, we have the following lemma:

Lemma 2. *The global cost for maintaining the data structures $\mathcal{D}_R$, $\mathcal{D}_L$, and $\mathcal{D}_I$ needed to record the r intersections in an in-place manner is $\mathcal{O}(r^2 \log n)$.*

4.3   Processing One Half of the Subarray

The algorithms for processing the halves of $A[0, \ldots, n-1]$ are symmetric, and thus we present the algorithm for processing the subarray $A[0, \ldots, \lfloor n/2 \rfloor - 1]$.

---

[1] We may need to adjust r by a constant factor such as to ensure that $16 \cdot r \log n \leq n/2$.

Counting Inversions. The algorithm for counting *all* inversions in $\langle b, e \rangle$ is an extension of the iterative *mergesort* algorithm: starting from the set of lines in sorted $<_b$-order, the algorithm iteratively merges the lines into $<_e$-order while counting inversions. During each *merge*-step of the algorithm, two subarrays already in sorted $<_e$-order are merged into a single $<_e$-sorted subarray. Each of these subarrays has been processed during the previous iteration, and thus all inversions involving lines from only one of these subarrays have been processed. Since all elements in $\mathtt{A_1}$ precede any element in $\mathtt{A_2}$ with respect to the $<_b$-order, the inversions induced by lines from different subarrays can be computed using the following observation [12, 13]: Each element $a_i \in \mathtt{A_1}$ induces an inversion with all elements in $\mathtt{A_2}$ preceding it in sorted $<_e$-order. Thus, to compute the number of inversions it is sufficient to maintain a counter that records how many elements from $\mathtt{A_2}$ have been written into sorted $<_e$-order. An obvious, yet crucial, fact guaranteeing the correctness of the inversion counting algorithm is that any two subarrays $\mathtt{A_1}$ and $\mathtt{A_2}$ that are merged in the j-th iteration of processing the m-th bottom-most level of the recursion tree are of the form $\mathtt{A_1} := \mathtt{A}[j \cdot 2^m, \ldots, (j+1) \cdot 2^m - 1]$ and $\mathtt{A_2} := \mathtt{A}[(j+1) \cdot 2^m, \ldots, (j+2) \cdot 2^m - 1].^2$

This leads to the following corollary which we can prove using induction and the fact that the algorithm starts with all lines in sorted $<_b$-order:

Corollary 2. *Let $\mathtt{A_1}$ and $\mathtt{A_2}$ be two subarrays of $\mathtt{A}$ that are merged in one step of the algorithm for inversion counting. Furthermore, define $\ell_{\min}$ to be the minimal line of $\mathtt{A_1} \cup \mathtt{A_2}$ with respect to the $<_b$-order and define $\ell_{\max}$ to be the maximal line of $\mathtt{A_1} \cup \mathtt{A_2}$ with respect to the $<_b$-order. Then, $\mathtt{A_1} \cup \mathtt{A_2}$ consists exactly of those lines ℓ in $\mathtt{A}$ for which $\ell_{\min} \leq_b \ell \leq_b \ell_{\max}$ holds.*

The algorithm COUNTANDRECORD (Algorithm 2) extends the above inversion counting algorithm to also record all intersections inside $\langle b, e \rangle$ whose ranks are recorded in $\mathcal{D}_R$.

Algorithm 2 can be implemented using constant extra space, and excluding the cost for recording the relevant intersections, the cost for running Algorithm 2 is linear in the size of the union of the two subarrays to be merged. Furthermore, Algorithm 2 can be simplified (by leaving out the code in Lines 6–9) to compute $|\mathcal{I}(b, e)|$ as required in Lines 3 and 9 of Algorithm 1. Afterwards, $\mathtt{A_1}$ and $\mathtt{A_2}$ can be merged using, e.g., the linear-time merging algorithm of Geffert *et al.* [9]. In combination with a bottom-up divide-and-conquer approach, this gives an in-place $\mathcal{O}(n \log n)$ implementation for the "inversion counting" in Algorithm 1.

Merging Two Subarrays into Sorted $<_e$-order. When actually merging $\mathtt{A_1}$ and $\mathtt{A_2}$ into sorted $<_e$-order, we also need to update the values stored in $\mathcal{D}_L$, since they reference the lines involved in the intersections found so far by directly indexing into $\mathtt{A}$. Merging $\mathtt{A_1}$ and $\mathtt{A_2}$ seems to corrupt the information recorded in $\mathcal{D}_L$, but fortunately the information of what goes where can be computed on the fly while running COUNTANDRECORD. To see this, observe that the

---

^{2} As noted earlier [1], the in-place divide-and-conquer scheme can be modified easily to correctly handle instances where the problem size is not a power of two.

Algorithm 2. Algorithm $\text{COUNTANDRECORD}(\text{A}_1, \text{A}_2, \langle b, e \rangle, c, \mathcal{D}_I)$ for incrementing the count c of intersections seen so far by the number of intersections induced by lines in A_1 and A_2 while recording all relevant intersections in $\mathcal{D}_I$. Only intersections that fall inside $\langle b, e \rangle$ are considered.

Require: A_1 and A_2 are sorted according to $<_e$. Each element in A_1 precedes all elements in A_2.

Ensure: A_1 and A_2 have not been modified, all intersections within $\langle b, e \rangle$ induced by lines from A_1 and A_2 have been recorded, and c reflects the number of intersections seen so far.

1: Let $i_1 := 0$ and $i_2 := 0$.   {Current position in subarray.}
2: **for** $i = 0$ to $\text{length}(\text{A}_1 \cup \text{A}_2) - 1$ **do**
3:    Let $\ell_1 := \text{A}_1[i_1]$ and $\ell_2 := \text{A}_2[i_2]$.   {The i-th element in sorted order is ℓ_1 or ℓ_2.}
4:    **if** $\ell_1 <_e \ell_2$ **then**
5:       Let $c_{i_1} := i_2$. {$\sharp$(inversions induced by ℓ_1) $= \sharp$(elements in A_2 preceding ℓ_1).}
6:       **for** each rank ρ in $\mathcal{D}_R \cap [c, \ldots, c + c_{i_1}]$ **do**
7:          Let ℓ be the line stored at $\text{A}_2[\rho - c]$.
8:          Update $\mathcal{D}_I$ to record the pair (ℓ_1, ℓ) as the intersection with rank ρ.
9:       **end for**
10:       Let $c := c + c_{i_1}$.   {Count intersections.}
11:       $i_1 := i_1 + 1$.   {Advance i_1.}
12:    **else**
13:       $i_2 := i_2 + 1$.   {Advance i_2 but do not do anything else.}
14:    **end if**
15: **end for**

---

algorithm COUNTANDRECORD is an iterative algorithm that computes, during the i-th iteration the element with rank i in the final sorted order (Line 3 of Algorithm 2). Thus we only need to be able to find out whether the line ℓ that will be the i-th element in sorted order is an element involved in an intersection. If this is the case, we simply update the reference pointing to ℓ to point to ℓ's position *after* the merge step: the i-th position in the union of A_1 and A_2. In order not to corrupt the values of $\mathcal{D}_L$ that are needed for correctly updating $\mathcal{D}_I$, we do not directly modify $\mathcal{D}_L$ to record the "new" position of each referenced entry. Instead, prior to processing each level of the recursion tree, we copy $\mathcal{D}_L$ into the scratch space that has been reserved earlier (see Section 4.2) and denote this copy as $\mathcal{D}_L'$. This copy will be used to record the updated values of $\mathcal{D}_L$.

Selecting the lines in A_1 and A_2 referenced from $\mathcal{D}_L$. Our algorithm for efficiently determining whether a line ℓ to be processed is one of the lines referenced from $\mathcal{D}_L$ is based upon Corollary 2. This corollary states that the entries in $\mathcal{D}_L$ that reference lines in the union of A_1 and A_2 are stored consecutively in a (possibly empty) sublist $\mathcal{D}_L[a_1, \ldots, a_2 - 1]$.

$\mathcal{D}_L$		References to $\text{A}_1 \cup \text{A}_2$	
	0	a_1	a_2	$2r - 1$

We obtain this sublist as follows: Prior to running the algorithm COUNTAND-RECORD on two subarrays A_1 and A_2, we scan $A_1 \cup A_2$ to determine (in linear time) the highest and lowest line of $A_1 \cup A_2$ with respect to the $<_b$-order. Using this information, we determine the (possibly empty) sublist $\mathcal{D}_L[a_1, \ldots, a_2 - 1]$ that contains all elements in $\mathcal{D}_L$ that reference any line in $A_1 \cup A_2$. Corollary 2 guarantees that the index a_1 used when merging the next two adjacent subarrays is obtained by setting $a_1 := a_2$ and that a_2 can be updated by scanning forward from $\mathcal{D}_L[a_2]$, i.e., a_1 and a_2 can be updated iteratively.

To efficiently use this sublist, we sort both $\mathcal{D}_L[a_1, \ldots, a_2 - 1]$ and its copy $\mathcal{D}'_L[a_1, \ldots, a_2 - 1]$ *according to the $<_e$-order*. This order allows us to process not only A_1 and A_2 but also all references in $\mathcal{D}_L[a_1, \ldots, a_2 - 1]$ during a synchronized scan. All of the above operations can be done in a (globally) efficient way:

Lemma 3. *The global cost for updating a_1 and a_2 is $\mathcal{O}(n \log n + r \log^2 n)$.*

Updating the relevant references in $\mathcal{D}_L$. Recall that the merging algorithm determines, during its i-th iteration, the line ℓ that is the i-th line of $A_1 \cup A_2$ in sorted $<_e$-order (Line 3 of Algorithm 2). During the synchronized scan, we keep track of the lowest line μ referenced from $\mathcal{D}_L[a_1, \ldots, a_2 - 1]$ *not* lower than ℓ; μ is the next line for which we need to record its new position. Let j be the index such that $\mathcal{D}_L[j]$ references μ. In case ℓ equals μ, we record that μ will reside at the i-th position of $A_1 \cup A_2$ in sorted $<_e$-order by updating $\mathcal{D}'_L[j]$ accordingly.

To make Algorithm 2 reflect these updates, we need to modify Lines 11 and 13 such as to compare the current line to μ and to update the index j as needed. This results in $\mathcal{O}(\log n)$ time spent per step in which the index j is updated, i.e., in $\mathcal{O}(r \log n)$ time per level of the recursion tree, since we can store the current value of μ using constant extra space and thus allow for constant-time access to μ. Thus, the overall extra time spent in these steps is in $\mathcal{O}(r \log^2 n)$.

Caveat: Updating $\mathcal{D}_I$ and $\mathcal{D}_L$. Adding a new intersection to $\mathcal{D}_I$ involves adding up to two (references to) lines to $\mathcal{D}_L$, and this is the sole reason for working with the copy $\mathcal{D}'_L$ of $\mathcal{D}_L$ and thus requiring extra scratch space. First of all, since all intersections recorded involve lines from $A_1 \cup A_2$, any (reference to a) line ℓ that is added to $\mathcal{D}_L$ will increase the value of a_2 by one. Furthermore, to keep $\mathcal{D}_L[a_1, \ldots, a_2 - 1]$ in sorted $<_e$-order, we need to change the order with respect to which we insert into $\mathcal{D}_L$ from the $<_b$-order to the $<_e$-order as soon as the insertion process reaches $\mathcal{D}_L[a_1, \ldots, a_2 - 1]$. These operations are simultaneously performed on $\mathcal{D}'_L$ as well.

For each component of a pair of indices that is added to $\mathcal{D}_I$ to record a newly-found intersection, we need to make sure that it encodes the correct position of the referenced line ℓ in $\mathcal{D}_L$, i.e., the correct position of ℓ w.r.t. the *sorted $<_b$-order*. To compute this position, we scan $\mathcal{D}_L[a_1, \ldots, a_2 - 1]$ to count the number of lines preceding ℓ in sorted $<_b$-order. (Note that this computation is only possible because, during the inversion counting step, we have updated the references in $\mathcal{D}'_L$ instead of the references in $\mathcal{D}_L$; thus, the entries in $\mathcal{D}_L$ still correctly reference the relevant lines at their *current* position.) If j lines precede ℓ, we update all indices in $\mathcal{D}_I$ referencing entries in $\mathcal{D}_L[a_1 + j, \ldots, |\mathcal{D}_L| - 1]$.

After having run the merging algorithm on A_1 and A_2, we reestablish the correct relation between entries in $\mathcal{D}_I$ and entries in $\mathcal{D}_L$: We replace the values encoded in $\mathcal{D}_L[a_1, \ldots, a_2 - 1]$ with the values encoded in $\mathcal{D}'_L[a_1, \ldots, a_2 - 1]$ and sort them according to $<_b$. This finishes processing the subarrays A_1 and A_2.

Lemma 4. *The global extra cost incurred by updating the references stored in the data structure $\mathcal{D}_L$ while merging subarrays is in $\mathcal{O}(r \log r \log^2 n)$.*

4.4   Finishing Up

After we have processed the first half of the input array using the second half to maintain the data structures $\mathcal{D}_R$, $\mathcal{D}_L$, and $\mathcal{D}_I$, we now reverse the roles of the two subarrays. To do so, we first need to copy the contents of the data structures to the first half of the array. The important detail to keep in mind is that, as a result of the inversion-counting algorithm, the first half of the array is sorted according to $<_e$. Thus, when copying the contents of the data structures to the first half of the array, the order according to which we have to decide whether two adjacent elements encode a binary zero or a binary one, is the $<_e$-order.

After we have run our subroutine on $A[\lfloor n/2 \rfloor, \ldots, n - 1]$, we need to finalize the algorithm by processing all intersections induced by lines stored in different halves of the array. As it turns out, however, we do not need to actually merge the lines in $A[0, \ldots, \lfloor n/2 \rfloor - 1]$ and $A[\lfloor n/2 \rfloor, \ldots, n - 1]$—it is sufficient to count the inversions and to construct the relevant intersections. This means that we can simply run the algorithm COUNTANDRECORD (Algorithm 2) without the modifications needed to record the "what-goes-where" information. As discussed in Section 2, the lines used to encode the data structures $\mathcal{D}_R$, $\mathcal{D}_L$, and $\mathcal{D}_I$ are at most one position off their correct position (in sorted $<_e$-order), and thus we can process them in sorted $<_e$-order with constant extra (look-ahead) space.

As a result of this final invocation of the algorithm COUNTANDRECORD, the data structure $\mathcal{D}_I$ will reference r pairs of entries in $\mathcal{D}_L$ which in turn reference pairs of lines in A. To select the intersections with rank κ, $\kappa_{b'}$, and $\kappa_{e'}$ (Line 7 of Algorithm 1), we could invoke a linear-time median-find algorithm. However, since we have chosen r small enough, we can simply use the algorithm implied by Corollary 1 to sort the pairs in $\mathcal{D}_I$ according to the x-coordinate of their intersection. The running time (including the time needed for resolving one level of indirection) is $\mathcal{O}(r \log r \log n)$. Combining this with Lemmas 2, 3, and 4 and the fact that $r \in \mathcal{O}(\sqrt{n})$, we obtain the following lemma:

Lemma 5. *A random sample $\mathcal{R}$ of $r = \lceil \sqrt{n} \rceil$ intersections inside a strip $\langle b, e \rangle$ can be constructed in-place and in $\mathcal{O}(n \log n)$ time.*

The same algorithm can be used to explicitly construct *all* of the at most r intersection points that are considered in the last iteration of Algorithm 1 (Line 13). This finishes the proof of Theorem 1.

5 Computing the Repeated Median Line Estimator

The *repeated median line estimator* is an estimator for line-fitting, which is even more robust with respect to data outliers—see [14] and the references therein. The repeated median line estimator is obtained by first computing, for each input point p_i the line m_i with median slope among all $n-1$ lines induced by p_i and another point in $\mathcal{P} \setminus \{p_i\}$. Among all such lines m_i, the line m that realizes the line with median slope is selected as the repeated median line estimator. In the dual setting this corresponds to finding, for each line ℓ, the intersection point on ℓ with the median x-coordinate and then to compute the median of these medians. Since the underlying algorithm is considerably more involved and maintains a much larger amount of statistical information, we are unable to match the expected running time of the original algorithm, and our in-place implementation exhibits an expected near-optimal running time.

Matoušek *et al.* [14] describe multiple variants of a randomized algorithm for computing a repeated median line estimator using $\mathcal{O}(n)$ extra space. The general approach is to proceed using randomized interpolation search in an only slightly different setting than the one described for slope selection (Section 3). In their case, the set $\mathcal{R}$ of samples drawn in each iteration is drawn from the set $\mathcal{P}$, i.e., from the set of input lines. Then, they compute the repeated median line estimator for all (points dual to) lines in $\mathcal{R}$ and a candidate strip $\langle b, e \rangle$ that is supposed to contain the (global) final result. To verify that $\langle b, e \rangle$ indeed contains the median-of-median intersection, the algorithm of Matoušek *et al.* then counts the number of medians that lie to the left, inside, and to the right of $\langle b, e \rangle$, respectively. They do so by computing all of these counts during a single run of a modification of the inversion-counting algorithm. During this run, counters L_i and I_i are maintained for each line ℓ_i that record the number intersections involving ℓ_i left of, resp. inside $\langle b, e \rangle$.

The main purpose of computing the counters L_i and I_i is to determine, for each line ℓ_i independently, whether its median intersection lies inside $\langle b, e \rangle$. Processing all n lines in one batch (as described above) is done for the sole purpose of efficiency. Since, in an in-place setting, we cannot accommodate $\mathcal{O}(n)$ counters in a subarray of size n, we need to process the lines in smaller batches. If we choose the batch size to be $\mathcal{O}(n/\log n)$, we can compute all counts in-place. This, however, increases the number of runs needed for checking whether $\langle b, e \rangle$ is a valid candidate from one to $\mathcal{O}(\log n)$. Additionally, each step of the algorithm for updating the counters now takes $\mathcal{O}(\log n)$ extra time for updating a binary-encoded value, and thus we loose another $\mathcal{O}(\log n)$-factor in the overall running time. Finally, the most efficient strategy for constructing the candidate strip $\langle b, e \rangle$ described in [14], computing the repeated median line by range searching and counting techniques, seems to be unavailable in an in-place setting. As a consequence, we have to resort to an in-place variant of a less efficient randomization technique, that (in contrast to the situation described in Section 3) results in an expected number of $\mathcal{O}(\log n)$ iterations of the global algorithm.

Lemma 6. *The repeated median line estimator can be computed in-place and in expected $\mathcal{O}(n \log^4 n)$ time.*

References

1. P. Bose, A. Maheshwari, P. Morin, J. Morrison, M. Smid, and J. Vahrenhold. Space-efficient geometric divide-and-conquer algorithms. *Computational Geometry: Theory & Applications*, 2006. To appear, accepted November 2004.
2. H. Brönnimann and T. M.-Y. Chan. Space-efficient algorithms for computing the convex hull of a simple polygonal line in linear time. *Computational Geometry: Theory & Applications*, 2006. In press.
3. H. Brönnimann, T. M.-Y. Chan, and E. Y. Chen. Towards in-place geometric algorithms. In *Proc. 20th Symp. Computational Geometry*, pp. 239–246, 2004.
4. H. Brönnimann and B. M. Chazelle. Optimal slope selection via cuttings. *Computational Geometry: Theory and Applications*, 10(1):23–29, 1998.
5. H. Brönnimann, J. Iacono, J. Katajainen, P. Morin, J. Morrison, and G. T. Toussaint. Space-efficient planar convex hull algorithms. *Theoretical Computer Science*, 321(1):25–40, 2004.
6. R. Cole, J. S. Salowe, W. L. Steiger, and E. Szemerédi. An optimal-time algorithm for slope selection. *SIAM J. Computing*, 18(4):792–810, 1989.
7. M. B. Dillencourt, D. M. Mount, and N. S. Nethanyahu. A randomized algorithm for slope selection. *Intl. J. Computational Geometry and Applications*, 2(1):1–27, 1992.
8. H. Edelsbrunner. *Algorithms in Combinatorial Geometry*. Springer, Berlin, 1987.
9. V. Geffert, J. Katajainen, and T. Pasanen. Asymptotically efficient in-place merging. *Theoretical Computer Science*, 237(1–2):159–181, 2000.
10. P. Huber. *Robust Statistics*. Wiley, New York, 1981.
11. M. J. Katz and M. Sharir. Optimal slope selection via expanders. *Information Processing Letters*, 47(3):115–122, 1993.
12. J. Kleinberg and É. Tardos. *Algorithm Design*. Addison-Wesley, Boston, 2006.
13. J. Matoušek. Randomized optimal algorithm for slope selection. *Information Processing Letters*, 39(4):183–187, 1991.
14. J. Matoušek, D. M. Mount, and N. S. Nethanyahu. Efficient randomized algorithms for the repeated median line estimator. *Algorithmica*, 20(2):136–150, 1998.
15. N. Megiddo. Applying parallel computation algorithms in the design of serial algorithms. *J. ACM*, 30(4):852–865, 1983.
16. J. I. Munro. An implicit data structure supporting insertion, deletion, and search in $O(\log^2 n)$ time. *J. Computer and System Sciences*, 33(1):66–74, 1986.
17. L. Shafer and W. L. Steiger. Randomizing optimal geometric algorithms. In *Proc. 5th Canadian Conference on Computational Geometry*, pp. 133–138, 1993.
18. J. Vahrenhold. Line-segment intersection made in-place. In *Proc. 9th Workshop Algorithms and Data Structures, Lecture Notes in Computer Science* 3608, pp. 146–157, 2005.
19. J. W. J. Williams. Algorithm 232: Heapsort. *Comm. ACM*, 7(6):347–348, 1964.

Quadratic Programming and Combinatorial Minimum Weight Product Problems

Walter Kern[1] and Gerhard Woeginger[2]

[1] Faculty of Electrical Engineering, Mathematics and Computer Science,
Department of Applied Mathematics,
University of Twente, P.O.Box 217, NL-7500 AE Enschede
kern@math.utwente.nl
[2] Department of Mathematics and Computer Science,
Eindhoven University of Technology,
P.O. Box 513, NL-5600 MB Eindhoven
gwoegi@win.tue.nl

Abstract. We present a fully polynomial time approximation scheme
(FPTAS) for minimizing an objective $(\mathbf{a}^T\mathbf{x} + \gamma)(\mathbf{b}^T\mathbf{x} + \delta)$ under linear constraints $\mathbf{Ax} \leq \mathbf{d}$. Examples of such problems are combinatorial
minimum weight product problems such as, e.g., the following: Given a
graph $G = (V, E)$ and two edge weights $\mathbf{a}, \mathbf{b} : E \to \mathbb{R}_+$ find an $s - t$ path
P that minimizes $\mathbf{a}(P)\mathbf{b}(P)$, the product of its edge weights relative to
$\mathbf{a}$ and $\mathbf{b}$.

Keywords: Quadratic Programming, approximation scheme, shortest
path.

AMS-Class: 90C20, 90C26, 90C27.

1 Introduction

The problem of minimizing a quadratic objective function $\mathbf{x}^T\mathbf{Qx} + \mathbf{c}^T\mathbf{x}$ under
linear constraints $\mathbf{Ax} \leq \mathbf{d}$ is well-known to be NP-hard ([4]), even when $\mathbf{Q}$ has
only a single negative eigenvalue ([9]). In case $\mathbf{Q}$ is positive semidefinite, the
problem can be solved efficiently ([6] or [12]). Here, we focus on the case where
the objective is the product of two affine functions:

$$z^* = \min \ (\mathbf{a}^T\mathbf{x} + \gamma)(\mathbf{b}^T\mathbf{x} + \delta) \tag{1.1}$$
$$\mathbf{Ax} \leq \mathbf{d}.$$

The complexity status of this (in general non-convex) problem is open (cf.
[9]). We present a *fully polynomial time approximation scheme* (FPTAS) for this
class. More precisely, we present an algorithm which correctly decides whether
$z^* < 0$, $z^* = 0$ or $z^* > 0$ holds and, in addition, computes for any given $\varepsilon > 0$
an ε-*approximate solution*, i.e., a feasible solution of (1.1) whose objective value
differs from the optimum z^* by at most $\varepsilon|z^*|$. (In case $z^* \leq 0$ we can even solve
the problem exactly in polynomial time.) The running time of the algorithm is
polynomially bounded in $1/\varepsilon$ and the size of (1.1).

T. Calamoneri, I. Finocchi, G.F. Italiano (Eds.): CIAC 2006, LNCS 3998, pp. 42–49, 2006.

In Section 3 we discuss possible applications of our result to combinatorial minimum weight product problem such as the following: Given a graph $G = (V, E)$ with two non-negative edge weights $a, b : E \to \mathbb{R}_+$, find an $s - t$ path P minimizing $\mathbf{a}(P)\mathbf{b}(P)$, the product of its edge weights relative to $\mathbf{a}$ and $\mathbf{b}$.

Remark. Vavasis ([11]) presents a FPTAS for (more general) quadratic objectives $q(\mathbf{x}) = \mathbf{x}^T \mathbf{Q} \mathbf{x} + \mathbf{c}^T \mathbf{x}$ with a bounded number of negative eigenvalues. His work, however, is based on a different concept of "ε-approximate solution": In [11], a feasible $\mathbf{x}$ is ε-approximate if its objective value differs from the optimum z^* by at most

$$\varepsilon(\max_{\mathbf{Ax} \leq \mathbf{d}} q(\mathbf{x}) - \min_{\mathbf{Ax} \leq \mathbf{d}} q(\mathbf{x})).$$

This concept of "ε-approximation" is not suited for the combinatorial applications that we discuss in Section 3. It is unclear whether our results can somehow be extended to the case of bounded number of negative eigenvalues.

2   The Algorithm

Relative to (1.1), we consider the related system

$$\begin{aligned}
\alpha - \mathbf{a}^T \mathbf{x} - \gamma &= 0 \\
\beta - \mathbf{b}^T \mathbf{x} - \delta &= 0 \\
\mathbf{Ax} &\leq \mathbf{d}
\end{aligned} \tag{2.1}$$

of $m + 2$ (in-)equalities in variables $(\alpha, \beta, \mathbf{x}) \in \mathbb{R}^{n+2}$.

Let $P \subseteq \mathbb{R}^{n+2}$ denote the polyhedron defined by (2.1) and let

$$\widehat{P} := \{(\alpha, \beta) \mid \exists x : (\alpha, \beta, \mathbf{x}) \text{ solves } (2.1)\} \subseteq \mathbb{R}^2$$

denote its projection into $\mathbb{R}^2$.

With $f(\alpha, \beta) := \alpha\beta$, our problem can thus be restated as

$$z^* = \min_{(\alpha, \beta) \in \widehat{P}} f(\alpha, \beta). \tag{2.2}$$

Note that $\widehat{P}$, being a projection of P, may have exponentially many describing inequalities. Yet we can clearly solve linear optimization problems over $\widehat{P}$, as these reduce to LP's over P. Indeed, for $\mathbf{c} \in \mathbb{R}^2$, we have

$$\min_{\binom{\alpha}{\beta} \in \widehat{P}} \mathbf{c}^T \binom{\alpha}{\beta} \quad \hat{=} \quad \min_{\binom{\alpha}{\beta}_{\mathbf{x}} \in P} (\mathbf{c}^T, \mathbf{0}^T) \begin{pmatrix} \alpha \\ \beta \\ \mathbf{x} \end{pmatrix} \tag{2.3}$$

Equivalently (cf., e.g., [10]), we can efficiently solve the separation problem for $\widehat{P}$. As a consequence of this, we may apply the ellipsoid method to check whether

$\widehat{P}$ is full-dimensional or not and – and in case it is not – determine the (possibly infinite) line segment that equals $\widehat{P}$. Thus, in case $\widehat{P}$ is not full-dimensional, (2.2) reduces to a 1-dimensional quadratic problem, which is readily solved.

In what follows, we therefore assume throughout that $\widehat{P}$ is full-dimensional. Next we can check (by means of linear programming) which of the four quadrants in $\mathbb{R}^2$ is (properly) intersected by $\widehat{P}$. This allows us to distinguish between the following three cases

$$z^* = 0, \ z^* < 0 \text{ and } z^* > 0,$$

which we treat separately.

$\underline{z^* = 0}$. This is tantamount to $\widehat{P} \subseteq \mathbb{R}^2_+$ or $\widehat{P} \subseteq \mathbb{R}^2_-$ and $\widehat{P}$ intersecting (touching) one of the coordinate axes. Assume, say, that $\widehat{P} \subseteq \mathbb{R}^2_+$ and

$$\min_{\begin{pmatrix} \alpha \\ \beta \end{pmatrix} \in \widehat{P}} \quad \beta = 0.$$

In this case an optimal solution x^* of (2.2) is obtained by solving the second problem in (2.3) with $\mathbf{c}^T = (0,1)$.

$\underline{z^* < 0}$, i.e., $\widehat{P}$ contains some $(\alpha, \beta) \in \widehat{P}$ with $\alpha\beta < 0$.

Let $\widehat{P}^{\pm} := \left\{ \begin{pmatrix} \alpha \\ \beta \end{pmatrix} \in \widehat{P} \mid \alpha \leq 0, \beta \geq 0 \right\}$ and $\widehat{P}^{\mp} := \left\{ \begin{pmatrix} \alpha \\ \beta \end{pmatrix} \in \widehat{P} \mid \alpha \geq 0, \beta \leq 0 \right\}$.

Then (2.2) basically reduces to two separate "convex" problems on $P^{\pm}$ resp. $P^{\mp}$. Indeed, for $z < 0$ let

$$L_z := \left\{ \begin{pmatrix} \alpha \\ \beta \end{pmatrix} \mid \alpha < 0, \beta > 0, \alpha\beta \leq z \right\}$$

and $C_z := L_z \cap \widehat{P}^{\pm}$ (cf. figure 2.1).

Clearly,

$$\min_{\begin{pmatrix} \alpha \\ \beta \end{pmatrix} \in \widehat{P}^{\pm}} \quad f(\alpha, \beta) = \min_{C_z \neq \emptyset} \ z \tag{2.4}$$

holds. Now C_z is a convex set for $z < 0$ and it is straightforward to design a separation algorithm for C_z (cf., e.g., [3], section 10.6). It is then routine wor k to verify that we may use the ellipsoid algorithm to determine (exactly) the optimum value $z^*_{\pm}$ in (2.4). (Note that the KKT conditions imply that the optimum is achieved in a rational point.)

Applying the same arguments to $\widehat{P}^{\mp}$ (in case this is non-empty), we obtain a corresponding $z^*_{\mp}$ and observe that $z^* = \min \left\{ z^*_{\pm}, z^*_{\mp} \right\}$ solves (2.2).

$\underline{z^* > 0}$. This case occurs when $\widehat{P} \subseteq \mathbb{R}^2_+$ (or $\widehat{P} \subseteq \mathbb{R}^2_-$) and $\widehat{P}$ does not touch any coordinate axes. This case may be considered as "essentially concave", as several local minima may exist (cf. figure 2). In what follows we assume w.l.o.g. that $\widehat{P} \subseteq \mathbb{R}^2_+$.

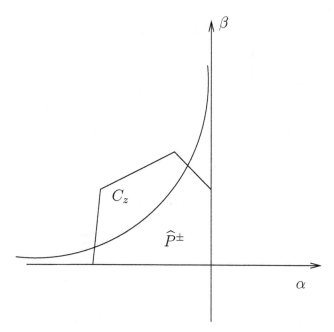

Fig. 1. L_z and C_z

Lemma 1. *The minimum in (2.2) is achieved at a vertex of $\widehat{P}$.*

Proof: This is an immediate consequence of the fact that $f(\alpha, \beta) = \alpha\beta$ is *quasi-concave* on $\mathbb{R}^2_+$ (cf. [2]), i.e., for any $\mathbf{x}_1, \mathbf{x}_2 \in \mathbb{R}^2_+$, f achieves its minimum on the line segment $[\mathbf{x}_1, \mathbf{x}_2]$ in one of the endpoints:

$$f(\lambda\mathbf{x}_1 + (1 - \lambda)\mathbf{x}_2) \geq \min\left\{f(\mathbf{x}_1), f(\mathbf{x}_2)\right\}, \qquad \lambda \in [0, 1]. \tag{2.5}$$

$\square$

It is well-known that the vertices of a polyhedron have components with size polynomially bounded in the size of the describing system of inequalities. Thus the vertices of P and, the more, the vertices (α, β) of $\widehat{P}$ satisfy $\alpha, \beta < 2^p$ with p polynomial in the size of the problem instance (1.1). In particular, we conclude that

$$\underline{z} := 2^{-2p} \leq z^* \leq 2^{2p} =: \overline{z}. \tag{2.6}$$

holds.

We seek to determine the value z^* approximately by binary search. Given $\underline{z} < z < \overline{z}$ we check whether $z^* < z$ holds (approximately) or not by approximating the level curve

$$\ell_z = \left\{\begin{pmatrix} \alpha \\ \beta \end{pmatrix} \mid \alpha\beta = z, \ \alpha, \beta \geq 0\right\}$$

by finitely many tangent lines at the points $(\alpha_k, \beta_k) = (\sqrt{z}(1 + \varepsilon)^k, \sqrt{z}(1 + \varepsilon)^{-k})$, $k = 0, \pm1, \ldots, \pm K$, where $K > 0$ is chosen so that $\alpha_K > 2^p$ (hence K is polynomially bounded in $1/\varepsilon$ and the size of (1.1)).

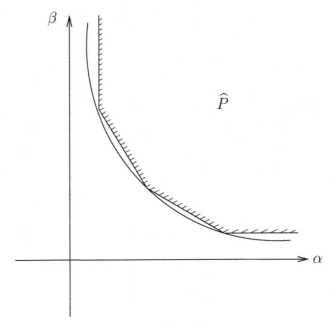

Fig. 2. $\widehat{P} \subseteq \mathbb{R}_+^2$

More precisely, to determine whether $z^* < z$ holds approximately, we solve polynomially (in $1/\varepsilon$ and the size of (1.1)) many linear optimization problems

$$z_k = \min_{x \in \widehat{P}} (\beta_k, \alpha_k)x.$$

(Note that (β_k, α_k) is the gradient of $f(\alpha, \beta) = \alpha\beta$ in (α_k, β_k).)

Lemma 2. If $z_k \le 2z$ for some k, $|k| \le K$, then $z^* \le z$. If $z_k > 2z$ for all k, $|k| \le K$, then $z \le (1+\varepsilon)z^*$.

Proof: The first claim is obvious.

As to the second claim, assume $z_k > 2z$ for all k. Let $z^* = \alpha\beta$, $(\alpha, \beta) \in P$ and assume w.l.o.g. that $\alpha \ge \beta$. Let k be the smallest k such that $\alpha \le \alpha_k$. If $k \le 0$, then $\beta \le \alpha \le \alpha_0 = \beta_0$ implies

$$z_0 \le \beta_0\alpha + \alpha_0\beta \le 2\alpha_0\beta_0 = 2z,$$

a contradiction. Hence $k \ge 1$. By assumption, we have

$$2z < z_k \le \beta_k\alpha + \alpha_k\beta \le \beta_k\alpha_k + \alpha_k\beta = z + \alpha_k\beta.$$

Hence $\alpha_k\beta > z$, i.e., $\beta > \beta_k$. But then

$$\alpha\beta > \alpha_{k-1}\beta_k = (1+\varepsilon)^{-1}z,$$

as claimed.   □

This enables us to perform a binary search for z^* on $[\underline{z}, \overline{z}]$, solving (2.2) approximately in time polynomially bounded in $1/\varepsilon$ and the size of (1.1).

3 Minimum Weight Product Problems

Every combinatorial minimum weight problem

$$\min \left\{ \mathbf{c}^T \mathbf{x} \mid \mathbf{x} \in D \right\} \tag{3.1}$$

where $D \subseteq \{0,1\}^n$ has a corresponding *minimum ratio* version, where the objective $\mathbf{c}^T\mathbf{x}$ is replaced by a quotient $\mathbf{p}^T\mathbf{x}/\mathbf{q}^T\mathbf{x}$ with $\mathbf{q} > \mathbf{0}$. Probably the best known example is the socalled "tramp steamer problem", where $D \in \{0,1\}^E$ is the set of directed circuits through a given node in a digraph $G = (V, E)$ (cf., e.g., [1]). Typically such minimum ratio problems seek to model *multicriteria objective functions* (e.g., "maximize profit versus time"). Such minimum ratio version are well-studied in the literature and it is known since long ([8]) that the minimum ratio version is (modulo polynomial time computation) at most as difficult as the original minimum weight problem.

In the context of multicriteria objectives it is often equally natural to consider other combinations of weights such as, e.g., *product versions* with objective $(\mathbf{a}^T\mathbf{x})(\mathbf{b}^T\mathbf{x})$. For example, if $D \subseteq \{0,1\}^E$ is the set of $s - t$ paths in a graph, then $\mathbf{a} \in \mathbb{R}_+^E$ may define failure probabilities and $\mathbf{b} \in \mathbb{R}_+^E$ may define edge costs ([7]). In contrast to minimum ratio problems, however, such product versions of minimum weight problems appear to be more difficult in general.

Our FPTAS from Section 2 can be used to approximately solve minimum weight product problems in case $D \subseteq \{0,1\}^n$ is the vertex set of a polyhedron $\mathbf{Ax} \leq \mathbf{d}$ and we are able to solve (3.1) efficiently. Thus, for example, our result applies when D is the set of $s - t$ paths, spanning trees or perfect matchings in a graph.(Note that our arguments in Section 2 rely only on the assumption that we can efficiently optimize a linear objective over $\mathbf{Ax} \leq \mathbf{d}$.)

For simplicity, we restrict our discussion to minimum weight product $s - t$ paths as a generic example. Consider a directed graph $G = (V, E)$ with two given edge weights $\mathbf{a}, \mathbf{b} : E \to \mathbb{R}_+$ and assume we are to find an $s - t$ path p minimizing the product $\mathbf{a}(p)\mathbf{b}(p)$ of its edge weights relative to $\mathbf{a}$ and $\mathbf{b}$. We first seemingly relax our problem, replacing the path p by an $s - t$ flow of value 1. Let $\mathbf{A} \in \mathbb{R}^{n \times m}$ denote the node-arc incidence matrix of G and let $\mathbf{d} \in \mathbb{R}^n$ have coordinates $d_s = 1$, $d_t = -1$ and $d_v = 0$ else. Then our relaxation can be written as

$$z^* = \min(\mathbf{a}^T\mathbf{x})(\mathbf{b}^T\mathbf{x})$$
$$\mathbf{Ax} = \mathbf{d} \tag{3.2}$$
$$\mathbf{x} \geq \mathbf{0}$$

Clearly $z^* \geq 0$ holds. Furthermore, $z^* = 0$ holds only in the trivial case where an $s - t$ path p with $\mathbf{a}(p) = 0$ or $\mathbf{b}(p) = 0$ exists. Hence we may assume $z^* > 0$. In this case, the minimum in (3.2) is achieved at a vertex of the feasible region (due to Lemma 1), which corresponds to an $s - t$ path. So (3.2) is an exact restatement of our original problem.

As our FPTAS from section 2 obtains the ε-approximate solution $\mathbf{x}$ of (3.2) via linear programming, we may assume w.l.o.g. that $\mathbf{x}$ is an $s - t$ path. (Alternatively, decompose $\mathbf{x} = \lambda_1 \mathbf{x}_1 + \ldots + \lambda_k \mathbf{x}_k$ into a convex combination of $s - t$ paths $\mathbf{x}_i$ and use (2.4) to exhibit one of the $s - t$ paths $\mathbf{x}_i$ as an approximately optimal solution.)

We like to remark that a similar approach also works for slightly different objective functions like, e.g., $\widetilde{f}(\mathbf{x}) = (\mathbf{a}^T \mathbf{x})\sqrt{\mathbf{b}^T \mathbf{x}}$. All we need is that the level curves of $\widetilde{f}$ can be nicely approximated by piecewise linear functions.

We conclude our discussion by commenting on the complexity of the (exact) problem (3.2). Its complexity status (P versus NP) is open and only two special cases are known to be efficiently solvable: $\mathbf{a} = \mathbf{b}$ (trivial) and $\mathbf{a} = \mathbf{1}$ ([7]). The latter also follows from our approach by observing that it suffices to approximate the level curves only in the points $\alpha_1 = 1, \ldots, \alpha_{n-1} = n - 1$, corresponding to the possible values $\alpha = \mathbf{a}^T \mathbf{x}$ for an $s - t$ path $\mathbf{x} \in \{0, 1\}^E$.

Alternatively, the case $\mathbf{a} = \mathbf{1}$ may also be settled directly by computing for each possible path length $k = 1, \ldots, n - 1$ the corresponding minimum b-weight b_k over all $s - t$ paths of length k (and observing that $z^* = \min\limits_k k b_k$). The computation of b_k can be accomplished as follows. Let $V_0, \ldots, V_k$ be $k + 1$ copies of V and let G^k denote the directed graph on $V_0 \cup \ldots \cup V_k$ with arcs going from V_t to V_{t+1}, joining vertices as in G. More precisely, the arc set E^k of G^k is given by

$$E^k = \{(i_t, j_{t+1}) \mid (i, j) \in E, \ 0 \le t \le k - 1\},$$

where i_t is the copy of i in V_t.

The edge weights $b : E \to \mathbb{R}_+$ give rise to edge weights $b : E^k \to \mathbb{R}_+$ by setting $b_{i_t, j_{t+1}} = b_{ij}$. Now b_k, the minimum b-weight of an $s - t$ path of length k in G is simply the minimum b-weight of an $s_0 - t_k$ path in G^k.

For general $a, b : E \to \mathbb{R}_+$, (3.2) can be solved by computing all vertices of $\widehat{P} \subseteq \mathbb{R}^2$ that minimize a linear function $(\alpha, \beta) \to \alpha + \lambda\beta, \ \lambda \in \mathbb{R}_+$, over $\widehat{P}$. (Each local minimizer of $f(\alpha, \beta) = \alpha\beta$ over P must be such a vertex.) These vertices can be determined successively: Let $\lambda > 0$ and consider the parametric minimum $s - t$ problem with edge costs $c_\lambda = a + \lambda b$. For $\lambda > 0$ sufficiently small, a min cost $s - t$ path relative to cost c_λ will be an $s - t$ path x_0 that is minimal relative to a and, among all such a-minimal paths, has minimum b-weight. Standard sensitivity analysis then allows us to exhibit a largest interval $[\lambda_0 = 0, \lambda_1]$ such that x_0 is optimal relative to c_λ for each $\lambda \in [\lambda_0, \lambda_1]$. We then proceed by chosing $\lambda > \lambda_1$ sufficiently small and a min cost path x_1 relative to $c_\lambda = a + \lambda b$ to determine the next interval $[\lambda_1, \lambda_2]$ for which x_1 is optimal etc.

The running time of this procedure basically equals the number of *breakpoints* $\lambda_1, \lambda_2, \ldots$ in the parametric min cost $s - t$ path problem with parametrized cost function $c = a + \lambda b, \ \lambda \ge 0$. Gusfield ([5]) has shown that this number has a subexponential bound $O(n^{\log n})$. For this reason, we do not expect (3.2) to be NP-hard.

To make our presentation selfcontained we briefly sketch the argument form [5]. Let B_n denote the number of breakpoints in the parameteric min cost path problem. Furthermore, we let B_n^k denote the number of breakpoints if only paths of length k are allowed, i.e., if we replace G by G^k as defined above.

For fixed k we estimate B_n^k as follows. Fix a node r in the middle layer $V_{\lfloor k/2 \rfloor}$ of G^k and let $B_n^k(r)$ denote the number of breakpoints if only $s - t$ paths through r are allowed. As λ varies, the costs of $s - r$ and $r - t$ paths in G^k vary independently. So we can conclude that

$$B_n^k(r) \leq 2B_n^{\lfloor k/2 \rfloor}.$$

This proves

$$B_n^k \leq \sum_r B_n^k(r) \leq 2nB_n^{\lfloor k/2 \rfloor}$$

and $B_n^k = O(n^{\log n})$ follows inductively. Hence also $B_n \leq \sum_k B_n^k = O(n^{\log n})$, as claimed.

References

1. R. Ahuja, T. Magnanti and J. Orlin: *Network Flows: Theory, Algorithms and Applications*, Prentice Hall, 1993.
2. M. Avriel, W.E. Dievert, S. Schaible and I. Zhang: *Generalized Convexity*, Plenum Press, New York, 1988.
3. U. Faigle, W. Kern and G. Still: *Algorithmic Principles of Mathematical Programming*, Kluwer, 2001.
4. M. Garey and D. Johnson: *Computers and Intractability*, A Guide to the Theory of NP-Completeness, Freeman, San Francisco, 1979.
5. D. Gusfield: *Sensitivity analysis for combinatorial optimization*, Memorandum UCB/ERL M80/22, Electronics Research Laboratory, Berkeley, 1980.
6. M. Kozlov, S. Tarasov and L. Hacijan: *Polynomial Solvability of Convex Quadratic Programming*, Soviet Math. Doklady 20, 1108–1111, 1979.
7. T. Kuno: *Polynomial algorithms for a class of minimum rank-two cost path problems*, Journal of Global Optimization 15, 405–417, 1999.
8. N. Megiddo: *Combinatorial Optimization with rational objective functions*, Mathematics of OR 4(4), 414–424, 1979.
9. P. Pardalos and S. Vavasis: *Quadratic Programming with One Negative Eigenvalue is NP-hard*, Journal of Global Optimization 1, 15–22, 1991.
10. A. Schrijver: *Theory of linear and Integer Programming*, Wiley, 1986.
11. S. Vavasis: *Approximation algorithms for indefinite quadratic programming*, Math. Prog. 57, 279–311, 1992.
12. S. Vavasis: *Nonlinear optimization: complexity issues*, Oxford University Press, 1991.

Counting All Solutions of Minimum Weight Exact Satisfiability

Stefan Porschen

Institut für Informatik, Universität zu Köln, Pohligstr. 1 D-50969 Köln, Germany
porschen@informatik.uni-koeln.de

Abstract. We show that the number of all solutions of minimum weight exact satisfiability can be found in $O(n^2 \cdot \|C\| + 2^{0.40567 \cdot n})$ time, for a CNF formula C containing n propositional variables equipped with arbitrary real-valued weights. In recent years merely the unweighted counterpart of this problem has been studied [2, 3, 7].

Keywords: Minimum weight exact satisfiability, minimum weight set partition, maximum weight independent set, counting problem.

1 Introduction

Besides decision and optimization problems, counting problems are interesting and important objects of computational complexity theory. For a search - or optimization problem S, its counting version denoted $\#S$ searches for the number of solutions of S (which is less than enumerating all solutions explicitly). In this paper we deal with the counting versions of two NP-hard optimization problems, namely minimum weight set partition (MINW-SP) and minimum weight exact satisfiability (MINW-XSAT). Both underlying decision problems, namely set partition (SP) [4] and exact satisfiability (XSAT) [9] are NP-complete. Therefore the counting versions are #P-complete [10]. MINW-SP takes as input a collection $\mathcal{M}$ of subsets of a finite set M, where each $T \in \mathcal{M}$ is equipped with a weight $w(T) \in \mathbb{R}$. A solution of MINW-SP is a subfamily $\mathcal{T} \subseteq \mathcal{M}$ of lowest total weight such that each $m \in M$ is contained in exactly one $T \in \mathcal{T}$. In other words, a solution $\mathcal{T}$, if existing, provides a partition of M of least possible weight. MINW-XSAT takes as input a conjunctive normal form (CNF) formula C, such that each Boolean variable $x \in \{0,1\}$ is equipped with a weight $w(x) \in \mathbb{R}$. MINW-XSAT searches for a truth assignment setting to 1 exactly one literal in each clause of C such that the total weight of all *variables* set to 1 is minimal (for a precise definition cf. Section 2).

Counting problems have attracted some attention during the last years. Dahlöf et al. in [2], e.g., construct an algorithm that solves the unweighted #XSAT problem in $O(2^{0.81131 \cdot n})$ time based on an $O(2^{0.40567 \cdot n})$ time algorithm for solving #MAXW-IS, i.e., for counting all maximum (positive integer) weighted independent sets in a finite graph (n is the number of Boolean variables, resp. vertices). Their bound for unweighted #XSAT has been improved to $O(2^{0.40567 \cdot n})$ in [7]. And in [3] the up to now best bound of $O(2^{0.2857 \cdot n})$ is shown for #XSAT. None

T. Calamoneri, I. Finocchi, G.F. Italiano (Eds.): CIAC 2006, LNCS 3998, pp. 50–59, 2006.
© Springer-Verlag Berlin Heidelberg 2006

of the algorithms solving #XSAT mentioned so far is able to enumerate all exact models of the input formula; each of them outputs the number of solutions only. Hence, there is no evidence whether or how these algorithms can be adapted for determining the number of all solutions of MINW-XSAT when arbitrary real-valued weights are assigned to the variables. However, in this paper, refining the techniques in [7], we provide an algorithm that also solves #MINW-XSAT in $O(n^2 \cdot \|C\| + 2^{0.40567 \cdot n})$ time, where $\|C\|$ is the length of input formula C. This algorithm, essentially benefits from a monotonization procedure that reduces #MINW-XSAT for arbitrary formulas to #MINW-XSAT for positive monotone formulas. The latter problem is attacked by an appropriate adaptation of the #MAXW-IS algorithm in [2]. #MINW-SP can be solved in the same manner as it turns out to be identifiable with the monotone #MINW-XSAT problem in a dual sense.

It should be noted that the weighted variant of a problem can increase its computational complexity considerably. For instance, in case of 2-SAT, i.e., the satisfiability problem for formulas only containing clauses of length at most two, a solution can be found in linear time [1]. Whereas, finding a minimum weight truth assignment for variable-weighted formulas is NP-hard. This can be seen by a straightforward reduction from the vertex cover problem to 2-SAT for monotone formulas via the *formula* graph (as defined below). Thus we are motivated to consider the weighted exact satisfiability counting problem. For the optimization problem MINW-XSAT recently the bound $O(2^{0.2441 \cdot n})$ has been found [8].

2   Basic Notions and Notation

A *literal* is a propositional variable $x \in \{0, 1\}$ or its negation $\overline{x} := \neg x$ (negated variable). The *complement* of a literal l is denoted as $\overline{l}$. A *clause* c is the disjunction of different literals and thus is represented as a finite set. A CNF formula C is a conjunction of different clauses and is thus represented as a finite clause set. For short, we throughout use the term *formula* meaning a clause set as defined. For a given formula C, clause c, by $V(C), V(c)$ we denote the set of variables contained in C, c, respectively. $V_\epsilon(C), V_\epsilon(c)$, for $\epsilon \in \{+, -\}$, denotes the set of variables occuring positive, resp. negative in C, c. We distinguish between the length $\|C\|$ of a formula C and the number $|C|$ of its clauses. Let CNF denote the set of all formulas, and let CNF_+ denote the set of all *positive monotone* formulas, i.e., no clause contains a negated variable. For a formula $C \in \text{CNF}$ and a literal l, we denote by $C(l) := \{c \in C : l \in c\}$ the set of all clauses in C containing l.

Exact satisfiability (XSAT) is a variant of the prominent satisfiability problem, and asks whether input formula $C \in \text{CNF}$ admits a truth assignment $t : V(C) \rightarrow \{0, 1\}$ mapping exactly one literal in each clause of C to 1. Such a truth assignment, for short, is called an *x-model (of C)*. The empty set also is a formula: $\varnothing \in \text{CNF}$ which is exactly satisfiable. However, a formula C containing the empty clause ($\varnothing \in C$) is never satisfiable. For solving #XSAT, one has to determine the number of all x-models of an input formula $C \in \text{CNF}$. The minimum weight XSAT problem (MINW-XSAT) is defined as follows: For $C \in \text{CNF}$ and weight function $w : V(C) \rightarrow \mathbb{R}$, find a x-model of minimal weight, *minimum* x-model for short. For any $X \subseteq V(C)$, we set $w(X) := \sum_{x \in X} w(x)$. The *weight* of a model t is defined as $w(t) := w(t^{-1}(1)) = \sum_{x \in V(C)} t(x)w(x)$. The counting problem #MINW-XSAT is to determine the number of all minimum x-models of a weighted input formula. For $C \in \text{CNF}$, let $X(C)$ denote the set of all x-models of C. Similarly, given $w : V(C) \to \mathbb{R}$, let $X_{\min}(C, w) \subseteq X(C)$ denote the set of all minimum x-models of C with respect to w.

We shall make use of a simple graph concept assigned to a formula. For a monotone formula $C \in \text{CNF}_+$, we define its *formula graph* G_C with vertex set $V(C)$. Two vertices are joined by an edge if there is a clause containing the corresponding variables.

3   Solving #MINW-XSAT Restricted to Monotone Formulas

This section is devoted to provide an algorithm solving #MINW-XSAT restricted to the class CNF_+, also called monotone #MINW-XSAT, in $O(n^2 \cdot \|C\| + 2^{0.40567 \cdot n})$ time, for formulas C containing n weighted variables. By a dualization argument presented below it will turn out that such an algorithm also solves #MINW-SP. In [2], Dahlöf and Jonsson proved an upper bound of $O(2^{0.40567 \cdot n})$ for calculating the number of all maximum weight independent sets in a graph of n vertices each equipped with a positive integer weight. For convenience, we refer to that algorithm as to the DJ-Algorithm. Recall that the maximum weight independent set problem (MAXW-IS) gets as input a finite (simple) graph $G = (V, E)$, and a vertex weight function $w : V \to \mathbb{N}$. It asks whether there is an independent set in G, i.e., a set of pairwise non-adjacent vertices of maximal weight. Observe that MAXW-IS is NP-hard even if all weights are equal to 1 which follows from the vertex cover problem [4], because a minimum cardinality vertex cover in G is the complement of a maximum cardinality independent set.

For attacking monotone #MINW-XSAT we reduce it to a conditional variant of #MAXW-IS called #MINW-MAXW-IS. The underlying optimization problem MINW-MAXW-IS is defined as follows, for arbitrary weight functions $f_1 : V \to \mathbb{N}, f_2 : V \to \mathbb{R}$:

Input: $G = (V, E)$, $f : V \to \mathbb{N} \times \mathbb{R}$ with $f(x) =: (f_1(x), f_2(x)), x \in V$.
Output: $X \subseteq V$ such that $f_2(X) = \min\{f_2(Y) : Y \in F_1^{\max}(G)\}$, where $F_1^{\max}(G)$ is the set of all independent sets Y in G such that $f_1(Y)$ is maximal, with $f_i(S) := \sum_{x \in S} f_i(x)$, $i = 1, 2$, for any $S \subseteq V(G)$.

Thus, an algorithm solving #MINW-MAXW-IS, has to count all independent sets X in G of minimal weight w.r.t. the second component under the condition that $f_1(X)$ is maximal.

Proposition 1. *Counting all solutions of* MINW-MAXW-IS *is possible in* $O(2^{0.40567 \cdot |V(G)|})$ *time, for input instances* $G, f = (f_1, f_2)$, *where* $f_1 : V(G) \to \mathbb{N}$, $f_2 : V(G) \to \mathbb{R}$.

Sketch of PROOF. In [2] an adaptation of the DJ-algorithm (not affecting its running time) is outlined for counting among all maximum weighted independent sets in a graph only those that have minimal cardinality (cf. [2], proof of Prop. 5.2). Clearly that variant corresponds to #MINW-MAXW-IS where $f_2(x) = 1$ for each vertex $x \in V(G)$. In this adapted version, the return function of the algorithm is employed by a separate component reserved for the cardinality of the current independent set under consideration. It is obvious that the mentioned adaptation carries over also to the generalization where f_2 is an arbitrary real-valued function: Simply take the value $f_2(X)$ in the corresponding component of the return function. It is not hard to see that the resulting recursive algorithm works as desired with this modification.   □

Now, Monotone #MINW-XSAT can be identified as a subproblem of #MINW-MAXW-IS in the following way. For $C \in \text{CNF}_+$ with formula graph $G := G_C$, consider variable weight function $w : V(C) \to \mathbb{R}$. Recall that each variable $x \in V(C)$ constitutes a vertex in G and that two vertices are joined by an edge if the corresponding variables occur together in a clause. Since we have positive literals only, $C(x) = \{c \in C | x \in c\}$ is the subformula of all clauses in C containing x. As vector-valued weight function $f = (f_1, f_2)$ we define $f : V(C) \to \mathbb{N} \times \mathbb{R}$, by $f_1(x) := |C(x)|$, $f_2(x) := w(x)$, for each $x \in V(C)$. Now, t is a minimum weight x-model of C if and only if vertex set $t^{-1}(1)$ is a solution of MINW-MAXW-IS for G, f. Indeed, let t be any x-model of C, then each $x \in t^{-1}(1)$ is the unique variable exactly satisfying subformula $C(x)$, hence the corresponding vertex contributes first component weight $f_1(x) = |C(x)|$ in G. Clearly, variables in $t^{-1}(1)$ must yield a partition $C = \bigcup_{x \in t^{-1}(1)} C(x)$, thus $f_1(t^{-1}(1)) = |C|$ which is maximum, because a larger weight meant that there are clauses in which more than one variable is set to 1. Conversely, it is easy to see that each independent set $X \subset V(G)$ of weight $w(X) = f_1(X) = |C|$ defines a x-model of C assigning 1 to exactly those variables corresponding to vertices in X, and 0 to the remaining variables. Observe that an independent set of weight larger than $|C|$ cannot exist, because otherwise there are two variables occuring in the same clause and corresponding vertices are adjacent in G. Hence such an independent set indeed has maximal first component weight. Therefore, t is a minimum x-model, if and only if it satisfies $w(t) = f_2(t^{-1}(1)) = \min\{f_2(\hat{t}^{-1}(1)) : \hat{t} \in \text{X}(C)\}$ which is equivalent to the fact that $t^{-1}(1)$ provides a solution of MINW-MAXW-IS for input instance $G, f = (f_1, f_2)$ as defined above.

Theorem 1. #MINW-XSAT *for positive monotone formulas C of n variables (resp. #MINW-SP for a collection C of n input sets c), where variables (resp. input sets) are equipped with arbitrary real-valued weights, can be solved in $O(n^2 \cdot \|C\| + 2^{0.40567 \cdot n})$ time, with $\|C\| = \sum_{c \in C} |c|$.*

PROOF. First, we show that the reduction provided above from #MINW-XSAT to #MINW-MAXW-IS can be executed in $O(n^2 \cdot \|C\|)$ time, where $n := |V(C)|$ and $C \in \text{CNF}_+$. This confirms the claim of the theorem regarding #MINW-XSAT relying on Proposition 1. So, for computing the weighted formula graph

(G_C, f), we first have to determine the vertex weights, for which an array W is maintained. Each position of W stores a variable occuring in C together with both weight components. Regarding the first component f_1, we have to determine the number of occurences $|C(x)|$ of each variable x in C. This can be done by running once through the formula. Each variable x found in C is compared to all variables already stored in W, for each of which we maintain a counter corresponding to the number of its occurences in C. If we find a match the counter for x is incremented by 1, otherwise the variable is stored in the next position of W, and its counter is initialized by value one, finally the second weight component $w(x)$ is assigned. Therefore a running time proportional to $n^2 \cdot \|C\|$ results. Next we have to form the edges of the formula graph. Clearly, this can be done by building a clique $K_{|c|}$ for each clause $c \in C$. As there are $|C| \leq \|C\|$ clauses and each clause contains at most n^2 variables, the time needed for constructing all edges of G_C is upper bounded by $O(n^2 \cdot \|C\|)$.

It remains to verify the claim of the theorem regarding #MINW-SP. First observe that MINW-SP and MINW-XSAT for monotone formulas essentially are the same, as the following dualization argument shows: Let $(M, \mathcal{M}, w)$ be an input instance of MINW-SP with weight function $w : \mathcal{M} \to \mathbb{R}$. Assigning to each $T \in \mathcal{M}$ a Boolean variable $x_T \in \{0, 1\}$ equipped with weight $w(T)$, and assigning to each $m \in M$ a clause c_m that contains variable x_T if and only if $m \in T$ yields a variable-weighted positive monotone input formula of MINW-XSAT. It is easy to see that solving MINW-XSAT for this formula is the same as solving MINW-SP for $(M, \mathcal{M}, w)$. The converse direction reducing monotone MINW-XSAT to MINW-SP proceeds analogously. For completing the proof it is left to verify that the reduction previously described can be done in at most $O(n^2 \cdot \|C\|)$ time: To that end we hold a table $A_m(T)$ of Boolean having size $|M| \cdot \|\mathcal{M}\|$ storing $A_m(i) = 1$ iff $m \in T_i$, where $\mathcal{M} := \{T_1, \dots, T_{|\mathcal{M}|}\}$ and $M = \{m_1, \dots, m_{|M|}\}$ are assumed to be ordered. After having filled this table we row-wise assign to each clause c_m all x_{T_i} with $A_m(i) = 1$ needing $O(|M| \cdot |\mathcal{M}|)$ time. For filling the table, we run once through $\mathcal{M}$ starting with T_1 working column-wise. In column i, we and assign value 1 one to entry $A_j(i)$ if we find $m_j \in T_i$. So, table filling needs $O(\|\mathcal{M}\|)$ time overall. Thus in summary we obtain $O(\|\mathcal{M}\| + |M| \cdot |\mathcal{M}|) = O(\|C\| + |C| \cdot n) = O(n \cdot \|C\|)$, where we took into account that M is in bijection to C, $\mathcal{M}$ is in bijection to $V(C)$, and $\|C\| = \|\mathcal{M}\| > |C| = |\mathcal{M}|$. □

Observe that monotone MINW-XSAT more directly can be identified with the *minimum weight exact hitting set problem*. Corresponding input instances consist of a base set S of arbitrarily weighted elements (which are the variables of a formula $C \in \text{CNF}_+$) and a collection $\mathcal{T}$ of subsets of S (corresponding to the clauses in C). Then one searches for a minimum weight subset $X \subseteq S$ such that X contains exactly one element of each $T \in \mathcal{T}$. Clearly a minimum weight x-model t of C via $t^{-1}(1) \subseteq V(C)$ yields a minimum weight hitting set and vice versa, correspondingly. We thus obtain:

Corollary 1. *Counting all solutions of minimum weight hitting set takes $O(n^2 \cdot \|C\| + 2^{0.40567 \cdot n})$ time, for a base set of n arbitrarily weighted elements and subset collection C.* □

4 Solving #MINW-XSAT for CNF

In this section we provide a polynomial time reduction from #MINW-XSAT for arbitrary CNF formulas to #MINW-XSAT restricted to the class CNF_+ of monotone formulas. This enables us also to solve the unrestricted #MINW-XSAT problem in $O(2^{0.40567 \cdot n})$ time. The main idea is, to establish a sequence of polynomial time computable mappings that, iteratively, transform an arbitrary input instance (C, w) into (C', w') where C' is positive monotone and such that the number of minimum x-models of the original instance is preserved, i.e., equals the number of minimum x-models of the transformed instance.

Since for the empty formula $\varnothing$ holds $V(\varnothing) = \varnothing$, we have $|X(\varnothing)| = 2^0 = 1$. The next lemma states a useful connection between the x-model spaces and the minimum x-model spaces of variable-weighted CNF formulas; it has been proven in [8].

Lemma 1. *For $C, C' \in \mathrm{CNF}$ and arbitrary real-valued weight functions w, w' defined on $V(C)$ resp. $V(C')$, assume that there exists a bijection*

$$F : X(C) \ni t \mapsto t' := F(t) \in X(C')$$

such that $(*)$: $w(t) = w'(t') + \alpha$, where $\alpha \in \mathbb{R}$ is a constant independent of t and t'. Then the restriction $F_{\min} := F|X_{\min}(C, w)$ is a bijection between $X_{\min}(C, w)$ and $X_{\min}(C', w')$; and we have $|X_{\min}(C, w)| = |X_{\min}(C', w')|$.

If a clause contains more than one complemented pairs, then it can never be exactly satisfiable, hence a formula containing such a clause has 0 x-models. However, clauses containing exactly one complemented pair can be removed from the formula such that the number of (minimum) x-models can be recovered, as stated in the following lemma:

Lemma 2. *For $C \in \mathrm{CNF}$ with weight function $w : V(C) \to \mathbb{R}$, let $c \in C$ contain exactly one complemented pair: $x, \overline{x} \in c$. Let C_c be the formula obtained from C by removing c and assigning all literals to 0 that occur in $c' := c - \{x, \overline{x}\}$ and finally removing all duplicate clauses. Let w_c be defined as the restriction of w to $V(C_c) = V(C) - V(c')$. Then the following holds true:*

(i) $|X(C)| = 2|X(C_c)|$ if $x \notin V(C_c)$ and $|X(C)| = |X(C_c)|$ if $x \in V(C_c)$,
(ii) $|X_{\min}(C, w)| = |X_{\min}(C_c, w_c)|$.

PROOF. Obviously $V(C_c) = V(C) - V(c')$, because by removing duplicate clauses no other variable can be removed from the formula. For proving (i) and (ii), first assume that $x \in V(C_c)$. Then a bijection $F : X(C) \to X(C_c)$ obviously is given by $F(t) := t|V(C) - V(c')$ if the reverse is defined by extending $t' \in X(C_c)$ to $V(C)$ by assigning all literals in c' to 0, which, clearly, is required for every truth assignment to be a x-model for C. So, we have (i) in that case. Moreover, one easily gets $w(t) = w_c(F(t)) + \alpha$, where $\alpha = \sum_{y \in V_-(c')} w(y)t(y)$ which is a constant since each $t \in X(C)$, if existing at all, assigns all literals in c' to 0. Thus (ii) follows by Lemma 1.

If $x \notin V(C_c)$, then x occurs in C in clause c' only. Let $X_i(C)$ be the set of x-models of C which assign $x = i \in \{0,1\}$. Clearly, $X(C) = X_0(C) \cup X_1(C)$ as disjoint union. And both $X_0(C)$ and $X_1(C)$ are in bijection via F_i to $X(C_c)$, as above, by restriction. Hence, we have (i) in this case. Obviously, $X_{\min}(C, w) \subset X_i(C)$, for either $i = 0$ (in case $w(x) > 0$) or $i = 1$ (in case $w(x) \leq 0$). Relation $(*)$ in Lemma 1 is easily seen to be satisfied, for $F_i, i \in 0, 1$, hence (ii) is proven. □

In the following, we call a formula *cp-free* if none of its clauses contains a complemented pair of variables. The transformation in the next lemma removes pure negative literals:

Lemma 3. *For a cp-free formula $C \in$ CNF with weight function $w : V(C) \to \mathbb{R}$, let $x \in V(C)$ be a variable that exclusively occurs negated in C. Let C_x be the formula obtained from C by replacing each occurence of $\overline{x}$ by x and let $w_x : V(C) \to \mathbb{R}$ be defined as w except for $w_x(x) := -w(x)$. Then:*

(i) $|X(C)| = |X(C_x)|$,
(ii) $|X_{\min}(C, w)| = |X_{\min}(C_x, w_x)|$.

PROOF. (i) follows, since we have $V(C) = V(C_x)$ and obviously every $t \in X(C)$ yields a $t' \in X(C_x)$ defined as t except for $t'(x) = 1 - t(x)$ and vice versa. In other words, the mapping $F : X(C) \ni t \mapsto F(t) := t' \in X(C_x)$ with t' as defined above, is a bijection of x-model spaces. To prove (ii), we assume that $C \in$ XSAT otherwise we are done. From $w(t) = \sum_{y \in V(C)} w(y)t(y)$, and the fact that t and $t' := F(t)$ as well as w and w_x are distinct at x only, we easily obtain $w(t) = w_x(t') + w(x) = w_x(t') + c$. Due to relation $(*)$ of Lemma 1 the assertion follows, as F is a bijection of x-model spaces. □

Next we state a transformation called *simple resolution* which in a different form was used in [6]. Given a formula C and a literal l, recall that $C(l)$ denotes the set of all clauses containing l. The following lemma is proven in [8]. The proof rests on the fact that the simple resolution transforming a CNF formula C into CNF formula C_{ij}, as defined below, provides a bijection between $X(C)$ and $X(C_{ij})$ as shown in [7]. Moreover, a corresponding transformation on the weight function w yielding $w_{ij} : V(C_{ij}) \to \mathbb{R}$, as defined below, satisfies relation $(*)$ of Lemma 1 finishing the proof sketch.

Lemma 4. *Let $C \in$ CNF be a cp-free formula and $w : V(C) \to \mathbb{R}$ be an arbitrary weight function. Let $c_i = \{x\} \cup u, c_j = \{\overline{x}\} \cup v \in C$ where $x \in V(C)$ and u, v are literal sets. Let C_{ij} be the formula obtained from C as follows:*

(1) Replace every clause $c \in C(x)$ by the clause $c - \{x\} \cup v$,
(2) replace every clause $c \in C(\overline{x})$ by the clause $c - \{\overline{x}\} \cup u$,

(3) set all literals in $u \cap v$ to 0,
(4) remove all duplicate clauses from the current clause set.
Let $w_{ij} := V(C_{ij}) \to \mathbb{R}$ be the weight function defined as follows: for each $y \in V(C_{ij}) - V(u \oplus v)$, set $w_{ij}(y) := w(y)$, and
(1') if $V_+(u \oplus v) \cap V_-(u \oplus v) = \{z\}$, then set $\forall y \in V(u \oplus v) - \{z\} : w_{ij}(y) := w(y)$
and

$$w_{ij}(z) := \begin{cases} w(z) + w(x), & \text{if } \overline{z} \in u, z \in v \\ w(z) - w(x), & \text{else} \end{cases}$$

(2') if $V_+(u \oplus v) \cap V_-(u \oplus v) = \varnothing$, then set $\forall y \in V(v - u) : w_{ij}(y) := w(y)$
and $\forall y \in V_+(u - v) : w_{ij}(y) := w(y) - w(x)$ and $\forall y \in V_-(u - v) : w_{ij}(y) := w(y) + w(x)$.
Then we have $V(C_{ij}) = V(C) - \{x\} - V(u \cap v)$, $|C_{ij}| \leq |C| - 1$ and:
$|X_{\min}(C, w)| = |X_{\min}(C_{ij}, w_{ij})|$.

Now we are ready to present Procedure Monotonization which as input gets a non positive monotone CNF formula C and recursively calls itself until C is positive monotone thereby it computes a multiplicator $N \in \{0, 1\}$ for C. N gets value 0 iff C turns out to be not exactly satisfiable during the monotonization process:

Procedure. Monotonization$(C, w; N)$
Input: $C \in \text{CNF}$, $w : V(C) \to \mathbb{R}$
Output: $C' \in \text{CNF}_+$, w' such that $|X_{\min}(C, w)| = |X_{\min}(C', w')|$, $N \in \{0, 1\}$
begin
(1) $N \leftarrow 1$
(2) **if** $\varnothing \in C$ **then return** $N \leftarrow 0$
(3) **if** $\exists c \in C$ containing ≥ 2 complemented pairs **then return** $N \leftarrow 0$
(4) **if** $\exists c \in C$ containing 1 complemented pair $\{x, \overline{x}\}\}$ **then**
       $C \leftarrow C_c, w \leftarrow w_c$, Monotonization$(C, w; N)$
(5) **if** $\exists x \in V(C)$ occuring only negated in C **then**
       $C \leftarrow C_x, w \leftarrow w_x$, Monotonization$(C, w; N)$
(6) **if** $\exists c_i = \{x\} \cup u, c_j = \{\overline{x}\} \cup v \in C, x \in V(C)$ **then**
       if $|V_+(u \oplus v) \cap V_-(u \oplus v)| > 1$ **then return** $N \leftarrow 0$
       $C \leftarrow C_{ij}, w \leftarrow w_{ij}$, Monotonization$(C, w; N)$
(7) **return** C, w, N
end

Theorem 2. *For $C \in \text{CNF}$, $w : V(C) \to \mathbb{R}$, Procedure Monotonization, in $O(n^2 \cdot \|C\|)$ time, correctly computes a monotone formula $C' \in \text{CNF}_+$ with $w' : V(C') \to \mathbb{R}$ such that $|X_{\min}(C, w)| = |X_{\min}(C', w')|$.*

PROOF. Correctness of Steps (1) to (3) is obvious. Correctness of Steps (4) to (6) follows by Lemmas 2 to 4, and by the fact that the current formula is cp-free when Step (5) is executed for the first time. Thus the current formula C returned in Step (7) is positive monotone and the weight function is such that the number of minimum weight x-models is preserved with respect to the original input.

Adressing the claim for the running time we assume that we can rely on appropriate data structures, such as doubly linked lists: For each variable x, we maintain a list containing pointers to all clauses containing x, carrying additional information whether x appears negative or not. Similarly, for each clause we hold a list, containing pointers to all variables contained, and assume that these lists are doubly linked. It is not hard to verify that these data structures, for given input instance, can be filled in $O(n^2 \cdot \|C\|)$ time, and that this bound also dominates the running time of Procedure Monotonization relying on these data structures. $\qquad\square$

Now we are ready for presenting the main algorithm solving #MINW-XSAT for arbitrary formulas with variable weights:

Algorithm. #MINW-XSAT$(C, w; |\mathrm{X}_{\min}(C, w)|)$
Input: $C \in \mathrm{CNF}$, $w : V(C) \to \mathbb{R}$
Output: number of all minimum weight x-models $|\mathrm{X}_{\min}(C, w)|$ of C, w
begin
(1) **if** C is not positive monotone **then**
(2)   Monotonization$(C, w; N)$
(3)   **if** $N = 0$ **then return** $|\mathrm{X}_{\min}(C, w)| \leftarrow 0$
(4)   **if** $C = \varnothing$ **then return** $|\mathrm{X}_{\min}(C, w)| \leftarrow 1$
(5) solve monotone #MINW-XSAT (* let r be its result *)
(6) **return** $|\mathrm{X}_{\min}(C, w)| \leftarrow r$
end

Theorem 3. *Algorithm* #MINW-XSAT *correctly calculates the number of all minimum weight x-models in time* $O(n^2 \cdot \|C\| + 2^{0.40567 \cdot |V(C)|})$, *for arbitrary input* $C \in \mathrm{CNF}$, $w : V(C) \to \mathbb{R}$.

PROOF. Theorem 2 establishes the correctness of statement (2) which needs to be executed only if the input formula does not be monotone. By the correctness of Prodecure Monotonization it is guaranteed that the multiplicator N is 0 iff $C \notin \mathrm{XSAT}$ in which case the number of x-models is 0, hence (3) is correct. Correctness of (4) is due to the fact mentioned above that the empty formula has only one x-model. Correctness of Step (5) is due to Theorem 1, based on the reduction of monotone #MINW-XSAT to #MINW-MAXW-IS via the formula graph of the weighted positive monotone formula as output by Procedure Monotonization.

Adressing the running time, observe that the test in (1) needs $O(\|C\|)$ time, and Procedure Monotonization can be executed $O(n^2 \cdot \|C\|)$ time due to Theorem 2. Finally, Step (5) solving monotone #MINW-XSAT performs in $O(n^2 \cdot \|C\| + 2^{0.40567 \cdot |V(C)|})$ time according to Theorem 1 completing the proof. $\qquad\square$

5   Concluding Remarks

We proposed an algorithm for #MINW-XSAT running in $O(n^2 \cdot \|C\| + 2^{0.40567 \cdot n})$ time, for input formulas $C \in \mathrm{CNF}$ of n real-valued weighted variables. Observe that testing all possible truth assignments in a brute-force manner needs $O(n^2 \cdot \|C\| \cdot 2^n)$ time.

An open question is whether all solutions can be enumerated explicitly with polynomial delay only, which is not provided by an algorithm merely counting solutions. Such an enumeration algorithm running with polynomial delay only, in the number of solutions, has been provided, e.g., by Johnson et al. [5] for enumerating all maximal independent sets in a finite graph.

Whether faster algorithms solving unweighted #XSAT like that in [3] can be adapted also to treat the weighted case without affecting the running time, is also an open problem.

References

1. B. Aspvall, M. R. Plass, and R. E. Tarjan, A linear-time algorithm for testing the truth of certain quantified Boolean formulas, Inform. Process. Lett. 8 (1979) 121-123.

2. V. Dahllöf, P. Jonsson, An Algorithm for Counting Maximum Weighted Independent Sets and its Applications, in: Proceedings of the 13th ACM-SIAM Symposium on Discrete Algorithms, pp. 292-298, 2002.

3. V. Dahllöf, P. Jonsson, and R. Beigel, Algorithms for four variants of the exact satisfiability problem, Theoretical Comp. Sci. 320 (2004) 373-394.

4. M. R. Garey and D. S. Johnson, Computers and Intractability: A Guide to the Theory of NP-Completeness, W. H. Freeman and Company, San Francisco, 1979.

5. D. S. Johnson, M. Yannakakis, and C. H. Papadimitriou, On Generating All Maximal Independent Sets, Inform. Process. Lett. 27 (1988) 119-123.

6. B. Monien, E. Speckenmeyer, and O. Vornberger, Upper Bounds for Covering Problems, Methods of Operations Research 43 (1981) 419-431.

7. S. Porschen, On Some Weighted Satisfiability and Graph Problems, in: "P. Vojtas, et al. (Eds.), Proceedings of the 31st Conference on Current Trends in Theory and Practice of Informatics", Lecture Notes in Comp. Sci., Vol. 3381, pp. 278-287, Springer-Verlag, Berlin, 2005.

8. S. Porschen, Solving Minimum Weight Exact Satisfiability in Time $O(2^{0.2441n})$, in: "X. Deng, et al. (Eds.), Proceedings of the 16th International Symposium on Algorithms and Computation (ISAAC 2005)", Lecture Notes in Comp. Sci., Vol. 3827, pp. 654-664, Springer-Verlag, Berlin, 2005.

9. T. J. Schaefer, The complexity of satisfiability problems, in: Proceedings of the 10th ACM Symposium on Theory of Computing, pp. 216-226, 1978.

10. L. Valiant, The complexity of enumeration and reliability problems, SIAM J. Comput. 9 (1979) 410-421.

Clause Shortening Combined with Pruning Yields a New Upper Bound for Deterministic SAT Algorithms

Evgeny Dantsin[1], Edward A. Hirsch[2,*], and Alexander Wolpert[1]

[1] Roosevelt University, 430 S. Michigan Av., Chicago, IL 60605, USA
{edantsin, awolpert}@roosevelt.edu
[2] Steklov Institute of Mathematics, 27 Fontanka, St. Petersburg 191023, Russia
hirsch@pdmi.ras.ru

Abstract. We give a deterministic algorithm for testing satisfiability of Boolean formulas in conjunctive normal form with no restriction on clause length. Its upper bound on the worst-case running time matches the best known upper bound for randomized satisfiability-testing algorithms [6]. In comparison with the randomized algorithm in [6], our deterministic algorithm is simpler and more intuitive.

1 Introduction

The problem of satisfiability of a propositional formula in conjunctive normal form (SAT) can be easily solved in 2^n polynomial-time steps, where n is the number of variables in the input formula. Since the early 1980s, this upper bound has been successively improved for k-SAT (the restricted case of SAT where clauses have at most k variables). The best bound to date for deterministic k-SAT algorithms is $(2 - 2/(k+1))^n$ up to a polynomial factor [3]. For randomized k-SAT algorithms, the currently best known bound is due to [8]; a close bound is given in [11]. These general bounds are improved for $k = 3$ in [2, 7].

The list of successive improvements for SAT (with no restriction on clause length) is shorter:

deterministic algorithms		randomized algorithms	
$2^{n\left(1 - \frac{2}{\sqrt{n \log n}}\right)}$	[4]	$2^{n\left(1 - \frac{1}{2\sqrt{n}}\right)}$	[10]
$2^{n\left(1 - \frac{1}{\log(2m)}\right)}$	[5]	$2^{n\left(1 - \frac{1}{\log(2m)}\right)}$	[12]
		$2^{n\left(1 - \frac{1}{\ln(m/n) + O(\ln \ln m)}\right)}$	[6]

Here n and m are respectively the number of variables and the number of clauses. For simplicity, we give the bounds above omitting polynomial factors; such a

---

* Supported in part by Russian Science Support Foundation, Russian Foundation for Basic Research, and INTAS grant 04-77-7173.

T. Calamoneri, I. Finocchi, G.F. Italiano (Eds.): CIAC 2006, LNCS 3998, pp. 60–68, 2006.

factor is typically linear in the length of the input formula (yet there are several exceptions).

In this paper we give a deterministic algorithm for SAT with no restriction on clause length. Its upper bound on the worst-case running time is

$$2^{n\left(1 - \frac{1}{\ln(m/n) + O(\ln \ln m)}\right)}$$

up to a polynomial factor. This bound matches the best known upper bound for randomized SAT algorithms [6]. In comparison with the randomized algorithm in [6], our deterministic algorithm is simpler and more intuitive.

Clause shortening approach. Our algorithm employs the *clause shortening* technique first used by Schuler [12] in his randomized algorithm. This technique is based on the following idea:

> For any "long" clause (longer than some k), either we can shorten this clause by choosing any k literals in the clause and dropping the other literals, or we can substitute false for these k literals in the entire formula.

Schuler's algorithm shortens every clause to its first k literals and applies the k-SAT algorithm [9] to the resulting k-CNF formula. If no satisfying assignment is found, Schuler's algorithm simplifies the initial formula by choosing a long clause at random and substituting false for its first k literals. This procedure is recursively applied to the simplified formula until no clause contains more than k literals. The upper bound in [12] is obtained when taking $k = \log(2m)$.

The derandomization [5] of Schuler's algorithm uses the same idea. Let F be an input formula consisting of clauses $C_1, \ldots, C_m$. Assume that the first m' clauses are longer than k and the other clauses have length $\leq k$. For each C_i where $i \leq m'$, let D_i be the clause that is made up from the first k literals of C_i. Then F is equivalent to the disjunction of the following $m' + 1$ formulas:

$$
\begin{aligned}
F_1 \quad &= F\left[D_1 = \mathsf{false}\right] \\
&\vdots \\
F_{m'} \quad &= F\left[D_{m'} = \mathsf{false}\right] \\
F_{m'+1} &= D_1 \wedge \ldots \wedge D_{m'} \wedge T
\end{aligned}
$$

where T is $C_{m'+1} \wedge \ldots \wedge C_m$, i.e., T is the "tail" consisting of "short" clauses. The derandomized algorithm first tests satisfiability of $F_{m'+1}$ using a k-SAT subroutine. If no satisfying assignment is found, the algorithm is recursively applied to each of $F_1, \ldots, F_{m'}$.

Clause shortening combined with pruning. There is some inefficiency in the derandomized version of Schuler's algorithm. Namely, when testing F_i, we may have to test its subformula corresponding to $D_j = \mathsf{false}$. On the other hand, when testing F_j, we may come to the same subformula. To eliminate this inefficiency, we prune the tree of recursively tested formulas as follows: for each formula F_i, we replace all clauses $C_1, \ldots, C_{i-1}$ by their counterparts $D_1, \ldots, D_{i-1}$. In other words, we use the fact that F is equivalent to the disjunction of the following formulas:

$$
\begin{aligned}
F_1 &= (C_1 \wedge C_2 \wedge C_3 \wedge \ldots \wedge C_{m'-1} \wedge C_{m'} \wedge T) \quad [D_1 = \mathsf{false}] \\
F_2 &= (D_1 \wedge C_2 \wedge C_3 \wedge \ldots \wedge C_{m'-1} \wedge C_{m'} \wedge T) \quad [D_2 = \mathsf{false}] \\
F_3 &= (D_1 \wedge D_2 \wedge C_3 \wedge \ldots \wedge C_{m'-1} \wedge C_{m'} \wedge T) \quad [D_3 = \mathsf{false}] \\
&\;\vdots \\
F_{m'} &= (D_1 \wedge D_2 \wedge D_3 \wedge \ldots \wedge D_{m'-1} \wedge C_{m'} \wedge T) \quad [D_{m'} = \mathsf{false}] \\
F_{m'+1} &= (D_1 \wedge D_2 \wedge D_3 \wedge \ldots \wedge D_{m'-1} \wedge D_{m'} \wedge T)
\end{aligned}
$$

Similarly to the derandomization above, our algorithm first tests $F_{m'+1}$ and then, if no satisfying assignment is found, it tests each of $F_1, \ldots, F_{m'}$. We give details of our algorithm in Sect. 3 and prove its worst-case upper bound in Sect. 4.

2   Definitions and Notation

We deal with Boolean formulas in conjunctive normal form (CNF). By a *variable* we mean a Boolean variable that takes truth values true or false. A *literal* is a variable x or its negation $\neg x$. A *clause* C is a set of literals such that C contains no complementary literals. A *formula* F is a set of clauses; n and m denote, respectively, the number of variables and the number of clauses in F. If each clause in F contains at most k literals, we say that F is a *k-CNF formula*.

An *assignment* to variables $x_1, \ldots, x_n$ is a mapping from $\{x_1, \ldots, x_n\}$ to $\{\mathsf{true}, \mathsf{false}\}$. This mapping is extended to literals: each literal $\neg x_i$ is mapped to the complement of the truth value assigned to x_i. We say that a clause C is *satisfied* by an assignment A if A assigns true to at least one literal in C. The formula F is *satisfied* by A if every clause in F is satisfied by A. In this case, A is called a *satisfying* assignment for F. We consider substitutions of truth values for some variables in a formula. If D is a set of literals, we write $F[D = \mathsf{false}]$ to denote the formula obtained from F as follows: any clause that contains the negation of a literal in D is removed from F, the literals occurring in D are deleted from the other clauses.

Here is a summary of the notation used in the paper.

- F denotes a CNF formula; n denotes the number of variables in F; m denotes the number of clauses in F.
- If C is a clause then $|C|$ denotes its length (the number of literals).
- We write $\log x$ to denote $\log_2 x$.
- $H(x)$ denotes the entropy function: $H(x) = -x \log x - (1-x) \log(1-x)$.

3   Algorithm

We describe an algorithm parameterized by a function $k(n, m)$. This function determines the length to which input clauses are to be shortened. The algorithm computes the value of $k(n, m)$ for particular n and m, then it runs a recursive procedure that implements the clause shortening approach combined with pruning. This recursive **Procedure $\mathcal{S}$** described below uses a k-SAT algorithm of [3] as a subroutine.

Lemma 1 ([3]). *There exists a deterministic algorithm that tests satisfiability of an input formula F in time at most*

$$m \cdot q(n) \cdot \left(2 - \frac{2}{k+1}\right)^n$$

where $q(n)$ is a polynomial in n, and k is the maximum length of clauses in F.

Procedure $\mathcal{S}$

Input: a CNF formula F and a positive integer k.

1. Assume F consists of clauses $C_1, \ldots, C_m$. Change each clause C_i to a clause D_i as follows: If $|C_i| > k$ then choose any k literals in C_i and drop the other literals; otherwise leave C_i as is, i.e., $D_i = C_i$. Let F' denote the resulting formula.
2. Test satisfiability of F' using the algorithm defined in Lemma 1.
3. If F' is satisfiable, output "satisfiable" and halt. Otherwise, for each i, do the following:
   (a) Convert F to F_i as follows:
       i. Replace C_j by D_j for all $j < i$;
       ii. Assign false to all literals in D_i.
   (b) Recursively invoke **Procedure $\mathcal{S}$** on (F_i, k).
4. Return "unsatisfiable".

Algorithm $\mathcal{A}_{k(n,m)}$

Parameter: a positive integer function $k(n, m)$
Input: a CNF formula F with m clauses over n variables $(n \le m)$

1. Compute $k = k(n, m)$.
2. Invoke **Procedure $\mathcal{S}$** on (F, k).

4 Upper Bound

First we give an upper bound for **Algorithm $\mathcal{A}_{k(n,m)}$**. Then we find a particular function $k(n, m)$ that approximately minimizes this upper bound.

Theorem 1. *Let $k(n, m)$ be an integer function such that:*

$$3 \le k(m, n) \le \log m. \tag{1}$$

Then Algorithm $\mathcal{A}_{k(n,m)}$ runs in time

$$O(\sqrt{m}) \cdot \tfrac{n}{k} \cdot q(n) \cdot 2^{n\left(1 - \frac{\log e}{k+1}\right) + O(m \cdot 2^{-k})}, \tag{2}$$

where $q(n)$ is the polynomial appearing in Lemma 1.

Proof. Let $t(F)$ be the running time of **Procedure** $\mathcal{S}$ on (F, k). It is not difficult to see that $t(F)$ can be estimated as follows:

$$t(F) \leq t_0(F') + \sum_{i=1}^{m} t(F_i) \tag{3}$$

where F' and F_i are as described in **Procedure** $\mathcal{S}$, and $t_0(F')$ is the running time of the k-SAT algorithm from Lemma 1 on F'. Let $T(n, m, m')$ denote the maximum of the running time of **Procedure** $\mathcal{S}$ on (G, k) where G is a formula with $\leq n$ variables and $\leq m$ clauses such that at most m' of its clauses contain $> k$ literals. For the k-SAT algorithm, we define $T_0(n, m)$ as the maximum running time on a different set of formulas, namely let $T_0(n, m)$ be the maximum running time of the algorithm from Lemma 1 on the set of formulas F' such that each F' has $\leq m$ clauses over $\leq n$ variables and the maximum length of clauses is not greater than k.

Then for any n and m, inequality (3) implies the following recurrence relation:

$$T(n, m, m') \leq T_0(n, m) + \sum_{i=0}^{m-1} T(n - k, m, m' - i). \tag{4}$$

If we iteratively substitute $T(n - L, m, m' - i)$ into this recurrence, we turn its right-hand side into the sum of terms of the form $T_0(n - lk, m)$ for $l \leq n/k$.

Our proof strategy is as follows. We consider the recursion tree of our algorithm and estimate the total amount T_l of work done at its l-th level (i.e., the sum of terms $T_0(n - lk, m)$). We then find l^* that maximizes this estimation. The total running time is then at most n/k times the estimation for the level l^*.

To estimate T_l, we note that the number of nodes at the l-th level

$$\sum_{i_1=1}^{m} \sum_{i_2=1}^{i_1} \cdots \sum_{i_l=1}^{i_{l-1}} 1$$

is the number of ways to choose l possibly equal elements out of m, i.e., $\binom{m+l-1}{l}$ (see, e.g., [13, Sect. 1.2]). Then

$$T_l \leq m \cdot q(n) \cdot \left(2 - \tfrac{2}{k+1}\right)^{n-lk} \cdot \binom{m+l-1}{l}. \tag{5}$$

Let E_l denote the right-hand side of the estimation (5). It is straightforward to see that $E_{l+1} \leq E_l$ if and only if

$$\tfrac{m+l}{l+1} \cdot \left(2 - \tfrac{2}{k+1}\right)^{-k} \leq 1,$$

which is equivalent to

$$\tfrac{m+l}{l+1} \cdot 2^{-k} \cdot \left(1 + \tfrac{1}{k}\right)^k \leq 1.$$

Therefore, the maximum of E_l over l is attained at the following integer l^*:

$$l^* = \tfrac{m\alpha - 2^k}{2^k - \alpha} + \delta,$$

where $\alpha = (1 + 1/k)^k$ and $-1 < \delta < 1$.

The next step is to give lower and upper bounds on l^*. We prove that

$$m \cdot 2^{-k} \leq l^* \leq 5.12 \cdot m \cdot 2^{-k} \qquad (6)$$

To prove the lower bound, we use $k \leq \log m$ and $\alpha \geq (1 + 1/3)^3 \approx 2.37$ (which follows from $k \geq 3$):

$$l^* = \frac{m\alpha - 2^k}{2^k - \alpha} + \delta$$
$$\geq m \cdot 2^{-k} \cdot \left(\frac{\alpha - 2^k/m}{1 - \alpha/2^k}\right) - 1$$
$$\geq m \cdot 2^{-k} \cdot \left(\frac{\alpha - 1}{1}\right) - 1$$
$$\geq m \cdot 2^{-k}.$$

The upper bound is proved using condition (1) and $\alpha < e$. Indeed,

$$l^* = \frac{m\alpha - 2^k}{2^k - \alpha} + \delta$$
$$\leq m \cdot 2^{-k} \cdot \left(\frac{\alpha - 2^k/m}{1 - \alpha/2^k}\right) + 1$$
$$\leq m \cdot 2^{-k} \cdot \left(\frac{e}{1 - e/8}\right) + 1$$
$$\leq m \cdot 2^{-k} \cdot \left(\frac{e}{1 - e/8} + 1\right)$$
$$\leq 5.12 \cdot m \cdot 2^{-k}.$$

Now we estimate the total amount of work done at the level l^*:

$$E_{l^*} = m \cdot q(n) \cdot 2^{n-kl^*} \cdot \left(1 - \frac{1}{k+1}\right)^{n-kl^*} \cdot \binom{m+l^*-1}{l^*}. \qquad (7)$$

The last factor in the right-hand side of (7) can be estimated using Stirling's approximation as in [1, page 4]:

$$\binom{m+l^*-1}{l^*} = O\left(\frac{1}{\sqrt{m+l^*}}\right) \cdot 2^{H\left(\frac{l^*}{m+l^*-1}\right)(m+l^*-1)}$$
$$= O\left(\frac{1}{\sqrt{m}}\right) \cdot e^{-l^* \ln \frac{l^*}{m+l^*-1} - (m-1) \ln \frac{m-1}{m+l^*-1}}.$$

Using $l^* - 1 < m$ and $\ln(1 + x) < x$, we have

$$\binom{m+l^*-1}{l^*} = O\left(\frac{1}{\sqrt{m}}\right) \cdot e^{l^* \ln \frac{m}{l^*} + l^* \ln\left(1 + \frac{l^*-1}{m}\right) + (m-1) \ln\left(1 + \frac{l^*}{m-1}\right)}$$
$$= O\left(\frac{1}{\sqrt{m}}\right) \cdot e^{l^*\left(\ln \frac{m}{l^*} + 2\right)}.$$

The factor $\left(1 - \frac{1}{k+1}\right)^{n-kl^*}$ in (7) can be estimated using the inequality $\ln(1 - x) < -x$:

$$\left(1 - \frac{1}{k+1}\right)^{n-kl^*} = e^{(n-kl^*) \ln\left(1 - \frac{1}{k+1}\right)} \leq e^{-\frac{n-kl^*}{k+1}} < e^{-\frac{n}{k+1} + l^*}.$$

Hence, we can estimate E_{l^*} as follows:

$$E_{l^*} \leq O(\sqrt{m}) \cdot q(n) \cdot 2^{n-kl^*} \cdot e^{-\frac{n}{k+1}+l^*} \cdot e^{l^*\left(\ln \frac{m}{l^*}+2\right)}$$
$$= O(\sqrt{m}) \cdot q(n) \cdot 2^n \cdot 2^{-\frac{n \log e}{k+1}} \cdot e^{-kl^* \ln 2} \cdot e^{l^*} \cdot e^{l^*\left(\ln \frac{m}{l^*}+2\right)}$$
$$= O(\sqrt{m}) \cdot q(n) \cdot 2^{n\left(1-\frac{\log e}{k+1}\right)} \cdot e^{\beta l^*},$$

where

$$\beta = 3 + \ln \frac{m}{l^*} - k \ln 2 = 3 + \ln \frac{m}{2^k \cdot l^*}.$$

The lower bound on l^* in (6) implies $\beta < 3$. Therefore, using the upper bound in (6), we have

$$E_{l^*} \leq O(\sqrt{m}) \cdot q(n) \cdot 2^{n\left(1-\frac{\log e}{k+1}\right)} \cdot e^{3l^*}$$
$$\leq O(\sqrt{m}) \cdot q(n) \cdot 2^{n\left(1-\frac{\log e}{k+1}\right)} \cdot e^{3 \cdot (5.12 \cdot m \cdot 2^{-k})}$$
$$\leq O(\sqrt{m}) \cdot q(n) \cdot 2^{n\left(1-\frac{\log e}{k+1}\right)} \cdot 2^{O(1) \cdot m \cdot 2^{-k}}.$$

Remark 1. What value of k minimizes bound (2)? Straightforward differentiation of the exponent

$$n\left(1 - \frac{\log e}{k+1}\right) + O(m \cdot 2^{-k})$$

gives the following equation:

$$k = \log(m/n) + 2\log(k+1) + O(1).$$

We can approximate a fix-point solution to this equation taking

$$k = \log(m/n) + d \cdot \log \log m$$

where $d > 1$ is a constant close to 1.

Theorem 2. *For any number $d > 1$, let $\mathcal{A}_d$ be an algorithm obtained from Algorithm $\mathcal{A}_{k(m,n)}$ by taking the following function $k(m,n)$:*

$$k(m,n) = \begin{cases} \lfloor \log(m/n) + d \cdot \log \log m \rfloor & \text{if } \log m < n^{1/d}, \\ \lfloor \log m \rfloor & \text{otherwise.} \end{cases}$$

Then $\mathcal{A}_d$ runs in time

$$O(\sqrt{m}) \cdot \frac{n}{k} \cdot q(n) \cdot 2^{n\left(1-\frac{1}{\ln(m/n)+d \cdot \ln \log m}+o\left(\frac{1}{k}\right)\right)} \tag{8}$$

on formulas such that $\log m < n^{1/d}$ and runs in time

$$O(\sqrt{m}) \cdot \frac{n}{k} \cdot q(n) \cdot 2^{n\left(1-\frac{1}{\ln(2m)}\right)} \tag{9}$$

on all other formulas, where $q(n)$ is the polynomial from Lemma 1.

Proof. We prove both bounds by applying Theorem 1. Note that the function $k(m,n)$ defined in the claim satisfies the inequality $k \leq \log m$ required by Theorem 1. This is obvious for $k = \lfloor \log m \rfloor$ and follows from $\log m < n^{1/d}$ for

$$k = \lfloor \log(m/n) + d \cdot \log \log m \rfloor. \tag{10}$$

To prove bound (8), we first write the upper bound given by Theorem 1 in the following form:

$$O(\sqrt{m}) \cdot \tfrac{n}{k} \cdot q(n) \cdot 2^{n(1-\gamma)}, \text{ where } \gamma = \tfrac{\log e}{k+1} - \tfrac{O(1) \cdot m}{n \cdot 2^k}.$$

Substituting the value of k from (10) in the second term of γ, we have

$$\gamma \geq \tfrac{\log e}{k+1} - \tfrac{O(1)}{(\log m)^d}$$

$$\geq \tfrac{\log e}{k} - \tfrac{\log e}{k(k+1)} - \tfrac{O(1)}{(\log m)^d}$$

$$\geq \tfrac{\log e}{k} - o\left(\tfrac{1}{k}\right) \quad \text{using } k \leq \log m \text{ and } d > 1$$

$$\geq \tfrac{1}{\ln(m/n) + d \cdot \ln \log m} - o\left(\tfrac{1}{k}\right).$$

Bound (9) is easily obtained from the upper bound given by Theorem 1 by substitution of $\lfloor \log m \rfloor$ for k.

Remark 2. Both bounds (8) and (9) hold for all formulas. Bound (8) is asymptotically better for formulas such that $\log m < n^{1/d}$, while bound (9) is better for all other formulas.

Remark 3. What is the best value of d? On the one hand, the smaller d is, the smaller k we have, which yields a better asymptotics of bound (8). In addition, the smaller d is, the weaker the $\log m \leq n^{1/d}$ restriction becomes. On the other hand, the smaller d we take, the slower $o(1/k)$ tends to zero (or, equivalently, the asymptotic behavior starts with lager values of m).

Remark 4. The randomized algorithm for SAT in [6] runs in time

$$2^{n\left(1 - \frac{1}{\ln(m/n) + O(\ln \ln m)}\right)}$$

up to a polynomial factor. It is straightforward to check that for any $d > 1$, the exponential part of the bound in Theorem 2 also can be written in this form, i.e., our upper bound for deterministic algorithms matches the best known upper bound for randomized algorithms.

Acknowledgement. We thank Natalia Tsilevich for her contribution to the proof of Theorem 1 and for helpful discussions.

References

1. B. Bollobás. *Random Graphs*. Cambridge University Press, 2nd edition, 2001.
2. T. Brueggemann and W. Kern. An improved local search algorithm for 3-SAT. *Theoretical Computer Science*, 329(1-3):303–313, December 2004.
3. E. Dantsin, A. Goerdt, E. A. Hirsch, R. Kannan, J. Kleinberg, C. Papadimitriou, P. Raghavan, and U. Schöning. A deterministic $(2 - 2/(k + 1))^n$ algorithm for k-SAT based on local search. *Theoretical Computer Science*, 289(1):69–83, 2002.
4. E. Dantsin, E. A. Hirsch, and A. Wolpert. Algorithms for SAT based on search in Hamming balls. In *Proceedings of the 21st Annual Symposium on Theoretical Aspects of Computer Science, STACS 2004*, volume 2996 of *Lecture Notes in Computer Science*, pages 141–151. Springer, March 2004.
5. E. Dantsin and A. Wolpert. Derandomization of Schuler's algorithm for SAT. In *Proceedings of the 7th International Conference on Theory and Applications of Satisfiability Testing, SAT 2004*, volume 3542 of *Lecture Notes in Computer Science*, pages 80–88. Springer, 2005.
6. E. Dantsin and A. Wolpert. A faster clause-shortening algorithm for SAT with no restriction on clause length. *Journal on Satisfiability, Boolean Modeling and Computation*, 1:49–60, November 2005.
7. K. Iwama and S. Tamaki. Improved upper bounds for 3-SAT. In *Proceedings of the 15th Annual ACM-SIAM Symposium on Discrete Algorithms, SODA 2004*, page 328, January 2004.
8. R. Paturi, P. Pudlák, M. E. Saks, and F. Zane. An improved exponential-time algorithm for k-SAT. In *Proceedings of the 39th Annual IEEE Symposium on Foundations of Computer Science, FOCS'98*, pages 628–637, 1998.
9. R. Paturi, P. Pudlák, and F. Zane. Satisfiability coding lemma. In *Proceedings of the 38th Annual IEEE Symposium on Foundations of Computer Science, FOCS'97*, pages 566–574, 1997.
10. P. Pudlák. Satisfiability – algorithms and logic. In *Proceedings of the 23rd International Symposium on Mathematical Foundations of Computer Science, MFCS'98*, volume 1450 of *Lecture Notes in Computer Science*, pages 129–141. Springer-Verlag, 1998.
11. U. Schöning. A probabilistic algorithm for k-SAT and constraint satisfaction problems. In *Proceedings of the 40th Annual IEEE Symposium on Foundations of Computer Science, FOCS'99*, pages 410–414, 1999.
12. R. Schuler. An algorithm for the satisfiability problem of formulas in conjunctive normal form. *Journal of Algorithms*, 54(1):40–44, January 2005. A preliminary version appeared as a technical report in 2003.
13. R. P. Stanley. *Enumerative Combinatorics*, volume 1. Wadsworth & Brooks/Cole, 1986.

Network Discovery and Verification
with Distance Queries[*]

Thomas Erlebach[1], Alexander Hall[2], Michael Hoffmann[1], and Matúš Mihaľák[1]

[1] Department of Computer Science, University of Leicester
{te17, mh55, mm215}@mcs.le.ac.uk
[2] Institute for Theoretical Computer Science, ETH Zürich
alex.hall@inf.ethz.ch

Abstract. The network discovery (verification) problem asks for a minimum subset $Q \subseteq V$ of queries in an undirected graph $G = (V, E)$ such that these queries discover all edges and non-edges of the graph. In the distance query model, a query at node q returns the distances from q to all other nodes in the graph. In the on-line network discovery problem, the graph is initially unknown, and the algorithm has to select queries one by one based only on the results of previous queries. We give a randomized on-line algorithm with competitive ratio $O(\sqrt{n \log n})$ for graphs on n nodes. We also show lower bounds of $\Omega(\sqrt{n})$ and $\Omega(\log n)$ on the competitive ratio of deterministic and randomized on-line algorithms, respectively. In the off-line network verification problem, the graph is known in advance and the problem is to compute a minimum number of queries that verify all edges and non-edges. We show that the problem is $\mathcal{NP}$-hard and present an $O(\log n)$-approximation algorithm.

1 Introduction

The growing interest in decentralized networks such as the Internet or peer-to-peer networks has introduced many new algorithmic challenges. A key property of these networks is that there is no central authority that maintains a map of the network. Obtaining an accurate map, usually represented as a graph, is not easy due to the dynamic growth of the network. A common approach to obtain a map of a network, or at least a good approximation, is to make some local measurements, which could be seen as local views of the network from selected nodes, and combine these in an appropriate manner. There is an extensive body of related work studying various aspects of this approach, see e.g. [14, 9, 15, 12, 13, 11, 3, 16, 8, 1, 6, 7].

As making measurements at a node is usually costly, the problem of minimizing the number of such measurements arises naturally. Nevertheless, it was proposed only recently to study this problem from a combinatorial optimization point of view: Beerliova et al. [4] introduce the network discovery and verification

---

[*] Work partially supported by European Commission - Fet Open project DELIS IST-001907 Dynamically Evolving Large Scale Information Systems, for which funding in Switzerland is provided by SBF grant 03.0378-1.

T. Calamoneri, I. Finocchi, G.F. Italiano (Eds.): CIAC 2006, LNCS 3998, pp. 69–80, 2006.

problems, which ask to find a map of a network with a small number of queries (measurements). In the on-line network discovery problem only the nodes V of a graph G are known in the beginning. An algorithm can make queries at nodes of the graph, and each query returns a local view of the graph. The task of the algorithm is to choose a minimum subset $Q \subseteq V$ of queries, such that the whole graph is discovered, i.e., all edges and non-edges are known. The network verification problem is the off-line version of the problem: The whole graph is known to the algorithm, and the task is to compute a minimum set Q of queries that verify all edges and non-edges. One motivation for the off-line version is checking with as few measurements as possible whether a given map is still correct.

In order to discover a graph, it may seem sufficient to discover only its edges. However, especially in view of the on-line setting, it is also necessary to have a proof (i.e., discover) for each unconnected node pair that indeed there is no edge between them. An on-line algorithm can only know that it has finished discovering the graph when both edges and non-edges have been discovered. Considering both also makes it possible to quantify how much knowledge about the network is revealed by a given set of queries. This could also be helpful e.g. when investigating the quality of previously published maps of the Internet.

In [4], a very strong query model was used: A query at a node v reveals all edges and non-edges whose endpoints have different distances from v. This model was motivated by the consideration that in certain scenarios one can identify all edges on shortest paths between the query node and all other nodes. In this paper, we study network discovery and network verification in the model where a query $q \in V$ gives all distances from q to any other node of the investigated graph G. We refer to the on-line problem as DIST–ALL–DISCOVERY and to the off-line problem as DIST–ALL–VERIFICATION. This *distance query model* is much weaker than the model used in [4], in the sense that typically a query reveals much less information about the network.

There are several reasons that motivate us to study the distance query model. First, in many networks it is realistically possible to obtain the distances between a node and all other nodes, while it is difficult or impossible to obtain information about edges or non-edges that are far away from the query node. For example, so-called distance-vector routing protocols work in such a way that each node informs its neighbors about upper bounds on the distances to all other nodes until these values converge; in the end, the routing table at a node contains the distances to all other nodes, and a query in our model would correspond to reading out the routing table. Another scenario is the discovery of the topology of peer-to-peer networks such as Gnutella [5]. With the Ping/Pong protocol it is possible to use a Ping command to ask all nodes within distance k (the TTL parameter of the Ping) to respond to the sender [2]. Repeated Pings could be used to determine the distances to all other nodes. Real peer-to-peer networks, however, are often so large that it becomes prohibitive to send Pings for larger values of k, and there are also many other aspects that make the actual discovery of the topology of a Gnutella network very difficult [2]. Nevertheless, we believe that our model is a good starting point for studying fundamental issues in the discovery of networks that support Ping/Pong-like protocols.

Related Work. There are several ongoing large-scale efforts to collect data representing local views of the Internet. The most prominent one is probably the RouteViews project [15] by the University of Oregon. It collects data (in the form of lists of paths) from a large number of so-called border gateway protocol routers. More recently, and due to good publicity very successfully, the DIMES project [9] has started collecting data with the help of a volunteer community. Users can download a client that collects paths in the Internet by executing successive traceroute commands. A central server can direct each client individually by specifying which routes to investigate. Data obtained by these or similar projects has been used with heuristics to obtain maps of the Internet, basically by simply overlaying the paths found by the respective project, see e.g. [13, 15, 9, 14]. Another line of research aims at inferring from such local views the types of the economic relationships between nodes in the Internet graph [11, 16, 8].

Beerliova et al. [4] propose the problem of network discovery (verification) and study it for the "layered graph" query model: A query $q \in V$ returns all edges and non-edges between nodes of different distance from q. They give an $o(\log n)$ inapproximability result for the off-line version and a randomized on-line algorithm with competitive ratio $O(\sqrt{n \log n})$. The on-line algorithm we present in this paper is based on a similar approach, but requires new ideas.

Our Results. In Sect. 2 we give basic definitions concerning network discovery and verification in the distance query model. We then characterize the queries that discover an individual non-edge and the sets of queries that together discover an individual edge. (At first sight, it may seem that the only way to discover an edge in the distance query model is to query one of its incident nodes. It turns out, however, that more intricate deductions are possible and edges at a larger distance from the query nodes can be discovered.) In Sect. 3 we show lower bounds on the number of queries needed to discover or verify a graph, based on the independence number α, clique number ω, and size of the edge set of the graph. For DIST–ALL–VERIFICATION we present in Sect. 4 polynomial-time algorithms for basic graph classes: chains, cliques, trees, cycles, and hypercubes. For general graphs, the problem turns out to be $\mathcal{NP}$-hard, and an $O(\log n)$-approximation algorithm is presented. For DIST–ALL–DISCOVERY we show in Sect. 5 that no deterministic on-line algorithm can be better than $O(\sqrt{n})$-competitive and no randomized on-line algorithm can be better than $O(\log n)$-competitive. Finally, we present our main result, a randomized on-line algorithm with competitive ratio $O(\sqrt{n \log n})$. Proofs omitted due to space restrictions can be found in [10].

2   Definitions and Preliminaries

Throughout this paper we assume graphs to be undirected and connected. For a given graph $G = (V, E)$, we denote the number of nodes by $n = |V|$ and the number of edges by $m = |E|$. For two distinct nodes $u, v \in V$, we say that $\{u, v\}$

is an *edge* if $\{u, v\} \in E$ and a *non-edge* if $\{u, v\} \notin E$. The set of non-edges is denoted by $\overline{E}$. By $\overline{G}$ we denote the complement of G, i.e., $\overline{G} = (V, \overline{E})$.

A *query* is specified by a node $v \in V$ and is called a query *at* v or simply the query v. In the *distance query model* the answer of a query at v consists of the distances from v to every node of G. We refer to sets of nodes with the same distance from v as *layers*. We use L_i or simply *layer i* to refer to the layer of nodes at distance i from the query node. By $d_G(u, v)$ we denote the distance from u to v in G. We may omit the subscript G if it is clear from the context to which graph the distance refers. Let $\mathbf{D}_G(Q)$, for $Q \subseteq V$, be a collection of distance vectors, one vector $d_G(Q, v)$ for each node $v \in V$. The vector $d_G(Q, v)$ has dimension $|Q|$, and each component gives the distance $d_G(q, v)$ of one of the (query) nodes $q \in Q$ to v; the i-th component corresponds to the i-th query node. Thus, we write $\mathbf{D}_G(Q) \neq \mathbf{D}_{G'}(Q)$, for $G' = (V, E')$, if there exists at least one query $q \in Q$ and a node $v \in V$ such that $d_G(q, v) \neq d_{G'}(q, v)$. Conversely, $\mathbf{D}_G(Q) = \mathbf{D}_{G'}(Q)$, if $d_G(q, v) = d_{G'}(q, v)$ holds for all queries $q \in Q$ and all nodes $v \in V$.

As opposed to the layered query model studied in [4], in the distance query model a query at node v does not explicitly return edges or non-edges. We shall show, however, how the information about the distances of nodes to (possibly a combination of several) queries can be utilized for discovering individual edges or non-edges of the graph. First we give a formal notion of what we mean by "discovering" a graph in this model. We use the two terms discover and verify to distinguish between the on-line and the off-line setting, they are otherwise equivalent (and we sometimes use the word "discover" also in the off-line setting). The following definitions hold for both terms but for simplicity are stated only for the network discovery setting.

A query set $Q \subseteq V$ for the graph $G = (V, E)$ *discovers the edge* $e \in E$ (*discovers the non-edge* $\overline{e} \in \overline{E}$), if for all graphs $G' = (V, E')$ with $\mathbf{D}_G(Q) = \mathbf{D}_{G'}(Q)$ it holds that $e \in E'$ ($\overline{e} \in \overline{E'}$). $Q \subseteq V$ *discovers the graph* G, if it discovers all edges and non-edges of G.

If Q discovers G, this implies that any graph G' with $\mathbf{D}_G(Q) = \mathbf{D}_{G'}(Q)$ must have the same edges and non-edges as G, i.e., $G' = G$. Conversely, if a query set Q for G yields $\mathbf{D}_G(Q) = \mathbf{D}_{G'}(Q)$ only for $G' = G$ and for no other graph, then Q discovers G. This gives an equivalent definition: A query set $Q \subseteq V$ *discovers the graph* $G = (V, E)$, if for every graph $G' = (V, E') \neq G$ at least one of the resulting distances changes, i.e., $\mathbf{D}_G(Q) \neq \mathbf{D}_{G'}(Q)$. Intuitively, the queries Q that discover a graph G can distinguish it from any other graph G' (sufficient and necessary condition).

Observation 1. *For $G = (V, E)$ the query set $Q \subseteq V$ discovers a non-edge $\{u, v\} \in \overline{E}$ if and only if there exists a query $q \in Q$ with $|d(q, u) - d(q, v)| \geq 2$.*

Proof. The implication "$\Leftarrow$" is obvious. To see the second implication "$\Rightarrow$", assume that $\{u, v\}$ is a non-edge and that (for a contradiction) every query node q gives $|d(q, u) - d(q, v)| \leq 1$. Then, if $\{u, v\}$ was an edge, the distances returned by Q would not change, as u and v are either in the same layer or in consecutive layers of each query $q \in Q$. □

For a query q and $\{u, v\} \in \overline{E}$ with $|d(q, u) - d(q, v)| \geq 2$, we say that q *discovers the non-edge* $\{u, v\}$.

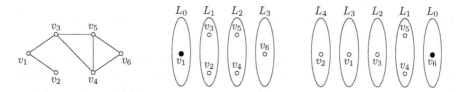

Fig. 1. Edge $\{v_3, v_4\}$ of a graph (left) is discovered by the combination of queries at nodes v_1 and v_6; the distances to the query node v_1 (middle) and v_6 (right) are depicted as layers of the graph

An edge may be discovered by a combination of several queries; this is a major difference to the layered graph query model of [4], where the set of edges and non-edges discovered by a set of queries is simply the union of the edges and non-edges discovered by the individual queries. If a node w is in layer $i + 1$ of a query q, this shows that w must be adjacent to at least one node from layer i. If layer i has more than one node, then it is not necessarily clear which node from layer i is adjacent to w. Figure 1 shows an example of how a combination of two queries can discover an edge even if each of the two queries alone does not discover the edge: The edge $\{v_3, v_4\}$ is neither discovered by a query at v_1 nor by a query at v_6 alone. The query at v_1 reveals that v_4 is connected to v_2 or to v_3 (or both). The query at v_6 identifies $\{v_2, v_4\}$ as a non-edge. From these two facts one can deduce that v_4 must be connected to v_3, i.e., $\{v_3, v_4\}$ is an edge. This discussion is generalized by the following observation [10].

Observation 2. *For $G = (V, E)$ the queries $Q \subseteq V$ discover an edge $\{u, v\} \in E$ if and only if there is a query $q \in Q$ with the following two properties:*

(i) *The nodes u and v are in consecutive layers of query q, say, u in the i-th layer L_i and v in the $(i + 1)$-th layer L_{i+1}, and $L_i \setminus \{u\}$ does not contain any neighbor of v.*

(ii) *The queries Q discover all non-edges between v and the nodes in $L_i \setminus \{u\}$.*

We say that a query for which (i) holds is a *partial witness* for the edge $\{u, v\}$. The word "partial" indicates that the query alone is not necessarily sufficient to discover the edge; additional queries may be necessary to discover the non-edges required by (ii).

We conclude that a set of queries discovers a graph G if and only if it discovers all non-edges and contains a partial witness for every edge.

3   Lower Bounds

In this section we show lower bounds on the number of queries needed to discover (or verify) a graph G, based on the independence number α, the clique number ω, and the number of edges m.

Lemma 1. *For any graph G with independence number α and diameter $diam >$ 2, at least $\log_{\lceil \frac{diam}{2} \rceil}(\alpha) - 1$ queries are needed to discover G. If $diam = 2$, we need at least $\alpha - 1$ queries.*

Lemma 2. *For any graph G with clique number ω at least $\omega - 1$ queries are necessary to discover G.*

Proof. Consider a clique K of size ω in G. Let q be the first query. The nodes of K appear in at most two consecutive layers i and $i + 1$ of query q. Observe that q is a partial witness of an edge from K if and only if there is exactly one node v from K in layer i and the remaining nodes of K are in layer $i + 1$. Moreover, q is a partial witness only for edges incident with v. After query q, there is still a clique K' of size $\omega - 1$ for which no query has been made that is a partial witness of any of its edges. Therefore, by induction (using the fact that one query is necessary for a clique of size 2 as the base case), it follows that we need at least $\omega - 1$ queries to discover G. □

Lemma 3. *Any graph G with n nodes and m edges needs at least $m/(n-1)$ queries to be discovered.*

Proof. Consider the layers of an arbitrary query $q \in V$. For each node v on layer i, q can be a partial witness for at most one edge $\{u, v\}$ with u in layer $i - 1$. Therefore, q can be a partial witness for at most $n - 1$ edges. Since a set of queries that discovers G must contain a partial witness for each of the m edges of G, the bound follows. □

4   Network Verification

Polynomially Solvable Cases. We discuss some classes of graphs for which the optimal number of queries for network verification can be determined in polynomial time.

Lemma 4. *G can be verified with 1 query if and only if G is a chain. A clique K_n on n vertices needs $n - 1$ queries to be verified.*

The example of the cycle with 4 nodes C_4 shows that there is a graph that needs $n - 1$ queries to be verified and is not a clique. The same holds for graphs that are obtained from K_n by deleting one edge, for $n \geq 4$. In general, for cycles the following lemma holds.

Lemma 5. *A cycle C_n, $n > 6$, can be verified optimally with 2 queries.*

Now we characterize the optimal query sets for trees. For this, we define a *leg* to be a maximal path in the tree starting at a leaf and containing only vertices of degree at most 2, see Fig. 2. If the tree is not a chain, there must be a node u of degree greater than 2 adjacent to the last vertex of the leg. We call u a *body* and we say that the leg is *adjacent* to its body u. The body u with all its adjacent legs is called a *spider*. Nodes that are not part of a spider are called *connectors* (i.e., nodes that are not in a leg and have no adjacent leg).

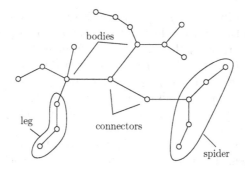

Fig. 2. Legs, bodies, spiders and connectors in a tree

Lemma 6. *Let $T = (V, E)$ be a tree that is not a chain. Denote by $B \subset V$ the set of bodies of the graph. Let l_b, for $b \in B$, be the number of legs adjacent to b. Let $T[B]$ be the induced subgraph of T on vertex set B. Let $VC(T[B])$ denote a minimum vertex cover of $T[B]$. Then the minimum number of queries to verify T is $\sum_{b \in B}(l_b - 1) + |VC(T[B])|$.*

Proof. We show first that we indeed need at least this many queries. Observe that if there is no query in two legs adjacent to a body, then we cannot verify the non-edges formed by vertices of the two legs at the same distance from the body. So, for each body, there must be at least one query in every leg except one. Moreover, if there are two legs of two different bodies which are connected by an edge then there must be at least one query in one of these legs. Otherwise we cannot verify the non-edge between vertices of the legs at the same distance from their bodies. Therefore, for any two bodies connected by an edge, at least one of them has a query in every leg. The bodies all of whose legs contain a query form a vertex cover of $T[B]$, and therefore a minimum vertex cover gives a lower bound on the number of spiders that have a query in every leg.

To prove that the claimed number of queries is sufficient, we construct a query set Q in the following way. We compute a minimum vertex cover of $T[B]$ (which can be done in polynomial time on trees). Let u be a body. We add the leaves of $l_u - 1$ of its legs to Q. If u is in the vertex cover, we add also the leaf of the last (the l_u-th) leg to Q.

We show now that Q verifies T. We start with non-edges. Let $\{v, w\}$ be a non-edge. We distinguish several cases. First, consider the case that both v and w are from legs. Consider the following subcases. If v and w are from the same leg, the non-edge is clearly verified by any query. If v and w are from different legs, and there is a query q in the leg where v or w is, then this query verifies the non-edge. (Note that there must be a query in the leg of v or w if they are in different legs of the same spider, or in legs of spiders whose bodies are adjacent.) Now assume that v and w are from different spiders with bodies u and u', which are not neighbors, and there is no query in the legs containing v and w. Let the path from u to u' be $u, x, \ldots, y, u'$, where $x = y$ is possible. Let q be a query from a leg adjacent to a body b such that the path from b to u does not contain x, possibly $b = u$. Let d_v be the distance from u to v, d_w be the distance from u' to w and let $d \geq 2$ be the distance between u and u'. If q does not verify the non-edge $\{v, w\}$ then $|d(q, v) - d(q, w)| = |d_v - (d + d_w)| \leq 1$. Then a query q' from a leg adjacent to a body b' such that the path from b' to u' does not contain y, possibly $b' = u'$, satisfies $|d(q', v) - d(q', w)| = |(d_v + d) - d_w| \geq 3$ and thus q' verifies the non-edge.

Now, consider the case that at least one of the two nodes, say, the node v, is not from a leg. Then any query in a tree of the forest $T \setminus \{v\}$ that does not contain w verifies the non-edge. Observe that such a query always exists.

Therefore Q verifies all non-edges. We claim now that Q verifies all edges. For this observe that for a tree T any query is a partial witness for every edge. To see this, imagine the tree rooted at the query node. So, Q verifies T. □

Lemma 7. *A query set that verifies a d-dimensional hypercube H_d is a vertex cover, and any vertex cover verifies a d-dimensional hypercube H_d for $d \geq 4$. A minimum vertex cover verifies H_3. Therefore, the optimal number of queries is 2^{d-1} (size of a minimum vertex cover in H_d) for $d \geq 3$.*

Complexity and Approximability. We can show that DIST–ALL–VERIFICATION is $\mathcal{NP}$-hard by a reduction from the VERTEX–COVER problem (see [10]).

Theorem 1. *The problem* DIST–ALL–VERIFICATION *is $\mathcal{NP}$-hard.*

An $O(\log n)$-approximation algorithm for DIST–ALL–VERIFICATION can be obtained using the well-known greedy algorithm for the set cover problem: Each vertex v corresponds to a set containing the non-edges a query at v verifies and the edges for which a query at v is a partial witness, and the goal is to cover all edges and non-edges.

Theorem 2. *For the problem* DIST-ALL-VERIFICATION, *there is an $O(\log n)$-approximation algorithm.*

5   Network Discovery

Lower Bounds for On-line Algorithms. We present lower bounds on the competitive ratio of on-line algorithms for DIST–ALL–DISCOVERY. Consider the graph G_k from Fig. 3. It is a tree built recursively from a smaller tree G_{k-1} as depicted in the figure. Alternatively, G_k can be described as follows. Start with a chain of length $2k-1$ from x to v_k. For $1 \leq i \leq k$, the node on the chain at distance $2i-1$ from x is labeled as v_i. To each such node v_i, $1 \leq i \leq k$, we attach another chain (which we call *arm*) of length $2i - 1$, starting at v_i. The number n_k of nodes of G_k satisfies $n_k = n_{k-1} + 1 + 2k$ for $k > 1$ and $n_1 = 3$. Hence, $n_k = k^2 + 2k$.

G_k is a non-trivial tree and, by Lemma 6, the optimum number of queries is 2. Now consider any deterministic algorithm A. As all vertices are indistinguishable

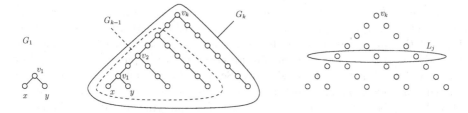

Fig. 3. Graph used in the proof of the lower bound $\Omega(\sqrt{n})$ for on-line algorithms (left and middle); layers after query at vertex v_k (right)

to A, we may assume that the initial query q_0 made by A is at v_k. This sorts the vertices into layers according to their distance from v_k. No non-edge is discovered within the layers. In particular, the non-edge $\{x, y\}$ in G_1 (see Fig. 3) is not discovered. We now show that A needs at least k additional queries to discover $\{x, y\}$. Observe that in the rightmost arm (attached to v_k) we have vertices from every layer. A picks a vertex from some layer j and, because all the vertices in this layer are indistinguishable for A, we may force A to pick the vertex from the rightmost arm. Such a query in the rightmost arm does not reveal any new information within G_{k-1}. The vertices within one layer of G_{k-1} remain indistinguishable for A. Thus, when A places its first query in G_{k-1}, we can force it to be at a node from G_{k-1}'s rightmost arm. We can continue recursively in this manner and therefore we can force A to query in every arm before it discovers $\{x, y\}$. Hence, A needs at least $1 + k$ queries to discover G_k.

Since $n_k = k^2 + 2k$, we have that $k = \Theta(\sqrt{n_k})$. Together with the fact that the optimum needs 2 queries, we get a lower bound of $\Omega(\sqrt{n})$ for deterministic algorithms. Furthermore, from the same construction we can also derive a lower bound for randomized on-line algorithms, see [10] for details.

Theorem 3. *For* DIST–ALL–DISCOVERY, *there is no* $o(\sqrt{n})$-*competitive deterministic and no* $o(\log n)$-*competitive randomized on-line algorithm.*

Randomized On-line Algorithm. We present a randomized algorithm for DIST–ALL–DISCOVERY. Its competitive ratio $O(\sqrt{n \log n})$ is very close to the lower bound $\Omega(\sqrt{n})$ for deterministic algorithms, but leaves a gap to the lower bound $\Omega(\log n)$ for randomized algorithms.

Theorem 4. *There is a randomized on-line algorithm with competitive ratio* $O(\sqrt{n \log n})$ *for* DIST–ALL–DISCOVERY.

Proof. The algorithm runs in two phases. In the first phase it makes $3\sqrt{n \ln n}$ queries at nodes chosen uniformly at random. In the second phase, as long as there is still an undiscovered pair $\{u, v\}$ (i.e., the queries executed so far have not discovered whether $\{u, v\}$ is an edge or non-edge), the algorithm executes the following. First, it queries both u and v. This discovers whether $\{u, v\}$ is an edge or non-edge. In case it is a non-edge, the algorithm then knows from the queries at u and v the set S of all queries that discover $\{u, v\}$: S is the set of vertices w for which $|d(u,w) - d(v,w)| \geq 2$. The algorithm then queries the whole set S. In case $\{u,v\}$ is an edge, the algorithm distinguishes three cases. First, if the queries at u and v discover a non-edge, say, $\{u,w\}$, that had not been discovered before, the algorithm proceeds with the pair $\{u,w\}$ instead of $\{u,v\}$ and handles it as described above. Second, if the number of neighbors of u and the number of neighbors of v is at most $\sqrt{n/\ln n}$, then the algorithm queries also all neighbors of u and v (notice that after querying u and v we know all their neighbors). With this information we know the set S of vertices that are partial witnesses for $\{u,v\}$: a vertex w is in S if and only if the two vertices are at distances i and $i+1$ from w and all the other neighbors of the more distant vertex are at distances $i+1$ or $i+2$. The algorithm then queries all vertices in S. Third, if the number of neighbors of u or the number of neighbors of v is larger than $\sqrt{n/\ln n}$, the algorithm does not do any further processing for this pair (i.e., this iteration of the second phase is completed) and proceeds with choosing another undiscovered pair $\{u',v'\}$ (if one exists).

The algorithm can be viewed as solving a HITTINGSET problem. For every non-edge $\{u,v\}$ let S_{uv} be the set of vertices that discover $\{u,v\}$. Similarly, for every edge $\{u,v\}$ let S_{uv} denote the set of all partial witnesses for $\{u,v\}$. The algorithm discovers the whole graph G if it hits all sets S_{uv}, for $\{u,v\} \in E \cup \overline{E}$. In the first phase, the algorithm aims to hit all the sets S_{uv} of size at least $\sqrt{n}\ln n$. Then, in the second phase, as long as there is an undiscovered pair $\{u,v\}$, the algorithm attempts to query the whole set S_{uv}; if $\{u,v\}$ is an edge, it also queries all the neighbors of u and v in order to determine S_{uv}, except in the case where the degree of u or v is too large. In the case that the undiscovered pair $\{u,v\}$ is an edge for which a partial witness has already been queried before, the query at u or v must discover a new non-edge, and the algorithm uses that non-edge instead of $\{u,v\}$ to proceed.

We analyze the algorithm as follows. Let OPT be the optimal number of queries. Consider a pair $\{u,v\}$ for which the set S_{uv} has size at least $\sqrt{n}\ln n$. In each query of the first phase, the probability that S_{uv} is not hit is at most $1 - \sqrt{n}\ln n/n = 1 - \sqrt{(\ln n)/n}$. Thus, standard calculations show that the probability that S_{uv} is not hit throughout the first phase is at most $1/n^3$. There are at most $\binom{n}{2}$ sets S_{uv} of cardinality at least $\sqrt{n}\ln n$. The probability that at least one of them is not hit in the first phase is at most $\binom{n}{2} \cdot \frac{1}{n^3} \leq \frac{1}{n}$.

Now consider the second phase, conditioned on the event that the first phase has indeed hit all sets S_{uv} of size at least $\sqrt{n}\ln n$. If the undiscovered pair $\{u,v\}$ is a non-edge, after querying u and v we know S_{uv}, and querying the whole set S_{uv} requires at most $\sqrt{n}\ln n$ queries (note that $|S_{uv}| \leq \sqrt{n}\ln n$ if $\{u,v\}$ is a non-edge that has not been discovered in the first phase). If the pair $\{u,v\}$ is an edge and the queries at u and v discover a new non-edge, the algorithm proceeds with that non-edge and makes at most $\sqrt{n}\ln n$ further queries (as above), hence at most $\sqrt{n}\ln n + 2$ queries in total for this iteration of the second phase. Otherwise, if the number of neighbors of u and of v is bounded by $\sqrt{n/\ln n}$, we query also all neighbors of u and v to determine the set S_{uv}, amounting to at most $2\sqrt{n/\ln n}$

queries, and then the set S_{uv}, giving another $\sqrt{n \ln n}$ queries (since S_{uv} has not been hit in the first phase). In total, we make at most $\sqrt{n \ln n} + 2\sqrt{n/\ln n}$ queries in this iteration of the second phase. Consider the remaining case, i.e., the case where the undiscovered pair $\{u, v\}$ is an edge, no partial witness for the edge has been queried before, and u or v has degree larger than $\sqrt{n/\ln n}$. Assume that there are k iterations of the second phase in which the undiscovered pair falls into this case. Note that no node can be part of an undiscovered pair in two such iterations. Hence, we get that $2|E| \geq k\sqrt{n/\ln n}$ and, by Lemma 3, $\text{OPT} \geq \frac{|E|}{n} \geq \frac{k\sqrt{n}}{2n\sqrt{\ln n}} = \frac{k}{2\sqrt{n \ln n}}$ and therefore $k \leq 2\sqrt{n \ln n} \cdot \text{OPT}$.

Let ℓ denote the number of iterations of the second phase in which the set S_{uv} was determined and queried (i.e., all iterations except the k iterations discussed above). We call such iterations *good* iterations. The overall cost of the second phase is at most $\ell\sqrt{n \ln n} + 2\ell\frac{\sqrt{n}}{\sqrt{\ln n}} + 2k$. Clearly, $\text{OPT} \geq \ell$, because no two undiscovered pairs $\{u, v\}$ considered in different good iterations can be discovered by the same query (or have the same partial witness). So the cost of the algorithm is at most $3\sqrt{n \ln n} + \ell\sqrt{n \ln n} + 2\ell\frac{\sqrt{n}}{\sqrt{\ln n}} + 2k = O(\sqrt{n \log n}) \cdot \text{OPT}$.

We have that with probability at least $1 - \frac{1}{n}$, the first phase succeeds and $O(\sqrt{n \log n}) \cdot \text{OPT}$ queries are made by the algorithm. If the first phase fails, the algorithm makes at most n queries (clearly, the algorithm need not repeat any query). This case increases the expected number of queries made by the algorithm by at most $\frac{1}{n} n = 1$. Thus, we have that the expected number of queries is at most $O(\sqrt{n \log n}) \cdot \text{OPT} + \frac{1}{n} n = O(\sqrt{n \log n}) \cdot \text{OPT}$.

$\square$

6   Conclusions and Future Work

In this paper, we have studied network discovery and network verification in the distance query model. The network verification problem is $\mathcal{NP}$-hard and admits an $O(\log n)$-approximation algorithm. For certain graph classes there exist polynomial optimal algorithms or easy characterizations of optimal query sets. For the network discovery problem, we have presented lower bounds of $\Omega(\sqrt{n})$ and $\Omega(\log n)$ on the competitive ratio of deterministic and randomized on-line algorithms, respectively, and designed a randomized on-line algorithm that achieves competitive ratio $O(\sqrt{n \log n})$.

The query model studied in this paper is motivated by real-world scenarios such as discovering the topology of a network that uses a distance-vector routing protocol by analyzing selected routing tables. An interesting direction for future work would be to consider a more realistic model where queries can only be executed at certain nodes of the network; this is motivated by the fact that only a rather small subset of the nodes in the Internet or in a network such as Gnutella can actually be used for queries. While our off-line results translate to such a model with forbidden query nodes in a straightforward way, it is not clear whether our on-line algorithm can be adapted to this model or a different approach needs to be employed.

References

1. D. Achlioptas, A. Clauset, D. Kempe, and C. Moore. On the bias of traceroute sampling; or, power-law degree distributions in regular graphs. In *Proc. 37th Ann. ACM Symp. Theory of Computing (STOC'05)*, pages 694–703, 2005.
2. V. Aggarwal, S. Bender, A. Feldmann, and A. Wichmann. Methodology for estimating network distances of Gnutella neighbors. In *Proceedings of the Workshop on Algorithms and Protocols for Efficient Peer-to-Peer Applications*, INFORMATIK 2004, 2004.
3. P. Barford, A. Bestavros, J. Byers, and M. Crovella. On the marginal utility of deploying measurement infrastructure. In *Proc. ACM SIGCOMM Internet Measurement Workshop*, November 2001.
4. Z. Beerliová, F. Eberhard, T. Erlebach, A. Hall, M. Hoffmann, M. Mihaľák, and L. S. Ram. Network discovery and verification. In *Proc. 31st International Workshop on Graph-Theoretic Concepts in Computer Science (WG'05)*, LNCS 3787, pages 127–138. Springer, 2005.
5. Clip2. The Gnutella protocol specification v0.4, 2001. http://www9.limewire.com/developer/gnutella_protocol_0.4.pdf.
6. L. Dall'Asta, I. Alvarez-Hamelin, A. Barrat, A. Vázquez, and A. Vespignani. Statistical theory of internet exploration. *Physical Review E*, 71, 2005.
7. L. Dall'Asta, I. Alvarez-Hamelin, A. Barrat, A. Vázquez, and A. Vespignani. Exploring networks with traceroute-like probes: theory and simulations. *Theoretical Computer Science*, 2006. To appear.
8. G. Di Battista, T. Erlebach, A. Hall, M. Patrignani, M. Pizzonia, and T. Schank. Computing the types of the relationships between autonomous systems. Submitted to IEEE/ACM Transactions on Networking, 2005.
9. DIMES. Mapping the Internet with the help of a volunteer community. http://www.netdimes.org/.
10. T. Erlebach, A. Hall, M. Mihal'ák, and M. Hoffmann. Network discovery and verification with distance queries. Research Report CS-06-002, Department of Computer Science, University of Leicester, March 2006.
11. L. Gao. On inferring autonomous system relationships in the internet. *IEEE/ACM Transactions on Networking*, 9(6):733–745, December 2001.
12. R. Govindan and A. Reddy. An analysis of internet inter-domain topology and route stability. In *Proc. IEEE INFOCOM 1997*, April 1997.
13. R. Govindan and H. Tangmunarunkit. Heuristics for Internet map discovery. In *Proc. IEEE INFOCOM 2000*, pages 1371–1380, Tel Aviv, Israel, March 2000.
14. Internet Mapping Project. Lucent Bell Labs. http://www.cs.bell-labs.com/who/ches/map/.
15. Oregon RouteViews Project. University of Oregon. http://www.routeviews.org.
16. L. Subramanian, S. Agarwal, J. Rexford, and R. Katz. Characterizing the internet hierarchy from multiple vantage points. In *Proc. IEEE INFOCOM 2002*, 2002.

Deciding the FIFO Stability of Networks in Polynomial Time

Maik Weinard

Institut für Informatik
Johann Wolfgang Goethe–Universität Frankfurt am Main
Robert-Mayer-Straße 11-15
60054 Frankfurt am Main, Germany
weinard@thi.informatik.uni-frankfurt.de

Abstract. FIFO is the most prominent queueing strategy due to its simplicity and the fact that it only works with local information. Its analysis within the adversarial queueing theory however has shown, that there are networks that are not stable under the FIFO protocol, even at arbitrarily low rate. On the other hand there are networks that are universally stable, i.e., they are stable under every greedy protocol at any rate $r < 1$.

The question as to which networks are stable under the FIFO protocol arises naturally. We offer the first polynomial time algorithm for deciding FIFO stability and simple-path FIFO stability of a directed network, answering an open question posed in [1, 4]. It turns out, that there are networks, that are FIFO stable but not universally stable, hence FIFO is not a worst case protocol in this sense. Our characterization of FIFO stability is constructive and disproves an open characterization in [4].

1 Introduction

The issue of queueing arises in communication networks. Communication networks can be naturally modeled as graphs with the vertices representing the access points of the network and the edges represent the established connections. Users insert data, organized in packets of roughly the same size. These packets travel along a path to their destination where they are absorbed. Natural goals are to keep the number of packets in the system at the same time small and to prevent long transportation times.

The connections have certain capacities and if the number of packets seeking to continue their journey along a specific edge exceeds this capacity, a choice must be made as to which packet may proceed immediately and which ones have to wait – the problem of queueing.

The adversarial queueing theory [5] was introduced to provide a formal framework that allows worst case analyses of queueing policies and network topologies: the insertion of packets and the assignment of a path and a destination is done by an adversary who is only restricted to not straight-forwardly overload an

T. Calamoneri, I. Finocchi, G.F. Italiano (Eds.): CIAC 2006, LNCS 3998, pp. 81–92, 2006.
© Springer-Verlag Berlin Heidelberg 2006

edge. A good queueing policy or a good network topology will keep the number of packets, that are in the system at the same time, limited. This ability will be referred to as *stability*.

There are queueing policies that perform well in all networks, like *Longest − In − System* that prefers packets that have so far spent the longest time in the system. FIFO however is by far the most prominent queueing policy due to its simplicity and its restriction to local information (as opposed to the packets age, that a packet must bring along and that must be trusted). Unfortunately it has been shown, that there are graphs (even planar ones) [3, 6], so that the FIFO protocol allows to pile up an unlimited number of packets even if the rate r at which packets are inserted is arbitrarily small. Of course, these are worst-case graphs.

On the other hand there are networks, that are stable under every queueing policy – universally stable graphs. Recently a characterization of these graphs together with an algorithm to decide universal stability has been introduced [1]. There the question was posed as to how FIFO stable networks can be characterized and whether it can be efficiently decided if a given network is FIFO stable. So far stability of FIFO has only been observed on universally stable graphs, giving rise to the question if FIFO is in this sense a worst case policy.

A partial answer has been provided in [4], where an open characterization is presented. These results namely prove, that one of two presented polynomial time algorithms correctly classifies FIFO stability. Hence the existence of a polynomial time algorithm is established.

We provide the missing piece of the classification and correct an erroneously classified set of graphs in [4]. Therefore we are able to present the first constructive classification and a polynomial time algorithm for both FIFO stability and FIFO simple-path stability of directed multigraphs. We show that FIFO is not a worst case queueing strategy, namely, that there are networks, that are FIFO stable but not universally stable and that there are networks, that are simple-path FIFO stable but not universally simple-path stable.

2   Preliminaries

Throughout this paper we consider directed graphs without loops but with multiple edges. As usual in adversarial queueing theory we assume, that the network operates in consecutive non-overlapping steps and that each step breaks down into three substeps. First new packets together with their assigned paths are inserted by the adversary, then packets chosen by the queueing strategy traverse edges, and finally packets that have traversed their last edge are absorbed. Hence packets that are inserted in a step t can traverse their first edge in the same step if they are chosen by the queueing strategy. We will use Q_e to denote the queue of edge e and $Q_e(t)$ as the set of *candidate*-packets that the queueing strategy chooses from in step t for edge e.

In this paper our focus is on the FIFO protocol. FIFO always picks a packet of maximum waiting time breaking ties arbitrarily.

We work with the leaky bucket adversarial model of [5, 2] and consider two types of adversaries: the (r, b)-adversary and the (r, b)-simple-path-adversary. During every interval of t consecutive steps and for every edge e both adversaries may insert at most $rt + b$ packets with e in the assigned path. Parameter $r \leq 1$ is called the rate and b is the burstiness. The (r, b)-simple-path-adversary is furthermore restricted to assign only simple paths, i.e., paths, in which no vertex appears more than once. The paths assigned by the (r, b)-adversary may have multiple appearances of a vertex, multiple appearances of edges however are forbidden.

Definition 1. *1. A network G is stable under a queueing strategy P against a (r, b)-[simple-path]-adversary, if, starting with an empty initial configuration, the number of packets in the system is upper bounded by some $B(r, b, G)$ that depends on the network topology and the parameters of the [simple-path]-adversary but not on the time that the network is exposed to the [simple-path]-adversary.*

2. A network G is [simple-path] FIFO stable if it is stable against every (r, b)-[simple-path]-adversary with $r < 1$ under the FIFO protocol.

We first note, that it suffices to classify strongly connected digraphs as to their stability.

Lemma 1. *Let $G = (V, E)$ be a digraph and let $V_1, V_2, \ldots, V_k$ constitute the strongly connected components of G. Set $E_i = \{e \in E | head(e), tail(e) \in V_i\}$ and $G_i = (V_i, E_i)$.*

1. G is FIFO stable if and only if every G_i is FIFO stable.
2. G is simple-path FIFO stable if and only if every G_i is simple-path FIFO stable.

The idea is, that an adversary equipped with extra burstiness can mimic the surrounding graph of a component. For a proof see [7]. From here on we may hence concentrate on strongly connected digraphs.

3   Sources of Instability

Instability is usually shown by providing an inductive sequence of insertions for an adversary: it is assumed, that an initial set S_0 of packets is already in the system, usually concentrated in one specific queue. Then a sequence of insertions is described, such that after these insertions a larger set S of packets takes the place of the original packets. As such a sequence of insertions can then be repeated arbitrarily often, the number of packets exceeds every bound and instability follows.

Often the initial set S_0 needs to have a certain minimal size. Observe, that for FIFO a set S_0 of arbitrary constant size can be inserted by an adversary with sufficiently high burstiness.[1]

---

[1] This argument is sometimes used without the appropriate care. Observe that it only applies for protocols, such as FIFO, that are ignorant of a packets history.

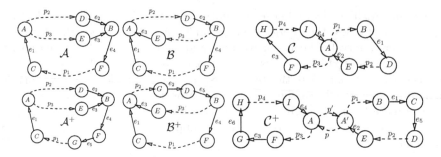

Fig. 1. Our families of unstable or simple-path unstable graphs. In all our diagrams an interrupted line indicates a simple path of an arbitrary number (including 0) of edges. If a path has 0 edges, the vertices at its ends coincide. We demand that within each graph the depicted paths p_i are edge disjoint. (p and p' in C^+ may have common edges.) We further demand, that none of the explicitly depicted edges e_i is a part of any simple path p_j.

All the instability results we need in the following sections have been previously obtained for example in [2] or [4]. We therefore only introduce our families of unstable graphs (Fig.1). The complete set of proofs can be found in [7].

Lemma 2. *A digraph from family A, B or C is not FIFO stable. A digraph from family A^+, B^+ or C^+ is not simple-path FIFO stable.*

4 Deciding FIFO Stability

We now concentrate on FIFO stability, before discussing simple-path stability in the next section.

4.1 The Special Case of 2-Vertex Components

Strongly connected digraphs with two vertices are a special case that needs specific arguments. Let $G = (\{X, Y\}, E)$ be a strongly connected digraph. Observe, that at least one edge from X to Y and one edge from Y to X must exist.

Lemma 3. *A strongly connected graph $G = (\{X, Y\}, E)$ is FIFO stable, if and only if at least one of the vertices has out-degree 1.*

Proof. First assume, that at least two edges in both directions are present. Then we have a graph of family C with $X = A = B = F$, $Y = D = E = H = I$ and empty p_1, p_2, p_3 and p_4. Observe that the sequence of insertions in the instability proof is still legal as we allow vertices to appear several times in a path, the edge disjointness of simultaneous insertions is not affected.

It remains to be shown, that graphs of the family D are FIFO stable.

We introduce $D_e(t)$ where e is an edge. $D_e(t)$ is the set of packets in the system after the insertions of step t and before the transportation of step t that have to traverse edge e and have not yet done so.

Fig. 2. Family $\mathcal{D}$ of FIFO stable networks with two vertices. $k \geq 1$ holds.

At first we verify, that it suffices to show, that the size of D_f never reaches a certain bound $B(r, b) < \infty$: If the network was unstable it could be forced to contain an arbitrary number of packets. Using further insertions we could move these packets to Q_f and only lose a constant fraction.

So from here on we assume, that T_0 is the first time, so that after step T_0 $D_f(T_0)$ holds at least $B(r, b)$ packets. We will see, that when $B(r, b)$ is picked sufficiently high, we reach a contradiction. Namely we see, that D_f must have consisted of more than $B(r, b)$ packets in an earlier time. Consequently $|D_f|$ is bounded and by the previous observation the system is stable. We choose $B(r, b) = \frac{2b}{(1-r)^2}$ and use the following properties of this choice

$$\frac{B(r, b) - b}{r} - \frac{b}{1 - r} > B(r, b) \tag{1}$$

$$(1 - r)\frac{B(r, b) - b}{r} - b \geq 0 \tag{2}$$

Let p be the oldest packet in $D_f(T_0)$ and let T_1 be the time, when p was inserted into the system. The packets in $D_f(T_0)$ have a common edge f. Consequently in the $T_0 - T_1 + 1$ steps of the period $[T_1, T_0]$, at most $r(T_0 - T_1 + 1) + b$ packets requiring edge f have been inserted. Our assumption thus leads to $r(T_0 - T_1 + 1) + b \geq B(r, b)$ and consequently

$$T_0 - T_1 + 1 \geq \frac{B(r, b) - b}{r}. \tag{3}$$

Let us assume p is ready to cross f at some time T' with $T_1 \leq T' \leq T_0$. Hence if p is inserted directly into Q_f, then $T' = T_1$ and if $T' > T_1$ then p is inserted into Q_{e_i} for some $1 \leq i \leq k$ at time T_1 and traverses e_i in step $T' - 1$. Of course p might also remain in Q_{e_i}, in this case we set $T' = T_0 + 1$.

In the period $[T', T_0]$ a packet traverses f in every step. At most $r(T_0 - T' + 1) + b$ packets join D_f by new insertions. As T_0 is picked minimally, we get $|D_f(T')| < |D_f(T_0)| \leq |D_f(T')| + r(T_0 - T' + 1) + b - (T_0 - T' + 1)$. This leads to

$$\frac{b}{1 - r} > T_0 - T' + 1. \tag{4}$$

Adding inequalities (3) and (4) we get the following bound for $T' - T_1$:

$$T' - T_1 > \frac{B(r, b) - b}{r} - \frac{b}{1 - r} \geq B(r, b) \tag{5}$$

The last bound is due to property (1) of $B(r, b)$. Thus we know, that packet p was inserted in Q_{e_i} and stayed in this queue for more than $B(r, b)$ steps. As p was present but not allowed to traverse e_i we know that another packet did in each of these steps. As FIFO is applied a set of packets of size more than $B(r, b)$ must have been in the queue of e_i when p was inserted in step T_1. Hence $|Q_{e_i}(T_1)| > B(r, b)$ holds.

Choose $T_2 < T_1$ maximal such that $Q_{e_i}(T_2) = \emptyset$. Between T_2 and T_1 at most $r(T_1 - T_2) + b$ packets requiring edge e_i were newly inserted into the system. As T_2 is chosen maximal, a packet traversed e_i in every step. We may conclude, that at least $|Q_{e_i}(T_1)| + (T_1 - T_2) - (r(T_1 - T_2) + b)$ of those packets that constitute $Q_{e_i}(T_1)$ were already in the system at time T_2 but not yet in Q_{e_i}. As f is the only possible predecessor of e_i in any path, these packets must have required f at time T_2. By definition they were in $D_f(T_2)$. We continue

$$|D_f(T_2)| \geq |Q_{e_i}(T_1)| + (T_1 - T_2) - (r(T_1 - T_2) + b)$$
$$= |Q_{e_i}(T_1)| + (1 - r)(T_1 - T_2) - b$$
$$> B(r, b) + (1 - r)(T_1 - T_2) - b$$

Finally we need a lower bound for $T_1 - T_2$. During the interval $(T_2, T_1]$ at most $r(T_1 - T_2) + b$ packets requiring e_i can be inserted newly and at most $T_1 - T_2$ *old* packets may arrive in Q_{e_i} via f. $T_1 - T_2$ packets leave Q_{e_i} via e_i. We thus get $r(T_1 - T_2) + b + (T_1 - T_2) - (T_1 - T_2) \geq B(r, b)$ yielding $T_1 - T_2 \geq \frac{B(r,b)-b}{r}$. We can now complete our lower bound of $|D_f(T_2)|$:

$$|D_f(T_2)| > B(r, b) + (1 - r)(T_1 - T_2) - b$$
$$\geq B(r, b) + (1 - r)\frac{B(r, b) - b}{r} - b \geq B(r, b)$$

The last inequality is due to property (2) and completes the proof.   □

This completes the classification of two node components. Lemma 3 contradicts a claim in [4]: there the network from family $\mathcal{D}$ with $k = 2$ (named $\mathcal{U}_1$ in [4]) is left unclassified, and the network with $k = 3$ (named $\mathcal{U}_1^1$) is claimed to be unstable. This leads to a faulty open classification of FIFO stability in [4]. As the networks from $\mathcal{D}$ with $k \geq 2$ are not universally stable [1], we have the following corollary.

Corollary 1. *Networks that are FIFO stable but not universally stable exist.*

4.2   The Case of Three and More Vertices

Lemma 4. *A strongly connected digraph $G = (V, E)$ with $|V| \geq 3$ is FIFO stable if and only if G is the simple directed cycle of k vertices.*

Proof. The simple directed cycle is universally stable under any greedy protocol as has been shown in [2]. It remains to show that all other digraphs of n vertices are FIFO unstable.

In a strongly connected graph of more than three vertices, there must exist a vertex B with an incoming edge from A and an outgoing edge to C with $A \neq C$. As A must be reachable from C, there is a simple path from C to A. If this path contains B, we have a subgraph of family $\mathcal{C}$. Hence the graph is unstable.

If the simple path from C to A does not contain B, we have a simple directed cycle with at least 3 vertices in our graph. Let us rename the vertices of this cycle $v_1, v_2, \ldots v_l$ so that edges are of the form (v_i, v_{i+1}) and (v_l, v_1). If this cycle is not the entire graph, one vertex of the cycle (w.l.o.g. v_1) must have a further outgoing edge to a node w that may or may not be another node of the cycle. The cycle however must be reachable from w and hence a simple path p from w to a vertex v_i of the cycle must exist. We choose p so that v_i is the first node of the cycle on p. If w itself is on the cycle, p is empty and we have $w = v_i$. Now the value of i yields three cases, that lead to a subgraph of family $\mathcal{A}, \mathcal{B}$ or $\mathcal{C}$ respectively.

For $i = 1$ we have $w \neq v_i$ since otherwise the graph contained a loop and hence p is not empty. We have a subgraph of family $\mathcal{C}$ with v_1 taking the role of A. The cycle p_1, e_1, p_2, e_2 is constituted by our cycle $v_1, \ldots, v_l$. p_3 is empty, hence $F = A = v_1$. (v_1, w) serves as e_3 and the simple path from w back to v_1 is p_4, e_e.

For $2 \leq i \leq l-1$ we get a subgraph of the family $\mathcal{A}$. Node v_1 is A and v_i is B. Path $(v_1, w)p$ is $p_2 e_2$. The path $(v_1, v_2), \ldots, (v_{l-1}, v_l)$ constitutes $p_3 e_3$. Finally F is v_{i+1}, C is v_l and $e_4 p_1 e_1$ is the path from v_i to v_1 on the cycle. $i < l$ guarantees, that this path consists of at least two edges (e_4 and e_1).

For $i = l$ we get a subgraph of family $\mathcal{B}$. v_1 serves as B, v_l is A. Path p_2 is empty, (v_l, v_1) is e_2 and $(v_1, w)p$ constitutes $p_3 e_3$. The path $e_4 p_1 e_1$ is the path along the cycle from v_1 to v_l. Observe that it consists of at least two edges. □

Theorem 1. *A directed graph with multiple edges but without loops is FIFO stable if and only if all of its strongly connected components with more than two vertices are simple directed cycles and in all of its strongly connected components with exactly two vertices X and Y there is at most one edge from X to Y or at most one edge from Y to X.*

Proof. Lemma 1 establishes, that a network is FIFO stable if and only if all of its strongly connected components are FIFO stable. A strongly connected component of one vertex does not contain an edge and is hence stable. Lemma 3 characterizes stability for digraphs with two nodes and Lemma 4 delivers the characterization for components with more than two nodes. □

For strongly connected graphs with at least three vertices, FIFO stability and universal stability coincide. Hence even though FIFO is not a worst case queueing strategy – it is close to being one. The algorithm to decide FIFO stability for a given network is now straightforward.

Theorem 2. *Algorithm 1 decides the FIFO stability of a directed multigraph in polynomial time.*

Algorithm 1. Deciding FIFO stability

1. INPUT: A directed multigraph G without loops
2. Find all graphs $G_1 = (V_1, E_1), \ldots, G_k = (V_k, E_k)$ induced by the strongly connected components of G. Let $outdeg_i(v)$ be the out-degree of vertex v within G_i.
3. FOR i from 1 to k DO
   (a) IF $|V_i| = 2$ THEN
      i. Let V_i={u,v}
      ii. IF $outdeg_i(u) > 1$ and $outdeg_i(v) > 1$ THEN RETURN(FALSE)
   (b) IF $|V_i| \geq 3$ DO
      i. FOR all $v \in V_i$ DO
         A. IF $outdeg_i(v) > 1$ RETURN(FALSE)
4. RETURN(TRUE)

Proof. Finding the strongly connected components can be done with two depth first searches, one with reversed edges. Counting nodes, edges and determining the out-degree of a node is standard. Correctness is established by Theorem 1. Observe, that the simple cycle of l vertices is the only strongly connected directed multigraph with l vertices in which every vertex has out-degree 1.   □

5   Deciding Simple-Path FIFO Stability

Crucial for deciding simple-path stability is the maximal size of a simple cycle in G – the *circumference* of the graph. We call u a neighbor of v, if there exists an edge from u to v or from v to u. For a directed strongly connected multigraph G we define $core(G)$ as the graph obtained from G by removing vertices with just one neighbor (and their incident edges) until no such vertex remains. We start with a simple observation.

Lemma 5. *A directed strongly connected graph is simple-path FIFO stable, if and only if $core(G)$ is simple-path FIFO stable.*

Proof. If G is simple-path FIFO stable, then $core(G)$ as a subgraph is simple-path FIFO stable.

Assume X is a vertex of G with only one neighbor Y. No simple path passes through X: all simple paths touching X either start in X or end in X. Hence X can be replaced by two nodes X' with only the incoming edges and X'' with only outgoing edges and the set of simple paths in the graph does not change. Lemma 1 now yields the claim, as X' and X'' each constitute a one vertex component that is trivially simple-path stable. Hence simple-path stability of the core graph implies simple-path stability of the original graph.   □

We may hence concentrate on deciding the simple-path FIFO stability of cores of strongly connected digraphs.

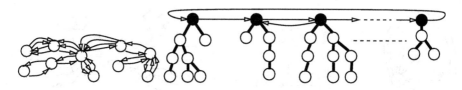

Fig. 3. A oriented multi-tree and a decorated cycle as discussed in Lemma 7. The thick edges are to represent a collection of at least one upward and one downward edge. The black vertices and the edges between them constitute the core of the graph.

5.1   Circumference 2

If a strongly connected graph has a maximal cycle length of 2, it holds that for every $(u, v) \in E$, (v, u) is also in E, since u must be reachable from v and a path from v to u with more than one edge completes a longer cycle. Hence strongly connected graphs with maximal cycle length two are trees with at least one edge from each father to his son and back. Following [1] we call such digraphs *oriented multitrees* (Fig.3). Their core is the empty graph and using Lemma 5 we get:

Lemma 6. *A strongly connected directed graph with circumference 2 is simple-path FIFO stable.*

5.2   Circumference ≥ 4

For strongly connected digraphs with circumference ≥ 4 it turns out, that here as well universal stability and FIFO stability coincide. We are lead to a set of graphs called *decorated directed cycles* in [1]. Figure 3 shows an example.

Lemma 7. *A strongly connected digraph G with circumference $k \geq 4$ is simple-path FIFO stable if and only if its core is the ring $v_0, \ldots, v_{k-1}$ with exactly one edge $(v_i, v_{i+1})^2$ for $0 \leq i \leq k - 1$. Furthermore there may be multiple backwards edges from v_i to v_{i-1} as long as the backwards edges do not complete a backward cycle themselves. If the backwards edges also constitute a cycle, there must be exactly one edge between all vertices v_i and v_{i-1}.*

Proof. Graphs of the described shape are universally simple-path stable by [1], hence we only need to show, that every FIFO stable strongly connected digraph G with circumference at least four is of the above shape. G is simple-path FIFO stable if and only if $core(G)$ is. The cycle of length at least four, that is in G by case assumption, is also in $core(G)$.

Assume there was a simple path (possibly a single edge) from v_i to v_j with $j \notin \{i - 2, i - 1, i\}$ that is edge disjoint to the edges of the cycle. Then we have a subgraph of family $\mathcal{A}^+$ with $v_i = A$, $v_j = B$, p_3, e_3 is the extra path and $p_2, e_2, e_4, e_5, p_1, e_1$ is the cycle.

---

^{2} We always assume index-arithmetics are performed modulo the cycle length.

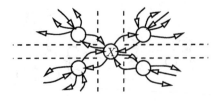

Fig. 4. An example of a graph from family $\mathcal{E}$, the graph S_2 considered in [1] and an illustration for the proof of Lemma 10

If there is a path (possibly a single edge) from v_i to v_{i-2} we have a subgraph of family $\mathcal{B}^+$. (p_2 is empty, p_3, e_e is this extra path and e_2, e_5, e_4, p_1, e_1 constitute the cycle.)

We also have a subgraph from family $\mathcal{B}^+$ if there is a path from v_i to v_{i-1} that consists of at least two edges. In this case p_3 is empty, e_3, p_2, e_2, e_5 is the cycle and e_4, p_1, e_1 the extra path.

So there are no connections between the vertices of the cycle except the edges along the cycle and possibly backward edges. Hence removing the edges of the cycle and the backward edges, breaks up the graph into k independent components. Assume there was a cycle of size at least three in any of these components, then this cycle and the cycle of the v_i-vertices share at most one vertex and no edges. We thus have a subgraph from family $\mathcal{C}^+$ yielding instability. Hence all these components have a maximal cycle length of two and are therefore oriented multitrees that vanish when constructing the core of G.   □

5.3   Circumference 3

For digraphs with circumference 3 there are simple-path FIFO stable networks, that are not universally simple-path stable. We start by introducing family $\mathcal{E}$.

Definition 2. *A directed multigraph* $G = (V, E)$ *with* $V = \{A, B, C_1, C_2, \ldots, C_k\}$ *is of family $\mathcal{E}$, if and only if*

- *there is exactly one edge from A to B,*
- *there is at least one edge from B to each C_i,*
- *there is at least one edge from each C_i to A,*
- *there are no edges between C_i and C_j for any i,j,*
- *for each i there are no edges from A to C_i or no edges from C_i to B. I.e., there is no path $A \to C_i \to B$.*

Figure 4 gives an example of a graph from $\mathcal{E}$.

Lemma 8. *Directed multigraphs from family $\mathcal{E}$ are simple-path FIFO stable.*

Lemma 9. *Directed multigraphs with just three vertices are simple-path FIFO stable.*

The proofs follow principal ideas similar to the proof of Lemma 3 and are presented in [7].

Corollary 2. *There are networks, that are simple-path FIFO stable but not universally simple-path stable.*

Proof. In [1] a graph S_2, that is not universally stable, is presented (see Fig. 4). This graph is in our set $\mathcal{E}$. □

Lemma 10. *A directed multigraph G with circumference three is simple-path FIFO stable if and only if $core(G)$ is in $\mathcal{E}$ or consists of only three vertices.*

Proof. If $core(G) \in \mathcal{E}$, Lemma 5 and 8 show stability. If the core consists of just three vertices Lemma 5 and 9 yield stability.

For the *only if* part, consider an arbitrary stable graph G with circumference 3. We need to show that its core is of the shape described in the lemma. We call a cycle of length three a triangle. We call two triangles *different*, if they do not share exactly the same vertices. If $core(G)$ does not contain different triangles, it consists of a ring with three vertices (possibly with backward edges) and $core(G)$ is of the shape described in the lemma.

$core(G)$ cannot contain two edge disjoint different triangles: they would yield a graph from $\mathcal{C}^+$ contradicting $core(G)$'s stability (If a vertex is shared by the triangles, we have $A = A'$ in $\mathcal{C}^+$ and if two vertices are shared we have $A = A'$ as well as $E = I$ in $\mathcal{C}^+$. Observe that the paths used in the instability proof remain simple under these assumptions.)

Hence all pairs of different triangles in $core(G)$ share an edge. Consequently one edge is shared by all different triangles. We call this edge (A, B) and the other vertices of the triangles in $core(G)$ C_i ($i \geq 2$ holds). There is exactly one edge from A to B (otherwise there would be edge disjoint different triangles). Furthermore there is at least one edge from B to each C_i and one from each C_i to A. There are no edges from C_i to C_j, since if there are, A, B, C_i, C_j, A is a cycle of length 4. Furthermore there is no i with edges from A to C_i and C_i to B (A, C_i, B, C_j, A with $i \neq j$ constitutes a cycle of length four otherwise.) So the set $S := \{A, B, C_1, \ldots, C_k\}$ and its internal edges are of the shape as described in the lemma.

Finally assume there was another vertex X in $core(G)$, that is not part of a triangle. Then removing X and its incident edges must split $core(G)$ into separate graphs. Only one of them can contain triangles, as there are no edge disjoint triangles in $core(G)$. The other graphs are hence oriented multitrees that cannot occur in $core(G)$ – a contradiction (see Fig. 4, right). Hence G is of the shape described in the lemma. □

Theorem 3 summarizes the previous lemmas and justifies the following algorithm whose tests can obviously be made in polynomial time.

Theorem 3. *A strongly connected directed multigraph is simple-path FIFO stable, if and only if its core has at most three vertices, is in family $\mathcal{E}$ or it is a simple directed cycle with backward edges, that either constitute another simple directed cycle or do not form a cycle at all.*

Algorithm 2. Deciding simple-path FIFO stability

1. INPUT: A directed multigraph G without loops
2. Find all graphs $G_1 = (V_1, E_1), \ldots, G_k = (V_k, E_k)$ induced by the strongly connected components of G.
3. FOR i from 1 to k DO
   (a) Compute $G' = (V', E')$ the core of G_i
   (b) IF $(|V'| > 3)$ AND $G' \notin \mathcal{E}$ THEN
       i. Test if G' is the simple cycle. Backward edges may exist arbitrarily if they do not form a cycle themselves. If the backward edges form a cycle, this cycle must be simple. If this test fails, RETURN(FALSE)
4. RETURN(TRUE)

---

6   Conclusion and Open Problems

We have classified directed multigraphs as to their FIFO stability and simple-path FIFO stability and provided efficient algorithms for these problems. It turned out, that FIFO is not a worst case policy, but rather close to being one. The problem remains open for undirected graphs and the different packet trajectories considered on them.

References

1. Àlvarez, C., Blesa, M., Serna, M., A Characterization of Universal Stability in the Adversarial Queuing Model, *SIAM J. Comput, Vol. 34, No 1, 2004*, pp. 41-66
2. Andrews, M., Awerbuch, B., Fernández, A., Leighton, T., and Liu, Z., Universal-Stability Results and Performance Bounds for Greedy Contention-Resolution Protocols, *Journal of the ACM, Vol. 48, No 1, January 2001*, pp. 39-69
3. Bhattacharjee, R. and Goel, A., Instability of FIFO at arbitrarily low rates in the adversarial queueing model, *Proc. of the 44th Symposium on Foundations of Computer Science, 2003*, pp. 160-167
4. Blesa, Maria J., Deciding Stability in Packet-Switched FIFO Networks Under the Adversarial Queuing Model in Polynomial Time, *Proc. of the 19th International Symposium on Distributed Computing, 2005*, LNCS Vol. 3724, pp 429-441
5. Borodin, A., Kleinberg, J., Raghavan, P., Sudan, M., and Williamson, D. P. Adversarial queueing theory, *J. of the ACM, Vol. 48, No 1, January 2001*, pp. 13-38
6. Koukopoulos, D., Mavronicolas, M., Spirakis, P., FIFO is Unstable at Arbitrarily Low Rates (Even in Planar Networks), *Electronic Colloq. on Computational Complexity, 2003*
7. Weinard, Maik, Deciding the FIFO Stability of Networks in Polynomial Time (full version), *Technical report: Frankfurter Informatik-Berichte, No 3/2005*, ISSN 1616-9107

Heterogenous Networks Can Be Unstable at Arbitrarily Low Injection Rates*

Dimitrios Koukopoulos[1] and Stavros D. Nikolopoulos[2]

[1] Department of Cultural Heritage Management & New Technologies,
University of Ioannina, GR-30100 Agrinio, Greece
koukopou@ceid.upatras.gr
[2] Department of Computer Science, University of Ioannina,
GR-45110 Ioannina, Greece
stavros@cs.uoi.gr

Abstract. A distinguishing feature of today's large-scale platforms for distributed computation and communication, such as the *Internet,* is their *heterogeneity,* predominantly manifested by the fact that a wide variety of *communication protocols* are simultaneously running over different distributed hosts. A fundamental question that naturally poses itself for such common settings of heterogeneous distributed systems concerns their ability to preserve or restore an acceptable level of performance during link failures. In this work, we address this question for the specific case of stability properties of greedy, contention-resolution protocols operating over a *packet-switched* communication network that suffers from *link slowdowns*. We focus on the *Adversarial Queueing Theory* framework, where an adversary controls the rates of packet injections and determines packet paths. In addition, the power of the adversary is enhanced to include the manipulation of *link slowdowns*. Within this framework, we show that the composition of LIS (*Longest-in-System*) with any of SIS (*Shortest-in-System*), NTS (*Nearest-to-Source*) and FTG (*Furthest-to-Go*) protocols is unstable at rates $\rho > 0$ when the network size and the link slowdown take large values. These results represent the current record for instability bounds on injection rate for compositions of greedy protocols over dynamic adversarial models, and also suggest that the potential for instability incurred by the composition of *two* greedy protocols may be *worse* than that of some *single* protocol.

1   Introduction

Motivation-Framework. Some of the most important features of contemporary large-scale platforms for distributed communication and computation, such as the *Internet,* is their *robustness* and *heterogeneity.* Robustness is the ability of

---

* This research was co-funded by the European Union in the framework of the program "Pythagoras II" of the "Operational Program for Education and Initial Vocational Training" of the 3rd Community Support Framework of the Hellenic Ministry of Education, funded by national sources and the European Social Fund (ESF).

T. Calamoneri, I. Finocchi, G.F. Italiano (Eds.): CIAC 2006, LNCS 3998, pp. 93–104, 2006.
© Springer-Verlag Berlin Heidelberg 2006

communication despite network link failures, while heterogeneity comes around in many different flavors. For example, the specifics of how the computers in different parts of the network are connected with each other, and the properties of the links that foster the interconnection, is difficult to characterize uniformly. Moreover, although, conceptually, the Internet uses a unified set of protocols, in practice each protocol has been implemented with widely varying features (and of course bugs) [9, 11]. As the Internet evolves into a ubiquitous communication infrastructure that supports multiple protocols running on different network hosts, its dependability in the presence of various failures becomes critical. These failures can degrade system performance and lead to service disruption. Thus, the study of performance and correctness properties of heterogeneous distributed systems which suffer from link failures becomes a necessity. This study could help on detecting, understanding and overcoming the conditions leading to these mentioned negative effects, as well as helping to their prevention.

Objectives. We are interested in the behavior of *packet-switched networks* in which packets arrive dynamically at the *nodes* and they are routed in discrete time steps across the *links*. Recent years have witnessed a vast amount of work on analyzing packet-switched networks under *non-probabilistic* assumptions (rather than stochastic ones); we work within a model of *worst-case* continuous packet arrivals, originally proposed by Borodin *et al.* [7] and termed *Adversarial Queueing Theory* to reflect the assumption of an *adversarial* way of packet generation and path determination. A major issue that arises in such a setting is that of *stability*– will the number of packets in the network remain bounded at all times? The answer to this question may depend on the *rate* of injecting packets into the network, the *slowdown* of the links, which is the time delay which is suffered by outgoing packets in order to be forwarded on a link, and the composition of *protocols* running on different network hosts in order to resolve packet conflicts. The underlying goal of our study is to establish the stability properties of heterogeneous networks when packets are injected by an adversary and the link slowdowns are chosen by the same adversary in a dynamic way.

Model of Quasi-Static Slowdowns. Most studies of packet-switched networks assume that one packet can cross a network link (an edge) in a single time step. This assumption is well motivated when we assume that all network links are identical. However, a packet-switched network can contain different types of links, which is common especially in large-scale networks like Internet. Also, a real network can suffer from link failures due to natural disasters (like hurricanes), human action (like hacker attacks) or by unintentional software failures. Then, it is well motivated to assign a slowdown to each link. Furthermore, if each link slowdown takes on values in the two-valued set of integers $\{1, D\}$ for $D > 1$, D takes on large values and each value remains fixed for a long time, then we can consider approximately as a link failure the assigning of slowdown D to a link, while the assigning of unit slowdown to a link can be considered as the proper service rate. Therefore, the study of the stability behavior of networks and protocols under our model of quasi-static slowdowns can be considered as an approximation of the fault-tolerance of a network where links can temporarily fail (infinite slowdown)[1]. The goal of this study is to provide an insight towards detecting, understanding, and overcoming the conditions leading to performance degradation and service disruption of today's communication networks during network attacks or failures.

In this work, we embark on a study of the impact of heterogeneity of distributed systems on their performance properties if the adversary can determine the paths of packet injections along with the slowdowns of network edges in each time step. More specifically, we wish to pose the general question of which performance properties of heterogeneous *packet-switched* networks (where compositions of protocols are running on different network hosts) are maintained and which are not in the presence of link failures. This subfield of study was initiated by Borodin *et al.* in [8] in the case of networks where a single protocol is responsible for the resolution of packet conflicts. Note that we continue to assume uniform packet sizes.

Stability. Roughly speaking, a protocol P is *stable* [7] on a network $\mathcal{G}$ against an adversary $\mathcal{A}$ of rate ρ if there is a constant B (which may depend on $\mathcal{G}$ and $\mathcal{A}$) such that the number of packets in the system is bounded at all times by B. On the other hand, *a protocol* P *is universally stable* [7] if it is stable against every adversary of rate less than 1 and on every network. Here, we consider four *greedy*, universally stable, contention-resolution protocols under the Adversarial Queueing Theory (Table 1).

Table 1. Greedy protocols considered in this paper (**US** stands for universally stable)

Protocol name	Which packet it advances:	US
SIS (*Shortest-In-System*)	The most recently injected packet	✓ [4]
LIS (*Longest-In-System*)	The least recently injected packet	✓ [4]
FTG (*Furthest-To-Go*)	The furthest packet from its destination	✓ [4]
NTS (*Nearest-To-Source*)	The nearest packet to its origin	✓ [4]

Contribution. We define here the *weakest* possible adversary of dynamically changing network *link slowdowns* in the context of Adversarial Queueing Theory (AQM) where the adversary may set link slowdowns to any of two integer values 1 and D (D > 1 is a parameter called high slowdown).[2] Moreover, once a link slowdown takes on a value, the value stays fixed for a continuous time period proportional to the number of packets in the system at the time of setting the slowdown to the value. We call this the *Adversarial, Quasi-Static Slowdown Queueing Theory* model (**AQSSQM**). In this framework, we establish that the

---

[1] However, infinite link slowdown is only an approximation of link failure, because in a slowdown the packet has left the queue and is being transmitted; however, when a failure occurs, the packet is not being transmitted but stored somewhere, and thus it participates later in the queue scheduling.

[2] In AQM only one slowdown value is available to the adversary.

composition of LIS with any of SIS, NTS and FTG protocols is unstable for arbitrarily low injection rates. We prove that increasing the network size along with dynamic changing of link slowdowns can drop to arbitrarily low values the lower bound on injection rate that guarantees instability for heterogeneous networks. To show this, we provide interesting combinatorial constructions of a *size-parameterized* network where we specify the contention-resolution protocol to be used to each queue. For purpose of completeness and comparison, we summarize, in Table 2, all results that are shown in this work and, also, in [16] (for AQM) and [18] (for the Adversarial Quasi-Static Queueing Model - AQSQM), concerning instability bounds on the injection rate for the composition pairs LIS-SIS, LIS-NTS and LIS-FTG.

Table 2. Instability bounds of the compositions of LIS with any of the SIS, NTS, and FTG protocols in AQM vs. AQSQM vs. AQSSQM

	Instability (AQM)	Instability (AQSQM)	Instability (AQSSQM)
LIS − SIS	$\rho > 0.5$ [16, Thm. 3.1]	$\rho > \sqrt{2} - 1$ [18, Thm. 2]	$\rho > 0$ [Thm. 1]
LIS − NTS	$\rho > 0.5$ [16, Thm. 3.1]	$\rho > \sqrt{2} - 1$ [18, Thm. 2]	$\rho > 0$ [Thm. 2]
LIS − FTG	$\rho > 0.5$ [16, Thm. 3.1]	$\rho > \sqrt{2} - 1$ [18, Thm. 2]	$\rho > 0$ [Thm. 3]

The combinatorial constructions of networks and adversaries that we have employed for showing that certain compositions of universally stable protocols can be unstable for arbitrarily low injection rates when link slowdowns can change dynamically, significantly extend ones that appeared before in [7, 15, 16, 18]. In more detail, some of the tools we devise in order to obtain constructions of networks and adversaries that imply improved bounds are the following:

- We employ combinatorial constructions of networks with multiple successively pairs of parallel queues; we judiciously use such paths for the simultaneous injection of various non-overlapping sets of packets. Also, this construction allows the adversary to inject a set of packets at a time period over a path with unit slowdown edges, while the previously injected sets of packets are delayed in another queue due to its high slowdown D.
- We use the technical notions of *investing flow* and *short flow*; these are some special cases of packet flows used in our adversarial constructions consisting of inductive *phases*. Roughly speaking, an investing flow injects packets in a phase some of which will remain in the system till the beginning of the next phase, in order to guarantee the inductive hypothesis for the next phase; on the other hand, short flows consist of packets injected on judiciously chosen links of the network and their role is to delay the investing flows.

Related Work. The issue of composing distributed protocols (resp., objects) to obtain other protocols (resp., objects), and the properties of the resulting (*composed*) protocols (resp., objects), has a rich record in Distributed Computing Theory (see, e.g., [20]). For example, Herlihy and Wing [13] establish that a composition of *linearizable* memory objects (possibly distinct), each managed by its own protocols, preserves linearizability. Robustness has been extensively studied in the context of fault-tolerant distributed systems. A landmark paper on failures in Tandem systems and the techniques to prevent them is [12]. In parallel and even earlier, a mathematical framework was developed in the Operations Research world to manage the robustness and risk in systems composed of various components [5].

Adversarial Queueing Theory [7] received a lot of interest in the study of stability and instability issues (see, e.g., [2, 4, 10, 15, 17, 21]). The universal stability of various natural greedy protocols (SIS, LIS, NTS and FTG) was established by Andrews *et al.* [4]. Also, several greedy protocols such as NTG (Nearest-To-Go) have been proved unstable at arbitrarily low rates of injection in [21]. The subfield of study of the stability properties of compositions of universally stable protocols was introduced by Koukopoulos *et al.* in [15, 16, 17] where lower bounds of 0.683, 0.519 and 0.5 on the injection rates that guarantee instability for the composition pairs LIS-SIS, LIS-NTS and LIS-FTG were presented.

Borodin *et al.* in [8] studied for the first time the impact on stability when the edges of a network can have capacities or slowdowns. They proved that many well-known universally stable protocols (SIS, NTS, FTG) do maintain their universal stability when the link capacity or slowdown is changing dynamically, whereas the universal stability of LIS is not preserved. This work was further extended by Koukopoulos *et al.* in [18, Theorems 2, 3] proving lower bounds of $\sqrt{2}-1$ on the injection rates that guarantee instability for the LIS protocol and its compositions with the SIS, NTS and FTG protocols under dynamically changing link capacities. Also, Koukopoulos in [14] studied the impact of link slowdowns on network stability when a single protocol is used or a forbidden subgraph for universal stability is induced. Moreover, in [1, 3] there have been generalizations of the adversarial queueing theory to networks with dynamic failures. Finally, in [6] it is proposed a generalization of the adversarial queueing theory where the network traffic flow is continuous in time and arbitrary packet lengths, link speeds and link propagation delays are allowed.

2   Preliminaries

The model definitions are patterned after those in [7, Section 3], adjusted to reflect the fact that the edge slowdowns may vary arbitrarily as in [8, Section 2], but we address the weakest possible model of changing slowdowns. We consider that a routing network is modelled by a directed graph $\mathcal{G} = (V, E)$. Each node $u \in V$ represents a communication switch, and each edge $e \in E$ represents a link between two switches. In each node, there is a buffer (queue) associated with each outgoing link. Time proceeds in discrete time steps. Buffers store packets that are injected into the network with a route, which is a simple directed path in $\mathcal{G}$. A *packet* is an atomic entity that resides at a buffer at the end of any step. It must travel along paths in the network from its *source* to its *destination,* both of which are nodes in the network. When a packet is injected, it is placed in the buffer of the first link on its route. When a packet reaches its destination, we say that it is *absorbed*. During each step, a packet may be sent from its current node along one of the outgoing edges from that node. Edges can have different integer slowdowns, which may or may not vary over time. Denote $D_e(t)$ the *slowdown* of the edge e at time step t. That is, we assume that if a packet p is scheduled to traverse the edge e at time t, then packet p completes the traversal of e at time $t + D_e(t)$ and during this time interval, no other packet can be scheduled on e.

Let $D > 1$ be an integer parameter. We demand that $\forall e$ and $\forall t$ $D_e(t) \in \{1, D\}$. We also demand for each edge e that $D_e(t)$ stays at some value for a continuous period of time at least equal to $f(\rho, D)s$ time steps, where s is the number of packets in the system at the time of setting the link slowdown to the value and $f(\rho, D)$ is a function of the injection rate ρ of the adversary in the network and the high link slowdown D. We call this the *Adversarial, Quasi-Static Slowdown Queueing Theory Model*. Our model is different from the failure model in [1, 3] because in our model a packet p is delayed after leaving the queue of the edge e, while in the failure model p waits in the queue of e.

Any packets that wish to travel along an edge e at a particular time step, but are not sent, wait in a queue for e. At each step, an *adversary* generates a set of requests. A *request* is a *path* specifying the route that will be followed by a packet.[3] We say that the adversary generates a set of packets when it generates a set of requested paths. Also, we say that a packet p *requires* an edge e at time t if e lies on the path from its position to its destination at time t.

Fix any arbitrary positive integer $w \geq 1$. For any edge e of the network and any sequence of w consecutive time steps, define $N(w, e)$ to be the number of paths that are injected by the adversary during the time interval of w consecutive time steps requiring to traverse the edge e. For any constant ρ, $0 < \rho \leq 1$, a (w, ρ)-*adversary* is an adversary that injects packets subject to the following *load condition*: For every edge e and for every sequence τ of w consecutive time steps, $N(\tau, e) \leq \rho \sum_{t \in \tau} \frac{1}{D_e(t)}$. We say that a (w, ρ)-adversary injects packets at rate ρ with *window size* w. The assumption that $\rho \leq 1$ ensures that it is not necessary a priori that some edge of the network is overloaded.

In order to formalize the behavior of a network, we use the notions of *system* and *system configuration*. A triple of the form $\langle \mathcal{G}, \mathcal{A}, \mathsf{P} \rangle$ where $\mathcal{G}$ is a network, $\mathcal{A}$ is an adversary and P is the used protocol (or list of protocols) on the network queues is called a system. In every time step t, the current configuration C^t of a system $\langle \mathcal{G}, \mathcal{A}, \mathsf{P} \rangle$ is a collection of sets $\{S_e^t : e \in \mathcal{G}\}$, such that S_e^t is the set of packets waiting in the queue of the edge e at the end of step t.

In the adversarial constructions we study here for proving instability, we split time into *phases*. In each phase, we study the evolution of the *system configuration* by considering corresponding *time rounds*. For each phase, we inductively prove that the number of packets of a specific subset of queues in the system increases in order to guarantee instability. This inductive argument can be applied repeatedly, thus showing instability. Furthermore, we assume that there is a sufficiently large number of packets s_0 in the initial system configuration. This

---

[3] In this work, it is assumed, as it is common in packet routing, that all paths are simple paths where edges cannot be overlapped, while vertices can be overlapped.

will imply instability results for networks with an *empty* initial configuration, as it was established in [4, Lemma 2.9]. For simplicity, and in a way similar to that in [4], we omit floors and ceilings from our analysis, and we, sometimes, count time steps and packets only roughly. This may only result to loosing small additive constants, while it implies a gain in clarity.

3   Unstable Compositions of Protocols

In this section, we prove that the composition of the LIS protocol with any of SIS, NTS and FTG protocols can become unstable for arbitrarily low injection rates. Before proceeding to the adversarial constructions for proving instability we give two basic definitions.

Definition 1. *We denote by X_i the set of packets that are injected into the system in the ith round of a phase. These packet sets are characterized as* investing flows *because only packets from these sets will remain in the system at the beginning of the next phase contributing in packet accumulation.*

Definition 2. *We denote by S_i the set of packets the adversary injects into the system in the ith round of a phase. These packet sets are characterized as* short flows *because they are injected on judiciously chosen links of the network for delaying investing flows.*

3.1   A Parameterized Network Family $\mathcal{G}_l$

We provide here a parameterized family of networks $\mathcal{G}_l$ (see Figure 1). The motivation that led us to such a parameterization in the network topology is *two-fold*: (a) The existence of many pairs of parallel queues in the network allows the adversary to inject an investing flow at a time round over a path with unit slowdown edges, while the previously injected investing flows are delayed in another queue due to its high slowdown D. Also, this structure permits the simultaneous injection of an investing flow on one queue of a pair, and a short flow on the other, without violating the rule of the restricted adversarial model. (b) Such a parameterized network topology construction, enables a parameterized analysis of the system configuration evolution into distinguished rounds whose number depends on the parameterized network topology. In LIS-FTG composition, the parameterization, besides the parallel edges, includes additional chains of queues for the exploitation of FTG in blocking investing flows.

3.2   Parameterized Adversarial Constructions

The main ideas of the adversarial constructions we present are: (a) the accurate tuning of the duration of each round of every phase j (as a function of the high slowdown D, the injection rate ρ and the number of packets in the system at the beginning of phase j, s_j) to maximize the growth of the packet population in the system, (b) the careful setting of the slowdowns of some edges to D for specified time intervals in order to accumulate packets, and (c) the careful injections of packets that guarantee that the load condition is satisfied.

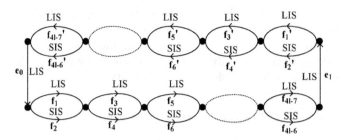

Fig. 1. The network $\mathcal{G}_l$

Theorem 1. *Let $\rho' = 0.0056$. For the network $\mathcal{G}_l$ where $l > 1000$ is a parameter linear to the number of network queues there is an adversary $\mathcal{A}_1$ of rate ρ that can change the link slowdowns of $\mathcal{G}_l$ between the two integer values 1 and $D > 1000$ such that the system $\langle \mathcal{G}_l, \mathcal{A}_1, \mathsf{LIS}, \mathsf{SIS} \rangle$ is unstable for every $\rho > \rho'$. When $\{D, l\} \to \infty$ the system $\langle \mathcal{G}_l, \mathcal{A}_1, \mathsf{LIS}, \mathsf{SIS} \rangle$ is unstable for $\rho > 0$.*

Proof. Consider an instance of the parameterized network family (network $\mathcal{G}_l$, see Figure 1). The edges $e_0, e_1, f_1, f_3, f_5, \ldots, f_{4l-7}$ $f_1', f_3', f_5', \ldots, f_{4l-7}'$ of $\mathcal{G}_l$ use the LIS protocol, while the remaining edges of $\mathcal{G}_l$ use the SIS protocol. The construction of the adversary $\mathcal{A}_1$ is broken into phases.

Inductive Hypothesis. At the beginning of phase j (suppose j is even), there are s_j packets that are queued in the queues f_{4l-9}', f_{4l-6}' (in total) requiring to traverse the edges e_0, f_1.

Induction Step. At the beginning of phase $j+1$, there will be $s_{j+1} > s_j$ packets that will be queued in the queues f_{4l-9}, f_{4l-6} (in total) requiring to traverse the edges e_1, f_1'.

We will construct an adversary $\mathcal{A}_1$ such that the induction step will hold. Proving that the induction step holds, we ensure that the inductive hypothesis will hold at the beginning of phase $j+1$ for the symmetric edges with an increased value of s_j, $s_{j+1} > s_j$. By the symmetry of the network, repeating the phase construction an unbounded number of times, we will create an unbounded number of packets in the network.

From the inductive hypothesis, initially, there are s_j packets (that constitute the set of packets S) in the queues f_{4l-9}', f_{4l-6}' requiring to traverse the edges e_0, f_1. In order to prove the induction step, it is assumed that the set S has a large enough number of $|S| = s_j$ packets in the initial system configuration. During phase j, the adversary plays l rounds of injections as follows:

Round 1: It lasts $|T_1| = s_j$ time steps. During this round the edge f_1 has high slowdown D, while all the other edges have unit slowdown. The adversary injects a set X_1 of $|X_1| = \rho|T_1|$ packets in the queue e_0 wanting to traverse the edges $e_0, f_2, f_3, f_6, f_7, f_{10}, \ldots, f_{4l-9}, f_{4l-6}, e_1, f_1'$.

Evolution of the system configuration. The packets of the set S delay the packets of the set X_1 in the queue e_0 that uses the LIS protocol because they are longer time in the system than the packets of the set X_1. At the same time, the packets of the set S are delayed in f_1 due to the high slowdown of the edge f_1. At the end of this round, the remaining packets of the set S in f_1 are $|S'| = |S| - |T_1|/D$.

Round 2: It lasts $|T_2| = |S'|$ time steps. During this round the edge f_2 has high slowdown D, while all the other edges have unit slowdown. The adversary injects a set X_2 of $|X_2| = \rho|T_2|$ packets in the queue f_1 requiring to traverse the edges $f_1, f_3, f_6, f_7, f_{10}, \ldots, f_{4l-9}, f_{4l-6}, e_1, f_1'$ and a set S_2 of $|S_2| = \rho|T_2|/D$ packets in the queue f_2 requiring to traverse the edge f_2.

Evolution of the system configuration. The packets of the set X_2 are delayed by the packets of the set S' in the queue f_1 that uses the LIS protocol because the packets of the set S' are longer time in the system than the packets of the set X_2. At the same time, the packets of the set X_1 are delayed in the queue f_2 that uses the SIS protocol due to its high slowdown D and the packets of the set S_2 that are shorter time in the system than the packets of the set X_1. Therefore, the remaining packets of the set X_1 in the queue f_2 are $|X_1| + |S_2| - |T_2|/D = |X_1| + (\rho - 1)|T_2|/D$.

Round 3: It lasts $|T_3| = |X_1| + |X_2| + (\rho - 1)|T_2|/D$ time steps. During this round the edge f_6 has high slowdown D, while all the other edges have unit slowdown. The adversary injects a set X_3 of $|X_3| = \rho|T_3|$ packets in the queue f_3 requiring to traverse the edges $f_3, f_5, f_7, f_{10}, \ldots, f_{4l-9}, f_{4l-6}, e_1, f_1'$ and a set S_3 of $|S_3| = \rho|T_3|/D$ packets in the queue f_6 requiring to traverse the edge f_6.

Evolution of the system configuration. The packets of the sets X_1, X_2 delay the packets of the set X_3 in the queue f_3 that uses the LIS protocol because they are longer time in the system than the packets of the set X_3. At the same time, the packets of the sets X_1, X_2 are delayed in f_6 that uses the SIS protocol due to the high slowdown of the edge f_6 and the packets of the set S_3 that are shorter time in the system than the packets of the sets X_1, X_2. Therefore, the remaining packets of the sets X_1, X_2 in the queue f_6 are $|X_1| + |X_2| + (\rho - 1)\frac{|T_2| + |T_3|}{D}$.

Round l: It lasts $|T_l| = \sum_{i=1}^{l-1} |X_i| - (\rho - 1)\sum_{i=2}^{l-1} |T_i|/D$ time steps. During this round the edge f_{4l-6} has high slowdown D, while all the other edges have unit slowdown. The adversary injects a set X_l of $|X_l| = \rho|T_l|$ packets in the queue f_{4l-9} requiring to traverse the edges $f_{4l-9}, f_{4l-7}, e_1, f_1'$ and a set S_l of $|S_l| = \rho|T_l|/D$ packets in the queue f_{4l-6} requiring to traverse the edge f_{4l-6}.

Evolution of the system configuration. The packets of the sets $X_1, \ldots, X_{l-1}$ delay the packets of the set X_l in the queue f_{4l-9} that uses the LIS protocol because they are longer time in the system than the packets of the set X_l. At the same time, the packets of the sets $X_1, \ldots, X_{l-1}$ are delayed in f_{4l-6} that uses the SIS protocol due to the high slowdown of the edge f_{4l-6} and the packets of the set S_l that are shorter time in the system than the packets of the sets $X_1, \ldots, X_{l-1}$.

Therefore, the remaining packets of the sets $X_1, \ldots, X_{l-1}$ in the queue f_{4l-6} are $\sum_{i=1}^{l-1} |X_i| + (\rho - 1) \sum_{i=2}^{l} |T_i|/D$.

Thus, the number of packets in the queues f_{4l-9}, f_{4l-6} requiring to traverse e_1, f_1' at the end of this round is $s_{j+1} = \rho s_j + (\rho + \frac{\rho-1}{D}) \sum_{i=2}^{l} |T_i|$. Moreover, $\sum_{i=3}^{l} |T_i| = (\rho + \frac{D+\rho-1}{D}) \sum_{i=3}^{l-1} |T_i| + (2\rho - \frac{1}{D} - \frac{\rho-1}{D^2})|T_1|$. Thus, $s_{j+1} = \rho s_j + (\rho + \frac{\rho-1}{D}) \frac{D-1}{D} s_j + (\rho + \frac{\rho-1}{D})(2\rho - \frac{1}{D} - \frac{\rho-1}{D^2}) \frac{1-(\rho+\frac{D+\rho-1}{D})^{l-2}}{1-\frac{(D+1)\rho}{D}} s_j$. In order to have instability, we must have $s_{j+1} > s_j$. Therefore, for instability it suffices $\rho + (\rho + \frac{\rho-1}{D}) \frac{D-1}{D} + (\rho + \frac{\rho-1}{D})(2\rho - \frac{1}{D} - \frac{\rho-1}{D^2}) \frac{1-(\rho+\frac{D+\rho-1}{D})^{l-2}}{1-\frac{(D+1)\rho}{D}} > 1$. If we let $\rho = 0.0056$, $D = 1000$ and $l = 1000$, the inequality holds. Thus, for $\{D, l\} > 1000$ the inequality holds, too.

When $D \to \infty$, it holds that $\frac{1}{D^k} \to 0$ for all $k \geq 1$. Then, our inequality becomes $2\rho(\rho + 1)^{l-2} > 1$. Thus, $\rho > \frac{1}{2(\rho+1)^{l-2}}$. When $l \to \infty$ and $x > 0$, it holds that $(1 + x)^{l-2} \to \infty$. Therefore, for $\{D, l\} \to \infty$ the inequality $\rho > \frac{1}{2(\rho+1)^{l-2}}$ holds for $\rho > 0$. Note that if we have a sequence of equations $f_{D,l}(\rho)$ and there exists the limit $\lim_{\{D,l\} \to \infty} f_{D,l}(\rho) = f_\infty(\rho)$, then it holds fundamentally by the theory of function limits that if $\rho(D, l)$ is the root of $f_{D,l}(\rho) = 0$, then $\lim_{\{D,l\} \to \infty} \rho(D, l)$ is the root of $f_\infty(\rho)$. Therefore, for $\rho > 0$ the system is unstable. This argument can be repeated for an infinite number of phases showing that the number of packets in the system increases forever for $\rho > 0$.   □

With a similar adversarial construction to Theorem 1, we show that the composition of the LIS and NTS protocols can become unstable for arbitrarily low injection rates considering an instance of the parameterized network family (network $\mathcal{G}_l$, see Figure 1). The network $\mathcal{G}_l$ is also used for proving the instability of the composition of the LIS and SIS protocols. However in this case, the edges $f_2, f_4, f_6, \ldots, f_{4l-6} f_2', f_4', f_6', \ldots, f_{4l-6}'$ of $\mathcal{G}_l$ use the NTS protocol instead of the SIS protocol, while the remaining edges of $\mathcal{G}_l$ use the LIS protocol. Thus, the following theorem, analogous to Theorem 1, holds.

Theorem 2. *Let $\rho' = 0.0056$. For the network $\mathcal{G}_l$ where $l > 1000$ is a parameter linear to the number of network queues there is an adversary $\mathcal{A}_2$ of rate ρ that can change the link slowdowns of $\mathcal{G}_l$ between the two integer values 1 and $D > 1000$ such that the system $\langle \mathcal{G}_l, \mathcal{A}_2, \text{LIS}, \text{NTS} \rangle$ is unstable for every $\rho > \rho'$. When $\{D, l\} \to \infty$ the system $\langle \mathcal{G}_l, \mathcal{A}_2, \text{LIS}, \text{NTS} \rangle$ is unstable for $\rho > 0$.*

Similarly, we show that the composition of the LIS and FTG protocols can become unstable for arbitrarily low injection rates considering an instance $\mathcal{G}_l'$ of the parameterized network family $\mathcal{G}_l$ (see Figure 2). The topology of the network $\mathcal{G}_l'$ has a significant difference with the networks that are used for proving Theorems 1, and 2. The network $\mathcal{G}_l'$ contains additional paths, comparing to the other three cases, that start at queues that use the FTG protocol. These paths have sufficient lengths, such that the injected short flows have the same blocking effects over the injected investing flows when they conflict in queues that use FTG, as happens in LIS-SIS and LIS-NTS cases. Thus, the following theorem, analogous to Theorem 1 and Theorem 2, holds.

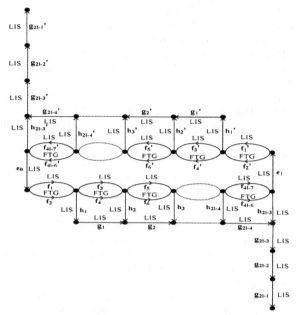

Fig. 2. The network $\mathcal{G}'_l$

Theorem 3. *Let $\rho' = 0.0056$. For the network $\mathcal{G}'_l$ where $l > 1000$ is a parameter linear to the number of network queues there is an adversary $\mathcal{A}_3$ of rate ρ that can change the link slowdowns of $\mathcal{G}'_l$ between the two integer values 1 and $D > 1000$ such that the system $\langle \mathcal{G}'_l, \mathcal{A}_3, \mathsf{LIS}, \mathsf{FTG} \rangle$ is unstable for every $\rho > \rho'$. When $\{D, l\} \to \infty$ the system $\langle \mathcal{G}'_l, \mathcal{A}_3, \mathsf{LIS}, \mathsf{FTG} \rangle$ is unstable for $\rho > 0$.*

4   Conclusions

In this work, we studied how the dynamic changing of link slowdowns affects the instability properties of compositions of contention-resolution protocols that include LIS. However, we do not have any clue what happens with compositions of protocols that do not include LIS. Also, our results suggest that, for every unstable network, its instability bound in the model of quasi-static slowdowns may be lower than for the classical adversarial queueing model or other dynamic adversarial model. Proving (or disproving) this remains an open problem.

References

1. C. Alvarez, M. Blesa, J. Diaz, A. Fernandez, M. Serna, Adversarial Models for Priority-Based Networks, *Proc. of the 28th Int'l Symposium on Mathematical Foundations of Computer Science*, 2003, LNCS. 2747, pp. 142–151.
2. C. Alvarez, M. Blesa, M. Serna, A Characterization of Universal Stability in the Adversarial Queuing model, SIAM Journal on Computing, 34 (2004) 41–66.

3. C. Alvarez, M. Blesa, M. Serna, The Impact of Failure Management on the Stability of Communication Networks, *Proc. of the 10th Int'l Conference on Parallel and Distributed Systems*, 2004, pp. 153–160.
4. M. Andrews, B. Awerbuch, A. Fernández, J. Kleinberg, T. Leighton, Z. Liu, Universal Stability Results for Greedy Contention-Resolution Protocols, Journal of the ACM, 48 (2001) 39–69.
5. R. Barlow and F. Proschan, Statistical Analysis of Reliability and LifeTesting Models, New York: Holt, Rinehart and Winston, 1975.
6. M. Blesa, D. Calzada, A. Fernández, L. López, A. Martínez, A. Santos, M. Serna, Adversarial Queueing Model for Continuous Network Dynamics, *Proc. of the 30th Int'l Symposium on Mathematical Foundations of Computer Science*, 2005, LNCS. 3618, pp. 144–155.
7. A. Borodin, J. Kleinberg, P. Raghavan, M. Sudan, D. Williamson, Adversarial Queueing Theory, Journal of the ACM, 48 (2001) 13–38.
8. A. Borodin, R. Ostrovsky, Y. Rabani, Stability Preserving Transformations: Packet Routing Networks with Edge Capacities and Speeds, *Proc. of the 12th Annual ACM-SIAM Symposium on Discrete Algorithms*, 2001, pp. 601–610.
9. D. Clark, The Design Philosophy of the DARPA Internet Protocols, ACM Computer Communication Reviews, 18 (1988) 106-114.
10. J. Diaz, D. Koukopoulos, S. Nikoletseas, M. Serna, P. Spirakis, D. Thilikos, Stability and Non-Stability of the FIFO Protocol, *Proc. of the 13th Annual ACM Symposium on Parallel Algorithms and Architectures*, 2001, pp. 48–52.
11. S. Floyd and V. Paxson, Difficulties in Simulating the Internet, IEEE/ACM Transactions on Networking, 9 (2001) 392–403.
12. J. Gary, Why do computers stop and what can be done about it?, *Symposium on Reliability in Distributed Software and Database Systems*, 1986.
13. M. P. Herlihy and J. Wing, Linearizability: A Correctness Condition for Concurrent Objects, *Proc. of the ACM Transactions on Programming Languages and Systems*, 1990, Vol. 12, No. 3, pp. 463–492.
14. D. Koukopoulos, The Impact of Dynamic Link Slowdowns on Network Stability, *Proc. of the 8th Int'l Symposium on Parallel Architectures, Algorithms and Networks*, 2005, pp. 340–345.
15. D. Koukopoulos, M. Mavronicolas, S. Nikoletseas, P. Spirakis, On the Stability of Compositions of Universally Stable, Greedy, Contention-Resolution Protocols, *Proc. of the 16th Int'l Symposium on DIStributed Computing*, 2002, LNCS. 2508, pp. 88–102.
16. D. Koukopoulos, M. Mavronicolas, S. Nikoletseas, P. Spirakis, The Impact of Network Structure on the Stability of Greedy Protocols, Theory of Computing Systems, 38 (2005) 425–460.
17. D. Koukopoulos, S. Nikoletseas, P. Spirakis, Stability Issues in Heterogeneous and FIFO Networks under the Adversarial Queueing Model, *Proc. of the 8th Int'l Conference on High Performance Computing*, 2001, LNCS. 2228, pp. 3–14.
18. D. Koukopoulos, M. Mavronicolas, P. Spirakis, Instability of Networks with Quasi-Static Link Capacities, *Proc. of the 10th Int'l Colloquium on Structural Information and Communication Complexity*, Carleton Scientific, 2003, pp. 179–194.
19. Z. Lotker, B. Patt-Shamir, A. Rosén, New Stability Results for Adversarial Queuing, SIAM Journal on Computing, 33 (2004) 286–303.
20. N. Lynch, Distributed Algorithms, Morgan Kaufmann, 1996.
21. P. Tsaparas, Stability in Adversarial Queueing Theory, M.Sc. Thesis, Computer Science Department, University of Toronto, 1997.

Provisioning a Virtual Private Network Under the Presence of Non-communicating Groups

Friedrich Eisenbrand[1] and Edda Happ[2]

[1] University of Dortmund, Otto Hahn Str. 14,
D-44221 Dortmund, Germany
`Friedrich.Eisenbrand@cs.uni-dortmund.de`
[2] Max-Planck-Institut für Informatik, Stuhlsatzenhausweg 85,
D-66123 Saarbrücken, Germany
`edda@mpi-sb.mpg.de`

Abstract. Virtual private network design in the hose model deals with the reservation of capacities in a weighted graph such that the terminals in this network can communicate with one another. Each terminal is equipped with an upper bound on the amount of traffic that the terminal can send or receive. The task is to install capacities at minimum cost and to compute paths for each unordered terminal pair such that each valid traffic matrix can be routed along those paths.

In this paper we consider a variant of the virtual private network design problem which generalizes the previously studied symmetric and asymmetric case. In our model the terminal set is partitioned into a number of groups, where terminals of each group do not communicate with each other.

Our main result is a 4.74 approximation algorithm for this problem.

1 Introduction

Suppose that a large globally operating company wants to connect all of its branch-offices into a common network to ensure communication between the offices. One approach to do so is to build the network on top of an existing public network by buying a certain amount of link capacities which is then reserved exclusively for the use of this company. In this way the company has established a *virtual private network*. The capacity reservation on links comes with certain costs which we assume to be linear in the amount of reserved capacity.

The network can be modeled as an undirected graph $G = (V, E)$ with edge costs $c : E \longrightarrow \mathbb{R}_+$ reflecting the cost of reserving one unit of capacity on an edge. The branch-offices are a subset $T \subseteq V$ of the nodes which are the *terminals* of this network design problem. A solution to the problem is an assignment of capacities to the edges and paths P_{ij} for each unordered pair $\{i, j\} \subseteq T$ of terminals such that all possible traffic between the terminals can be routed along those paths over the network.

T. Calamoneri, I. Finocchi, G.F. Italiano (Eds.): CIAC 2006, LNCS 3998, pp. 105–114, 2006.

Predicting the amount of traffic that pairs of terminals exchange is often illusive. In the so-called hose model [1, 2] the knowledge of the exact amount of traffic which is exchanged between the terminal pairs is relaxed into a prediction of how much traffic occurrs at each terminal. Here, each terminal $v \in T$ has an threshold $b(v) \in \mathbb{Z}_{\geq 0}$ which is an upper bound on the amount of network traffic that this terminal can interchange with other terminals.

A *traffic matrix* $D \in \mathbb{Q}_{\geq 0}^{T \times T}$ is a symmetric rational matrix which represents the amount of communication between terminals. The traffic matrix is *valid*, if it respects the upper bounds, i.e., if the following holds for each terminal $i \in T$

$$\sum_{j \in T, j \neq i} D(i, j) \leq b(i) . \tag{1}$$

Virtual private network design is the optimization problem that searches a minimum cost assignment of capacities to the edges and specifies for each unordered terminal pair $i, j \in T$ a path P_{ij} in the network such that each valid traffic matrix can be routed along these paths without exceeding the capacities.

This virtual private network design problem has received a considerable amount of attention. Gupta et al. [3] provided a 2-approximation algorithm for this problem and showed that it can be solved in polynomial time when the graph stemming from the edges with nonzero capacity reservation is supposed to form a tree. It is a well known conjecture that there always exists an optimal tree reservation. Hurkens, Keijsper and Stougie [4] have recently shown that this is the case in ring networks. Computational evidence that it also holds in arbitrary networks is for example presented in [5, 6].

In the *asymmetric* variant of virtual private network design, one distinguishes between traffic which is *sent* and traffic which is *received* by a terminal. A traffic matrix then has to respect these upper bounds on each vertex in order to be valid. Via duplicating each terminal into two terminals, where one copy can only send and the other can only receive traffic, the asymmetric variant can be formalized as follows. The terminal set T is partitioned into two sets $\mathcal{R}$ and $\mathcal{S}$, representing *receivers* and *senders*, respectively. The terminals are equipped with upper bounds $b(v)$ as above. A traffic matrix is valid, if it satisfies (1) and if $D(i, j) = 0$ whenever i and j are both senders or both receivers.

The asymmetric virtual private network design problem is NP-hard [3] which follows from a reduction to the steiner tree problem. Gupta et al.[3] gave the first constant factor approximation algorithm for this problem. Gupta, Kumar and Roughgarden [7] presented a randomized approximation algorithm. Their algorithm samples terminals which are then connected into a high bandwidth core. The remaining terminals are connected along their shortest paths to this core. The approximation ratio of this algorithm is 5.55. This result was refined to a 4.74 approximation [8] which also finds a tree solution. The first non-tree approximation algorithm achieves an approximation factor of 3.55 [9]. Italiano, Leonardi and Oriolo [10] consider the setting in which the sums of the sender and receiver thresholds are equal.

1.1   A Setting in Which Some Terminals Do Not Communicate

In this paper we consider a variant of the virtual private network design problem which generalizes both the symmetric and asymmetric version of virtual private network design.

The terminals T are partitioned into disjoint sets $T_1, \ldots, T_k$. Network traffic only occurs between terminals i and j if i and j are in different sets $T_i \neq T_j$. This means that a traffic matrix Q is now valid if $D(i, j) = 0$ for all $i, j \in T_\ell$ and all $1 \leq \ell \leq k$ and

$$\sum_{j \in T, j \neq i} D(i, j) \leq b(i) \text{ for all } i \in T \ .$$

The goal is now to determine paths between each unordered pair of terminals belonging to different sets and to reserve capacities on the edges such that each valid traffic matrix can be routed and the capacity reservation has minimum cost. In the following we refer to this combinatorial optimization problem as *virtual private network design* (VPND).

If the terminal sets $T_1, \ldots, T_k$ are singletons, then we are dealing with the symmetric virtual private network design problem. If the terminals are partitioned into two sets only, then this is the setting of the asymmetric case. Thus, our model is flexible enough to capture both variants of network design which have previously been studied in the literature.

A possible application scenario where this more general model is relevant is as follows (Fig. 1). Some companies want to cooperate and to connect all their branch-offices via a common virtual private network. The companies themselves are already connected.

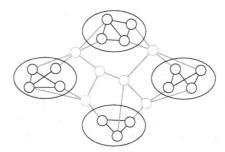

Fig. 1. Companies having internal networks have established a joint network

One possible approach would be to use these connections and to treat the existing small networks as one terminal. Thus, all communication leaving a company network would have to be collected in one selected node and then sent outside. This might cause congestion in the small networks and might lead to a necessary renegotiation for new contracts with the providers of the small networks. It could be cheaper to consider the VPND problem in which the terminal sets correspond to the companies which are already connected.

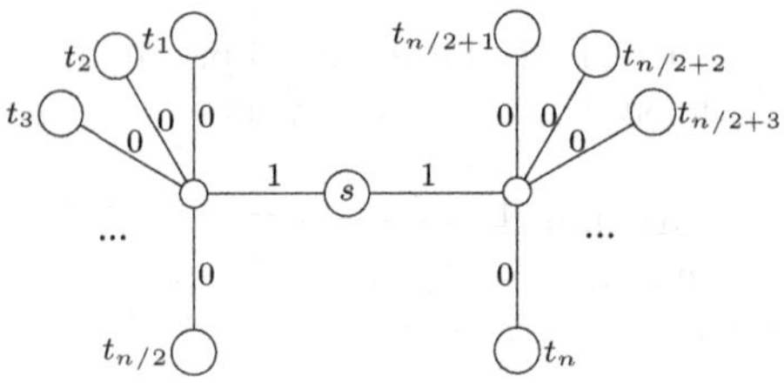

Fig. 2. A network which demonstrates that the optimal solution can differ considerably depending on the partitioning of the terminals. The edges are labeled with their costs.

We would also like to mention that the network design problem of building a minimum cost virtual private network connecting one terminal of each company is hard to approximate with a factor of less than $\log(n)$ since it is a generalization of the group steiner tree problem [11] which is known to have this bound [12, 13].

The following example (Fig. 2) shows that the optimal solution can differ considerably on the same graph and the same terminals depending on the partitioning of the terminals. The terminals are the set $\{s, t_1, \ldots, t_n\}$. The threshold on each node is one.

A solution to the corresponding symmetric problem requires a reservation of $\frac{n}{2}$ on the edges adjacent to s which has cost n while a solution to the asymmetric problem where the set of senders is $\{s\}$ and the set of receivers is $\mathcal{R} = \{t_1, \ldots, t_n\}$ requires only a reservation of 1. It is easy to see there exists a partitioning of the terminals for each even natural number $i \in \{1, \ldots, n\}$ such that the optimum value of the corresponding VPND problem is exactly i.

Contribution of This Paper

The above example shows that an arbitrary reduction to the symmetric or asymmetric case does not yield a constant factor approximation. Our main result however is a proof that ignoring the terminal partitions, i.e. solving the corresponding symmetric case, yields a constant factor approximation to VPND unless the problem is unbalanced. This is the case if the size of one terminal partition is larger than the sum of the remaining partitions. In this case we show that an optimal tree solution of the asymmetric problem stemming from identifying the large terminal partition as the set of receivers and collecting the remaining terminals from the other partitions into a set of senders yields a constant factor approximation.

Assume without loss of generality that the terminal sets are ordered in decreasing cardinality, i.e., $|T_i| \geq |T_j|$ for $i < j$. We call the VPND instance *unbalanced* if $|T_1| \geq \sum_{i>1} |T_i| - 1$.

We show that the following algorithm is a 4.74 approximation algorithm for VPND.

Algorithm 1. VPND $(G, \bigcup_{i=1}^{k} T_i, c)$

---

1. If the VPND-instance is unbalanced then return an approximate tree solution for the asymmetric problem with senders T_1 and receivers $T_2 \cup \ldots \cup T_k$.
2. Otherwise output an approximate solution of the symmetric VPND-instance with terminal set $T_1 \cup \ldots \cup T_k$.

We show that if we use the randomized approximation algorithm [8] in step 1 and the algorithm [3] in step 2, then we achieve an overall approximation ratio of 4.74 which coincides with the approximation ratio of the algorithm in [8].

2 Subinstances and Their Optimal Solutions

By duplicating terminals we can assume that $b(i) = 1$ for all $i \in T$. Suppose that the paths $\mathcal{P}$ are given along which the flow has to be routed. We can compute the corresponding necessary capacity assignment as follows. Consider the complete k-partite graph $B = (T_1 \cup \cdots \cup T_k, E^B)$ and the set of matchings $\mathcal{M}$ of B. Each $M \in \mathcal{M}$ corresponds to a valid traffic matrix. We have to make sure that for all M all paths can be packed. Therefore we compute the capacity $u(e)$ of an edge e as

$$u(e) = \max_{M \in \mathcal{M}} |\{P_{rs} \in \mathcal{P} \mid e \in P_{rs} \text{ and } rs \in M\}| \ . \tag{2}$$

The following is a generalization of a similar statement for the asymmetric case [9].

Lemma 1. Let $H_1, \ldots, H_\ell$ be a partitioning of the terminals T. We denote the VPND-instance on graph G with Terminals $T \cap H_i$ and corresponding partitioning $T_1 \cap H_i, \ldots, T_k \cap H_i$ by I_i. Then one has

$$\sum_{i=1}^{k} \text{OPT}_i \leq \text{OPT} \ ,$$

where OPT_i is the optimum cost of instance I_i.

Proof. Let $\mathcal{P}$ be an optimal set of paths for the original VPND-instance with resulting capacity reservation $u : E \to \mathbb{Z}_+$. The subset $\mathcal{P}_i \subseteq \mathcal{P}$ of paths with both endpoints in H_i defines a solution to instances I_i with the corresponding capacity reservation $u_i : E \to \mathbb{Z}_+$. It suffices to show that $\sum_{i=1}^{k} u_i(e) \leq u(e)$ for each edge $e \in E$.

It follows from (2) that for each $i = 1, \ldots, k$

$$u_i(e) = \max_{M_i \in \mathcal{M}_i} |\{P_{rs} \in \mathcal{P}_i \mid e \in P_{rs} \text{ and } rs \in M_i\}| \ .$$

Let $\tilde{M}_i$ denote the matching for which the maximum is attained. Then, the disjoint union $\tilde{M} := \bigcup_{i=1}^{k} \tilde{M}_i$ is a matching of B. It thus follows from (2) that

$$
\sum_{i=1}^{k} u_i(e) = \sum_{i=1}^{k} \left| \{ P_{rs} \in \mathcal{P}_i \mid e \in P_{rs}, rs \in \tilde{M}_i \} \right|
$$

$$
= \left| \{ P_{rs} \in \mathcal{P} \mid e \in P_{rs}, rs \in \tilde{M} \} \right| \leq u(e)
$$

for each edge $e \in E$. This concludes the proof.   □

3   An Unbalanced Terminal Set

Let us first consider unbalanced instances of VPND with $|T_1| \geq \sum_{i>1} |T_i| - 1$ which is the case in step 1 of the algorithm.

Theorem 1. *Let $(G, \bigcup_{i=1}^{k} T_i, c)$ be an unbalanced VPND instance. Then any tree solution to the corresponding asymmetric virtual private network design problem with $\mathcal{R} = T_1$ and $\mathcal{S} = \bigcup_{i>1} T_i$ is a valid solution to the VPND-instance.*

Proof. Assume the opposite. Then there exists a valid traffic matrix corresponding to a k-partite matching M that cannot be routed on the tree solution to the asymmetric virtual private network design problem. Since any bipartite matching on $\mathcal{S} \cup \mathcal{R}$ can be routed, M contains matched pairs $t_i t_j$ with $t_i, t_j \notin T_1$. Let M^* be a non-routable matching having the *minimal number* of such pairs. Consider matching $M' := M^* \setminus \{t_i t_j\}$ where $t_i, t_j \notin T_1$. Since $|T_1| \geq \sum_{i>1} |T_i| - 1 > |\bigcup_{i>1} T_i \setminus \{t_i, t_j\}| = \sum_{i>1} |T_i| - 2$ there is at least one terminal $t^* \in T_1$ which is idle in M'. So $M' \cup \{t_i t^*\}$ and $M' \cup \{t_j t^*\}$ must be routable since the number of pairs with neither terminal in T_1 is smaller than in M^*. That means that on the path from t_i to t^* and on the path from t_j to t^* one unit of capacity must be free, and therefore also on the unique path from t_i to t_j. So M^* is routable.   □

We have to require tree solutions to guarantee an unambiguous path from t_i to t_j independent of t^*. Figure 3 shows a non-tree solution to the asymmetric virtual private network design problem with $\mathcal{R} = \{t_1^*, t_2^*\}$ and $\mathcal{S} = \{t_1, t_2, t_3\}$. If

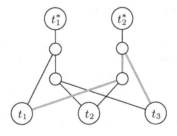

Fig. 3. Non-tree solution of asymmetric problem version not sufficient for VPND

we consider the corresponding VPND with $T_1 = \mathcal{R}$, $T_2 = \{t_1, t_2\}$, and $T_3 = \{t_3\}$, then the condition $|T_1| \geq |T_2| + |T_3| - 1$ holds, but however we fix the path between t_1 and t_3 (e.g. as given in gray in Fig. 3) there is a valid traffic matrix (e.g. $t_2^* t_2, t_1 t_3$) that is not routable even though there is a path between t_1 and t_3 with free capacity (but it is not the fixed path).

The current best tree approximation algorithm to the asymmetric virtual private network design problem has an approximation factor of 4.74 [8]. Since any solution to the VPND $(G, \bigcup_{i=1}^{k} T_i, c)$ is also a solution to the problem when we replace some sets by their union we have $\text{OPT}_{\text{asym}} \leq \text{OPT}_{\text{VPND}}$ implying that the above is also a 4.74 approximation to VPND for unbalanced instances.

4 A Balanced Terminal Set

In the following we denote the shortest path distance between i and j in the Graph $G = (V, E)$ with edge costs $c : E \longrightarrow \mathbb{R}_+$ by $\ell(i, j)$. Finding the terminal t that minimizes $\sum_{t' \in T} \ell(t, t')$ and adding one unit of capacity along each shortest path gives a valid solution to the symmetric virtual private network design problem (G, T, c) where every terminal can communicate with any other [3]. Obviously, it is also a solution to VPND where we restrict the communication to be only between terminals of different sets.

Let us now consider VPND instances where $|T_1| \leq \sum_{i>1} |T_i| - 2$ which we will call *balanced*. We use the following theorem which is proven in [9, Theorem 2] to show that in this case the cheapest shortest path tree is a factor 3 approximation to the optimum solution.

Theorem 2 ([9]). *Consider an instance of VPND with two terminal sets T_1, T_2 with $|T_1| = |T_2|$. Let M be an arbitrary matching of the complete graph on $T_1 \cup T_2$. Then*

$$\sum_{uv \in M} \ell(u, v) \leq \text{OPT} .$$

To prove the approximation factor for VPND we use the following lower bound.

Theorem 3. *Let OPT be the optimal cost of a balanced VPND and M an inclusion-wise maximal matching of the terminals. Then*

$$\sum_{t_1 t_2 \in M} \ell(t_1, t_2) \leq \frac{|M|}{|M| - 1} \cdot \text{OPT} .$$

Proof. Recall that $|T_1| \leq \sum_{i>1} |T_i| - 2$. If the number of terminals is odd, then one terminal node is free and we have $|T_1| \leq \sum_{i>1} |T_i| - 3$. Thus, if we discard the one possibly free node and the endpoints of the lightest edge in the matching from the terminal set, we obtain a new instance where $|T_1| \leq \sum_{i>1} |T_i|$. Thus, we can assume that $|T_1| \leq \sum_{i>1} |T_i|$ holds, $|M|$ is even and all terminals are matched. We will now show that we have

$$\sum_{t_1 t_2 \in M} \ell(t_1, t_2) \leq \text{OPT} .$$

The assertion then follows since we removed the lightest edge from the matching.

We proceed by showing that the edges of M can be paired in such a way that each pair of edges t_1t_2, t_3t_4 satisfies one of the following conditions.

(i) There exists a terminal set T_i such that $t_1, t_2 \in T_i$ and $t_3, t_4 \notin T_i$, or
(ii) there does not exist a terminal set T_i such that $t_1, t_2 \in T_i$ or $t_3, t_4 \in T_i$.

Let $M_i \subseteq M$ be the set of those edges of M having both endpoints in T_i. In other words

$$M_i = \{uv \in M \mid u, v \in T_i\}$$

and assume that the cardinality of M_l is maximal. The endpoints of the remaining edges of M belong to different sets. We partition them into the set $\overline{M}$ containing edges having one node in T_l and the set $\widetilde{M}$ containing edges that do not comprise nodes of T_l.

We now distinguish two cases. Suppose first that $|M_l| > \sum_{j \neq l} |M_j|$. Since $|T_l| \leq \sum_{j \neq l} |T_j|$ this implies

$$|M_l| \leq \sum_{j \neq l} |M_j| + |\widetilde{M}| \ .$$

This allows us to pair each edge of M_l with an edge from $\bigcup_{j \neq l} M_j \cup \widetilde{M}$ such that all edges of $\bigcup_{j \neq l} M_j$ are paired. These pairs will satisfy Condition (i). The remaining edges of $\widetilde{M}$ and the edges of $\overline{M}$ are then paired arbitrarily and satisfy Condition (ii).

In the second case one has

$$|M_l| \leq \sum_{j \neq l} |M_j| \ . \tag{3}$$

Consider an ordering $e_1, \ldots, e_\mu$ of the edges of $M_1 \cup \cdots \cup M_k$ in which the elements of M_i precede the elements of M_j whenever $i < j$. If μ is odd, we can find an edge e_i that can be paired with an edge from $\overline{M} \cup \widetilde{M}$ such that Condition (i) holds and (3) still holds. So assume that μ is even. The pairings $\{e_i, e_{\mu/2+i}\}$ for $1 \leq i \leq \mu/2$ satisfy Condition (i) and are of the left type in Figure 4. The remaining edges can be paired arbitrarily.

Let $\{t_1t_2, t_3t_4\}, \ldots, \{t_{4s+1}t_{4s+2}, t_{4s+3}t_{4s+4}\}$ be such a pairing where each pair satisfies either condition (i) or condition (ii). We now partition T into the sets $H_i = \{t_{4i+1}, t_{4i+2}, t_{4i+3}, t_{4i+4}\}$ for $i = 0, \ldots, s$ and show that

$$\ell(t_{4i+1}t_{4i+2}) + \ell(t_{4i+3}t_{4i+4}) \leq \text{OPT}_i \ , \tag{4}$$

where we use the terminology of Lemma 1.

If the paired edges satisfy condition (ii), then the edges correspond to a valid traffic matrix and (4) clearly holds.

There are two possibilities on how the terminals in a paired set of edges satisfying Condition (i) can be colored, see Fig. 4. Here membership to a terminal set is interpreted as a color. In the first case, the endpoints of the first edgeshare

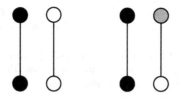

Fig. 4. The possible colorings of nodes in paired edges satisfying condition (i)

the same color as well as the endpoints of the second edge. The assertion then
follows from Theorem 2. We further constrain the problem of the second case
by recoloring the gray node white. In other words, we forbid communication
between the gray and white node. The optimal solution to this problem is at
most as expensive as the optimal solution to the original one. This settles (4).

Applying Lemma 1 concludes the proof.   □

Consider the complete graph $K = (T, E^K)$ on the terminals T with edge costs
equal to the shortest path distances between the terminals in the original graph
G. The cost of the shortest path tree of a terminal t in G is equal to the cost
of the star of t in the graph K. The edges E^K can be covered by at most $|T|$
matchings. Therefore there exists a star whose cost is bounded by $2 \cdot \ell(M)$, where
M is a maximum weight matching of K. By Theorem 3 one has

$$\ell(M) \leq \frac{\lfloor \frac{|T|}{2} \rfloor}{\lfloor \frac{|T|}{2} \rfloor - 1} \cdot \mathrm{OPT} \leq \frac{3}{2} \cdot \mathrm{OPT}$$

for $|T| \geq 6$ which is the case whenever any $|T_i| \geq 2$. If $T_i = 1$ for all i it is a
symmetric virtual private network design problem for which this tree is known
to be a 2 approximation.

This implies the following theorem.

Theorem 4. *Let* $(G, \cup_{i=1}^k T_i, c)$ *be a* balanced VPND-*instance. The cheapest
shortest path tree yields a tree-solution whose cost is at most* $3 \cdot \mathrm{OPT}$.

By combining Theorem 1 and Theorem 4 we obtain our main result.

Theorem 5. *There exists a 4.74 randomized approximation algorithm for* VPND.

References

1. Duffield, N.G., Goyal, P., Greenberg, A., Mishra, P., Ramakrishnan, K.K., van der
   Merive, J.E.: A flexible model for resource management in virtual private networks.
   In: Proceedings of the conference on Applications, technologies, architectures, and
   protocols for computer communication, ACM Press (1999) 95–108
2. Fingerhut, J.A., Suri, S., Turner, J.S.: Designing least-cost nonblocking broadband
   networks. Journal of Algorithms **24**(2) (1997) 287–309

3. Gupta, A., Kleinberg, J., Kumar, A., Rastogi, R., Yener, B.: Provisioning a virtual private network: a network design problem for multicommodity flow. (2001) 389–398
4. Hurkens, C., Keijsper, J., Stougie, L.: Virtual private network design: A proof of the tree routing conjecture on ring networks. In: Proceedings of the eleventh Conference on Integer Programming and Combinatorial Optimization, IPCO XI. (2005) to appear.
5. Erlebach, T., Regg, M.: Optimal bandwidth reservation in hose-model vpns with multi-path routing. In: INFOCOM. (2004)
6. Altin, A., Amaldi, E., Belotti, P., Pinar, M.: Provisioning virtual private networks under traffic uncertainty. In: Proceedings of the 3rd Cologne Twente Workshop on Graphs and Combinatorial Optimization (CTW'04). Volume 17 of Electronic Notes in Discrete Mathematics. (2004) 19–22
7. Gupta, A., Kumar, A., Roughgarden, T.: Simpler and better approximation algorithms for network design. In ACM, ed.: Proceedings of the Thirty-Fifth ACM Symposium on Theory of Computing, San Diego, CA, USA, June 9–11, 2003, New York, NY, USA, ACM Press (2003) 365–372
8. Eisenbrand, F., Grandoni, F.: An improved approximation algorithm for virtual private network design. In: Proceedings of the sixteenth annual ACM-SIAM symposium on Discrete algorithms, SODA 05. (2005) 928–932
9. Eisenbrand, F., Grandoni, F., Oriolo, G., Skutella, M.: New approaches for virtual private network design. (to appear in the proceedings of ICALP 05)
10. Italiano, G., Leonardi, S., Oriolo, G.: Design of networks in the hose model. In: Proceedings of ARACNE 2002. (2002) pp 65–76
11. Reich, G., Widmayer, P.: Beyond Steiner's problem: A VLSI oriented generalization. In: Graph-Theoretic Concepts in Computer Science WG-89. Volume 411 of Lecture Notes in Computer Science. (1990) 196–210
12. Ihler, E.: The complexity of approximating the class Steiner tree problem. Technical report, Institut für Informatik, Albert-Ludwigs-Universität Freiburg (1991)
13. Feige, U.: A threshold of ln n for approximating set cover. In: Proc. of the ACM sympository on the Theory of Computing. (1996) 314–318

Gathering Algorithms on Paths Under Interference Constraints

Jean-Claude Bermond[1,*], Ricardo Corrêa[2,**], and Minli Yu[3,***]

[1] MASCOTTE, joint project CNRS-INRIA-UNSA, 2004 Route des Lucioles, BP 93,
F-06902 Sophia-Antipolis, France
bermond@sophia.inria.fr

[2] Universidade Federal do Ceará, Departamento de Computação, Campus do Pici,
Bloco 910, 60455-760 Fortaleza, CE, Brazil
correa@lia.ufc.br

[3] University College of the Fraser Valley, Department of Mathematics and Statistics,
Abbotsford, BC, Canada V2S 4N2
joseph.yu@ucfv.ca

Abstract. We study the problem of gathering information from the nodes of a multi-hop radio network into a pre-determined destination node under interference constraints which are modeled by an integer $d \geq 1$, so that any node within distance d of a sender cannot receive calls from any other sender. A set of calls which do not interfere with each other is referred to as a round. We give algorithms and lower bounds on the minimum number of rounds for this problem, when the network is a path and the destination node is either at one end or at the center of the path. The algorithms are shown to be optimal for any d in the first case, and for $1 \leq d \leq 4$, in the second case.

Keywords: Gathering, interference, multi-hop radio network, path.

1 Introduction

1.1 Problem Statement

The problem that we consider in this paper was motivated by a question asked by FRANCE TELECOM about "how to provide Internet to villages" (see [3]) and is related to the following scenario. Suppose we are given a set of communication devices (for instance, network interfaces that connect computers to the Internet) which are placed in houses in a village. They require access to a gateway (for instance, a satellite antenna) to send and receive data through a multi-hop wireless network. The nodes communicate exclusively by means of radio transmissions,

* Partially supported by the CRC CORSO with France Telecom, by the european FET project AEOLUS, by the cooperation with Brazil project REGAL and by the INRIA associated team RESEAUXCOM with S.F.U.
** Partially supported by the Conselho Nacional de Desenvolvimento Científico e Tecnológico, CNPq, Brazil.
*** Partially supported by the INRIA associated team RESEAUXCOM.

T. Calamoneri, I. Finocchi, G.F. Italiano (Eds.): CIAC 2006, LNCS 3998, pp. 115–126, 2006.

referred to as *calls*. A call involves two nodes, the sender and the receiver, and is subject to the following constraints:

Reachability constraint: since every node has limited transmission power, the receiver must be close enough to the sender.

Interference constraint: unlike wired networks, a call can interfere with reception at certain nodes beyond the receiver. A node that is within interference distance of one call cannot be the receiver of another call.

Considering these two constraints, a message transmitted in a call can only be properly received if the receiver is reachable from the sender and there is no interference by another message being simultaneously transmitted. In this context, we study the following problem:

t-gathering problem: suppose each node of the network has a piece of information. The *t*-gathering problem consists of collecting (gathering) all these pieces of information into a special node *t*, called the *gathering node*.

In this paper, we propose solutions to this problem for the particular case of a path. Before going into details about our results, let us introduce the mathematical formulation of the problem.

1.2   Model and Assumptions

According to the model adopted in [1], the network described above is represented by an undirected graph $G = (V, E)$, where V is the set of nodes, each of them representing a communication device that is able to send and receive messages, and E is the set of edges, representing the possible communications. Let $d_G(s, r)$ indicate the distance in G, defined as the length of a shortest path between s and r. We model the reachability and the interference constraints by two positive integers $d_T \geq 1$ and $d_I \geq d_T$. A node $r \in V$ is reachable from $s \in V$ if and only if $d_G(r, s) \leq d_T$. An important case is $d_T = 1$, which means that a node is able to communicate only with its neighbors in the graph. The second parameter d_I models the interference constraint as follows: if s sends a message to r, then no node $w \in V$ such that $d_G(s, w) \leq d_I$ can receive another message.

Denote by $X_{s,r}$ a call where a node $s \in V$ sends message X to node $r \in V$. We assume that every call takes one unit of time (or one slot) to transmit one unit-length message. Two calls are said to be *compatible* if they do not interfere with each other. More precisely, two calls X_{s_1,r_1} and Y_{s_2,r_2} are compatible if $d_G(s_1, r_2) > d_I$ and $d_G(s_2, r_1) > d_I$. Observe that a consequence of the interference constraint is that $s_1 \neq r_2$ and $s_2 \neq r_1$, which implies that a node is not able to send and receive messages simultaneously. A *round* is a set of compatible calls, whereas an *algorithm* is a sequence of rounds.

In this paper, our aim is to find a *t*-gathering protocol using a minimum number of rounds in the specific case where G is a path. In fact, this stems from the assumption that the village consists of one main street. To our great surprise, the gathering problem is not so simple in this case, if one wants to obtain an exact optimal algorithm.

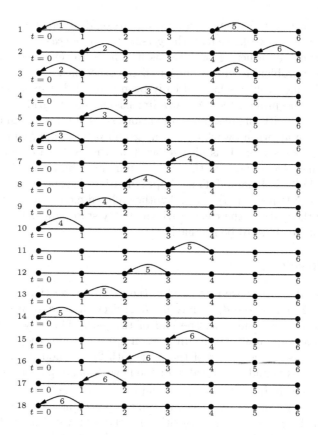

Fig. 1. Algorithm for a graph of 7 nodes and $d_I = 2$

In the algorithm shown in Figure 1 (where $d_T = 1$ and $d_I = 2$), the call $1_{1,0}$ interferes with $4_{4,3}$ because $d_G(1,3) \leq 2 = d_I$. This is the reason why they do not appear in the same round. On the other hand, the calls $1_{1,0}$ and $5_{5,4}$ are compatible. All the rounds shown in the figure consist of a single call or two compatible calls. It will be shown later that the algorithm consisting of this sequence of 18 rounds is in fact optimal.

A final remark with respect to the model adopted in this paper is that another possibility would be to represent the radio devices as nodes in the plane, and to state the reachability and interference constraints according to the euclidean distances. However, since we only consider paths, the two models are equivalent.

1.3   Related Work

The broadcasting and gossiping problems in radio networks with $d_T = d_I = 1$ are studied in [6, 8] and [4, 5, 7], respectively. Note that, in a broadcast, the same information has to be transmitted to all the other nodes and therefore flooding techniques can be used. When a node needs to send different messages to the other nodes of the network, we have the *personalized broadcasting problem*, which is equivalent to the gathering problem as it suffices to reverse the calls in the solution of one problem to get a solution of the other one.

Some gathering problems have already been studied. For example, in [2] optimal solutions are provided for the two-dimensional square grid. In [1], general results are given (with the possibility of various sizes of messages in each node); in particular, an algorithm working on any graph with an approximation factor of at most 4 is presented. It is also shown that the problem of finding an optimal gathering algorithm (one that uses a minimum number of rounds) does not admit a Fully Polynomial Time Approximation Scheme if $d_I > d_T$, unless P=NP, and is NP-HARD if $d_I = d_T$. Another related model can be found in [9], where the authors study the case in which steady-state flow demands between each pair of nodes have to be satisfied.

1.4 Our Results

The results of this paper are presented in the remaining sections as follows. We assume $d_T = 1$ and denote d_I simply by d. In Section 2, we deal with the case where the gathering node is at one end of the path. This case is simple and we describe an optimal algorithm. In Section 3, we consider the case where the gathering node is at the center of the path with $2p + 1$ nodes. We first give a lower bound (this bound is also valid for the flow model of [9]). Then, we design an algorithm which meets the lower bound for $p \leq p_1 = d + 1 + \frac{k(k+1)}{2}$. In the next subsection, we show how to strengthen the preceding lower bound. In fact, we show that, for $p \geq d + 2$, any algorithm for the path with the gathering node at the center needs $2\lfloor (d-1)/2 \rfloor + 1$ more rounds than that for the path of length p with the gathering node at one end. Our algorithm meets this strengthened lower bound for $d = 1, 2, 3, 4$ (which correspond to the practical cases). We close the paper with some concluding remarks in Section 4.

2 Paths with the Gathering Node at One End

Let Π_p be the path of length p (consisting of p edges and $p+1$ nodes). The nodes are denoted $0, 1, 2, \cdots, p$, and the edges are of the type $(i, i-1)$. Assume that the gathering node is $t = 0$. To simplify the notation, we denote the call $X_{i,i-1}$ by X_i and the minimum number of rounds by $g_d(p)$. The recursive scheduler depicted in Algorithm 1 is used to prove the result below (see Figure 1 for an example with $p = 6$ and $d = 2$).

Theorem 1. *For the path Π_p and $d \geq 1$,*

$$g_d(p) = \begin{cases} p(p+1)/2, & \text{if } p \leq d + 1 \\ (d+2)(2p - d - 1)/2, & \text{otherwise} \end{cases}$$

Algorithm 1. Gathering scheduler on Π_p

1: **if** $p > 0$ **then**
2:     Call recursively the gathering scheduler on Π_{p-1}
3:     **for** $j \leftarrow p, \ldots, d+3$ **do**
4:         Let $x = p - (d+2)$ and $i = j - (d+2)$
5:         Schedule P_j in the same round as X_i
6:     **for** $j \leftarrow \min\{p, d+2\}, \ldots, 1$ **do**
7:         Schedule P_j in a new round

Proof. The upper bound is given by Algorithm 1. Suppose that all calls involving messages smaller then P are scheduled in existing rounds as indicated in line 2. The calls involving the message P leaving a node $j \geq d+3$ are scheduled as indicated in lines 3-5. New rounds are then created for the remaining calls. Hence, proceeding by recurrence, we find that

$$g_d(p) \leq \sum_{i=1}^{p} \min\{i, d+2\},$$

which gives the upper bound of the theorem.

To show the lower bound, note that the information X of a node x must be transmitted via the calls X_j, $1 \leq j \leq x$. Furthermore, the interference constraint implies that at most one call X_j, for $1 \leq j \leq d+2$, can occur in a round. So, to send X, for $1 \leq x \leq d+1$, from node x to the gathering node, we need at least x rounds, all containing a call in the interval $[0, d+2]$. It follows that the rounds used for two distinct nodes x and x', $1 \leq x, x' \leq d+1$, are disjoint. Therefore, if $p \leq d+1$, then at least $1 + 2 + \cdots + p = p(p+1)/2$ rounds are required.

Now, consider $x \geq d+2$. To bring X to the gathering node, all the $d+2$ calls X_j, $1 \leq j \leq d+2$, for X must occur at different rounds. Moreover, these rounds must be different from those used for Y_j, $X \neq Y$ and $1 \leq j \leq d+1$. Consequently, at least $(d+2)[p-(d+1)]$ calls are required for X, $d+2 \leq x \leq p$, thus we have the lower bound for the case $p \geq d+2$.   □

3   Paths with the Gathering Node at the Center

3.1   Preliminaries

Let us denote by $\Pi_{-p}\Pi_p$ the path of length $2p$ with the $2p+1$ nodes $-p, -(p-1), \cdots, -1, 0, 1, 2, \cdots, p$, and with edges $(-i, -(i-1))$ and $(i, i-1)$. Assume that the gathering node is $t = 0$. We write $d = 2k+1$ or $d = 2k+2$, depending whether d is odd or even, respectively, and denote the minimum number of rounds by $g_d(p, p)$. Clearly, $g_d(p, p) \geq g_d(p)$ since $\Pi_{-p}\Pi_p$ is composed by two symmetric paths of length p. However, in order to attain any tight lower bound, it often requires the calls on one side of the paths to be paired with calls on the other side. When p is small, all the calls are incompatible and every algorithm is optimal.

Proposition 1. *If $p \leq k + 1$, then $g_d(p, p) = 2g_d(p) = p(p + 1)$.*

In the sequel, we consider $p > k+1$, in which case an optimal algorithm requires some compatible calls to be appropriately paired. Special attention needs to be devoted to the *critical calls*, that is the calls in the *critical interval* $[-(d+2), d+2]$ of nodes. A round is called an *obstruction* if it contains only one critical call. Like in the previous section, write X_i and $-X_i$ for the calls $X_{i,i-1}$ and $-X_{-i,-(i-1)}$, respectively.

In the critical interval, two calls X_i and Y_j interfere, and so do two calls $-X_i$ and $-Y_j$. Moreover, two calls $-X_i$ and Y_j interfere if and only if $i + j \leq d + 1$ because the distance between $-i$ and $j - 1$ is $i + j - 1$. For example, a call $-X_1$ can be paired only with calls Y_{d+1} or Y_{d+2}. Consequently, every round contains at most two critical calls and, in addition, a round contains two critical calls $-X_i$ and Y_j only if $i + j \geq d + 2$.

Let

$$A^+ = \bigcup_{i=1}^{k+1} \{X_i \mid i \leq x \leq p\} \text{ and } A^- = \{-X_i \mid X_i \in A^+\}. \tag{1}$$

Observe that these two sets are such that a call in A^+ cannot be paired with any call in A^-. The remaining critical calls define the sets

$$B^+ = \bigcup_{i=d-k+1}^{p'} \{X_i \mid i \leq x \leq p\} \text{ and } B^- = \{-X_i \mid X_i \in B^+\}, \tag{2}$$

where $p' = \min\{p, d + 2\}$. When d is odd, these sets partition the set of possible calls. But when d is even, there are also all the calls $-X_{k+2}$ and X_{k+2}. Observe that two critical calls can be paired only if one of them belongs to neither A^+ nor A^-.

3.2   A Lower Bound When $p \geq k + 2$

Let us turn our attention to a lower bound which will turn to be optimal when p is not too large.

Theorem 2. $g_d(p, p) \geq p(k + 2) + \lfloor d/2 \rfloor (p - k - 1)$.

Proof. To obtain the lower bound, we count the maximum number M of pairs $\{-X_i, Y_j\}$ which can be formed and we get $g_d(p, p) \geq 2g_d(p) - M$. It can be checked from (1) and (2) that $|A^-| \geq |B^+|$ and $|A^+| \geq |B^-|$. Since the calls of A^- (resp. A^+) can only be paired with calls in B^+ (resp. B^-), the maximum number of pairs involving calls in A^+, A^-, B^+ and B^- occurs when all calls in B^+ and B^- are paired with A^- and A^+, respectively. In addition, if d is even, we can also pair $-X_{k+2}$ with X_{k+2}, for $k + 2 \leq x \leq p$. Thus, $M = |B^+| + |B^-| + (d - 2k - 1)(p - k - 1)$.

First consider the case $p \leq d + 1$. Then $g_d(p) = \frac{p(p+1)}{2}$ by Theorem 1. If d is odd, then $|B^+| = |B^-| = \sum_{k+2 \leq i \leq p} p - i + 1 = \frac{(p-k-1)(p-k)}{2}$ and so $2g_d(p) - M = p(p + 1) - (p - k - 1)(p - k) = p(k + 2) + k(p - k - 1)$. Otherwise, d is even, and $|B^+| = |B^-| = \sum_{k+3 \leq i \leq p} p - i + 1 = \frac{(p-k-2)(p-k-1)}{2}$. This leads to $M = (p-k-1)^2$ and $2g_d(p) - M = p(p+1) - (p-k-1)^2 = p(k+2) + (k+1)(p-k-1)$.

For the case $p \geq d + 2$, we have to use the value of $g_d(p)$ given in Theorem 1 and observe that $g_d(p)$ increases by $d + 2$ as p increases by 1. To compute M, we also observe that now $p' = d + 2$. So, when p increases by 1, $|B^+|$ increases by $k + 2$ and M by $d + 3$. Therefore, $2g_d(p) - M$ increases by $d + 1 = k + \lfloor d/2 \rfloor$, ending the proof. □

3.3   An Optimal Algorithm

In this subsection, we present an algorithm whose number of rounds meets the lower bound described in the previous subsection. This algorithm corresponds to the sequence of rounds obtained with Algorithm 2. In the next subsection, we will show that this algorithm also gives optimal solution for larger values of p and $1 \leq d \leq 4$.

Algorithm 2. Gathering scheduler for $\Pi_{-p}\Pi_p$

1: **if** $p > 0$ **then**
2:      Call recursively the gathering scheduler for $\Pi_{-(p-1)}\Pi_{p-1}$
3:      **for** $j \leftarrow p, \ldots, d + 3$ **do**
4:          Let $x = p - (d + 2)$ and $i = j - (d + 2)$
5:          Schedule P_j in the same round as X_i
6:          Schedule $-P_j$ in the same round as $-X_i$
7:      **for** $j \leftarrow \min\{p, d + 2\}, \ldots, k + 2$ **do**
8:          **if** There is obstruction compatible with P_j **then**
9:              Schedule P_j in the smallest round that is compatible with P_j
10:         **else**
11:             Schedule P_j in a new round
12:         Schedule $-P_j$ in the smallest round that is compatible with $-P_j$
13:     **for** $j \leftarrow \min\{p, k + 1\}, \ldots, 1$ **do**
14:         Schedule P_j in a new round
15:         Schedule $-P_j$ in a new round

Algorithm 2 schedules the calls in a sequence of pairs of symmetric rounds in such a way that, if a pair of compatible critical calls $\{X_i, -Y_j\}$, with $x \neq y$, is scheduled in a certain round, then the round immediately after consists of the symmetric counterpart $\{-X_i, Y_j\}$. Similarly, if a round consists of a single positive call X_i, the next round consists of the single negative call $-X_i$. The algorithm for $d = 3$ and $d = 4$ are illustrated in Table 1 and 2, respectively.

The rounds in Algorithm 2 are scheduled recursively in the sense that the rounds involving the calls P_j and $-P_j$, for all $j \in \{1, 2, \ldots, p\}$, are scheduled after all the calls associated with the path consisting of $p-1$ positive and negative nodes are scheduled in line 2. This is done without modifying the order of rounds, but only by including the new calls in existing rounds, when possible, or creating

Table 1. Pairs and obstructions in the rounds derived from Algorithm 2 for $d = 3$. For every round shown in the table but those between horizontal lines, the algorithm also includes its symmetric counterpart.

Round	$p = 1$	$p = 2$	$p = 3$	$p = 4$	$p = 5$	$p = 6$	$p = 7$	$p = 8$	$p = 9$	$p = 10$	$p = 11$
1	1_1			-4_4		6_6			-9_9		11_{11}
3		2_2	-3_3				7_7	-8_8			
5		2_1			-5_5		7_6			-10_{10}	
7			3_2	-4_3				8_7	-9_8		
9			3_1		-5_4			8_6		-10_9	
11				4_2	-5_3				9_7	-10_8	
13				4_1		-6_5			9_6		-11_{10}
15					5_2	-6_4				10_7	-11_9
17					5_1		-7_5			10_6	
19						$\{-6_3, 6_3\}$					$\{-11_8, 11_8\}$
20						6_2	-7_4				11_7
22						6_1		-8_5			11_6
24							$\{-7_3, 7_3\}$				
25							7_2	-8_4			
27							7_1		-9_5		

new pairs and obstructions. In addition, the new calls are scheduled greedily in lines 9 and 12. More precisely, for the new calls outside the critical interval, they are included in existing rounds. Then, the critical calls are handled. First, the call $P_{\min\{p,d+2\}}$ is paired with the first available and compatible obstruction $-X_i$. Next, the symmetric counterpart is created by pairing $-P_{\min\{p,d+2\}}$ with X_i. The call $P_{\min\{p,d+2\}-1}$ is then paired with the next available and compatible positive obstruction and so on, until $-P_{k+2}$ is paired with either the first available and compatible obstruction (which will turn to be $(P-1)_{d-k}$) or with $-P_{k+2}$.

Algorithm 2 leads to the following upper bound for $g_d(p,p)$.

Theorem 3. $g_d(p, p) \leq p(k+2) + \lfloor d/2 \rfloor (p-k-1) + \max\{0, p-p_1\}$, for $p \geq k+1$, where $p_1 = d + 1 + \frac{k(k+1)}{2}$.

Sketch of the proof. To prove the theorem, we count the number $r_d(p)$ of rounds scheduled with Algorithm 2. First let us consider the odd case $d = 2k + 1$. The proof is by induction on p. We indicate only half of the rounds (the other being obtained by symmetry) which consist either of an obstruction or of a pair of critical calls exclusively. The calls outside the critical interval are easily handled in lines 5 and 6.

If $p = k+1$, all rounds are scheduled in lines 13-15. So, $r_d(k+1) = (k+1)(k+2)$. Next, we consider $p \geq k + 2$ and give the sequence A_d^p of obstructions left after line 2 which have to be paired with the sequence

$$\langle -P_\ell, -P_{\ell-1}, \ldots, -P_{k+2} \rangle, \ell = \min\{p, d + 2\}.$$

The first element of A_d^p is paired with $-P_\ell$, the second with $-P_{\ell-1}$ and so on. When p is large enough (as can be seen in Table 1) the last element of

Table 2. Similar to Table 1, but for $d = 4$

Round	$p=1$	$p=2$	$p=3$	$p=4$	$p=5$	$p=6$	$p=7$	$p=8$	$p=9$
1	1_1				-5_5		7_7		
3		2_2		-4_4				8_8	
5		2_1				-6_6		8_7	
7			$\{-3_3,3_3\}$						$\{-9_9,9_9\}$
8			3_2		-5_4				9_8
10			3_1			-6_5			9_7
12				$\{-4_3,4_3\}$					
13				4_2		-6_4			
15				4_1			-7_6		
17					$\{-5_3,5_3\}$				
18					5_2		-7_5		
20					5_1			-8_6	
22						$\{-6_3,6_3\}$			
23						6_2	-7_4		
25						6_1		-8_5	
27							$\{-7_3,7_3\}$		
28							7_2	-8_4	
30							7_1		-9_6
32								$\{-8_3,8_3\}$	
33								8_2	-9_5
35								8_1	
37									$\{-9_4,9_4\}$
38									$\{-9_3,9_3\}$
39									9_2
41									9_1

A_d^p is not an obstruction at step $p - 1$ but, in fact, P_{k+2} which is paired with $-P_{k+2}$.

First assume that $p \leq 2k + 2 = d + 1$. It turns out that

$$r_d(p) = r_d(p - 1) + 2\left(p - |A_d^p|\right). \tag{3}$$

Let $p = k + 1 + i$, where $1 \leq i \leq k + 1$. Then,

$$A_d^{k+1+i} = \langle (k - i + 2)_{k-i+2}, (k - i + 4)_{k-i+3}, ..., (k + i)_{k+1}\rangle,$$

which gives $|A_d^p| = p - k - 1$ and $r_d(p) = r_d(p - 1) + 2(k + 1)$.

When $p \geq d + 2$, write $s = k(k - 1)/2$ and let $\oplus$ denote a concatenation of sequences. In the two cases considered in the sequel, we replace $2\left(p - |A_d^p|\right)$ by $2\left(d + 2 - |A_d^p|\right)$ in (3) and obtain $|A_d^p| = k + 2$, as follows:

1. $2k + 3 \leq p \leq 2k + 2 + s$. Let $p = 2k + 2 + i$, where $1 \leq i \leq s$. The definition of A_d^p in this case is recursive:

$$A_d^{2k+2+i} = \begin{cases} \langle 2_1, 3_1, 4_2 \rangle, & \text{if } p = 5 \text{ and } d = 3 \\ \langle (P-2)_1, (P-1)_2, P_3 \rangle, & \text{if } p \geq 6 \text{ and } d = 3 \\ A_{d-2}^{2k+i} \oplus \langle (2k + 1 + i)_{k+1}\rangle, & \text{otherwise} \end{cases}$$

The recurrence for the number of elements in A_d^{2k+2+i} is

$$|A_d^p| = \begin{cases} 3, & \text{if } p \geq 5 \text{ and } d = 3 \\ |A_{d-2}^{p-2}| + 1, & \text{otherwise} \end{cases}$$

whose solution gives the desired result.

2. $2k + 3 + s \leq p \leq 3k + 2 + s = p_1$. Let $p = 2k + 2 + s + i$, where $1 \leq i \leq k$. Then, the cardinality of A_d^p follows directly from

$$A_d^{2k+2+i} = \langle (s + k + i)_1, (s + k + 1 + i)_2, ..., (s + 2k)_{k-i+1} \rangle \oplus \\ \langle (s + 2k + 1)_{k-i+1}, (s + 2k + 2)_{k-i+2}, ..., (s + 2k + i + 1)_{k+1} \rangle.$$

Finally, assume $p \geq 3k + 3 + s$, in which case we write $p = 3k + 2 + s + i$, where $i \geq 1$. In this case, the last element of A_d^p is P_{k+2}, which is paired with $-P_{k+2}$. Then, $r_d(p) = r_d(p - 1) + 2(d + 2 - |A_d^p|) + 1$ and

$$A_d^p = \langle (s + 2k + i + 1)_1, (s + 2k + i + 2)_2, ..., (s + 3k + i + 2)_{k+2} \rangle.$$

Putting the pieces together, we get the recurrence

$$r_d(p) \leq \begin{cases} (k+1)(k+2), & \text{if } p = k + 1 \\ r_d(p-1) + d + 1, & \text{if } k + 1 < p \leq p_1 \\ r_d(p-1) + d + 2, & \text{if } p > p_1. \end{cases} \qquad (4)$$

When d is even ($d = 2k+2$, illustrated in Table 2), we obtain the result from the case $d-1$ odd. First, observe that, if $p < 3k+3+s$, then $A_{2k+2}^p = A_{2k+1}^{p-1} \oplus \langle P_{k+2} \rangle$. Otherwise, A_d^p includes P_{k+2} and pairs kept from A_{2k+1}^{p-1} and A_{2k+1}^p depending on the call X_{k+2}, where $x = 3k + 3 + s$. A call Y_j is kept from A_{2k+1}^{p-1} if $y < x$ or ($y = x$ and $j > k + 2$), and from A_{2k+1}^p otherwise. The recurrence is then the same as above. The solution of (4) concludes the proof. $\qquad \square$

Combining Theorem 2 and Theorem 3, we get

Theorem 4. $g_d(p,p) = p(k+2) + \lfloor d/2 \rfloor (p - k - 1)$, for $k + 1 \leq p \leq p_1$, where $p_1 = d + 1 + \frac{k(k+1)}{2}$.

3.4   A Lower Bound for $p \geq d + 2$

Both Algorithm 1 and Algorithm 2 have a common property: X_i (resp. $-X_i$) appears in a round occurring before that of X_j (resp. $-X_j$) if $i > j$, and X_i (resp. $-X_i$) appears after Y_i (resp. $-Y_i$) if $x > y$. Indeed, one can easily modify any algorithm in order to satisfy such a property. For this reason, and without loss of generality, we suppose that the obstructions are maximal in the following sense.

Assumption 1. *If X_i (resp. $-X_i$) is an obstruction, then the following conditions hold:*

1. *either $x = p$ or $(X + 1)_i$ (resp. $-(X + 1)_i$) is an obstruction; and*
2. *either $i = 1$ or $-X_{i-1}$ (resp. $-X_{i-1}$) is an obstruction.*

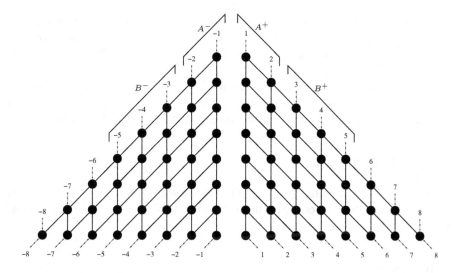

Fig. 2. Partial order $\preceq$ on the calls of an algorithm for $p = 8$ and $d = 3$

Moreover, the property mentioned above naturally defines a partial order $\preceq$, illustrated in Figure 2, in which $X_i \preceq Y_j$ if $x \leq y$ and $i \geq j$ and $-X_i \preceq -Y_j$ if $x \leq y$ and $i \geq j$. We may use the notation $X_i \prec Y_j$ when $x \neq y$ or $i \neq j$.

In the rest of this subsection, we present a lower bound for $p \geq d + 2$. This lower bound is based on the minimum number of obstructions that are induced by $\preceq$. The proofs are omitted due to space limitations.

Lemma 1 (Non-Crossing Lemma). *An algorithm cannot have two different pairs $\{-X_i, W_j\}$ and $\{-Y_k, Z_\ell\}$ with either $-Y_k \prec -X_i$ and $W_j \prec Z_\ell$, or $-X_i \prec -Y_k$ and $Z_\ell \prec W_j$.*

The previous lemma is applied in the results that follow.

Lemma 2. *P_1 and $-P_1$ are obstructions.*

An immediate consequence is the optimality of Algorithm 2 for $d = 1, 2$. In addition, we have

Lemma 3. *If $p \geq d + 2$, then every algorithm has at least $2k + 1$ positive and $2k + 1$ negative obstructions.*

This leads to our final result

Theorem 5. *If $p \geq d + 2$, then $g_d(p, p) = g_d(p) + 1$, for $d = 1, 2$ and $g_d(p, p) = g_d(p) + 3$, for $d = 3, 4$.*

4 Concluding Remarks

We presented algorithms for gathering information from nodes of a path with $2p+1$ nodes to its center or one end node, satisfying reachability and interference constraints. Optimal solutions are given for the first case, and for the second case when the interference distance d is at most 4. We conjecture that the algorithm for the second case is also optimal for larger values of d, leading to $g_d(p,p) = g_d(p) + \frac{(k+1)(k+2)}{2}$, for $p \geq p_1$.

The results in this paper can be extended for more general cases; for instance, when the gathering node is placed anywhere in the path. However, the choice of the center of the path is the one that minimizes the number of rounds.

Acknowledgements

The two last authors thank the MASCOTTE project where some of the research was done during their visits. Also, the authors would like to thank the anonymous referees for the useful suggestions.

References

1. J-C. Bermond, J. Galtier, R. Klasing, N. Morales, and S. Pérennes. Hardness and approximation of gathering in static radio networks. In *FAWN06*, Pisa,Italy, March 2006.
2. J.-C. Bermond and J. Peters. Efficient gathering in radio ids with interference. In *AlgoTel'05*, pages 103–106, Presqu'le de Giens, May 2005.
3. P. Bertin, J-F. Bresse, and B. Le Sage. Accs haut dbit en zone rurale: une solution "ad hoc". *France Telecom R&D*, 22:16–18, 2005.
4. M. Christersson, L. Gasieniec, and A. Lingas. Gossiping with bounded size messages in ad-hoc radio networks. In *Proceedings of ICALP'02*, volume 2380 of *LNCS*, pages 377–389. Springer-Verlag, 2002.
5. M. Chrobak, L. Gasieniec, and W. Rytter. Fast broadcasting and gossiping in radio networks. *Journal of Algorithms*, 43(2):177–189, 2002.
6. M. L. Elkin and G. Kortsarz. Logarithmic inapproximability of the radio broadcast problem. *Journal of Algorithms*, 52(1):8–25, 2004.
7. I. Gaber and Y. Mansour. Centralized broadcast in multihop radio networks. *Journal of Algorithms*, 46(1):1–20, 2003.
8. L. Gasieniec and I. Potapov. Gossiping with unit messages in known radio networks. In *Proceedings of the IFIP 17th World Computer Congress*, pages 193–205. Kluwer, B.V., 2002.
9. R. Klasing, N. Morales, and S. Pérennes. On the complexity of bandwidth allocation in radio networks with ste ady traffic demands. Technical report, INRIA Research Report RR-5432 and I3S Research Report I3S/RR-2 004-40-FR, 2004.

On the Hardness of Range Assignment Problems

Bernhard Fuchs

Zentrum für Angewandte Informatik Köln,
Universität zu Köln, Weyertal 80, 50931 Köln, Germany
bfuchs@zpr.uni-koeln.de

Abstract. We investigate the computational hardness of the CONNEC-
TIVITY, the STRONG CONNECTIVITY and the BROADCAST type of Range
Assignment Problems in $\mathbb{R}^2$ and $\mathbb{R}^3$. We present new reductions for the
Connectivity problem, which are easily adapted to suit the other two
problems. All reductions are considerably simpler than the technically
quite involved ones used in earlier works on these problems. Using our
constructions, we can for the first time prove NP-hardness of these problems for *all* real distance-power gradients $\alpha > 0$ (resp. $\alpha > 1$ for BROAD-
CAST) in 2-d, and prove APX-hardness of all three problems in 3-d for
all $\alpha > 1$. Our reductions yield improved lower bounds on the approximation ratios for all problems where APX-hardness was known before
already. In particular, we derive the overall first APX-hardness proof for
Broadcast. This was an open problem posed in earlier work in this area,
as was the question whether (Strong) Connectivity remains NP-hard for
$\alpha = 1$. Additionally, we give the first hardness results for so-called well-
spread instances.

1 Introduction

1.1 Problem Definition

Let S be a finite set of n points in the Euclidean space $\mathbb{R}^d$. A *range assignment*
for S is a function $r : S \to \mathbb{R}_+$. For convenience, we write r_v instead of $r(v)$. The
cost of a range assignment is defined as

$$cost(r) = \sum_{v \in S} r_v^\alpha$$

for some real constant $\alpha > 0$.

The underlying intuition is that the elements of S are given radio stations,
and one can choose for each station $v \in S$ a corresponding data transmission
range (radius) r_v. Sending data at radius r_v consumes energy proportional to
r_v^α. A prominent example for this setting is a wireless network. The parameter
α is called the *distance-power gradient*, and realistic values for α range from 1
to more than 6, according to [13], where the reader is directed to for further
reading on the background behind this model.

Let $dist(u, v)$ denote the Euclidean distance between two points $u, v \in \mathbb{R}^n$. A
range assignment r directly defines two kinds of graphs, which reflect communication properties of r:

T. Calamoneri, I. Finocchi, G.F. Italiano (Eds.): CIAC 2006, LNCS 3998, pp. 127–138, 2006.
© Springer-Verlag Berlin Heidelberg 2006

- Let $\vec{G}_r = (S, A_r)$ be the *directed communication graph* of range assignment r, where an arc (u, v) is contained in A_r iff u can send to v, i.e., the radius of u is at least as large as the distance between u and v. In short, $(u, v) \in A_r \Leftrightarrow r_u \geq dist(u, v)$.
- Let $G_r = (S, E_r)$ be the *undirected communication graph* of range assignment r. Here, an edge $\{u, v\}$ is included in E_r iff u can send to v and v can send to u, i.e., $\{u, v\} \in E_r \Leftrightarrow \min\{r_u, r_v\} \geq dist(u, v)$. In other words, G_r contains exactly the antiparallel arcs of $\vec{G}_r$.

We can now define which kinds of requirements we might demand of a range assignment r, and define the according optimization problems, i.e. the problem to find a range assignment r with the respective property that has minimal $cost(r)$ among all range assignments satisfying this property.

- CONNECTIVITY (C): G_r must be connected.
- STRONG CONNECTIVITY (SC): $\vec{G}_r$ must be strongly connected.
- BROADCAST (B): given a source node s, $\vec{G}_r$ must contain an arborescence rooted at s.

One can also consider each of these problems with an additional integer parameter h, which indicates the maximal number of hops a message may make on its way from one station to another, i.e. there must be a path of length at most h from s (resp. all nodes) to each other. We do not address this problem here, which can also be regarded as setting $h = n - 1$.

In this context, the notion of *well-spread* instances in 2-d was defined in [7], and approximation results for constant h for these special instances were shown therein. Define

$$D(S) = \max\{dist(u, v) \mid u, v \in S\}$$
$$\delta_s(S) = \min\{dist(s, v) \mid v \in S \setminus \{s\}\}$$
$$\delta(S) = \min\{\delta_s(S) \mid s \in S\}$$

As in [7], we say that a family $\mathcal{S}$ of 2-dimensional instances is *well-spread* if there exists some positive constant c such that, for any $S \in \mathcal{S}$, $\delta(S) \geq cD(S)/\sqrt{|S|}$ holds. A possible generalization for other dimensions is to call a family $\mathcal{S}$ of d-dimensional instances well-spread if there exists some positive constant c such that, for any $S \in \mathcal{S}$, $\delta(S) \geq cD(S)/\sqrt[d]{|S|}$ holds. Orthogonal regular grids of full dimension are the prototypical well-spread instances. In the following, we sometimes omit the specific set of stations S if it is clear from the context which S is meant.

We are going to use reductions from Vertex Cover problems in graphs with low degree. A *vertex cover* of a graph $G = (V, E)$ is a subset $V' \subseteq V$ of its nodes that contains at least one endpoint of each edge $e \in E$. The VERTEX COVER problem (VC for short) is to find a vertex cover of minimum cardinality. VC is among Karp's original NP-complete problems [11]. It remains hard on various restricted graph classes: Let k-VERTEX COVER (or k-VC for short) denote the Vertex Cover problem on graphs with maximum degree k. k-VC is APX-hard for

$k \geq 3$ [14]. Planar VC remains NP-complete (in fact even Planar k-VC for $k \geq 3$) [9], but admits a PTAS [1]. In [2], Berman and Karpinski gave the first explicit inapproximability results for low degree VC. Quite some progress has been made since then, and we are going to use the inapproximability results for 3-VC, 4-VC and 5-VC by Chlebík and Chlebíková [4], which are (to our knowledge) currently the best such results for these problems.

1.2 Previous Work and Our Contribution

For an introduction to this subject, we refer the reader to the survey [6], which to our knowledge is still the current state of affairs on this subject, at least the hardness results presented there. In table 1, previously known and our results for these problems for certain ranges of α and d are listed. New results in this paper are listed in bold print. Note that in the literature, the distance-power gradient α is often implicitly assumed to be an integer. In this paper, we allow α to be any positive real number.

All APX-hardness results, both in [7] and this article, are obtained by reductions from low degree Vertex Cover problems. An entry ρ means that no ρ-approximation algorithm can exist for the respective optimization problem, unless P = NP. The actual numbers given here are obtained by combining our resp. the construction in [7] with the inapproximability results for 3-, 4- resp. 5-VC by Chlebík and Chlebíková [4].

The first NP-hardness result for 3-d Range Assignment Problems was given by Kirousis et al in [12]. Clementi et al [7] showed NP-hardness in 2-d and APX-hardness in 3-d of SC for $\alpha \geq 2$. It was noticed by Calinescu et al [3] that their constructions also work for C. In [5], these construcions were modified to prove NP-hardness of B.

In this paper, we prove NP-hardness of C and SC for all constants $\alpha > 0$, and APX-hardness of C, SC and B for all $\alpha > 1$. Note that B is trivially in P for $0 < \alpha \leq 1$, as setting $r_s = \max_{v \in S}\{dist(s, v)\}$ and $r_v = 0$ for $v \neq s$ is an optimal range assignment for this problem.

The case $\alpha = 1$ for SC, i.e. SC with the Euclidean distance, was formulated as an open problem in [6].[1] Another open question in the same survey was if there exists a PTAS for B in general dimension, which we answer negatively here by giving the first APX-hardness result for this problem in 3-d. When $\alpha \gtrsim 1$, the term for ρ in the APX-hardness result for SC remains larger than $1 + \frac{1}{702}$. We can also for the first time provide NP-hardness results for well-spread instances for these problems, and, for $\alpha > d$, the first APX-hardness results for these problems.

The paper is organized as follows: We first review the techniques of Garey and Johnson, and adapt their reduction for our purposes. In the sections thereafter, we describe the particular reduction for the different Range-Assignment problems that we address here.

---

[1] In the PhD-thesis of G. Rossi [15], it is stated without proof that the results in [7] can be applied also when $\alpha = 1$.

Table 1. List of previous and new results for different Range Assignment Problems

	$d=2$ (old)	$d \geq 3$ (old)	$d=2$ (new)	$d \geq 3$ (new)
C, $0 < \alpha \leq 1$	—	—	NP-hard (also for w.s.i.)	NP-hard (also for w.s.i.)
C, $\alpha > 1$	—*	—*	NP-hard (also for w.s.i.)	APX-hard ($\rho = 1 + \frac{1}{260}$)
C, $\alpha > d$				(also for w.s.i.)
SC, $0 < \alpha < 1$	—	—		NP-hard (also for w.s.i.)
SC, $\alpha = 1$	—		NP-hard (also for w.s.i.)	NP-hard (also for w.s.i.)
SC, $1 < \alpha < 2$	—	NP-hard [12]		APX-hard ($\rho = 1 + \frac{\sqrt{2}^{\alpha} - 1}{(7 - \sqrt{2}^{\alpha}) \cdot 52}$)
SC, $\alpha \geq 2$	NP-hard [7]	APX-hard [7] ($\rho = 1 + \frac{1}{495}$)	NP-hard (also for w.s.i.)	APX-hard ($\rho = 1 + \frac{1}{260}$)
SC, $\alpha > d$				(also for w.s.i.)
B, $1 < \alpha < 2$	—	—	NP-hard (also for w.s.i.)	APX-hard ($\rho = 1 + \frac{1}{50}$)
B, $\alpha \geq 2$	NP-hard [5]	NP-hard [5]	NP-hard (also for w.s.i.)	
B, $\alpha > d$				(also for w.s.i.)

(* NP-hardness for $\alpha \geq 2$ is implicit in [7].)

2   The Reduction of Garey and Johnson

Given a finite set of points S lying in the real plane $\mathbb{R}^2$, the Rectilinear Steiner Tree problem seeks to find a tree interconnecting S using only horizontal and vertical lines of shortest possible total length. In 1977, Garey and Johnson published the first proof of the NP-hardness of this problem [9]. They also needed to employ the result of Garey, Johnson and Stockmeyer that the planar version of Vertex Cover still remains NP-hard, which was published one year earlier [10].

The line of reductions in [9] is the following:

Planar VC $\rightarrow$ Planar 3-VC $\rightarrow$ Planar Connected 4-VC $\rightarrow$ Rectilinear Steiner Tree Problem in $\mathbb{R}^2$

where a connected vertex cover is a vertex cover whose node set induces a connected graph.

In the last reduction of this line, an orthogonal drawing of a planar graph with maximum degree 4 is needed; quite efficient methods for this have been developed in graph drawing. Roughly speaking, in an orthogonal drawing of a graph G, all vertices have integer coordinates, all edges are represented by orthogonal polylines (i.e. piecewise axis-parallel lines between points with integer coordinates), and polylines intersect only at their end-vertices.

Lemma 1. *It is possible to efficiently construct polynomial orthogonal draw-ings of planar graphs with maximum degree 4 in 2-d and arbitrary graphs with maximum degree 6 in 3-d with maximum edge-length $O(n)$.*

In fact, in [8] a 3-d orthogonal drawing with maximum edge length $O(\sqrt{n})$ is constructed, and further references in this field can be found therein.

Let us now take a closer look at the reduction from Planar 3-Vertex Cover to Planar Connected 4-Vertex Cover. Variations of this step will be included in all later proofs.

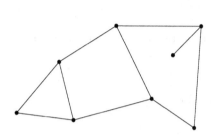

Fig. 1. An instance D of 3-VC

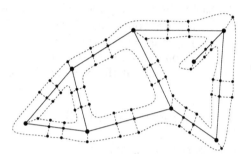

Fig. 2. The reduced Connected 4-VC in-stance $\bar{D}$. Backbone edges are dashed.

Maybe this construction is best explained by a picture. Figures 1 and 2, taken from [9], show a planar drawing D of an example instance of 3-VC and our reduced Connected 4-VC instance $\bar{D}$, which is almost but not entirely the same as D_{GJ}, which we call the graph constructed in [9]. $\bar{D}$ is constructed in the following way (for a rigorous proof, the reader is of course invited to refer to the original proof in [9]):

Let $D = (V, E)$ be a planar graph with maximum degree 3 with a fixed planar embedding. All through this paper, $n = |V|$ and $m = |E|$ are the number of vertices resp. edges of the original VC-instance D.

- First, split each edge $e = \{x, y\} \in E$ into three edges $\{x, x_e\}$, $\{x_e, y_e\}$ and $\{y_e, y\}$ by adding two new vertices x_e, y_e per edge. Call those new vertices C ("connectors") and the split edges E'. We call this intermediate graph $D' = (V \cup C, E')$.
- For each vertex $c \in C$, place one new vertex $b_{c,R}$ in each adjacent region R (one or two), and connect $b_{c,R}$ to c. For each vertex $v \in V$ of the original graph, place one new vertex $b_{v,R}$ in any neighboring region R, and connect $b_{v,R}$ to v.
- In each region R, connect all vertices $b_{\cdot,R}$ by a walk along the border of the region, like in figure 2. Collect the additional edges of this and the previous step in the set $\bar{E}$, and the nodes in the set B ("backbone").

Now we have described the construction of the planar graph $\bar{D} = (V \cup C \cup B, E' \cup \bar{E})$ with a fixed embedding.

The following lemma is implicit in the construction of Garey and Johnson:

Lemma 2. *D has a vertex cover of size $k \Leftrightarrow D'$ has a vertex cover of size $k+m$.*

The constructed graph $\bar{D}$ that we present here is in fact slightly different from the graph D_{GJ} constructed in [9]: In D_{GJ}, each backbone node $b \in B$ has a "spike", i.e., it is additionally connected to a copy of itself which has no other neighbor. The single purpose of these spikes is to ensure that, wlog., the whole set of backbone nodes B is included in every vertex cover of $\bar{D}$. Additionally, one connector per edge in D must lie in every vertex cover, providing connectivity of the backbone. As every node has a backbone neighbor, all feasible vertex covers for $\bar{D}$ are wlog. connected. So the size of a minimum vertex cover for D is k iff a minimum connected vertex cover of D_{GJ} has size $k + m + |B|$.

In the context of range assignments for radio stations, we will not need these spikes to ensure that the backbone is connected.

3 Hardness Results for CONNECTIVITY

3.1 NP-Hardness of CONNECTIVITY in 2-d

After this preparatory work, which is exactly the same as in [9] (except for leaving out the spikes), we complete the construction for our first Range Assignment Problem:

- As in [9], construct an orthogonal drawing of $\bar{D}$ in the plane.
- Scale the whole drawing by factor 3.
- Replace each line in the drawing by a set of equidistant points in the following way: Place one station at one end of the line, and:
  - For each polyline representing an edge originally in E', place stations on every point with integer coordinates, i.e. points at distance one.
  - For each line representing some part of a backbone edge (those in $\bar{E}$), place stations at distance $\frac{3}{4}$.

By scaling by a factor c we mean multiplying the coordinates of each point with c. By a *vertex (or node) station* we mean a station representing a node of the graph D'. By an *edge-end* we mean the last station on an edge before the vertex station.

Note that because of the scaling step, the first and last station on a straight line segment always have integer coordinates, and the minimum distance between two vertex stations is 3. See figures 3 and 4 for an illustration of this reduction.

Originally given a planar instance D of 3-Vertex Cover with a fixed embedding, we have constructed a blown-up version $\bar{D}$ and, by our last step, associated a set of points S in the plane with D. We now claim that a solution for the CONNECTIVITY problem for S also automatically yields a minimal vertex cover for D.

Theorem 1. *For any $\alpha > 0$, CONNECTIVITY in $\mathbb{R}^2$ is NP-hard.*

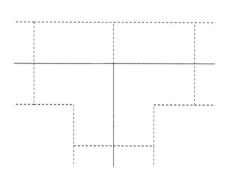

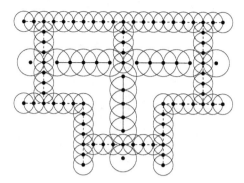

Fig. 3. A small part of an orthogonal drawing of a graph $\bar{D}$

Fig. 4. The resulting set of stations, with a minimal configuration and the induced communication graph

Before we begin to sketch the proof of this theorem, we formulate a simple lemma, which is fundamental in the hardness proofs for the CONNECTIVITY and STRONG CONNECTIVITY types of problems.

Lemma 3. *In a range assignment r for a set of stations S that satisfies the* CONNECTIVITY *or the* STRONG CONNECTIVITY *property, $r_s \geq \delta_s(S)$ must hold for all $s \in S$.*

Proof. If there was a station s with $r_s < \delta_s(S)$, it could not send data to any other station, meaning that there cannot be an arc/edge leaving s. So G_r cannot be connected, and $\vec{G}_r$ cannot be strongly connected, a contradiction.

As a direct consequence of Lemma 3, it makes sense to define a *minimal configuration* r_{min} in the following way: For all $s \in S$, set $r_{min}(s) = \delta_s$. A feasible range assignment r for CONNECTIVITY or STRONG CONNECTIVITY will satisfy $r \geq r_{min}$.

Proof. of Thm. 1: *(Sketch)* Let us look at the minimal configuration r_{min} for S: All stations on edges in E' have $r_{min} = 1$, and all stations on edges in $\bar{E}$, the backbone edges, have $r_{min} = \frac{3}{4}$. As all vertex stations have at least one adjacent edge from the backbone, all those stations also have $r_{min} = \frac{3}{4}$.

Observe that the undirected communication graph of the minimal configuration, $G_{r_{min}}$, already has quite large connected components (cf. fig. 4): There is one connected component corresponding to each edge in E', and the backbone is one connected component already. For notational convenience, we refer to a component corresponding to an edge $e \in E'$ as an *edge-component*. We also use the terms incident or adjacent with edge-components or vertex stations when we mean that the corresponding edges or vertices have this property.

Let $M = cost(r_{min})$ be the cost of the minimal configuration, and let k be the number of vertices in a minimal vertex cover for D. One has to argue that a minimal range assignment with property CONNECTIVITY has cost $M + \gamma(m + k)$,

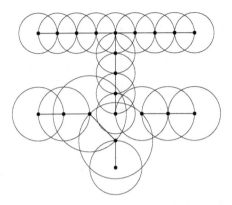

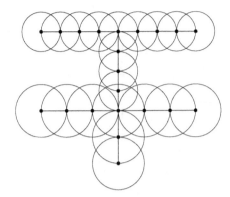

Fig. 5. An unwanted connection szenario

Fig. 6. The cheapest way to connect edges. Automatically, all incident edges are attached.

where $\gamma = 1 - \left(\frac{3}{4}\right)^{\alpha}$. Roughly speaking, one has to convince him-/herself that it is always best to use vertex stations to get edge components connected to the backbone (Fig. 6), and in particular, it never pays to connect adjacent edge components directly (Fig. 5).

3.2   NP-Hardness for Well-Spread Instances

Notice that the construction above is already "nearly" well-spread: The minimal distance is a constant. So each set of stations on a straight line segment already is a well-spread instance in $\mathbb{R}$. But for S being well-spread (in $\mathbb{R}^2$), there may be "too few" points in S. In order to ensure that the construction is well-spread we fill it up in such a way that the additional stations play no role in the reduction, i.e. they are never increased above their minimal distance in a minimal solution, and so only contribute to the cost of the minimal configuration.

To do so, we use the following trick: By construction, the outer face is bounded completely by backbone edges, all lying on a grid with mesh distance $\frac{3}{4}$. We now simply "fill up" the outer face with stations of this grid in such a way that the outer stations of this grid form a square of some desired size. For instance, twice the maximum of the width and height of the drawing would suffice to produce well-spread instances with constant $c \geq \frac{3}{4}\frac{1}{\sqrt{2}}$. Obviously, in a minimal configuration the additional vertices all have radius $\frac{3}{4}$ and all are in the same connected component as the backbone, and it does not make sense to increase any of their radii. It is obvious how to fill up such a well-spread square to a well-spread cube.

Theorem 2. *For any $\alpha > 0$, CONNECTIVITY on well-spread instances in $\mathbb{R}^2$ and $\mathbb{R}^3$ is NP-hard.*   □

3.3   APX-Hardness of CONNECTIVITY in 3-d

The construction in the NP-hardness proof for CONNECTIVITY is far from being approximation-preserving: The fixed cost M of the minimal configuration is far larger than the variable cost of the vertex cover. More precisely, M would have to be bounded by some (preferably low) constant factor times the number of vertices n. To the best of our knowledge, no orthogonal 3-d graph drawing method is known that uses only $O(n)$ total length, so we cannot hope to achieve this goal with this construction when $\alpha \leq 1$. However, the situation changes when $\alpha > 1$: Because the power function is now strictly convex, smaller radii cost far less than big radii. This means we can make the power of the internal radii on the edges negligible by inserting a large number of stations on every edge at a very small distance. A polynomial number of stations suffices. This fact has already been exploited by Clementi et al in their constructions [7], and is not proved here.

Lemma 4. *For $\alpha > 1$, we can achieve that for the total power consumption of a minimal configuration, we have $M = r_{min} < c$ for any constant $c > 0$ with a polynomial number s of stations.*

Before we begin with the construction, we state the results of Chlebík and Chlebíková [4] for the non-approximability of 3–, 4– and 5–Vertex-Cover, which are to our knowledge currently the best results for these problems.

Lemma 5 (Chlebík and Chlebíková, [4]). *It is NP-hard to approximate the solutions of:*

- *3–Vertex Cover to within $1 + \frac{1}{99}$,*
- *4–Vertex Cover to within $1 + \frac{1}{52}$,*
- *5–Vertex Cover to within $1 + \frac{1}{50}$,*

even on 3-, 4- resp. 5-regular graphs.

Given a low-degree instance of Vertex Cover $D = (V, E)$, we describe now how to build the graph $\bar{D}$ which later gets drawn in the Euclidean space $\mathbb{R}^3$. Note that we cannot use our reduction to prove APX-hardness of Range Assignment Problems in 2-d, because Planar Vertex Cover is not APX-hard, but there exists a PTAS for this problem [1]. This time, as we do not have to observe planarity, the construction of the backbone becomes very simple: Let the vertices of V be given in some arbitrary order $V = \{v_1, \ldots, v_n\}$. The backbone vertices B contain one copy of each original vertex: $B = \{v'_1, \ldots, v'_n\}$, and the backbone edges consist of one edge between each original node and its copy, and a cycle through all backbone nodes: $\bar{E} = \{\{v_i, v'_i\} \mid 1 \leq i \leq n\} \cup \{\{v'_i, v'_{i+1}\} \mid 1 \leq i \leq n - 1\} \cup \{\{v'_n, v'_1\}\}$. Call $\bar{D} = (V \cup B, E \cup \bar{E})$. Given a constant $0 < \varepsilon < 1$, choose s according to Lemma 4 and construct a polynomial set of stations S in the following way:

- Construct an orthogonal drawing of $\bar{D}$ in $\mathbb{R}^3$.
- Scale the drawing by factor 3.

- For all polylines representing original edges $e \in E$, remove the first and last open unit interval of the polyline (i.e. do not erase any integer points).
- Replace all remaining unit line segments with $s + 1$ stations along this line at distance $1/s$.

Here, the scaling step is needed to ensure that at least one length unit of each edge remains. Note that when the maximum degree in D is Δ, the maximum degree of $\bar{D}$ will be $\Delta + 1$. So according to Lemma 1, we must have $\Delta \leq 5$. We use this set of stations in order to prove

Theorem 3. *For any $\alpha > 1$, it is NP-hard to approximate* CONNECTIVITY *in* $\mathbb{R}^3$ *within* $1 + \frac{1}{260}$. *For $\alpha > d \geq 3$, approximating* CONNECTIVITY *within* $1 + \frac{1}{260}$ *remains NP-hard even when restricted to well-spread instances.*

Proof. The components of the minimal configuration are still the backbone and the original edges. Additionally to the vertex stations of a VC in D, one edge-end per edge has to have radius 1. As the VC is at least as big as m/Δ, we have an L-reduction with factor $\Delta + 1$. Setting $\Delta = 4$ produces the claimed result.

We can embed a reduced instance in a cube, containing a grid of small enough mesh distance. One can check that for $\alpha > d$, the power consumption of such a cube can be made arbitrarily small using a polynomial number of stations.

4   Hardness Results for STRONG CONNECTIVITY

In the remainder of this article, we will adapt our reductions for CONNECTIVITY for the STRONG CONNECTIVITY and BROADCAST problems. In fact, the reduction for NP-hardness will be exactly the same as for CONNECTIVITY, but the proof will slightly differ. The main difference is that now directed links are established already when one of the two stations has a large enough radius. Despite this difference, for $\alpha \geq 1$ the proof is very similar to the connected case. For $\alpha < 1$, some additional scaling will be needed.

Theorem 4. *For any $\alpha > 0$,* STRONG CONNECTIVITY *in* $\mathbb{R}^2$ *is NP-hard, already for well-spread instances.*

Concerning APX-hardness, the only difference to the construction for CONNECTIVITY is that now the outgoing arc of an edge does not have to be parallel to the ingoing arc. If some edge-end is increased to 1, and the incident vertex is not, it could indeed be cheaper to increase the border station further to $\sqrt{2}$ and send to another adjacent edge. This could indeed be cheaper if $\alpha < 2$, and as this is the only thing we have to worry about, let us assume in the following that $1 < \alpha < 2$.

We will now present a slightly changed reduction: To the original Vertex Cover instance $D = (V, E)$, no new vertices but only new edges will be added. For vertex set $V = \{v_1, \ldots, v_n\}$, add a directed Hamiltonian cycle $\bar{E} = \{(v_i, v_{i+1}) \mid 1 \leq i \leq n - 1\} \cup \{(v_n, v_1)\}$ as the backbone, already completing the construction of $\bar{D} = (V, E \overset{.}{\cup} \bar{E})$. The direction of the added arcs is needed only for notational convenience later on. Note that $\bar{D}$ may contain parallel backbone and original edges, which is not a problem.

The construction is now similar to the one before; the new thing is we also erase some part of backbone lines. We begin with the same construction as in Theorem 3, and choose $\beta = \sqrt[\alpha]{2 - \sqrt{2}^{\alpha}}$. We additionally erase the first[2] open interval of length β on every backbone edge. This ensures that all vertex stations have, wlog., at least radius β. The parameter β is chosen so that it is no longer cheaper to use non-vertex stations to connect components. Using this set of stations, one can prove

Theorem 5. *It is NP-hard to approximate* STRONG CONNECTIVITY *in* $\mathbb{R}^3$ *within* $1 + \frac{1}{260}$, *if* $\alpha \geq 2$, *and within* $1 + \frac{\sqrt{2}^{\alpha} - 1}{(7 - \sqrt{2}^{\alpha}) \cdot 52}$, *if* $1 < \alpha < 2$.

5   Hardness Results for BROADCAST

As the optimal solution when $0 < \alpha \leq 1$ is to have s directly broadcast to all stations, and all other stations have radius 0, we assume $\alpha > 1$. To prove NP-hardness, we use the Garey-Johnson backbone construction, draw it in the plane, and replace the lines with stations at a small enough distance as in the APX-type construction. This ensures that the overhead is small enough without arguing via a minimal configuration. APX-hardness is proven by the same construction as for APX-hardness of CONNECTIVITY. Exactly the vertex nodes of a vertex cover will have a significant radius.

Theorem 6. *For any* $\alpha > 1$, BROADCAST *in* $\mathbb{R}^2$ *is NP-hard, already for well-spread instances.*

Theorem 7. *For any* $\alpha > 1$, *approximating* BROADCAST *in* $\mathbb{R}^3$ *better than* $1 + \frac{1}{50}$ *is NP-hard. For* $\alpha > 3$, *approximating* BROADCAST *in* $\mathbb{R}^3$ *within* $1 + \frac{1}{50}$ *remains NP-hard even when restricted to well-spread instances.*

6   Open Problems and Acknowledgements

It is still an intriguing question if the above problems remain APX-hard in the plane or allow a PTAS. A similar reduction as in this paper from a planar APX-hard problem might produce such a result.

An interesting question in Graph Drawing would be whether graphs of low degree could be drawn with total length in $O(n)$ in 3-d. If so, this would imply APX-hardness of C and SC in 3-d even for $\alpha = 1$. On the other hand, a PTAS for this case would imply that such drawings cannot exist, unless P = NP.

The author would like to thank Walter Kern for suggesting this problem, Dominique Andres and Christoph Buchheim for helpful discussions, Therese Biedl for reference [8] and Marek Karpinski for reference [4].

---

[2] i.e. the interval starting at the tail of e.

References

1. B. Baker, *Approximation Algorithms for NP-Complete Problems on Planar Graphs*, Journal of the ACM **41** (1): 153–180, 1994.
2. P. Berman and M. Karpinski, *On Some Tighter Inapproximability Results*, in: Proc. 26th ICALP, pp. 200–209, 1999. Also available as ECCC Report TR98-065 at http://eccc.uni-trier.de/eccc/.
3. G. Călinescu, I. Măndoiu and A. Zelikovsky, *Symmetric Connectivity with Minimum Power Consumption in Radio Networks*, Proc. IFIP TCS **223**, pp. 119–130, 2002.
4. M. Chlebík and J. Chlebíková, *Inapproximability results for bounded variants of optimization problems*, Proc. FCT (2003), LNCS **2751**: 27–38, 2003. Also available as ECCC Report TR03-026.
5. A. Clementi, P. Crescenzi, P. Penna, G. Rossi and P. Vocca, *A Worst-Case Analysis of an MST-based Heuristic to Construct Energy-Efficient Broadcast Trees in Wireless Networks*. Technical Report 010, University of Rome "Tor Vergata", Math. Department, 2001.
6. A. Clementi, G. Huiban, P. Penna, G. Rossi and Y. Verhoeven, *Some Recent Theoretical Advances and Open Questions on Energy Consumption in Ad-Hoc Wireless Networks*, in: Proc. 3rd Workshop on Approximation and Randomization Algorithms in Communication Networks (ARACNE): 23–38, 2002.
7. A. Clementi, P. Penna and R. Silvestri, *On the Power Assignment Problem in Radio Networks*, Mobile Networks and Applications **9** (2): 125–140, April 2004. Also available as ECCC Report TR00-054.
8. P. Eades, C. Stirk and S. Whitesides, *The techniques of Kolmogorov and Bardzin for three-dimensional orthogonal graph drawings*, Information Processing Letters **60** (2): 97–103, 1996.
9. M. Garey and D. Johnson, *The Rectilinear Steiner Tree Problem is NP-Complete*, SIAM Journal of Applied Mathematics **32** (4): 826–834, 1977.
10. M. Garey, D. Johnson and L. Stockmeyer, *Some simplified NP-Complete Problems*, Theoretical Computer Science **1**: 237–267, 1976.
11. R. Karp, *Reducibility among combinatorial problems*, in: R. Miller and J. Thatcher (eds.), Complexity of Computer Computations, 85–103, 1972.
12. L. Kirousis, E. Kranakis, D. Krizanc and A. Pelc, *Power consumption in packet radio networks*, Theoretical Computer Science **243**: 289–305, 2000.
13. K. Pahlavan and A. Levesque, *Wireless Information Networks*, Wiley-Interscience, 1995.
14. C. H. Papadimitriou and M. Yannakakis, *Optimization, approximation, and complexity classes*, Journal of Computer and System Sciences **43**, 425–440, 1991.
15. G. Rossi, *The Range Assignment Problem in Static Ad-Hoc Wireless Networks* (PhD-thesis), University of Siena, 2003.

Black Hole Search in Asynchronous Rings Using Tokens

S. Dobrev[1], R. Královič[2], N. Santoro[3], and W. Shi[3]

[1] School of Information Technology and Engineering, University of Ottawa,
Ottawa, K1N 6N5, Canada
[2] Dept. of Computer Science, Comenius University, Mlynska dolina,
84248 Bratislava, Slovakia
[3] School of Computer Science, Carleton University,
Ottawa, K1S 5B6, Canada

Abstract. A *black hole* is a highly harmful host that disposes of visiting agents upon their arrival. It is known that it is possible for a team of mobile agents to locate a black hole in an asynchronous *ring* network if each node is equipped with a *whiteboard* of at least $O(\log n)$ dedicated bits of storage. In this paper, we consider the less powerful *token model*: each agent has has available a bounded number of tokens that can be carried, placed on a node or removed from it. All tokens are identical (i.e., indistinguishable) and no other form of communication or coordination is available to the agents. We first of all prove that a team of two agents is sufficient to locate the black hole in finite time even in this weaker coordination model. Furthermore, we prove that this can be accomplished using only $O(n \log n)$ moves in total, which is optimal, the same as with whiteboards. Finally, we show that to achieve this result the agents need to use only $O(1)$ tokens each.

1 Introduction

1.1 The Framework

Whereas exploration problems by mobile agents have been extensively studied in the context of *safe* networks, the reality of networked systems supporting mobility agents is that these systems are highly *unsafe*. Indeed, the most pressing concerns are all about security issues and mainly in regards to the presence of a *harmful host* (i.e., a network node damaging incoming agents) or of a *harmful agent* (e.g., a mobile virus infecting the network nodes) [2, 10, 11, 13, 14].

The computational and algorithmic research has just recently started to consider these issues. The computational issues related to the presence of a harmful agent have been explored in the context of intruder capture and network decontamination; in the case of harmful host the focus has been on the *black hole* (BH), a node that disposes of any incoming agent without leaving any observable trace of this destruction [3, 5, 6, 7, 12]. In this paper, we continue the investigation of the black hole search problem.

T. Calamoneri, I. Finocchi, G.F. Italiano (Eds.): CIAC 2006, LNCS 3998, pp. 139–150, 2006.
© Springer-Verlag Berlin Heidelberg 2006

As mentioned, a *black hole* is a network site that disposes of any incoming agent without leaving any observable trace of this destruction. It models e.g. a node where a resident process (e.g., an unknowingly installed virus) deletes visiting agents or incoming data; furthermore, any undetectable crash failure of a site in an asynchronous network transforms that site into a black hole. In presence of a black hole, the first important goal is to determine its location. To this end, a team of mobile system agents is deployed; their task is completed if, within finite time, at least one agent survives and knows the links leading to the black hole. The research concern is to determine under what conditions and at what cost mobile agents can successfully accomplish this task, called the *black hole search* (BHS) problem. The main complexity parameter is the *size* of the team; i.e., the number of agents used in the search. Another important measure is the amount of *moves* performed by the agents in their search.

Both solvability and complexity of BHS depend on a variety of factors, first and foremost on whether the system is *asynchronous* [4, 5, 6, 7] or *synchronous* [3, 12]. Indeed the nature of the problem changes drastically and dramatically. For example, both in synchronous and asynchronous systems, with enough agents it is possible to locate the black hole if we are aware of its existence; however, if there is doubt on whether or not there is a black hole in the system, in absence of synchrony this doubt can *not* be removed. In fact, in an asynchronous system, it is *undecidable* to determine if there is a black hole [6]. The consequences of this fact are numerous and render the asynchronous case considerably difficult. In this paper we continue the investigation of the asynchronous case.

Other important factors influencing solvability and complexity are the amount of a priori knowledge held by the agents (e.g., number n of nodes, map of network, etc.) and the means offered by the system for agents communication and coordination (e.g., whiteboard, blackboard, reliable message passing, etc). In particular, all existing investigations on BHS in asynchronous systems have assumed the presence of a powerful inter-agent communication mechanism, *whiteboards*, at all nodes. In the whiteboard model, each node has available a local storage area (the whiteboard) accessible in fair mutual exclusion to all incoming agents; upon gaining access, the agent can write messages on the whiteboard and can read all previously written messages. This mechanism can be used by the agents to communicate and mark nodes or/and edges, and has been commonly employed in several mobile agents computing investigations (e.g. see [1, 8, 9]).

Although many research questions are still open, the existing investigations have provided a strong characterization of the asynchronous BHS problem using whiteboard. In particular, it is possible for two agents with a map of the network to determine the location of the black hole in any bi-connected graph[1] using $O(n \log n)$ moves, provided there are $O(\log^2 n)$ bits whiteboards [6].

In this paper we consider an asynchronous *ring* network. The ring is the sparsest bi-connected graph and the one for which the cost (in terms of number of moves) for black hole search with whiteboards is the worst: $\Omega(n \log n)$. Using

---

[1] Edge bi-connectivity is required for BHS in asynchronous systems [6].

quite a different protocol, two agents can however locate the black hole with $O(n \log n)$ moves using $O(\log n)$ bits whiteboards [5].

We consider ring networks in the less powerful *token* model, often employed in the exploration of safe graphs. In this model, each agent has available a bounded number of tokens that can be carried, placed on a node or on a port or removed from it. All tokens are identical (i.e., indistinguishable) and no other form of communication or coordination is available. Some natural questions immediately arise: is the BHS problem is still solvable with this weaker mechanism, and if so under what conditions and at what cost. Notice that the use of tokens introduces another complexity measure: the number of tokens. Indeed, if the number of tokens is unbounded, it is possible to simulate a whiteboard environment; hence the question immediately arises of how many tokens are really needed.

1.2   Our Results

The network under consideration is an asynchronous ring network with a black hole. In such a network, in presence of whiteboards, the black hole search problem can be solved with team of just *two* agents, and performing only $\Theta(n \log n)$ moves [5]. We consider the same topology and examine the BHS problem using tokens.

We first of all prove that a team of *two* agents is sufficient to locate the black hole in finite time even in this weaker coordination model. Furthermore, we prove that this can be accomplished using only $O(n \log n)$ moves in total, which is optimal, the same as with whiteboards. Finally, we show that the agents need to use only $O(1)$ tokens. These results are established constructively: we do present protocols that allow a team of two agents to correctly locate the BH with that number of moves and with those few tokens. The first protocol uses a total of ten tokens, while in the second that number is reduced to 3.

Hence we show that, although tokens are a weaker means of communication and coordination, their use does not negatively affect solvability and it does not even lead to a degradation of performance. On the contrary, whereas the protocols using whiteboards assumed at least $O(\log n)$ dedicated bits of storage at each node, the ones proposed here use only three tokens in total.

2   Model and Basic Tools

2.1   The Model and Basic Observations

Let $\mathcal{R}$ be a anonymous ring of n nodes (i.e. all the nodes look the same, they do not have distinct identifiers). Operating on $\mathcal{R}$ is a set of k agents $a_1, a_2, ..., a_k$. The agents are *anonymous* (do not have distinct identifiers), *mobile* (can move from a node to a neighbouring node) and *autonomous*; each has computing and limited memory capabilities ($O(\log n)$ bits suffice for all our algorithms). All agents have the same behaviour, i.e. follow the same protocol, and start at the same node (however, they may start at different and unpredictable times), called *homebase* (HB for brevity). The agents do not know k, nor do they know how many agents have been awaken before. Since all agents start at the same node, we can assume that the ring is *oriented*, i.e. all ports are labelled left and right consistently in the whole ring[2].

The agents can interact with their environment and with each other only through the means of *tokens*. A token is an atomic entity that the agents can see, place it in the middle of a node or on a port and/or remove it. Several tokens can be placed on the same place. The agents can detect the multiplicity, but the tokens themselves are undistinguishable from each other. Initially, there are no tokens placed in the network, and each agent starts with some fixed number of tokens (depending on the algorithm).

The basic computational step of an agent (executed either when the agent arrives to a node, or upon wake-up) is to *examine* the node (returns a triple of non-negative integers - multiplicity of tokens at the middle of the node, on the right port and on the left port, respectively), *modify* the tokens (by placing/removing some of the tokens at the current node) and either *fall asleep* or *leave* the node through either left or right port. The whole computational step is performed as an atomic action, i.e. as if it took no time to execute it.

The computation is asynchronous in the sense that the time an agent sleeps or is on transit is finite but unpredictable. The links obey FIFO rule, i.e. the agents do not overtake each other when traveling over the same link in the same direction.

Note that the tokens are the only means of inter-agent communication we consider. There is no read/write memory (whiteboards) in the nodes the agents can access, nor is there face-to-face communication. In fact, the agents do not even need to be capable of seeing each other - they only see the tokens.

One of the nodes of the ring $\mathcal{R}$ is highly harmful – it disposes of every agent that enters it, without leaving any trace of this destruction observable from the outside. Due to this behaviour, this node is called *Black Hole* (or BH for brevity). All the agents are aware of the presence of the BH, but at the beginning the location of the BH is unknown. The goal is to locate the BH, i.e. at the end there must be at least one agent that has not entered the BH and knows the location of the BH.

The primary complexity measure is *size*: the number of agents needed to locate the BH. Other complexity measures we are interested in are *token size*: the number of tokens each agent starts with, *cost*: the total number of moves executed by the agents (worst case over all possible timings) and *time*: the time it takes to locate the black hole, from the moment the *second*[3] agent wakes-up until the black hole is located, assuming transiting an edge takes at most one time step.

---

[2] The agents can simply use the notions of left/right according to the port labelling at the HB, remembering as they move how does the labelling at the current node relate to the labelling at the HB.

[3] The first agent might immediately enter the black hole, and the second agent might wake-up arbitrarily late, resulting in unbounded time complexity if we measure from the time the first agent wakes up.

Since the first move of an agent can end up in the BH, we immediately get:

Lemma 1. *[5] At least two agents are needed to locate the black hole.*

Because of the asynchrony, the agents can not distinguish a slow node from the BH. From this we get:

Lemma 2. *[5] It is impossible to find the Black Hole if the size of the ring is not known.*

2.2   Basic Tools and Techniques

Cautious Walk with Token. At any moment of execution, the ports can be classified as *unexplored* – no agent has exited or arrived via this port yet, *dangerous* – an agent has exited via this port, but no agent has arrived yet via it, or *safe* – an agent has arrived via this port.

As our primary complexity measure is the number of agents needed, we aim to prevent unnecessary agent disappearances by making sure that no two agents enter the BH over the same link. This is achieved by forbidding the agents to leave a node via a dangerous port. This means the agents have to mark the dangerous ports. In order to facilitate progress, the agents are also required to remove the dangerous marks whenever they learn that a port marked dangerous in fact leads to a non-BH node: When an agent that left dangerous mark on a port of node u arrives at node v, it immediately returns to u, and removes the dangerous mark to signal that the link from u to v has became *safe*. Afterwards (provided it is not interrupted by a message in u) the agent returns to v and proceeds from there.

This technique has been introduced in [5] and named *Cautious Walk*. In [5] the whiteboards have been used to mark dangerous and safe ports. In this paper, the tokens placed on a port (one or two, depending on context) mean the port is dangerous (however, there will be cases where tokens might appear also on non-dangerous ports); by default, there are no tokens on safe and unexplored ports and the agents distinguish them from the context. (As we aim to use $O(1)$ tokens overall, while $O(n)$ ports will eventually become safe, we cannot afford to mark the safe ports by tokens.)

Elimination Technique. In order to minimize the *cost*, i.e. the total number of moves executed by all agents, we use elimination technique to effectively reduce the number of active agents to two – the first two agents to wake-up. The idea is to have the first two agents mark the HB by their tokens, the agents waking up later check the marks at the HB and become passive if they see there are already two agents active.

3   Algorithm *Divide with Token*

3.1   General Description and Ideas

A node is called *explored* if it has been visited by an agent, *unexplored* otherwise. The set of the explored nodes is called *explored region*; the part of the ring consisting of unexplored nodes the *unexplored region*. As all agents start at the HB, the explored region (and thus also the unexplored region) is connected. The two extremal explored nodes having an unexplored neighbour are called Last-Safe-Place (LSP) (at the very beginning, the HB is the sole LSP). The LSPs keep changing while the explored region is getting larger and larger.

The main technique for locating the BH is borrowed from [5]: The two agents logically partition the unexplored region into two connected parts of (almost) equal size, and then each agent goes to explore its part using Cautious Walk. Since there is only one BH, exactly one agent (say agent a) will finish exploring its part. The agent a then traverses the explored part until it reaches the LSP of agent b. At this moment, a knows the distance between LSPs, i.e. the size of the explored part. Since n is known, a also knows the size of the unexplored part. If the unexplored part consists of a single node v, a determines that the BH is located at v and terminates the algorithm (b has already been terminated by entering the BH). Otherwise, a logically divides the remaining unexplored region into two connected, almost-equally sized work assignments W_a and W_b for a and b, such that W_b contains the node to which b is currently heading. Afterwards, a leaves a message for b informing it about W_b and goes on to explore W_a. The process is repeated until the unexplored region contains single node – the BH.

The above general description omits details of how the agents communicate – the format of the message, identifying the LSP, implementing Cautious Walk. In [5] whiteboards were used to store the message (and the status of the ports). In this paper, we aim to implement the above approach using only tokens, and in fact using three tokens in total.

The first step in this direction is to use tokens only for dangerous ports, not for safe or unexplored ones. This can be easily achieved, as all ports between the dangerous ports leaving from LSPs are implicitly safe. The second step – using tokens to encode the message is trickier. Note that it is sufficient to leave as a message for b not the size of W_b, but the number of times a has finished its part before b has made any progress. b can re-compute the size of its part by that many times halving the size of its last work assignment (and being careful to use the same rounding as a used when the unexplored area had odd size). This means the message needs only to be of size at most $\log n$ and allows a to update its message by simply incrementing it[4].

The basic technique we use is to encode the message by the distance between the LSP of b and a token (*end-of-message* marker) left by a in the explored area. In fact, in order to minimize the number of tokens used, a leaves a message x by moving b's cautious walk token x nodes in the direction opposite to b's exploring. This way, when/if b returns to move its token and does not find it there, it knows there is a message waiting for it. The basic technical difficulty

---

[4] An alternative approach of leaving W_b has synchronization problems: Updating W_b means making it smaller, which would be unsafe if the other agent is reading the message in the same time. This can be solved by leaving $n/2 - W_b$ as a message, as W_b is always at most $n/2$.

lies in the fact that the tokens are undistinguishable and seeing a token might mean very different things, depending on the context. Careful case analysis is applied, moreover the algorithm crucially relies on the FIFO requirement and the atomicity of actions (so the agents cannot overtake each other on a link or in a node).

3.2   Detailed Description

The following variables(local to the agent) are maintained by each agent throughout the execution:

- *Steps* – the size of the remaining work assignment
- *HBpos* and *LSPpos* – relative position with respect to the HB and the LSP of the agent; this allows the agent to know when it is at the HB or its LSP without using tokens for marking them
- *MsgILeft* – the last message left by this agent
- *DistC* – used to calculate the distance between LSPs
- *Msg4Me* – the message left for me by the other agent

The tokens are used to mark dangerous ports during cautious walk (a token on a port), as *end-of-message* marks (a token in the middle of a node), and in the HB to limit the number of active agents to two. As the number of possible situations at the HB is quite high, the following careful encoding at the HB is used to limit the number of tokens used: (the triplet represents the number of tokens on the left port, middle, and right port of the HB, respectively):

- $(0,0,2)$ – the right port is dangerous (i.e. the HB is the LSP of the right agent), the left port is unexplored (the second agent has not woken-up yet)
- $(0,0,1)$ – the right port is safe (the LSP of the right agent already moved to the right), the left port is unexplored
- $(2,1,0)$ – both the right and the left port are dangerous, there is a message 0 waiting for the right agent (the second agent woke up before the first agent explored its first port, in the initial step it would have changed $(0,0,2)$ to $(2,0,1)$, but then immediately in SEEKING transformed that to $(2,1,0)$),
- $(1,0,1)$ – the right port is dangerous, the left port is safe
- $(2,0,0)$ – the left port is dangerous, the right port is safe
- $(1,0,0)$ – both ports are safe
- $(1,1,0)$ – if seen by the right agent whose LSP is the HB: the left port is safe, there is message 0 waiting for me; if seen by the left agent whose LSP is the HB: the right port is safe, there is message 0 waiting for me
- other configurations do not occur

Note that although configuration $(1,1,0)$ has two possible meanings, there will never be confusion as it does not occur in the case where the HB is LSP for both agents (at least one agent must have finished its assignment in order to leave a message).

The algorithm is described for the right agent a. The algorithm for the left agent b is almost identical, the only differences are using opposite directions, and using the floor function instead of ceil when calculating the work assignment.

In the initialization step, the elimination technique is used to limit the number of active agents to two, as well as to choose the right agent a and the left agent b. Note that unlike in [5], the initial work assignment of the first agent is the whole unexplored part – we do not want to deal with the case that the first agent explored its part, while the second agent has not started the algorithm yet. When the second agent starts the algorithm, it will seek the first agent to divide the workload based on what remains unexplored at that moment.

The procedure Checking is executed by an agent that finds a notice that a message has been left for it. The agent reads the message by traversing leftward until an end-point marker of the message (a token in the middle of a node) is found. Note that for the first message this means zero leftward moves. Moreover, a message is left only when both agents are active, therefore there is no confusion if the end-of-message marker is found at the HB.

Algorithm 1. INITIALIZATION and CHECKING

1: INITIALIZATION:(upon initial wake-up in the HB)
2: **if** the right port has no token on it **then**
3:     put two tokens on the right port
4:     execute procedure EXPLORE$(n-1)$ as the right agent a
5: **else if** the left port has no token on it **then**
6:     move one token from the right port to the left port and add additional token to the left port
7:     execute procedure SEEKING() as the left agent b
8: **else** become Passive immediately
9: **end if**

10: **procedure** CHECKING
11:     go left until a token is found in the middle of a node u, counting in *Msg4Me* the number of steps
12:     remove the token, return to your LSP and put the token on the right port
13:     **for** $(i = 0; i \leq Msg4Me; i++)$ **do**   // compute the new work assignment
14:         $Steps = \lceil Steps/2 \rceil$ // again floor by the left agent
15:     **end for**
16:     execute EXPLORE(*Steps*)
17: **end procedure**

In procedure EXPLORE an agent explores its work assignment using Cautious Walk, checking in the progress for messages from the other agent.

---

[5] The number of tokens there might have changed from 2 to 1 if u is the HB and the second agent had waken-up meanwhile, but that still does not mean a message notification. Note that this is a place where the code for the left agent is not simple translation of the code for the right agent. The test by the left agent will be: there is a token on the left port of u, or u is the HB and there are two tokens on the left port of u.

Algorithm 2. EXPLORE, SEEKING and CHECK&SPLIT

1: **procedure** EXPLORE(*Steps*)
2:    **while** true **do**
3:        // *might enter the* BH *in this step*
4:        go from the current node u to its right neighbour v
5:        return back to node u // *doing the Cautious Walk here*
6:        **if** there is a token on the right port of u^5 **then**   // *no message for me yet*
7:            remove one token from the right port of u
8:            move to v, put a token on the right port of v and decrement *Steps*.
9:            **if** *Steps* $= 0$ **then**
10:               // *finished exploring my assignment, now find the other agent*
11:               exit the loop and execute procedure SEEKING()
12:           **end if**
13:       **else** // *there is no token on the right port, i.e. message waiting for me*
14:           exit the loop and execute procedure CHECKING()
15:       **end if**
16:   **end while**
17: **end procedure**

18: **procedure** SEEKING
19:    go left until a token is found at node u, counting in *DistC* the distance travelled
20:    **if** found a token on the left port of a non-HB node, or two tokens on the left port of the HB **then**
21:        // u *is the LSP of the other agent, leave a message* 0
22:        move a token from the left port to the middle of u
23:        execute CHECK&SPLIT(*DistC*)
24:    **else if** found one token in the middle of a node u **then**
25:        // *this is the endpoint of my last message, agent b did not read it yet*
26:        // *update/increment the message*
27:        move the token one step to the right and increment *MsgILeft*
28:        execute CHECK&SPLIT(*DistC* + *MsgILeft* -1)
29:    **else if** found one token on the left port of the HB **then**   // HB, *ignore*
30:        ignore and continue on l.19 as if nothing found
31:    **end if**
32: **end procedure**

33: **procedure** CHECK&SPLIT(*Dist*)
34:    **if** *Dist* $= n - 2$ **then**   // *single unexplored node remaining*
35:        the BH is in the remaining unexplored node, **terminate**
36:    **else**
37:        return to your LSP and execute Explore($\lceil (n - Dist)/2 \rceil$)
38:    **end if**
39: **end procedure**

In procedure SEEKING the agent determines the distance between the LSPs and either locates the BH (if there is single unexplored node remaining) or leaves/updates the message for the other agent.

3.3   Correctness and Complexity Analysis

Lemma 3. *At most two agents will become active and put a token somewhere.*

Proof. By construction: The second agent to wake-up places two tokens on the left port of the HB (line 6 of INITIALIZATION) and one of these tokens stays there for the whole execution. Due to line 8 no other agent becomes active.

Lemma 4. *If there is a message left for an agent, the agent detects the presence of the message and correctly computes its contents.*

Proof. We prove the lemma for the agent a exploring to the right. The proof for agent b is analogous. When a finds out that there is a message waiting for it (by not finding its token when returning from cautious walk, lines 13-14 of EXPLORE), according to line 11. of CHECKING it travels to the left to locate the end-of-message token placed in the middle of a node. Due to FIFO and atomicity, it cannot overtake the agent b laying the message, i.e. it will always find the end-of-message token (even if b is concurrently incrementing the message). Moreover, the searching for the end-of-message token is safe, i.e. a will not travel past b's LSP: A non - 0 message is left only by an agent that already explored its assignment, i.e. the distance between LSPs must be at least $\lfloor n/2 \rfloor$.

Let us define the *work assignment* of the right/left agent as follows:

- If the agent is traversing the dangerous link from its LSP and there is a message waiting for it, the work assignment of the agent is only the node on the other side of the dangerous port
- Otherwise, the work assignment of the agent is the *Steps* nodes to the right/left of agent's LSP.

Lemma 5. *At any moment, the work assignments of the right and left agents are disjoint. Moreover, if there is no message waiting for an agent then the work assignments form a partition of the part of the ring delimited by the LSPs (or the LSP of the right agent and the HB, if there is single agent active) and not containing the HB.*

Proof. By induction over the execution. According to line 4 in INITIALIZATION, the initial work assignment of the right agent a covers all nodes between the HB and a's LSP (which is the HB). This property is maintained by construction of EXPLORE, until the second agent b leaves a message for a.

If there is a message waiting for an agent (say a), the agent b at the moment it left the message computed its work assignment as half of the part remaining unexplored between the LSPs. Since this part contains at least two nodes (otherwise b would terminate), half of it (= b's assignment) does not contain the node a is currently heading to.

Finally, if there are no messages waiting, it is either at the very beginning when there is only one agent (already dealt with) or after one agent (say a) read a message m and recomputed its *Steps* according to lines 13-14 of CHECKING.

Consider the value d of variable $DistC$ of agent b at the moment when it left the first message (containing 0) for agent a. Since b has just finished its assignment, by induction hypothesis the current value of a's $Steps$ equals to $n - d$ (i.e. the value b started with halving). At the moment a reads the message m, b had halved its Steps $m + 1$ times (line 37 in CHECK&SPLIT), and that is exactly what a does in lines 13-14 of CHECKING (also applying Lemma 4). The partitioning works properly even when halving odd-sized workloads because one agent uses ceiling and the other uses floor. Afterwards, the invariant of the lemma is maintained by construction of $Explore$ until another message is left.

Since, by construction, the only previously unexplored nodes an agent enters are in its work assignment, we immediately get:

Corollary 1. *At most one agent enters the* BH.

Another consequence of Lemma 5 is that at any moment there is at most one message present, and at most one agent executing SEEKING or CHECK&SPLIT.

From construction (line 37 of CHECK&SPLIT and lines 13-16 or CHECKING) it follows that the parameter STEPS is at least halved in each consecutive call to EXPLORE (i.e. the unexplored area is at least halved). Since no waiting is specified in any place of the algorithm; SEEKING, CHECK&SPLIT and CHECKING each take at most $n - 2$ moves and either terminate or are followed by a call to EXPLORE, the algorithm terminates in $O(n \log n)$ steps. As the only way to terminate is to detect there is a single unexplored node, this node must contain the BH.

Note that at any moment at most three tokens are present in the system: one remaining at the HB and one by each agent used for marking the dangerous port and messaging. This follows from:

- the first agent puts two tokens (line 3 of INITIALIZATION), the second agent adds one more token (line 6) and no more agents become active (line 8)
- at any other place of the algorithm an agent puts a token that it has previously removed from some other place (moving the dangerous port mark, leaving a message, incrementing a message or reading a message).

Putting all together we obtain:

Theorem 1. *Algorithm Divide with Token correctly locates the* BH *employing two agents and three tokens in total, spending* $O(n \log n)$ *moves.*

4   Conclusions

In this paper we answered the following question: *Is it possible to locate a black hole in an asynchronous ring network using tokens instead of whiteboards?*

We answered this question affirmatively, showing that this can actually be achieved without any loss of performance. In fact, we present two algorithms, both using $O(1)$ tokens, that allow a team of *two* agents to determine the location of the black hole with $\Theta(n \log n)$ moves.

This result constitutes is a significant improvement over previous results, which used both stronger communication media (whiteboards) and more memory in the network ($O(n)$, resp. $O(\log n)$ after some modifications).

The second protocol presented here uses only three tokens in total, while still maintaining the optimal cost of $\Theta(n \log n)$ moves. An intriguing question is whether it is possible to further reduce the number of tokens to 2, and what would be the cost in such case.

References

1. L. Barriere, P. Flocchini, P. Fraigniaud, and N. Santoro. Election and rendezvous in fully anonymous systems with sense of direction. *Theory of Computing Systems*, 2006. To appear. Preliminary version in Proc. of SIROCCO 2003.
2. D. M. Chess. Security issues in mobile code systems. In *Proc. Conf. on Mobile Agent Security*, LNCS 1419, pages 1–14, 1998.
3. J. Czyzowicz, D. Kowalski, E. Markou, and A. Pelc. Searching for a black hole in tree networks. In *Proc. 8th International Conference on Principles of Distributed Systems (OPODIS 2004)*, pages 35–45, 2004.
4. S. Dobrev, P. Flocchini, R. Kralovic, G. Prencipe, P. Ruzicka, and N. Santoro. Optimal search for a black hole in common interconnection networks. *Networks*, 2006. To appear. Preliminary version in Proc. of OPODIS 2002.
5. S. Dobrev, P. Flocchini, G. Prencipe, and N. Santoro. Mobile search for a black hole in an anonymous ring. *Algorithmica*, 2006. To appear. Preliminary version in Proc. of DISC 2001.
6. S. Dobrev, P. Flocchini, G. Prencipe, and N. Santoro. Searching for a black hole in a arbitrary networks: optimal mobile agent protocols. *Distributed Computing*, 2006. To appear. Preliminary version in Proc. of PODC 2002.
7. S. Dobrev, P. Flocchini, and N. Santoro. Improved bounds for optimal black hole search in a network with a map. In *Proc. of 10th International Colloquium on Structural Information and Communication Complexity*, pages 111–122, 2004.
8. P. Fraigniaud, L. Gasieniec, D. Kowalski, and A. Pelc. Collective tree exploration. In *6th Latin American Theoretical Informatics Symp.*, pages 141–151, 2004.
9. P. Fraigniaud and D. Ilcinkas. Digraph exploration with little memory. In *21st Symp. on Theoretical Aspects of Computer Science*, pages 246–257, 2004.
10. M.S. Greenberg, J.C. Byington, and D. G. Harper. Mobile agents and security. *IEEE Commun. Mag.*, 36(7):76 – 85, 1998.
11. F. Hohl. A model of attacks of malicious hosts against mobile agents. In *Proc. of the ECOOP Workshop on Distributed Object Security and 4th Workshop on Mobile Object Systems*, LNCS 1603, pages 105 – 120, 1998.
12. R. Klasing, E. Markou, T. Radzik, and F. Sarracco. Hardness and approximation results for black hole search in arbitrary graphs. In *Proc. 12th Coll. on Structural Information and Communication complexity (SIROCCO'05)*, pages 200–215, 2005.
13. R. Oppliger. Security issues related to mobile code and agent-based systems. *Computer Communications*, 22(12):1165 – 1170, 1999.
14. T. Sander and C. F. Tschudin. Protecting mobile agents against malicious hosts. In *Proc. of Conf on Mobile Agent Security*, LNCS 1419, pages 44–60, 1998.

On Broadcast Scheduling with Limited Energy

Christian Gunia*

University of Freiburg, Georges-Köhler-Allee 79,
79110 Freiburg, Germany
gunia@informatik.uni-freiburg.de

Abstract. Given a set of requests, we tackle the problem of finding 'good' broadcast schedules aiming at the minimization of their total flow time. While running at a fixed speed, in the considered model the server is only allowed to use a certain amount of energy to perform these broadcasts. For this task we present optimal and approximation algorithms, respectively, depending on the number of distinct request types and their transmission lengths. The problem is solvable within polynomial time in the offline setting if the transmission lengths of all request types are identical and the number of distinct request types is constant. The presented algorithm can be generalized to obtain an approximation on instances without identical transmission lengths. Regarding the online version, we show lower and upper bounds on the competitive ratio of an optimal algorithm, including randomized algorithms and algorithms using resource augmentation. These lower and corresponding upper bounds match (at least asymptotically).

Keywords: Algorithms, Computational complexity, Mobile and net computing.

1 Introduction

In recent years, taking nature as archetype multi-agent systems consisting of dozens of autonomous agents are used to solve various kinds of problems. These range from problems of industrial manufacturing [13] to distributed constraint satisfaction problems [12]. Typically, the agents require a solid and dependable communication among each other in order to work efficiently. This is often provided by means of a wireless network. Using this medium communication is implicitly done via broadcasts. As battery power of these agent is strongly limited and sending a message requires a significant amount of energy, strategies to realize energy savings are inevitable. We consider situations in which such a system has to offer its service at least for a given period of time; the system's operation time. Furthermore, we focus on situations where agents request information from other agents. Have a look at one single agent, which can be seen as a server. Since many other agents potentially want to cooperate with it, usually the same piece of information—called a page—is requested by a large

---

* The author was supported by the German Research Foundation (DFG) by the research training program (Graduiertenkolleg) No 1103 'Embedded Microsystems'.

T. Calamoneri, I. Finocchi, G.F. Italiano (Eds.): CIAC 2006, LNCS 3998, pp. 151–162, 2006.

number of agents. One single broadcast could answer all of these requests at a single blow. Unfortunately, the requests are typically not issued at the same point of time. One could envisage accumulating all requests and broadcast only once afterwards. As the energy consumption for broadcasting the page is independent of the number of clients waiting for it, the energy saving compared to many single broadcasts can be significant. Therefore, the lifetime of an agent can be prolonged or its manufacturing costs and weight can be reduced due to a smaller battery. However, by following this strategy requests that arrived early have to wait an unreasonable amount of time until being satisfied. Likely, this slows down solving the task the multi-agent system was designed for. In other words, the provided Quality of Service (QoS) decreases. Hence, the server tries to find a tradeoff between the obtained QoS and the energy consumption needed to provide it.

We obtain a similar situation by observing the current developments concerning mobile devices, like handheld computers (e. g., PDAs). In recent years, their computational power, storage capacity and possibilities of communication (via bluetooth, wireless lan, etc.) significantly extended. One could imagine, running a peer-to-peer application like eDonkey[1] or BitTorrent[2] on it, sharing small files like pictures or articles. The communication between them is established via a wireless connection to the Internet. Requests for the same file can be answered via broadcasts. Therefore, a file transmitted through the Internet could be tagged with multiple addresses. Hence, it is only sent once to an access-point and accordingly multiplicated on its way through the Internet. Consequently, the energy consumption within the PDA is (almost) independent of the actual number of receivers. Once again, a tradeoff between the QoS and the energy consumption within the PDA has to be found. A quite common and mostly adequate measure of the QoS is the flowtime: the accumulated waiting time of the requests until they are satisfied, which ought to be minimized.

Both presented examples show that we have to deal with additional problems when designing algorithms for mobile systems. Probably the most outstanding one is the significantly limited energy capacity. Reducing the energy consumption can be seen as an architectural design constraint for stationary applications ([8],[11]) but it becomes a crucial issue for mobile applications.

Notations. We will have a look at the model described in the following: Given a set of requests for a server, minimize their total flow time while not exceeding the server's predetermined amount of energy C; there exists one broadcast channel, i. e., only one page can be transmitted simultaneously, thereby satisfying all requests for that page issued so far. The server is allowed to select its broadcast schedule but always transmits at a fixed speed. For the rest of this work m defines the number of pages stored on the server while n denotes the number of requests. We refer to w_i as the transmission length of the i-th page and define $\mathcal{W} := (w_1, \ldots, w_m)$ and $\dim(\mathcal{W}) = m$. Since it consumes one energy unit to broadcast a page of unit length, $\mathcal{W}$ also identifies the vector of the pages' energy

---

[1] http://www.edonkey.com

[2] http://www.bittorrent.com

consumption. W denotes $\sum_{i=1}^{m} w_i$ and C the server's energy capacity. The j-th request is abbreviated by $r_j := (t_j, p_j)$, meaning it arrives at time t_j and is requesting page p_j. We assume all requests to be issued within the time interval $[0, T]$ which represents the system's operation time (otherwise, all requests not issued within $[0, T]$ would be neglected). A schedule $S := (\sigma_1, \ldots, \sigma_k)$ with $\sigma_i := (\tau_i, \pi_i)$ broadcasts page π_i at time τ_i. We consider only feasible schedules, i. e., schedules respecting the following three conditions:

1. The broadcasts do not overlap, i. e., $\forall i \in \{1, \ldots, k-1\} : \tau_{i+1} - \tau_i \geq w_{\pi_i}$.
2. Each request is answered, i. e., $\forall i \in \{1, \ldots, n\} \exists j \in \{1, \ldots, k\} : t_i \leq \tau_j \wedge p_i = \pi_j$.
3. Respect the energy capacity, i. e., $\sum_{i=1}^{k} w_{\pi_i} \leq C$.

Let γ_i denote the beginning of the broadcast that answers r_i. Then S yields a total flow time of

$$f(S) := \sum_{i=1}^{n} \gamma_i - t_i + w_{p_i}. \tag{1}$$

Since we tackle only feasible schedules, $f(S)$ is defined properly. In both the offline and online version we provide the algorithm with the time interval $[0, T]$. It is not difficult to verify that this parameter does not present any additional information to an offline algorithm—remember that all requests are issued within $[0, T]$. However, it can contain worthwile information for an online algorithm. Therefore, we denote the given offline problem by $BC_C^{\text{Off}, \mathcal{W}}$, where C denotes the available energy capacity. In the online version, abbreviated by $BC_{T,C}^{\text{On}, \mathcal{W}}$, the algorithm knows the latest point of time at which a request can occur. Since it is not clairvoyant, it can only take T, the requests issued till time t and its past behavior into account when deciding what to do at time t.

Previous Work. While Clementi et al. [3] deal with topological aspects of broadcast problems and present an approximation algorithm for the energy consumption by choosing appropriate transmission ranges, we focus on the timing aspect of the problem: When to send which page to ensure a certain QoS? Results of empirical studies on this as well as valuable background information can be found in [1]. Bansal, Coppersmith and Swiridenko presented an $O(\log^2(T + n))$-approximation algorithm for this problem without regarding an energy limit. Erlebach and Hall [5] showed the $\mathcal{NP}$-hardness of this problem even if all stored pages require the same transmission time. However, since they use a linear number of distinct pages this does not rule out the possibility to find an optimal schedule if the number of distinct pages is constant. Kalyanasundaram, Pruhs and Velauthapillai presented an $O(1)$-speed $O(1)$-approximation algorithm for the offline version of this problem in [7] using the principle of resource-augmentation (see, [6]). They also considered the online version and provided a worst-case analysis for different QoS measures. In this context, the term s-speed means that the online algorithm is allowed to send s pages within one time unit while the offline algorithm has to get by on one page per time unit. In [10] various strategies of client-side caching and server-side ordering are discussed, if—according to our terminology—the pages consist of different segments. The paper shows that broadcasting segments out of order does not improve any reasonable QoS measure, while the capability of receiving them out of order has a great impact on them. The aspects of preemption, i. e., interrupting the broadcast of a page and later return to that point, are discussed in [4] by providing a $O(4 + \varepsilon)$-speed $O(1 + 1/\varepsilon)$-approximation algorithm.

Our Results. To our best knowledge none of the previous works on timing aspects of broadcast problems directly tackle the trade-off between QoS and energy consumption. This is the first comprehensive analysis of timing aspects on broadcast problems considering energy consumption. The paper is divided into two main parts: offline and online versions of the introduced problem. Throughout each of both sections, we start by dealing with an easy version of the problem and extend the scenario step by step. Therefore, we introduce three different input classes: We have a look at situations when there exists a) just one page, b) m pages with equal transmission length and c) m pages with variable transmission lengths. For the offline version, we present a polynomial time algorithm computing the optimal solution for any instance of $BC_C^{\text{Off},\mathcal{W}}$ with $\dim(\mathcal{W}) = 1$, i. e., just one page. Afterwards, this algorithm is extended to provide the same result on instances with $\mathcal{W} = (w, \ldots, w)$ and, finally, give an approximation algorithm for general $BC_C^{\text{Off},\mathcal{W}}$, i. e., input class c) with variable transmission length.

In Section 3 we turn to the online version. After starting by analyzing the competitive ratio of deterministic algorithms, we will also look at the influence of randomization and resource-augmentation. The table in Figure 1 gives an overview of our results for the online case (for the definition of the augmentation factor λ we refer to Section 3). Thereby, each line corresponds to

		deterministic		randomized
$m = 1$	UB	$\frac{T}{\lambda C} + 1$	UB	$\frac{5}{8} \cdot \frac{T}{\lambda C - w} + 2$
	LB	$\frac{T}{\lambda C} + 1$	LB	$\frac{1}{4} \cdot \frac{T}{\lambda C - w} + 1$
$m > 1$	UB	$m \cdot \frac{T}{\lambda C - (m-1)w} + 1$	UB	$m \cdot \frac{T}{\lambda C - (m-1)w} + 1$
	LB	$\frac{m}{3} \cdot \frac{T}{\lambda C - mw}$	LB	$\frac{m}{32} \cdot \frac{T}{\lambda C - mw}$
variable	UB	$3m \cdot \frac{T}{\lambda C - W} + 1$	UB	$3m \cdot \frac{T}{\lambda C - W} + 1$
	LB	$\frac{m}{3} \cdot \frac{T}{\lambda C - W}$	LB	$\frac{m}{32} \cdot \frac{T}{\lambda C - W}$

Fig. 1. Overview of lower (LB) and upper bounds (UB) for the competitive ratio

one of the introduced input classes. Once more first taking care of instances with $\dim(\mathcal{W}) = 1$, we obtain lower and upper bounds on the competitive ratio of an optimal, deterministic, resource-augmented online algorithm that match up to a constant additive. For randomized online algorithms on this problem as well as for resource-augmented, randomized online algorithms on $BC_{T,C}^{\text{On},\mathcal{W}}$

with $\mathcal{W} = (w, \ldots, w)$ these bounds match asymptotically. Finally, generalizing the ideas used for the former results leads to asymptotically matching lower and upper bound on the competitive ratio of an optimal, randomized, resource-augmented online algorithm on $BC_{T,C}^{\mathrm{On},\mathcal{W}}$ for general $\mathcal{W}$. As broadcasts are typically used in wireless communication environments receiving a message is almost as expensive (in terms of energy) as sending it. Therefore, it is important for the clients to (roughly) know the point of time their request will be answered, so that they can deactivate their receiver until then and save energy. Hence, we also point out in Section 3 that this can be guaranteed for the online version of the problem without negatively affecting the provided QoS. Section 4 concludes and presents problems left open.

2   Offline-Algorithms

We suppose the reader to be familiar with the concept of dynamical programming. Some proofs within this paper are omitted due to space limitations. We start by constructing an optimal schedule, i. e., a schedule that provides a minimal total flow time while sticking to the energy limit. Note that in $BC_C^{\mathrm{Off},w}$ one broadcast of the page keeps the channel busy for w time units and consumes w energy units. Hence, this induces a number of available broadcasts, namely C/w. We assume this to not exceed n; otherwise, we round it down to n. Since an optimal schedule at most broadcasts once per request, this does not change the situation significantly. Therefore, the following theorem proves the problem to be in $\mathcal{P}$.

Input: input instance for $BC_C^{\mathrm{Off},w}$
Result: flowtime of optimal schedule

1 **create** a table holding the optimal flowtime for each feasible (i, j, b, c, k).
2 **initialize** table acccording to
3    $(i, j, b, c, 0) := \sum_{\alpha=i}^{j}((t_{\alpha+1} - t_\alpha) \cdot \sum_{\beta=b}^{\alpha} h_\beta)$ and $(i, i, b, c, k) := (i, i, b, c, 0)$
4 **for** increasing $j - i$ and k **do**
5    $T(l, d', b', k') := (i, l - 1, b, b', k') + \sum_{\alpha=b'}^{m-1} f_l(t_\alpha) \cdot h_\alpha + (m, j, c', c, k - k' - d')$
6   where $c' := \min\{\beta \mid t_l + (d' - 1) \cdot w < t_\beta\}$
7   and $m := \min\{\beta \mid t_l + d' \cdot w \le t_\beta\}$
8   and $f_l(t_\alpha) := \min_\beta\{f := t_l + \beta \cdot w - t_\alpha \mid f \ge 0\} - t_\alpha$ holds.
9    $(i, j, b, c, k) := \min\{T(l, d', b', k') \mid$ feasible $l, d', b', k'\}$
10 **end**
11 **return** the minimum of:
12 $\{(1, n - 1, 0, b, C/w - 1) + \sum_{\alpha=b}^{n} f_n(t_\alpha) \cdot h_\alpha\} \cup$
    $\{(1, i - 1, 0, b, C/w - \lceil(t_n - t_i)/w\rceil - 1) + \sum_{\alpha=b}^{n} f_i(t_\alpha) \cdot h_\alpha\}$ for all feasible b, i.

Fig. 2. The algorithm BCOPT

Theorem 1. *Let $n \in \mathbb{N}$ and $C, w \in \mathbb{R}^+$. The optimal solution of $BC_C^{\mathrm{Off},w}$ for n requests is computable in time $O((C/w)^3 n^7)$ and space $O((C/w)n^4)$.*

Proof. (sketched) It is not difficult to see that an optimal schedule consists of blocks of broadcasts; each one of these blocks starts at a request time t_i. The main idea of algorithm BCOPT is to partition a given problem into three parts: the broadcast block, one subproblem 'before' the block and one 'after' the block; by using dynamical programming the optimal solutions of these 'smaller' subproblems are already known. We have to ensure that we obtain an optimal solution for the given problem from the the concatenation of the optimal solutions of these parts. Therefore, we specify the interfaces between the three parts.

Thinking this out, we define (i, j, b, c, k) to represent the subproblem where broadcasts are only valid within $[t_i, t_j]$, the requests $r_b, r_{b+1}, \ldots, r_{c-1}$ have to be regarded and k broadcasts are available. We call such a tuple feasible if it represents a 'senseful' subproblem (details are omitted from this sketch). However, there are $O(kn^4)$ feasible tuples. For $i = j$ and $k = 0$ the subproblems can be solved directly as seen in Step 3 of algorithm BCOPT (Figure 2). The other entries of the table in Step 1 can be computed iteratively: if we fix the broadcast block and the first unanswered request $r_{b'}$ and $r_{c'}$ at its beginning and ending, respectively, we obtain the situation shown in Figure 3. Hence, we can compute the solution for (i, j, b, c, k) from smaller instances, i. e., instances that contain less requests or have less broadcasts available, by computing the solution for each possible broadcast block with each possible carryover and choosing the best one (Step 4). The value we are interested in is either $(i, n-1, 0, b, C/w - 1) + \sum_{j=b}^{n} f_n(t_j)$ if the last broadcast is done at time t_n, or the second term in Step 11. The latter is the case if the last broadcast is performed after T because it constitutes the ending of a broadcast block beginning at t_α and, consequently, containing $\lceil (t_n - t_\alpha)/w \rceil + 1$ broadcasts.

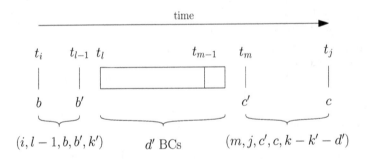

Fig. 3. Partition of the problem (i, j, b, c, k) into subproblems

The runtime of BCOPT derives directly from the number of feasible tuples and the fact that each iteration in Step 4 can be done in $O(k^2 n^2)$ time steps. By storing not only the flowtime for (i, j, b, c, k) but also the choice made in Step 9 we use backtracking to obtain the actual schedule.   □

By presenting an algorithm that solves $BC_C^{\text{Off},w}$ optimally, the previous theorem shows the problem is in $\mathcal{P}$. The next theorem shows properties of an approximation algorithm for that problem, although the actual algorithm is omitted from this paper due space limitations. We cite it since the upper bounds on its running time and memory requirements differ by a factor $\Omega(n^2)$ compared to BcOpt.

Theorem 2. *Let $n \in \mathbb{N}$ and $C, w \in \mathbb{R}^+$. A 2-approximation of $BC_C^{\text{Off},w}$ for n requests is computable in time $O((C/w)^2 n^3)$ and space $O((C/w)^2 n)$.*

We aim at generalizing Theorem 1 to situations where more than one page exists. Its main idea was the creation of optimal broadcast blocks of 'right lengths' at 'right times'. This was quite simple in the case $m = 1$ since fixing the starting point, the carryover and the length of a broadcast block completely determines it. Specifying these values in the situation of $m > 1$, fixes the starting points of the broadcasts within the broadcast block. However, the question remains *which* page to send at which time point.

In the general situation, there are requests $r_1, \ldots, r_n$ for the m pages given. The broadcast block starts at t_l and spends C energy units. Let b' be the vector consisting of the number of unsatisfied requests at time t_l and c' is the corresponding vector at time t_{l+C}. Both vectors, the starting point of the broadcast and its length are given. Therefrom, the starting points of the broadcasts, namely $\tau_1, \ldots, \tau_{C/w}$, fulfill $\tau_i = t_l + i \cdot w$. The task is to find an assignment of $\sigma_1, \ldots, \sigma_{C/w}$ minimizing the total flow time while satisfying all given constraints.

Lemma 1. *Considering the above stated problem, it is possible to decide whether there exists a schedule satisfying all constraints and to compute the one minimizing the total flow time within time $O((C/w)mn^{3m+2})$ and space $O(n^{2m+2})$.*

Theorem 3. *Let $m, n \in \mathbb{N}, C \in \mathbb{R}^+$ and $\mathcal{W} := (w, \ldots, w)$ with $w \in \mathbb{R}^+$ and $\dim(\mathcal{W}) = m$. The optimal solution of $BC_C^{\text{Off},w}$ for n requests is computable in time $O((C/w)^3 n^{4m+3})$ and space $O((C/w)mn^{2m+2})$.*

Proof. (sketched) A slight modification of algorithm BcOpt will do the trick. We still partition the problem into three parts: two subproblems P_L and P_R, and a broadcast block B between them. By combining the proofs of Lemma 1 and Theorem 1 it is possible to identify the subproblems by 5-tuples, namely (i, j, b, c, k), whereas i and j define the valid interval for broadcasts, b and c specify the carryover at the beginning and the end of the subproblem, and k identifies the number of broadcasts available. In comparison to BcOpt the carryovers b and c are vectors containing one component for each page type. Let b' and c' denote the carryover at the broadcast block's beginning and end, respectively.

According to Lemma 1, the optimal schedule and flow time is computable given the block's starting point, length, b' and c'. Hence, the following equation holds:

$$(i, j, b, c, k) = \min_{B, k'} \left\{ (i, l - 1, b, b', k') + \text{flow}(B) + \right.$$

$$\left. (m, j, c', c, k - k' - d') \right\}, \tag{2}$$

whereas flow(B) denotes the flow time of the requests satisfied by B. The problem's solution is computed in two steps. First, the optimal schedules for all broadcast blocks are determined using Lemma 1. This is done in time $O((C/w)^2mn^{5m+3})$ by using a table of size $O((C/w)n^{2m+1})$. Afterwards, we compute all 5-tuples in time $O((C/w)^2n^{4m+4})$ with a table of size $O((C/w)n^{2m+3})$ using (2). The values we are interested in are very similar to the ones of algorithm BcOpt but are omitted here.   □

Finally we show that for general instances at least an approximation can be obtained within polynomial time.

Theorem 4. *Let* $m, n \in \mathbb{N}, C \in \mathbb{R}^+$ *and* $\mathcal{W} = (w_1, \ldots, w_m) \in (\mathbb{R}^+)^m$. *A* $(1 + 3m + 3m \cdot w_{\max}/w_{\min})$-*approximation of* $BC_C^{\mathrm{Off},\mathcal{W}}$ *for* n *requests is computable in pseudopolynomial time, i. e.,* $\mathrm{poly}(C/w_{\min}, n)$.

3   Online-Algorithms

Definition 1. *For* $\lambda \in \mathbb{N}$ *we call an online algorithm* λ-*augmented if it does at most* $\lambda \cdot C$ *broadcasts on a problem* $BC_{T,C}^{\mathrm{On},\mathcal{W}}$.

The idea behind resource augmentation is to compensate for the loss of clairvoyance an online algorithm experiences in comparison with an offline algorithm by allowing it to use more resources. Note that an 1-augmented online algorithm is not augmented at all. In our considered model, giving an algorithm more resources, obviously results in a better competitive ratio. Have a look at the probably easiest algorithm for this problem: Independent of the requests, broadcast each $T/(wC)$ time units. Even this algorithm improves its competitive ratio nearly by a factor of λ^{-1} when provided with λC instead of C energy units. However, the next theorem proves this factor λ^{-1} to be the most any optimal algorithm can gain from resource augmentation.

Input: input instance for $BC_C^{\mathrm{On},(w,\ldots,w)}$

1 $\alpha := 1 - m + \lambda C/w$
2 **for** $i = 1, \ldots, \alpha$ **do**
3 $\quad p_i := (i \bmod m) + 1$
4 $\quad t_i := iT/\alpha$
5 $\quad$ broadcast page p_i at time t_i
6 **end**
7 broadcast page $i = 1, \ldots, m$ at time $T + (i - 1) \cdot w$.

Fig. 4. The algorithm BcDet$_m$

Theorem 5. *Let* $\lambda \in \mathbb{N}$ *and* $C, T, w \in \mathbb{R}^+$ *with* $C \leq T$. *The competitive ratio of each deterministic,* λ-*augmented online algorithm for* $BC_{T,C}^{\mathrm{On},w}$ *is at least* $T/(\lambda C) + 1$. *The competitive ratio of* BcDet$_1$ *is* $T/(\lambda C) + 1$.

As already mentioned in the introduction, broadcasts are often done via wireless communication devices where sending and receiving of a message is nearly equally expensive in terms of energy usage. Regarding this, Theorem 5 shows that using algorithm BcDet$_1$ (Figure 4)—whose predictable broadcasts allow clients to turn off their receivers within the 'idle-periods'—is worst-case optimal.

Since resource augmentation for itself does not yield a major advantage, let us have a look at randomization.

Input: input instance for $BC_C^{\text{On},w}$

1 $\alpha := \lambda C/w - 1$
2 **for** $i = 1, \ldots, \alpha$ **do**
3 $\quad$ draw t_i uniformly at random from $[w + iT/\alpha \, , \, (i+1)T/\alpha)$
4 $\quad$ broadcast at time t_i
5 **end**
6 broadcast at time T

Fig. 5. The algorithm BcRand

Theorem 6. *Let $\lambda \in \mathbb{N}$ and $C, T, w \in \mathbb{R}^+$ with $C \leq T$. The competitive ratio of each randomized, λ-augmented online algorithm for $BC_{T,C}^{\text{On},w}$ is at least $(T/(\lambda C - w) + 1)/4$. The competitive ratio of algorithm BcRand is at most $(5T/(\lambda C - w) + 2)/8$.*

Proof. The algorithm BcRand (Figure 5) obviously creates a feasible schedule. We identify the flow time of the i-th request with the random variable X_i. If we show that $E(X_i) \leq \alpha$ holds, we will get $n \cdot \alpha/(n \cdot w) = \alpha/w$ as an upper bound for the competitive ratio. Hence, it is sufficient to show $E(X_i) \leq w \cdot (5T/(\lambda C - w) + 2)/8$.

Let us have a look at some arbitrary request i. This request is located in some interval I with length $|I|$ and let x denote the request's distance from the beginning of I. With probability $(|I| - x)/|I|$ the request is answered by I's broadcast, since it takes place after x with this probability. The expected flow time created in this case is at most $(|I| - x)/2 + w$, where we added w to compensate for the exclusion of the first w time units. With probability $x/|I|$ it is answered by the broadcast of the next interval yielding an expected flow time of $3/2|I| - x + w/2$. According to the Total Probability Theorem [9], we can compute the expected flow times of these two cases separately and afterwards construct their weighted sum. Hence, we get

$$E(X_i) = \frac{|I| - x}{|I|} \cdot \left(\frac{|I| - x}{2} + w \right) + \frac{x}{|I|} \cdot \left(\frac{3}{2}|I| - x + \frac{w}{2} \right) \leq \frac{|I|^2 + x|I| - x^2}{2|I|} + w.$$

This reaches for $x = |I|/2$ its maximum of $|I| \cdot 5/8 + w$. Substituting $|I|$ by $T/((\lambda C/w) - 1)$ and adding another w to regard the transmission time omitted up to now, reveals a total flow time of at most $((5/8) \cdot T/(\lambda C - w) + 2) \cdot w$ as desired.

To show the lower bound we consider an arbitrary, feasible, randomized algorithm A. We partition T into intervals of length $Tw/(2\lambda C - w)$, i.e., into $(2\lambda C - w)/w$ parts—which we, w. l. o. g., assume to be a natural number. At the beginning of the i-th interval we request the page a_i times. Next, we specify the a_i one by one. Since we know A, we can compute its probability of broadcasting within each interval if we fix the worst-case instance constructed so far. Let $p_i^{a_i}$ denote the probability of a broadcast within the i-th interval and p_i^∞ its limit[3] for growing a_i. If $p_i^\infty < 1/2$ holds, it is possible to find an a_i guaranteeing that (i) the influence of the first $i-1$ intervals on the total flow time is bounded by an arbitrarily small constant and (ii) A performs no broadcast within this interval with probability at least $1/2$. Stopping the construction after a_i leads to

$$\frac{a_i \cdot (Tw/(2\lambda C - w) + w) + O(1)}{a_i \cdot w + O(1)}$$

as a lower bound on the competitive ratio since the a_i requests are waiting a complete interval to be satisfied. However, in the optimal schedule they can be satisfied immediately yielding a total flow time of $a_i \cdot w + O(1)$. This shows the claim if a_i is chosen sufficiently large and $p_i^\infty < 1/2$ holds.

Now we show that there has to be an interval with $p_i^\infty < 1/2$. Assume all p_i to be at least $1/2$ and define random variables $X_i \in \{0,1\}$ with $X_i = 1 \Leftrightarrow$ A broadcasts in the i-th interval. The number of broadcasts performed by A is obviously $\sum X_i$ and, hence, at least $(1/2) \cdot (2\lambda C - w)/w$. Therefore, the energy used by the algorithm before time T exceeds the limit $\lambda C - w$—recall that one broadcast has to be possible at T to guarantee a feasible schedule. This leads to a contradiction and completes the proof of the theorem.   □

We have seen that randomization and resource augmentation do not provide a major effect on the competitive ratio of an optimal online algorithm as far as problem instances with just one page are concerned. Furthermore, we have seen that algorithms BcRand and BcDet$_m$ are asymptotically worst case optimal for $m = 1$. Next, we want to generalize these results to instances that contain $m \geq 1$ pages.

Theorem 7. *Let $\lambda \in \mathbb{N}$ and $C, T \in \mathbb{R}^+$ with $\lambda C \leq T/3$. Further assume $W = (w, \ldots, w) \in (\mathbb{R}^+)^m$ and $W < C$. The competitive ratio of each deterministic, λ-augmented online algorithm for $BC_{T,C}^{\mathrm{On},W}$ is at least $(mT/(\lambda C - mw))/3$. The competitive ratio of BcDet_m is at most $1 + mT/(\lambda C - (m-1)w)$.*

As already observed in the case of one available page resource-augmentation does only yield a trivial impact on the competitive factor in this situation as well. Therefore, we have a look at the effect of randomization.

Theorem 8. *Let $\lambda \in \mathbb{N}$ and $C, T \in \mathbb{R}^+$ with $\lambda C \leq T/8$. Further assume $W = (w, \ldots, w) \in (\mathbb{R}^+)^m$ and $W < C$. The competitive ratio of each randomized, λ-augmented online algorithm for $BC_{T,C}^{\mathrm{On},W}$ is at least $(mT/(\lambda C - mw))/32$.*

---

[3] We assume $p_i^{a_i}$ to be monotonically decreasing with respect to a_i. However, this limitation is due to technical reasons and can be removed easily.

Theorem 9. *Let $\lambda \in \mathbb{N}$ and $C, T \in \mathbb{R}^+$ with $\lambda C \leq T$. Further assume $\mathcal{W} = (w_1, \ldots, w_m) \in (\mathbb{R}^+)^m$ with $w_1 \leq w_2 \leq \cdots \leq w_m, W < C$ and $w_1 \geq W(\lambda C - W)/(mT)$. The competitive ratio of each deterministic, λ-augmented online algorithm for $BC_{T,C}^{On,\mathcal{W}}$ is at least $(mT/(\lambda C - W))/3$. There exists a deterministic online algorithm A whose competitive ratio is at most $1 + 3mT/(\lambda C - W)$.*

Proof. Once again, we start by showing the upper bound first. Instead of writing the algorithm A down in pseudo-code, we explain its behavior. In order to make sure the presented schedule is feasible A will use at most $\lambda C - W$ energy units until T. To perform the broadcasts of page i it will use $C' := (\lambda C - W)/m$ energy and spend it 'equally' distributed. Assume for one moment, there would be m distinct channels so that m broadcasts could be done simultaneously. Hence, the algorithm dedicates each page one channel, and broadcasts each Tw_i/C' time units page i on channel i, starting at time Tw_i/C'. Observe, that consequently the last broadcast of each page takes place after T. When merging all broadcasts canonically to one channel, there could be conflicts as two broadcasts might overlap. Nevertheless, we are merging them canonically to one channel and describe how to resolve these conflicts. We use the broadcasts of page 1, i. e., the page broadcasted most frequently, as alignments for the others and enumerate the arising intervals chronologically by $I_1, \ldots, I_a$. Each of these intervals has at least length W according to our preconditions and the choice of the original schedule. Imagine a group of conflicting broadcasts that are all starting within $I \in \{I_1, \ldots, I_a\}$. Since the interval's size is at least W, they all can be relocated within I solving their conflicts. Particularly, they all can finish within I. Hence, we can solve all conflicts sequentially starting in interval I_1. Thus, we get a feasible schedule.

Let us regard the resulting schedule's QoS. Each request for page 1 starts receiving its answer within $mTw_1/(\lambda C - W)$ time units. On the other hand, each request for page $i > 1$ starts receiving its answer within $3 \cdot mTw_i/(\lambda C - W)$ time units since the distance between two broadcasts of page type i is at most trebled during the conflict resolution. This bounds $\mathcal{A}$'s total flow time from above by

$$\sum_{i=1}^{m} 3n_i \cdot \left(\frac{mTw_i}{\lambda C - W} + w_i \right) = 3 \left(\frac{mT}{\lambda C - W} + 1 \right) \cdot \sum_{i=1}^{m} (n_i \cdot w_i),$$

whereas n_i denotes the number of requests for page i. The optimal flow time is trivially bounded below by at least $\sum_{i=1}^{m} n_i \cdot w_i$ and, therefore, the proof of the upper bound is completed. The proof of the lower bound of Theorem 7 can be adapted to show this lower bound as well by replacing mw by W. $\qquad\square$

4   Conclusions and Outlook

We saw that the problem of finding an optimal broadcast schedule with respect to minimizing the flow time using a limited number of available energy lies in $\mathcal{P}$, if the number of pages is constant. It would be interesting to figure out if this still holds when the number of distinct page types is superconstant or when their transmission times are allowed to vary. Up to now, only an approximation algorithm for distinct transmission times is known due to Theorem 4.

Furthermore, asymptotically optimal algorithms for the online versions of this problem can be obtained by using the probably most simple ones as Theorems 5 to 9 have shown. Furthermore, their predictability provides the clients with the opportunity to save energy by shutting down their receivers until needed.

Finally, average case analyses are quite common to represent the behavior of an algorithm somewhat better than worst-case analyses (see, e. g., the simplex algorithm for solving linear programs [2]). Therefore, it would be interesting to know more about the average case behavior of broadcast algorithms in our model for 'typical' input distributions (e. g., web server accesses, online libraries, etc.).

References

1. S. Acharya and S. Muthukrishnan. Scheduling on-demand broadcasts: new metrics and algorithms. In *Proceedings of the 4th Annual ACM/IEEE international conference on Mobile computing and networking*, 1998.
2. I. Adler and N. Meggido. A simplex algorithm whose average number of steps is bounded between two quadratic functions of smaller dimension. *Journal of the ACM*, 32:871–895, 1985.
3. A. Clementi, P. Crescenzi, P. Penna, G. Rossi, and P. Vocca. On the complexity of computing minimum energy consumption broadcast subgraphs. In *Proceedings of the 18th Annual Symposium on Theoretical Aspects of Computer Science*, volume 2010 of *LNCS*, pages 121–131, 2001.
4. J. Edmonds and K. Pruhs. Multicast pull scheduling: When fairness is fine. *Algorithmica*, 36, 3:315–330, 2003.
5. T. Erlebach and A. Hall. NP-hardness of broadcast scheduling and inapproximability of single-source. In *Proceedings of the 13th Annual ACM-SIAM symposium on Discrete algorithms*, pages 194–202, 2002.
6. B. Kalyanasundaram and K. Pruhs. Speed is as powerful as clairvoyance. In *IEEE Symposium on Foundations of Computation*, pages 214–221, 2000.
7. B. Kalyanasundaram, K. Pruhs, and M. Velauthapillai. Scheduling broadcasts in wireless networks. In *Proceedings of the 8th Annual European Symposium on Algorithm*, volume 1879 of *LNCS*, pages 290–301, 2000.
8. M. Mudge. Power: A first-class architectural design constraint. *IEEE Computer Magazine*, 34(4):52–58, 2001.
9. A. Papoulis. *Probability, Random Variables, and Stochastic Processes, 2nd ed.* New York: McGraw-Hill, 1984.
10. K. Pruhs and P. Uthaisombut. A comparison of multicast pull models. In *Proceedings of the 10th Annual European Symposium on Algorithms*, volume 2461 of *LNCS*, pages 808–819, 2002.
11. K. Pruhs, P. Uthaisombut, and G. Woeginger. Getting the best response for your erg. In *Scandanavian Workshop on Algorithms and Theory*, 2004.
12. R. Mailler and V. Lesser. Using cooperative mediation to solve distributed constraint satisfaction problems. In *Proceedings of Third International Joint Conference on Autonomous Agents and MultiAgent Systems*, volume 1 of *IEEE Computer Society*, pages 446–453, 2004.
13. W. Shen and D.H. Norrie. Agent-based systems for intelligent manufacturing: A state-of-the-art survey. *Knowledge and Information Systems*, 1(2):129–156, 1999.

A Near Optimal Scheduler
for On-Demand Data Broadcasts

Hing-Fung Ting[*]

Department of Computer Science, The University of Hong Kong,
Pokfulam Road, Hong Kong
hfting@cs.hku.hk

Abstract. In an on-demand data broadcast system, clients make requests for data such as weather forecasts, stock prices and traffic information. The server of the system broadcasts the requested data at some time, and all pending requests on this data are satisfied with this single broadcast. All requests have deadlines. The system can abort the current broadcast for more valuable requests and a preempted broadcast may be restarted from the beginning later. In this paper, we design and analyse online scheduler for scheduling broadcasts in such system. The best previously known upper and lower bounds on the competitive ratio of such schedulers are respectively $\Delta + 2\sqrt{\Delta} + 2$ and $\sqrt{\Delta}$, where Δ is the ratio between the length of the longest and shortest data pages. In this paper, we design a scheduler that has competitive ratio $\frac{6\Delta}{\log \Delta} + O(\Delta^{5/6})$. We also improve the lower bound of the problem to $\frac{\Delta}{2\ln \Delta} - 1$, and hence prove that our scheduler is optimal within a constant factor.

1 Introduction

With the advances in satellite broadcasting, wireless networks and mobile computing, on-demand data broadcasting becomes an important technique for information dissemination and has already been widely adopted in daily life. NHK digital broadcast company is a good example for providing such services, and YESTV, TIVO and DTV-Plus are other examples. In an on-demand data broadcast system, clients make requests for data such as weather forecasting, stock prices, traffic information and sports results using various mobile devices such as notebooks, personal digital assistants (PDAs) and GPRS-enabled cellular phones. The server broadcasts the requested data at some time, and all pending requests on this data are satisfied with this single broadcast.

On-demand broadcasting has been studied extensively, both empirically [1, 6] and theoretically [3, 7, 12, 11]. Most of these studies focus on schedules that minimize the average or maximum respond time. They assume that once a request is generated by a user, the requests will be held until it is satisfied. As pointed out by Jiang and Vaidya [5], this assumption is not always true; clients are impatient and they may leave with their requests unserved after waiting too long.

---

[*] This research was supported partially by the Hong Kong RGC Grant HKU-7045/02E.

T. Calamoneri, I. Finocchi, G.F. Italiano (Eds.): CIAC 2006, LNCS 3998, pp. 163–174, 2006.
© Springer-Verlag Berlin Heidelberg 2006

To take this kind of behaviour into consideration, Kim and Chwa [9], and independently Kalyanasundaram and Velauthapillai [8], have recently proposed two similar models for on-demand broadcasting in which every request has a deadline. Kim and Chwa proposed a *restart* model, while Kalyanasundaram and Velauthapillai proposed a *resume* model.

The models. In both models, the input to the system is a sequence of page requests, which arrive at arbitrary time. The pages may have different lengths. At any time, the server can broadcast only one page. All pending requests for that page can be satisfied with this single broadcast. If a request is satisfied on or before its deadline, the system earns the profit specified by the request. Otherwise, the system does not earn any profit from the request. The objective is to maximize the total profit earned. Both models allow preemption; the server can abort the current broadcast for more valuable requests. In the restart model, the server is allowed to start a preempted broadcast later from the beginning. The resume model allows the server to resume a previously preempted broadcast later from the point of preemption.

In this paper, we focus on the restart model. We refer to [8, 12] for more details on the resume model and interesting upper and lower bound results on the performance of their schedulers.

Previous results. For ease of reference, we call the problem of scheduling on-demand data broadcast system BROADCAST. Let Δ be the ratio between the length of the longest and shortest pages. In [9], Kim and Chwa observed that for the special case when $\Delta = 1$ and all requests have tight deadlines, BROADCAST is closely related to the interval scheduling problem studied in [10, 13], where a set of requests for a page arriving at one time can correspond to a job with weight equal to the total value of the requests. They noted that the results in [13] imply that in this special case, there is a 4-competitive scheduler for BROADCAST, and this is best possible for any deterministic online scheduler. For the case of arbitrary deadlines, they proposed a 5.828-competitive scheduler. The competitive ratio was subsequently reduced to 5 by Chan *et al.* [2] and further to 4.56 by Zhang *et al.* [14].

For the general case when $\Delta > 1$, Kim and Chwa [9] derived a lower bound of $\min\{\sqrt{\Delta}, r\}$ on the competitive ratio where r is the maximum number of requests for a page arriving at a time. The bound was later improved to $\sqrt{\Delta}$ by Chan *et al.* [2], who also derived the first non-trivial upper bound. They observed that for $\Delta > 1$, BROADCAST is closely related to another job scheduling problem, namely the Job Scheduling with Cancellation problem (JS-CANCELLATION), in which users can send requests to cancel waiting jobs[1]. In [2], Chan *et al.* described a competitive-ratio preserving reduction from BROADCAST to JS-CANCELLATION. Then, they gave a $(4\delta + 3)$-competitive scheduler for JS-CANCELLATION where δ is the ratio between the length of the longest and shortest jobs. Together with the reduction, they concluded that there is a $(4\Delta + 3)$-competitive scheduler for

---

[1] A real-life example for JS-CANCELLATION is scheduling printing jobs on a printer; a printing job can be cancelled when they are waiting.

BROADCAST. The bound was subsequently reduced to $e\Delta + e + 1$ and further to $\Delta + 2\sqrt{\Delta} + 2$ (see [4]).

Note that there is a significant gap between the currently known upper bound (i.e., $\Delta + 2\sqrt{\Delta} + 2$) and lower bound (i.e., $\sqrt{\Delta}$) on the competitive ratio. We note that all known online schedulers for BROADCAST are conservative; they will not make any preemption that reduces the profit of the schedule. We observe that such conservative schedulers cannot have competitive ratio smaller than $\Delta - \epsilon$ for any $\epsilon > 0$. Let us consider the following input request sequence: There is a request on page Q arriving at time 0 where Q has length Δ and the request has value 1, and for $0 \leq i \leq \Delta - 1$, there is a request on page P_i arriving at time i where P_i has length 1 and the request has value $1 - \epsilon/\Delta$. All requests have tight deadlines. Note that any conservative scheduler will choose to broadcast Q to earn a profit of only 1, while the optimal scheduler broadcasts P_i at time i for $0 \leq i < \Delta$ and earns a profit of $\Delta - \epsilon$.

Our results. In this paper, we describe a novel online scheduler BRAVE for BROADCAST, which breaks the Δ barrier for conservative schedulers; it has competitive ratio $\frac{6\Delta}{\log \Delta} + O(\Delta^{5/6})$ where log is of base 2. BRAVE is greedy in nature, but it has the courage to make some preemptions that reduce the profit currently earned. To ensure that enough profit will be earned, BRAVE follows a set of rather complicated rules for making preemptions. In particular, when BRAVE decides whether to preempt a current broadcast B_{cur} by a new broadcast B, it does not compare the value of B with that of B_{cur}, nor compare it with the value of the current schedule, but compare it with the value of a carefully chosen broadcast that may be made much earlier than B. Furthermore, the decision for making a preemption depends not only on their values, but also their "distance".

From the preemption rules of BRAVE, it is easy to find difficult input instances for BRAVE and show that its competitive ratio is at least $\Omega(\Delta/\log \Delta)$. More importantly, these difficult input instances suggest an adversary that generates difficult inputs for any scheduler. Based on this adversary, we prove that any online scheduler for BROADCAST has competitive ratio at least $\frac{\Delta}{2\ln \Delta} - 1$ where ln is the natural logarithm. Hence, BRAVE is optimal within a constant factor.

Organization of the paper. In Section 2, we give the definitions and notations that are necessary for our discussion. We describe the online scheduler BRAVE in Section 3. In Section 4, we prove that the competitive ratio of BRAVE is at most $\frac{6\Delta}{\log \Delta} + O(\Delta^{5/6})$, and in Section 5, we prove that any deterministic online scheduler for BROADCAST has competitive ratio at least $\frac{\Delta}{2\ln \Delta} - 1$.

2 Definitions and Notations

This section gives the definitions and notations that are necessary for our discussion. For simplicity, we assume that the minimum length of a page is 1. Thus, Δ is the length of the longest page.

Given any page P, let $\ell(P)$ denote its length, which is the time needed for a complete broadcast of P. A *schedule* is a sequence of *broadcasts*

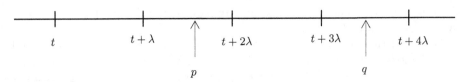

Fig. 1. An example on λ-distance

$S = \langle (P_1, t_1), (P_2, t_2), \ldots, (P_{\text{last}}, t_{\text{last}}) \rangle$ where $t_1 < t_2 < \cdots < t_{\text{last}}$ and (P_i, t_i) specifies that a broadcast on page P_i starts at time t_i. We say that (P_i, t_i) is a *complete* broadcast if it can be broadcast completely before the next broadcast (P_{i+1}, t_{i+1}) starts, or equivalently, $t_i + \ell(P_i) \leq t_{i+1}$. Otherwise, (P_i, t_i) is *incomplete* and we say that it is *preempted* by (P_{i+1}, t_{i+1}), and (P_{i+1}, t_{i+1}) is a *preempting* broadcast. Regardless of whether (P_i, t_i) is complete or not, we say that it has *supposed completion time* $t_i + \ell(P_i)$.

A request r is specified by a tuple (P, a, d, v) where P is the page that it requests, a is its arrival time, d is its deadline and v is its value. We say that r is *alive* at any time during $[a, d]$. We say that it is *satisfied* by the schedule S if it can listen to a complete broadcast on P from S before its deadline. More precisely, we say that the request (P, a, d, v) is satisfied by S before time t_o if there is a complete broadcast (P, t) in S such that $a \leq t < t + \ell(P) \leq \min\{t_o, d\}$.

Now, we define the profit of schedule S on serving a sequence of requests σ. For any request $r \in \sigma$, let $v(r)$ denote its value. For any page P and any time t, define $R_{S,\sigma}(P, t)$ to be the set of requests in σ that (i) ask for page P, (ii) are alive during $[t, t + \ell(P))$, and (iii) have not been satisfied by the schedule S at or before t. Note that if S has the broadcast (P, t), $R_{S,\sigma}(P, t)$ is exactly the set of requests in σ that are served by (P, t). Define *the value of P at t with respect to (S, σ)* to be $v_{S,\sigma}(P, t) = \sum_{r \in R_{S,\sigma}(P,t)} v(r)$. Note that when using S to serve σ, we will earn a profit of $v_{S,\sigma}(P, t)$ from broadcast $(P, t) \in S$ if it is complete, and earn nothing otherwise; we also say that the value of the broadcast (P, t) is $v_{S,\sigma}(P, t)$. Define the *profit* $\rho(S, \sigma)$ of S on serving σ to be the total value of the complete broadcasts in S for serving σ, i.e.,

$$\rho(S, \sigma) = \sum_{\substack{(P,t) \in S \\ P \text{ is complete}}} v_{S,\sigma}(P, t).$$

For any deterministic online scheduler A for BROADCAST, we say that A is *c-competitive*, and it has *competitive ratio* c, if given any input σ, A produces a schedule S for σ such that $\rho(O, \sigma) \leq c\rho(S, \sigma)$ where O is the optimal offline schedule for σ. Finally, we need the notion of λ-distance in order to describe BRAVE. Let λ be any positive number. For any two numbers $0 \leq x \leq y$, we say that y *is at a λ-distance of ℓ from* x, and their *λ-distance* $d_\lambda(x, y)$ is ℓ if

$$x + \ell\lambda \leq y < x + (\ell+1)\lambda.$$

For example, in Figure 1, p and q are at a λ-distance of 1 and 3 from t, respectively.

3   A BRAVE Scheduler

In this section, we describe our scheduler BRAVE. Essentially, we describe the criteria for BRAVE to make a preemption.

BRAVE is greedy in nature; whenever it has completed a broadcast and there are still unsatisfied requests, it immediately broadcasts the page that has the highest value at that time. However, unlike previous schedulers, BRAVE makes two kinds of preempting broadcasts: the profit-gaining broadcasts, whose completion will increase the profit of the current schedule, and the finish-trimming broadcasts, whose completion may not increase the profit, but will reduce the completion time of the current scheduler. Following are the details.

Let $\lambda_o = 3\Delta/\log \Delta$. The behaviour of BRAVE depends on λ_o. Suppose that σ is the sequence of requests arrived so far. Let $S = \langle B_1, \ldots, B_{\mathrm{nft}}, \ldots, B_{\mathrm{last}} \rangle$ be the sequence of broadcasts that BRAVE scheduled for σ. Suppose that B_{nft} is the last broadcast in S that is not a finish-trimming broadcast. (Such broadcast must exist because, as can be verified from the definition given below, the first broadcast B_1 is not a finish-trimming broadcast.) Note that B_{nft} may be equal to B_{last}. BRAVE decides whether to preempt the current broadcast $B_{\mathrm{last}} = (P_{\mathrm{last}}, t_{\mathrm{last}})$ based on the value of $B_{\mathrm{nft}} = (P_{\mathrm{nft}}, t_{\mathrm{nft}})$ (not the value of B_{last}).

Let t be any time during the broadcast of B_{last}.

- If there is a page P such that the value of P at t is higher than $\sqrt{\Delta}$ times the value of B_{nft}, i.e.,

$$v_{S,\sigma}(P,t) > \sqrt{\Delta} v_{S,\sigma}(P_{\mathrm{nft}}, t_{\mathrm{nft}}),$$

  BRAVE preempts the current broadcast B_{last} and broadcast P at t. We say that the preempting broadcast (P, t) is *profit-gaining*.
- Suppose that BRAVE cannot make any profit-gaining broadcast at t. Let Γ be the set of pages Q satisfying the following two conditions:
  (i) the supposed completion time of (Q, t) is smaller than that of the current broadcast B_{last}, i.e., $t + \ell(Q) < t_{\mathrm{last}} + \ell(P_{\mathrm{last}})$, and
  (ii) the value of (Q, t) is large enough when compared with that of B_{nft}, or more precisely,

$$v_{S,\sigma}(Q,t) > \Big(\frac{2^d}{\Delta^{1/3}}\Big) v_{S,\sigma}(P_{\mathrm{nft}}, t_{\mathrm{nft}})$$

  where $d = d_{\lambda_o}(t_{\mathrm{nft}}, t + \ell(Q))$ is the λ_o-distance between the starting time of B_{nft} and the supposed completion time of (Q, t).
  If Γ is not empty, BRAVE preempts B_{last} and broadcasts the page $Q \in \Gamma$ with the smallest page length (such that (Q, t) has the smallest supposed completion time). We say that the preemptive broadcast (Q, t) is *finish-trimming*.

In the rest of the paper, we derive an upper bound on the competitive ratio of BRAVE, and prove that there is no deterministic online scheduler for BROADCAST that has asymptotically smaller competitive ratio.

4   Brave Is ($\frac{6\Delta}{\log \Delta} + O(\Delta^{5/6})$)-Competitive

Let σ be any input request sequence. Let S be the schedule decided by Brave for σ, and let O be the optimal offline schedule for σ. In this section, we derive an upper bound on the competitive ratio of Brave, i.e., the ratio between $\rho(O, \sigma)$ and $\rho(S, \sigma)$. In our analysis, we assume that there is no preemption in O and all broadcasts in O are complete[2].

Recall that $R_{S,\sigma}(P, t)$ is the set of requests on P that are alive during $[t, t + \ell(P))$ and are not satisfied by S at or before t, and $v_{S,\sigma}(P, t)$ is the value of P at t with respect to (S, σ). The key step of our analysis is to prove that the following inequality:

$$\sum\nolimits_{(P,t)\in O} v_{S,\sigma}(P, t) \leq \left(\frac{6\Delta}{\log \Delta} + O(\Delta^{5/6}) \right) \rho(S, \sigma). \tag{1}$$

Intuitively, the inequality asserts that from the view point of S, the value of the broadcasts in O is not too much. We can derive an upper bound on the competitive ratio of Brave from Inequality (1) easily.

Theorem 1. We have $\rho(O, \sigma) \leq (\frac{6\Delta}{\log \Delta} + O(\Delta^{5/6}))\rho(S, \sigma)$ and hence the competitive ratio of Brave is at most $(\frac{6\Delta}{\log \Delta} + O(\Delta^{5/6}))$.

Proof. Note that $\rho(O, \sigma) = \sum_{(P,t)\in O} v_{O,\sigma}(P, t)$ because all broadcasts in O are complete. Therefore,

$$\rho(O, \sigma) - \sum\nolimits_{(P,t)\in O} v_{S,\sigma}(P, t) = \sum\nolimits_{(P,t)\in O} v_{O,\sigma}(P, t) - \sum\nolimits_{(P,t)\in O} v_{S,\sigma}(P, t)$$
$$= \sum\nolimits_{(P,t)\in O} \left(\sum\nolimits_{r\in R_{O,\sigma}(P,t)} v(r) - \sum\nolimits_{r\in R_{S,\sigma}(P,t)} v(r) \right),$$

which is no greater than the summation

$$\sum\nolimits_{(P,t)\in O} \left(\sum\nolimits_{r\in R_{O,\sigma}(P,t)\setminus R_{S,\sigma}(P,t)} v(r) \right) \tag{2}$$

where $R_{O,\sigma}(P, t) \setminus R_{S,\sigma}(P, t)$ denotes the difference of the two sets.

Note that for any $(P, t) \in O$, the requests in $R_{O,\sigma}(P, t) \setminus R_{S,\sigma}(P, t)$ are not satisfied by O at or before t, and are satisfied by S at or before t. Observe that these requests will be satisfied by O after (P, t) completes, and this implies that no request will be in more than one $R_{O,\sigma}(P, t) \setminus R_{S,\sigma}(P, t)$ in Summation (2). We conclude that a request can contribute its value $v(r)$ at most once to (2) and only those requests that are satisfied by S will make contribution. Therefore, (2) is no greater than $\rho(S, \sigma)$. Together with Inequality (1), the theorem follows.

In the rest of this section, we prove Inequality (1). Note that the schedule $S = \langle B_1, B_2, \ldots, B_n \rangle$ constructed by Brave on serving σ can be divided naturally into a collection Π_S of disjoint subsequence of broadcasts H where H may contain only a single broadcast, which is not preempting and is complete, or $H = \langle B_i, B_{i+1}, \ldots, B_{j-1}, B_j \rangle$ where

---

[2] We can make this assumption because O is offline and it does not need to make any broadcast that is going to be preempted.

1. B_i is not a preempting broadcast; it does not preempt any earlier broadcast;
2. $B_{i+1}, B_{i+2}, \ldots, B_j$ are all preempting broadcasts; and
3. the last broadcast B_j is also a complete broadcast.

We call such H a *subschedule* of S. For any subschedule $H = \langle B_i, B_{i+1}, \ldots, B_j \rangle \in \Pi_S$, define the *duration* $I(H)$ of H to be $[t_s, t_c]$ where t_s is the time the first broadcast B_i starts and t_c is the time the last broadcast B_j completes. For any broadcast $B = (P, t) \in O$ and any time interval I, we say that I covers B if O starts broadcasting B during I, or equivalently, $t \in I$. Let $\text{Cover}(O, I)$ be the set of broadcasts in O that are covered by I. The following simple lemma is our basic tool to relate S and O.

Lemma 1. *We have the following equality:*

$$\sum_{(P,t)\in O} v_{S,\sigma}(P,t) = \sum_{H\in\Pi_S} \left(\sum_{(P,t)\in\text{Cover}(O,I(H))} v_{S,\sigma}(P,t) \right).$$

Proof. It suffices to prove that for any broadcast $(P, t) \in O$, if $v_{S,\sigma}(P,t) > 0$, then there is a subschedule $H \in \Pi_S$ such that $(P, t) \in \text{Cover}(O, I(H))$, which is equivalent to proving that BRAVE is not idle at t. However, this is obvious because BRAVE can at least broadcast P at t to gain a profit of $v_{S,\sigma}(P,t) > 0$.

Below, we focus on one subschedule $H \in \Pi_S$ and derive an upper bound on $\sum_{(P,t)\in\text{Cover}(O,I(H))} v_{S,\sigma}(P,t)$. In our discussion, we suppose that

- the first broadcast in H is (P_1, t_1);
- there are exactly $k-1$ profit-gaining preempting broadcasts $(P_2, t_2), (P_3, t_3),$ $\ldots, (P_k, t_k)$ in H where $t_2 < t_3 < \cdots < t_k$,[3] and
- $(P_{\text{last}}, t_{\text{last}})$ is last broadcast in H.

Note that $(P_{\text{last}}, t_{\text{last}})$ may be equal to (P_k, t_k). Furthermore, $(P_{\text{last}}, t_{\text{last}})$ is the only broadcast in H that is complete. Let t_c be the completion time of this broadcast. Note that between t_i and t_{i+1} $(1 \leq i < k)$, and between t_k and t_c, BRAVE may make other preempting broadcasts, which are all finish-trimming.

The following lemma asserts that for any $1 \leq i < k$, the total value of the broadcasts in O that start during $[t_i, t_{i+1})$ (i.e., between the time BRAVE makes the two profit-making preempting broadcasts (P_i, t_i) and (P_{i+1}, t_{i+1})) is not too high. Intuitively, this is true because except the last broadcast, all these broadcasts complete before t_{i+1} and thus they could be finish-trimming broadcasts in S. However, we observe that these broadcasts must not be in S, and thus we conclude that they do not satisfy the second requirement for a finishing-trimming preemeption. In other words, their values are not high when compared with that of the profit-making broadcast (P_i, t_i).

Lemma 2. *For $1 \leq i < k$, we have*

$$\sum_{(P,t)\in\text{Cover}(O,[t_i,t_{i+1}))} v_{S,\sigma}(P,t) < (\tfrac{6\Delta}{\log\Delta} + \sqrt{\Delta})v_{S,\sigma}(P_i,t_i).$$

---

[3] k may be equal to 1.

Proof. Consider any $1 \leq i < k$. Let G be the sequence of broadcasts made by BRAVE during $[t_i, t_{i+1})$. Note that except (P_i, t_i), all broadcasts in G are finish-trimming and thus the supposed completion time of any of them is smaller than that of the previous one. It follows that the last broadcast $X \in G$ has the smallest supposed completion time, which is no greater than $t_i + \ell(P_i)$ and is larger than t_{i+1} (because it is preempted by (P_{i+1}, t_{i+1})). Therefore, we have

$$t_{i+1} < t_i + \ell(P_i) \leq t_i + \Delta. \tag{3}$$

Now we consider $\text{Cover}(O, [t_i, t_{i+1}))$, the set of broadcasts in O that are covered by $[t_i, t_{i+1})$. Suppose $\text{Cover}(O, [t_i, t_{i+1})) = \{(Q_1, s_1), (Q_2, s_2), \ldots, (Q_h, s_h)\}$ where $s_1 < s_2 < \cdots < s_h$. Recall that O does not have any preemptive broadcast and thus all the (Q_j, s_j)'s are complete. We have the following useful facts:

(i) $t_i \leq s_1 < s_2 < \cdots < s_h < t_{i+1}$.

(ii) The supposed completion time of any two (Q_j, s_j)'s differ by at least 1 because the minimum page length is 1.

(iii) Except (Q_h, s_h), all broadcasts (Q_j, s_j) complete before t_{i+1}.

From Fact (iii), Inequality (3) and the fact that $\lambda_o = 3\Delta/\log \Delta$, we conclude that for any $1 \leq j < h$,

$$s_j + \ell(Q_j) < t_{i+1} < t_i + \Delta < t_i + (\lfloor (\log \Delta)/3 \rfloor + 1) \lambda_o$$

and $s_j + \ell(Q_j)$ is at a λ_o-distance of at most $\lfloor (\log \Delta)/3 \rfloor$ from t_i. Therefore,

$$\sum_{\substack{(Q,s) \in \text{Cover}(O, [t_i, t_{i+1})), \\ (Q,s) \neq (Q_h, s_h)}} v_{S,\sigma}(Q, s) = \sum_{0 \leq d \leq \lfloor (\log \Delta)/3 \rfloor} \sum_{\substack{1 \leq j < h, \\ d_{\lambda_o}(t_i, s_j + \ell(Q_j)) = d}} v_{S,\sigma}(Q_j, s_j). \tag{4}$$

Fact (ii) implies that for any d, there are at most λ_o different $1 \leq j < h$ such that $t_i + d\lambda_o \leq s_j + \ell(Q_j) < t_i(d+1)\lambda_o$, or equivalently,

$$|\{j \mid d_{\lambda_o}(t_i, s_j + \ell(Q_j)) = d, \ j \neq h\}| \leq \lambda_o = 3\Delta/\log \Delta. \tag{5}$$

Consider any such j. Fact (iii) says that the supposed completion time of (Q_j, s_j) is smaller than t_{i+1}, which is smaller than that of the last broadcast $X \in G$, which has the smallest supposed completion time among the broadcasts in G. This implies (Q_j, s_j) is not in G, even though its supposed completion time is smaller than that of any broadcast in G and satisfied the first requirement of being a finish-trimming broadcast. We conclude that it violates the second requirement, and thus $v_{S,\sigma}(Q_j, s_j) \leq \left(\frac{2^d}{\Delta^{1/3}} \right) v_{S,\sigma}(P_i, t_i)$. Combining with (4), (5), we conclude that $\sum_{((Q,s) \in \text{Cover}(O, [t_i, t_{i+1}))} v_{S,\sigma}(Q, s)$ is equal to

$$\left(\sum_{\substack{0 \le d \le \lfloor (\log \Delta)/3 \rfloor}} \sum_{\substack{1 \le j < h, \\ d_{\lambda_o}(t_i, s_j + \ell(Q_j)) = d}} v_{S,\sigma}(Q_j, s_j) \right) + v_{S,\sigma}(Q_h, s_h)$$

$$\le \left(\sum_{\substack{0 \le d \le \lfloor (\log \Delta)/3 \rfloor}} \lambda_o \left(\frac{2^d}{\Delta^{1/3}} \right) v_{S,\sigma}(P_i, t_i) \right) + v_{S,\sigma}(Q_h, s_h)$$

$$< \left(\frac{6\Delta}{\log \Delta} \right) v_{S,\sigma}(P_i, t_i) + v_{S,\sigma}(Q_h, s_h).$$

Finally, note that $v_{S,\sigma}(Q_h, s_h) < \sqrt{\Delta} v_{S,\sigma}(P_i, t_i)$; otherwise BRAVE would have made a profit-making preemptive broadcast at $s_h < t_{i+1}$. The lemma follows.

Since the only complete broadcast $(P_{\text{last}}, t_{\text{last}})$ in H is covered by $[t_k, t_c]$, we need a tighter bound for this interval, which is given in the following lemma.

Lemma 3. *For the interval* $[t_k, t_c]$, *we have*

$$\sum_{(P,t) \in \text{Cover}(O, [t_k, t_c])} v_{S,\sigma}(P, t) \le \left(\frac{6 \cdot 2^w \Delta}{\Delta^{1/3} \log \Delta} + \sqrt{\Delta} \right) v_{S,\sigma}(P_k, t_k) \qquad (6)$$

where $w = d_{\lambda_o}(t_k, t_c)$ *is the* λ_o*-distance between* t_k *and* t_c. *Furthermore,* $w \le \lfloor (\log \Delta)/3 \rfloor$.

Proof. The proof is almost identical to that of Lemma 2, but we have to be more careful about the boundary condition. Recall that $(P_{\text{last}}, t_{\text{last}})$ is the last broadcast in H, which completes at t_c. Note that either $(P_{\text{last}}, t_{\text{last}}) = (P_k, t_k)$ or $(P_{\text{last}}, t_{\text{last}})$ is a finish-trimming broadcast[4]. In both cases, we have $t_c \le t_k + \ell(P_k)$. Suppose that $\text{Cover}(O, [t_k, t_c]) = \{(Q_1, s_1), \dots, (Q_h, s_h)\}$ and (Q_h, s_h) is the last broadcast in $\text{Cover}(O, [t_k, t_c])$. Note that for all the $(Q_j, s_j) \ne (Q_h, s_h)$, they complete before $t_c \le t_k + \ell(P_k)$, but they are not finish-trimming broadcasts in H; by the design of BRAVE, we conclude $v_{S,\sigma}(Q_j, s_j) \le \left(\frac{2^d}{\Delta^{1/3}} \right) v_{S,\sigma}(P_k, t_k)$. where d is the λ_o-distance between t_k and $s_j + \ell(Q_j)$. From the fact that t_c is at a λ_o-distance of w from t_k, we have

$$\sum_{((Q,s) \in \text{Cover}(O, [t_k, t_c]))} v_{S,\sigma}(Q, s) = \sum_{\substack{0 \le d \le w}} \sum_{\substack{1 \le j < h, \\ d_{\lambda_o}(t_i, s_j) = d}} v_{S,\sigma}(Q_j, s_j) + v_{S,\sigma}(Q_h, s_h)$$

$$(7)$$

Together with (5), we can derive Inequality (6) in exactly the same way as we did for Lemma 2.

For the second part of the lemma, note that t_c is the completion time of $(P_{\text{last}}, t_{\text{last}})$. If $(P_{\text{last}}, t_{\text{last}}) = (P_k, t_k)$, then $t_c = t_k + \ell(P_k) \le t_k + \Delta$; otherwise $(P_{\text{last}}, t_{\text{last}})$ is a finishing-trimming preemption, and as in the proof of Lemma 3, we conclude that $t_c < t_k + \ell(P_k) \le t_k + \Delta$. In both cases, we conclude that $w = d_{\lambda_o}(t_k, t_c) \le \lfloor (\log \Delta)/3 \rfloor$.

Now, we combine the bounds on those intervals $[t_i, t_{i+1})$ in $I(H)$. Define $v_{S,\sigma}(H)$ to be $v_{S,\sigma}(P_{\text{last}}, t_{\text{last}})$, the value of the only complete broadcast in H.

---

[4] It cannot be profit-making because (P_k, t_k) is the last profit-making broadcast.

Lemma 4. $\sum_{(P,t)\in \text{Cover}(O,I(H))} v_{S,\sigma}(P,t) \leq \left(\frac{6\Delta}{\log \Delta} + O(\Delta^{5/6})\right) v_{S,\sigma}(H).$

Proof. Recall that (P_1, t_1) is the first broadcast in H, and $(P_2, t_2), (P_3, t_3), \ldots,$ (P_k, t_k) are the sequence of profit-making broadcasts in H. Thus, for any $1 < i \leq k$, (P_{i-1}, t_{i-1}) is the nearest non finish-trimming broadcast before (P_i, t_i), and by the design of BRAVE, it makes the profit-making broadcast (P_i, t_i) because $v_{S,\sigma}(P_i, t_i) > \sqrt{\Delta} v_{S,\sigma}(P_{i-1}, t_{i-1})$. Together with Lemma 2, we have

$$
\begin{aligned}
S_o &= \sum_{1 \leq i < k} \sum_{(P,t)\in \text{Cover}(O,[t_i, t_{i+1}))} v_{S,\sigma}(P,t) \\
&\leq (\frac{6\Delta}{\log \Delta} + \sqrt{\Delta})(v_{S,\sigma}(P_{k-1}, t_{k-1}) + \cdots v_{S,\sigma}(P_1, t_1)) \\
&< (\frac{6\Delta}{\log \Delta} + \sqrt{\Delta})v_{S,\sigma}(P_k, t_k)\left(\frac{1}{\sqrt{\Delta}} + \frac{1}{(\sqrt{\Delta})^2} + \cdots\right)
\end{aligned}
\tag{8}
$$

Note that

- if $(P_k, t_k) = (P_{\text{last}}, t_{\text{last}})$, then by the fact that $w \leq \lfloor (\log \Delta)/3 \rfloor$, we have $v_{S,\sigma}(P_{\text{last}}, t_{\text{last}}) = v_{S,\sigma}(P_k, t_k) \geq (\frac{2^w}{\Delta^{1/3}}) v_{S,\sigma}(P_k, t_k)$;
- if $(P_k, t_k) \neq (P_{\text{last}}, t_{\text{last}})$, then $(P_{\text{last}}, t_{\text{last}})$ is a finish-trimming preempting broadcast and it satisfies the second requirement of the preempting rule, i.e., $v_{S,\sigma}(P_{\text{last}}, t_{\text{last}}) > (\frac{2^w}{\Delta^{1/3}})v_{S,\sigma}(P_k, t_k)$.

Thus, in both cases, $v_{S,\sigma}(P_k, t_k) \leq (\frac{\Delta^{1/3}}{2^w})v_{S,\sigma}(P_{\text{last}}, t_{\text{last}})$, and together with (8), we have

$$
S_o < (\frac{6\Delta}{\log \Delta} + \sqrt{\Delta})\frac{\Delta^{1/3}}{(\sqrt{\Delta} - 1)2^w}v_{S,\sigma}(P_{\text{last}}, t_{\text{last}}) = O(\Delta^{5/6})v_{S,\sigma}(H).
\tag{9}
$$

Furthermore, let $S_1 = \sum_{(P,t)\in \text{Cover}(O,[t_k, t_c])} v_{S,\sigma}(P,t)$, which is no more than

$$
(\frac{6 \cdot 2^w \Delta}{\Delta^{1/3} \log \Delta} + \sqrt{\Delta})v_{S,\sigma}(P_k, t_k) \leq (\frac{6 \cdot 2^w \Delta}{\Delta^{1/3} \log \Delta} + \sqrt{\Delta})\frac{\Delta^{1/3}}{2^w}v_{S,\sigma}(P_{\text{last}}, t_{\text{last}}).
$$

It can be verified that the last term is equal to $(\frac{6\Delta}{\log \Delta} + O(\Delta^{5/6}))v_{S,\sigma}(H)$. Finally, note that $\sum_{(P,t)\in \text{Cover}(O,I(H))} v_{S,\sigma}(P,t) = S_o + S_1$, the lemma follows.

Note that our main Inequality (1) is just a corollary of Lemma 4.

Corollary 1. $\sum_{(P,t)\in O} v_{S,\sigma}(P,t) \leq \left(\frac{6\Delta}{\log \Delta} + O(\Delta^{5/6})\right) \rho(S, \sigma)$

Proof. By Lemmas 1 and 4, we have

$$
\sum_{(P,t)\in O} v_{S,\sigma}(P,t) = \sum_{H \in \Pi_S} \sum_{(P,t)\text{Cover}(O,I(H))} v_{S,\sigma}(P,t),
$$

which is no more than $\sum_{H \in \Pi_S}(\frac{6\Delta}{\log \Delta} + O(\Delta^{5/6}))v_{S,\sigma}(H) = (\frac{6\Delta}{\log \Delta} + O(\Delta^{5/6}))\rho(S, \sigma)$.

5   BRAVE Is Optimal

In this section, we prove that when $\Delta \geq 2$, any deterministic online scheduler for BROADCAST must have competitive ratio at least $\frac{\Delta}{2 \ln \Delta} - 1$. Note that for any $1 \leq \Delta < 2$, the $(\Delta + 2\sqrt{\Delta} + 2)$-competitive scheduler of Fung [4] has constant competitive ratio. Furthermore, our lower bound shows that BRAVE is optimal within a constant factor because BRAVE is $(6 \frac{\Delta}{\log \Delta} + O(\Delta^{5/6}))$-competitive,

Let A be any deterministic online scheduler for BROADCAST. We now describe an adversary that generates difficult input request sequence for A. The adversary generates the input from two groups of requests:

- Group 1 has only one request $(Q, 0, \Delta, W)$, which arrives at time 0 and asks for a page Q of length Δ. The request has tight deadline and its value is W.
- Group 2 has $\lceil \Delta \rceil$ requests; for $0 \leq i \leq \lceil \Delta \rceil - 1$, the ith request is $(P_i, i - 1, i, v_i)$, which arrives at time $i - 1$, asks for a distinct page P_i of length one, has tight deadline and its value v_i is $c^{i-1} \frac{\ln \Delta}{\Delta} W$ where $c = 1 + \frac{\ln \Delta - \ln \ln \Delta}{\Delta}$.

Since each request r has tight deadline, r has to be served immediately, or it cannot be served at all. The adversary will present the requests to A one by one, and as soon as A decides to serve a request in Group 2, no more requests are issued. Let σ be the input sequence issued according to this strategy, S be the schedule generated by A for σ, and O be the optimal offline schedule for σ.

Theorem 2. *We have $\rho(O, \sigma) > (\frac{\Delta}{2 \ln \Delta} - 1)\rho(S, \sigma)$ and hence the competitive ratio of A is greater than $\frac{\Delta}{2 \ln \Delta} - 1$.*

Proof. According to the behavior of the adversary, A can earn profit from at most one request; either the single request Q in Group 1, or a request P_i in Group 2. We consider two cases.

Case 1. A serves the ith request in Group 2. Then, $\rho(S, \sigma) = c^{i-1} \frac{\ln \Delta}{\Delta} W$. Based on the value of c^{i-1}, we estimate the ratio between $\rho(O, \sigma)$ and $\rho(S, \sigma)$ as follows. If $c^{i-1} \leq 2$, we can serve Q and gain a profit of W. Thus $\rho(O, \sigma) \geq W$ and $\frac{\rho(O,\sigma)}{\rho(S,\sigma)} \geq \frac{W}{c^{i-1} \frac{\ln \Delta}{\Delta} W} = \frac{\Delta}{c^{i-1} \ln \Delta} \geq \frac{\Delta}{2 \ln \Delta}$.

Suppose that $c^{i-1} > 2$. Note that we can serve all the requests in Group 2 arriving at or before time $i - 1$ and earn a total profit of $(1 + c + \cdots + c^{i-1}) \frac{\ln \Delta}{\Delta} W$. Since $\rho(S, \sigma) = c^{i-1} \frac{\ln \Delta}{\Delta} W$, we have $\rho(O, \sigma)/\rho(A, \sigma)$ is greater than

$$\frac{1 + c + \cdots + c^{i-1}}{c^{i-1}} = \frac{c^i - 1}{c^{i-1}(c - 1)} = (c - \frac{1}{c^{i-1}}) \frac{1}{c - 1} > \frac{\Delta}{2(\ln \Delta - \ln \ln \Delta)} > \frac{\Delta}{2 \ln \Delta}.$$

Case 2. A earns the profit from the request $(Q, 0, \Delta, W)$ in Group 1, and thus does not serve any request in Group 2. Then, we have $\rho(A, \sigma) = W$. Note that in this case, the adversary has a chance to issue all requests in Group 2; we can serve all of them and gain a profit of $(1 + c + \cdots + c^{\lceil \Delta \rceil - 1}) \frac{\ln \Delta}{\Delta} W$. Since $c = 1 + \frac{\ln \Delta - \ln \ln \Delta}{\Delta}$, we have $\rho(O, \sigma)/\rho(S, \sigma)$ is no smaller than

$$(1 + c + \cdots + c^{\lceil \Delta \rceil - 1}) \frac{\ln \Delta}{\Delta} = \frac{\ln \Delta (c^{\lceil \Delta \rceil} - 1)}{\Delta(c - 1)} = \frac{\ln \Delta (c^{\lceil \Delta \rceil} - 1)}{\Delta \frac{\ln \Delta - \ln \ln \Delta}{\Delta}} > c^{\Delta} - 1. \quad (10)$$

Recall that for any $x > 0$, we have $\ln(1 + x) > x/(1 + x)$ and $1/(1 + x) > 1 - x$. Thus, $\ln c^{\Delta} = \Delta \ln \left(1 + \frac{\ln \Delta - \ln \ln \Delta}{\Delta}\right) > \left(\Delta \frac{\ln \Delta - \ln \ln \Delta}{\Delta}\right) / \left(1 + \frac{\ln \Delta - \ln \ln \Delta}{\Delta}\right) > (\ln \Delta - \ln \ln \Delta)\left(1 - \frac{\ln \Delta - \ln \ln \Delta}{\Delta}\right) = (\ln \Delta - \ln \ln \Delta) - ((\ln \Delta - \ln \ln \Delta)^2/\Delta)$, or equivalently,

$$c^{\Delta} > e^{(\ln \Delta - \ln \ln \Delta) - (\ln \Delta - \ln \ln \Delta)^2/\Delta} = \frac{\Delta}{\left(e^{(\ln \Delta - \ln \ln \Delta)^2/\Delta}\right) \ln \Delta}.$$

It can be verified that the function $f(\Delta) = e^{(\ln \Delta - \ln \ln \Delta)^2/\Delta}$ is monotonically decreasing, and thus $f(\Delta) \leq e^{(\ln 2 - \ln \ln 2)^2/2} = 1.75319 < 2$ for any $\Delta \geq 2$. It follows that $c^{\Delta} > \frac{\Delta}{2 \ln \Delta}$. Together with (10), we have $\frac{\rho(O,\sigma)}{\rho(S,\sigma)} > \frac{\Delta}{2 \ln \Delta} - 1$.

References

1. D. Aksoy and M. Franklin. R×W: a scheduling approach for large-scale on-demand data broadcast. *IEEE/ACM Transactions on networking*, 7(6):846–860, 1999.
2. W.T. Chan, T.W. Lam, H.F. Ting, and W.H. Wong. New results on on-demand broadcasting with deadline via job scheduling with cancellation. In *Proceedings of the 10th Annual International Conference on Computing and Combinatorics*, 210–218.
3. J. Edmonds and K. Pruhs. A maiden analysis of longest wait first. In *Proceedings of the fifteenth annual ACM-SIAM symposium on Discrete algorithms*, pages 818–827, 2004.
4. S.P.Y. Fung. *Online algorithms for the provision of quality of service in networks*. PhD thesis, The University of Hong Kong, 2005.
5. S. Jiang and N.H. Vaidya. Scheduling data broadcast to "impatient" users. In *Proceedings of the 1st ACM international workshop on Data engineering for wireless and mobile access*, pages 52–59, 1999.
6. S. Jiang and N.H. Vaidya. Response time in data broadcast systems: Mean, variance and tradeoff. *Mobile Networks and Applications*, 7(1):37–47, 2002.
7. B. Kalyanasundaram, K. Pruhs, and M. Velauthapillai. Scheduling broadcasts in wireless networks. *Journal of Scheduling*, 4(6):339–354, 2001.
8. B. Kalyanasundaram and M. Velauthapillai. On-demand broadcasting under deadline. In *Proceedings of the 11th Annual European Symposium on Algorithms, volume 2832 of* Lecture Notes in Computer Science, pages 313–324, 2003.
9. J.H. Kim and K.Y. Chwa. Scheduling broadcasts with deadlines. *Theoretical Computer Science*, 325(3):479–448, 2004.
10. R. Lipton and A. Tomkins. Online interval scheduling. In *Proceedings of the Fifth Annual ACM-SIAM Symposium on Discrete Algorithms*, pages 302–311, 1994.
11. S. Muthukrishnan and S. Acharya. Scheduling on-demand broadcasts: new metrics and algorithms. In *The Fourth Annual ACM/IEEE International Conference on Mobile Computing and Networking*, pages 43–54, 1998.
12. K. Pruhs and P. Uthaisombut. A comparison of multicast pull models. In *Proceedings of the 10th Annual European Symposium on Algorithms*, pages 808–819, 2002.
13. G.J. Woeginger. On-line scheduling of jobs with fixed start and end times. *Theoretical Computer Science*, 130:5–16, 1994.
14. F. Zhang, P.Y. Fung, F. Chin, C.K. Poon, and Y. Xu. Improved on-line broadcast scheduling with deadlines. Technical report, Submitted for publication.

Fair Cost-Sharing Methods for Scheduling Jobs on Parallel Machines*

Yvonne Bleischwitz[1,2] and Burkhard Monien[1]

[1] Faculty of Computer Science, Electrical Engineering and Mathematics,
University of Paderborn, Fürstenallee 11, 33102 Paderborn, Germany
{yvonneb, bm}@upb.de
[2] International Graduate School of Dynamic Intelligent Systems

Abstract. We consider the problem of sharing the cost of scheduling n jobs on m parallel machines among a set of agents. In our setting, each agent owns one job and the cost is given by the makespan of the computed assignment. We focus on α-budget-balanced cross-monotonic cost-sharing methods since they guarantee the two substantial mechanism properties α-budget-balance and group-strategyproofness and provide fair cost-shares. For identical jobs on related machines and for arbitrary jobs on identical machines, we give $(m + 1)/(2m)$-budget-balanced cross-monotonic cost-sharing methods and show that this is the best approximation possible. As our major result, we prove that the approximation factor for cross-monotonic cost-sharing methods is unbounded for arbitrary jobs and related machines. We therefore develop a cost-sharing method in the $(m + 1)/(2m)$-core, a weaker but also fair solution concept. We close with a strategyproof mechanism for the model of arbitrary jobs and related machines that recovers at least $3/5$ of the cost. All given solutions can be computed in polynomial time.

1 Introduction

Motivation and Framework. We consider the scenario, in which a service provider owns a set of machines and receives requests from agents to execute their jobs. Each agent has a non-publicly observable preference for his job to be processed. He submits a bid to the service provider that indicates the amount of money he is willing to pay. If his job is processed, he has to make a payment to the service provider. We refer to a payment as *cost-share*. The utility of an agent expresses his valuation of receiving the service at a certain cost-share. The aim of an agent is to maximize his utility. We assume that agents are selfish. Therefore the provider can generally not rely on receiving *truthful bids*, i.e. bids that equal the private preferences.

In our model, the provider's cost of assigning jobs to his machines is given by the *makespan*, i.e. the time needed until all machines have processed their assigned jobs. The provider's problem is to determine the set of served agents, their cost-shares, and a valid assignment for the served agents. He would like to recover as much of the cost as

* This work has been partially supported by the German Science Foundation (DFG) priority program 1126 *Algorithms of Large and Complex Networks* under grant MO 285/15-3, and by the European Union within the 6th Framework Programme under contract 001907 (DELIS).

T. Calamoneri, I. Finocchi, G.F. Italiano (Eds.): CIAC 2006, LNCS 3998, pp. 175–186, 2006.

possible. Furthermore, he aims to minimize the makespan for reasons of efficiency and he wants to prevent being manipulated by the agents. To be practicable, his problem has to be computable in polynomial time. Since his scheduling problem is NP-hard in general, he has to apply approximation algorithms. The proposed scenario is of particular importance for commercial computing centers as well as for the evolving commercial grid computing offerings.

The problem of scheduling a set of n jobs on a set of m parallel machines with the objective of minimizing the makespan is an extensively studied problem. The most commonly used models are the models of *related* and *unrelated* machines. In the model of related machines, the completion time of a job on a machine does only depend on its workload and on the speed of the machine, where in the model of unrelated machines, machines have player-specific completion times. Recently, these models have been considered in the context of game theory. In this branch of research, there is no central authority that assigns the jobs, but selfish agents themselves assign their jobs to machines. The objective is to obtain an assignment in Nash equilibrium in which no agent can profit by assigning his job to another machine, given that all other agents leave the assignment of their jobs unchanged.

We recall the provider's problem, which is to determine a set of served agents, their cost-shares, and an assignment for the set of served agents. We can utilize assignment algorithms to compute the assignment but need different tools for determining the set of service-receiving agents and their cost-shares.

The theory of mechanism design proposes *cost-sharing mechanisms* that provide a solution to the problem of choosing the set of served agents and their payments. These mechanisms apply *cost-sharing methods* to determine the cost-shares. Two important fairness properties of cost-sharing methods are *cross-monotonicity* and the *core property*. Cross-monotonic cost-sharing methods require that the cost-share of an agent can only decrease if more agents receive the service. The weaker core property assures, that a coalition is always charged not more than the optimal cost of exclusively assigning the jobs of the coalition. This implies that no coalition is overcharged. Furthermore, a cost-sharing method can be α-*budget-balanced*, which guarantees that the service provider covers an α-fraction of his cost and assures the serviced agents that their collective cost-share is not larger than the cost of an optimal solution. If it additionally satisfies the core-property, we say that it is in the α-*core*. Significant properties of cost-sharing mechanisms are *strategyproofness* and *group-strategyproofness*, demanding that an agent or a group of agents can not improve their utility by submitting untruthful bids. This keeps them from manipulating the service provider. Cross-monotonic cost-sharing methods play a very important role in the design of cost-sharing mechanism, since they can be applied to design group-strategyproof mechanisms [25, 18].

Contribution and Significance. The main contributions of this paper are results on cost-sharing methods that are both α-budget-balanced and cross-monotonic. To the best of our knowledge, this paper is the first to introduce cross-monotonic cost-sharing methods for scheduling jobs on parallel machines. We prove that cross-monotonic cost-sharing methods that are α-budget-balanced do not exist for $\alpha > (m + 1)/(2m)$, not even for identical jobs and machines, and give cross-monotonic methods with factor $\alpha = (m+1)/(2m)$ for arbitrary jobs and identical machines or identical jobs and related machines. For arbitrary jobs and related machines, cross-monotonicity is impracticable. Our results show, that it is impossible to obtain cross-monotonic cost-sharing methods that recover more than a $1/d$-fraction of the cost, and that it is possible to recover a $1/(2d)$-fraction, where d is the number of different workloads.

In order to achieve a better approximation, we design a weaker but also fair cost-sharing method that is in the $(m + 1)/(2m)$-core. In addition, we propose a strategyproof mechanism that recovers at least $3/5$ of the cost and makes no agent pay more than if his job were solely processed. All proposed methods run in polynomial time and compute Nash equilibria.

Related Work. The assignment problem for the model of unrelated [30, 22, 16] and related [8, 10, 12, 14, 15] machines has been extensively studied in the past. We focus on the model of related machines. Hochbaum and Shmoys [15] give a PTAS for this model. In this paper, we frequently apply the LPT algorithm proposed by Graham [12]. LPT is optimal for identical jobs, achieves an approximation ratio of $4/3 - 1/(3m)$ for identical machines [12] and an approximation ratio of $5/3$ for related machines [9]. It is explained in Section 2. For results on computing Nash equilibria, we refer to the surveys of Gairing et al. [11] and Czumaj [4].

Cost-sharing mechanisms have mainly been designed for multicast [7, 6, 1], set cover [5], facility location [5, 27, 23], Steiner trees [19, 18], Steiner forests [20, 21], multicommodity rent-or-buy [3], and single-source rent-or-buy [27, 13]. Penna and Ventre [28] study algorithmic properties of cost-sharing mechanisms that among other properties satisfy group strategyproofness and budget-balance.

Cross-monotonic cost-sharing methods have been investigated for facility location [27, 23], single-source rent-or-buy [27, 13], and Steiner trees and forests [19, 18, 20]. Impossibility results are given by Immorlica et al. [17]. Moulin and Shenker [25] study the relations between group-strategyproofness and cross-monotonicity. One of their central results is a mechanism that is group-strategyproof if it applies a cross-monotonic cost-sharing method. The core is a well studied solution concept that stems from coalitional games with transferable payoffs and has for example been considered by Shapley [29].

Results on scheduling in the mechanism design context exist for other scheduling models. With regard to the fairness concept of the Shapley value, Mishra et al. [24] investigate the case in which there is only one server that can serve only one job at a time. Nisan and Ronen [26] consider unrelated parallel machines. In contrast to our model, machines are owned by agents that submit bids on execution times. They give a strategyproof mechanism that computes an assignment with makespan smaller than m times the optimal makespan and conjecture that this is the best possible. They prove that there is no strategyproof mechanism that computes an assignment with makespan smaller than 2 times the optimal makespan. Archer and Tardos [2] consider the scenario in which agents own related machines and give a strategyproof mechanism whose computed assignment yields a makespan that is smaller than 3 times the optimal makespan.

Road Map. Section 2 gives the basic definitions from mechanism design and defines the scheduling problem. Our results on cross-monotonicity are given in Section 3. Finally, Section 4 mainly focuses on results for the core and gives a strategyproof mechanism.

2 Definitions

Let N be the set of potential customers with $|N| = n$. The set of machines owned by the service provider is denoted by M. Agent $i \in N$ has a private preference $v_i \in \mathbb{R}_{\geq 0}$ for his job to be processed. If his job is processed at a certain cost-share $x_i \in \mathbb{R}_{\geq 0}$ his utility is defined as $u_i = v_i - x_i$. Otherwise, his cost-share and his utility are zero. In his request of being served, he submits a bid b_i that corresponds to the amount of money he is willing to pay. Since he is guided by self-interest, he chooses his bid such as to maximize his utility. The provider experiences a certain cost by scheduling a set of jobs. We assume that this cost is given by the makespan as defined in Section 2.2. His problem is to determine a set of served agents $U \subseteq N$, their cost-shares $x_i(U) \in \mathbb{R}_{\geq 0}$, that recover as much of his cost as possible, and a valid assignment for U.

We give the basic definitions and results on cost sharing methods and mechanisms for an unspecified service in Section 2.1. This section provides solutions on how the provider can extract the agents' real preferences while recovering a certain fraction of his cost. Section 2.2 specifies the service of scheduling the agents' jobs. Throughout the paper, we use $[k], k \in \mathbb{N}$, to denote the set $\{1, \ldots, k\}$ of integers.

2.1 Mechanism Design for Cost-Sharing

Let $c_{\mathcal{A}}(U)$ be the cost of a solution computed by some algorithm $\mathcal{A}$ to provide the service to $U \subseteq N$. In many cases, this algorithm is an approximation algorithm to assure polynomial time. We write $opt(U)$ for the cost of an optimal solution to provide the service to U. For a given set $U \subseteq N$, a *cost-allocation function* $\xi : N \to \mathbb{R}$ *for the set* $U \subseteq N$ specifies the cost-shares of each $i \in U$. It satisfies $\xi(i) \geq 0$ for all $i \in U$ and $\xi(i) = 0$ for all $i \notin U$. Let $\xi(U) = \sum_{i \in U} \xi(i)$. A *cost-sharing method* is a collection of cost-allocation functions:

Definition 1 (cost-sharing method). *A cost-sharing method x is defined as a function* $x : N \times 2^N \to \mathbb{R}$ *satisfying for all $U \subseteq N$, that $x(i, U) \geq 0$ for all $i \in U$ and* $x(i, U) = 0$ *for all $i \notin U$. We will denote $x(i, U)$ by $x_i(U)$. Let $x(U) = \sum_{i \in U} x_i(U)$.*

Ideally, we would like to have *budget-balance*, i.e. $c_{\mathcal{A}}(U) = x(U) = opt(U)$ for all $U \subseteq N$. In many cases it is not possible to achieve budget-balance if the cost-sharing method is to meet other properties as well, or it is computationally hard to compute. Therefore, this condition is relaxed. A cost-sharing method $x(\cdot)$ is α-*budget-balanced* for $\alpha \leq 1$ if it satisfies $\alpha c_{\mathcal{A}}(U) \leq x(U) \leq opt(U)$ for all $U \subseteq N$. A cost-allocation function for $U \subseteq N$ is α-*budget-balanced*, if the above condition holds for U. Observe, that dividing the cost-shares by α results in cost-shares that guarantee the full coverage of the actual cost and an overall cost-share of less than α^{-1} times the optimal solution. Although this is the more intuitive definition we use the definition given first for reasons of clearness.

Both α-budget-balanced cost-sharing methods and cost-allocation functions can have the property to be in the α-core. Intuitively, no coalition is overcharged:

Definition 2 (the α-core property). *A cost-allocation function $\xi(\cdot)$ for $U \subseteq N$ is in the α-core iff it is α-budget-balanced and for all $U' \subseteq U : \xi(U') \leq opt(U')$. A cost-sharing method $x(\cdot)$ is in the α-core iff for all $U \subseteq N, x(\cdot, U)$ is in the α-core.*

The provider's problem can be solved by a *cost-sharing mechanism* and it's underlying cost-sharing method. A cost sharing mechanism is an algorithm that is given the agents' bids $\{b_i\}_{i \in N}$. It outputs the set of agents $U \subseteq N$ that receive the service and cost-shares $x_i(U) \in \mathbb{R}$ with $0 \leq x_i(U) \leq b_i$ for all $i \in U$ and $x_i(U) = 0$ for all $i \notin U$. Furthermore, it outputs a solution with cost $c_{\mathcal{A}}(U)$ to provide the service to U. We focus on assuring the following mechanism properties:

- **strategyproofness:** Agent $i \in N$ maximizes his utility by bidding $b_i = v_i$.
- **group-strategyproofness:** A coalition $U \subseteq N$ of users cannot collude and submit untruthful bids such that as a result, each of them has at least the same utility and at least one of them has a strictly larger utility compared to the outcome that results if each of them bids truthfully.
- **α-budget-balance**, $\alpha \leq 1 : \alpha c_{\mathcal{A}}(U) \leq x(U) \leq opt(U)$ holds for the set U of service-receiving agents.

Cross-monotonic cost-sharing methods play a crucial role in the context of how to achieve α-budget-balance and group-strategyproofness.

Definition 3 (cross-monotonicity). *A cost-sharing method $x(\cdot)$ is cross-monotonic if for all $U, U' \subseteq N, U' \subseteq U : x_i(U') \geq x_i(U) \; \forall i \in U'$.*

If the underlying cost sharing method is cross-monotonic and α-budget-balanced, a simple mechanism given by Moulin and Shenker [25] is α-budget-balanced and group-strategyproof [18].

It is easy to see that each α-budget-balanced cross-monotonic cost sharing method is in the α-core. From this we can conclude that if there is no cost-allocation function for some set $U \subseteq N$ in the α-core, then no α-budget-balanced cross-monotonic cost-sharing method can exist. On the other hand, there can be cost-sharing methods that are in the α-core and are not cross-monotonic.

2.2 The Scheduling Problem

Let N be the set of n agents with $|N| = n$. Each agent $i \in N$ owns exactly one job of workload $w_i \in \mathbb{N}$. Therefore, we will use $U \subseteq N$ to denote agents and jobs interchangeably. For $U \subseteq N$, let $W(U) = \sum_{i \in U} w_i$ and $w_{max}(U) = \max_{i \in U} w_i$. Let $d(U)$ denote the number of different workloads in U. Moreover, there is a set M of m machines. Each machine $j \in M$ has speed $s_j \in \mathbb{N}$. We assume that $s_1 \geq \ldots \geq s_m$. For $M' \subseteq M$, let $S(M') = \sum_{j \in M'} s_j$. If all speeds are the same, we say that the machines are identical. Otherwise we call them related. Jobs are identical, if all workloads are the same. Without loss of generality we assume that identical machines and jobs have speeds and workloads of one respectively.

An assignment allocates each job to exactly one machine. For a given assignment, let δ_j be the sum of the workloads of the jobs assigned to machine j. Then the completion time of a job assigned to machine j is (δ_j / s_j). The makespan is defined as $\max_{j \in M}(\delta_j / s_j)$. We call the machines whose completion time is equal to the makespan *makespan machines*. The optimal solution for a set of jobs $U \subseteq N$ is an assignment with minimal makespan, denoted by $opt(U)$.

To compute an assignment, we apply Graham's LPT algorithm [12]. LPT processes the jobs in decreasing order and assigns each job to a machine on which it experiences the smallest completion time (taking into account the jobs that have been assigned already). For a set $U \subseteq N$ we use $lpt(U)$ to denote the makespan resulting from LPT, i.e. $lpt(U) = c_{LPT}(U)$. For an assignment for jobs $U \subseteq N$ computed by LPT, let $m(U)$ be the set of machines that jobs are assigned to. The running time of LPT is $O(n)$ for identical jobs and identical machines, $O(n \log m)$ for identical jobs and related machines, and $O(n \log n)$ otherwise. Even though there are better approximation algorithms for the assignment problem [8, 10, 14, 15], our main results cannot be improved by switching to another algorithm. A nice additional property of LPT that we exploit in most proofs is that in each iteration, the current assignment is in Nash equilibrium. An assignment is in Nash equilibrium, if no agent can improve by deviating from the current assignment, i.e. for each job i from the set of served agents $U \subseteq N$ that is assigned to machine $j \in M$ it holds that $(\delta_k + w_i)/s_k \geq \delta_j/s_j$ for all $k \in M \setminus \{j\}$.

There are three LPT specific assignment properties that we will utilize in our proofs. Lemma 1 states these well-known properties.

Lemma 1. *Let $U \subseteq N$ and let $\hat{U} \subseteq U$ be the jobs assigned by LPT until the makespan first occurs, and let $\tau = |m(\hat{U})|$. Then it holds, that:*

1. *For identical machines, $W(U)/m \leq opt(U)$.*
2. *For related machines, $W(\hat{U})/S(m(\hat{U})) \leq opt(\hat{U})$.*
3. *If machines are identical and there are at least two jobs assigned to a makespan machine, then $lpt(U) \leq \frac{2m}{m+1} \frac{W(U)}{m}$.*
4. *If there are at least two jobs assigned to some machine, then*
   - *for related machines: $lpt(U) \leq \frac{2\tau}{\tau+1} \frac{W(\hat{U})}{S(m(\hat{U}))}$.*
   - *for identical jobs: $lpt(U) \leq \frac{2m(U)}{m(U)+1} \frac{|U|}{S(m(U))}$.*

3   Results on Cross-Monotonicity

In this section, we give α-budget-balanced cross-monotonic cost-sharing methods that yield α-budget-balanced group-strategyproof mechanisms if used as input for the mechanism by Moulin and Shenker [25]. All proposed methods rely on solving the assignment problem via LPT. The property that LPT computes a Nash equilibrium is utilized frequently. We say that an algorithm *computes a cost-sharing method* $x(\cdot)$ in time $f(m, n)$ if for each set $U \subseteq N$ the cost-shares $\{x_i(U)\}_{i \in U}$ are computed in time $f(m, n)$. The mechanism by Moulin and Shenker runs in time $O(nf(m, n) + g(m, n))$, where $g(m, n)$ is the running time of LPT. Proofs omitted due to space restrictions are provided in the full version of this paper.

Theorems 1 and 2 propose $(m + 1)/(2m)$-budget-balanced cross-monotonic cost-sharing methods for the scheduling problem with identical jobs and for the scheduling problem with identical machines. Due to Theorem 5 in Section 4, these cross-monotonic methods achieve the best budget-balance factor possible.

Theorem 1. *There is an $\frac{m+1}{2m}$-budget-balanced cross-monotonic cost-sharing method for the scheduling problem with arbitrary jobs and identical machines computable in time $O(n)$.*

Theorem 2. *There is an $\frac{m+1}{2m}$-budget-balanced cross-monotonic cost-sharing method for the scheduling problem with identical jobs and related machines computable in time $O(n \log m)$.*

The central Theorem 4 states, that the approximation factor for cross-monotonic cost-sharing methods is unbounded for arbitrary jobs and related machines. It depends on $d(N)$, the number of different workloads in the set N. By Theorem 3, it is possible to achieve $(2d(N))^{-1}$-budget-balance:

Theorem 3. *There is a $(2d(N))^{-1}$-budget-balanced cross-monotonic cost-sharing method for the scheduling problem with arbitrary jobs and related machines computable in time $O(n \log n)$.*

Theorem 4. *For the scheduling problem with arbitrary jobs and related machines, there is no α-budget-balanced cross-monotonic cost-sharing method for the factor α with $\alpha > (d(N) + \epsilon)^{-1}$, $\forall \epsilon > 0$.*

Proof. We proceed as follows: we fix a set of machines and consider classes of scheduling instances in which the job workloads equal their speeds. Classes are defined by specifying the number of agents and jobs respectively of a certain job workload. For average cost-shares on these instances, we derive properties that are met by all cross-monotonic cost-sharing methods. Afterwards, we derive a bound on α-budget balance where α will be determined later.

Instances. The considered classes consist of instances with $d = d(N)$ different workloads $1, a, \ldots, a^{d-1}$, with $a \in \mathbb{N}_{>1}$. There are m_j machines having a speed of a^{d-j} with $j \in [d]$. Let $m_1 = 1$ and $m_j = (a - 1) \sum_{l=1}^{j-1} m_l a^{j-l}$ for $j \geq 2$. It holds that $m_j = a^2 m_{j-1}$ for $j \geq 3$, which can easily be proved by induction. We use the more complicated formulation that simplifies later arguments. For $j \in [d]$, let N_j be the set of all agents with jobs of workload a^{d-j} and $n_j = |N_j|$. Then, $N = \cup_{j \in [d]} N_j$. For $U \subseteq N$, let $U_j = U \cap N_j$. U_j extracts from U all jobs with workload a^{d-j}. Let the profile $(u_1, \ldots, u_d)$ denote the class of all sets U with $u_j = |U_j|$ for all $j \in [d]$.

Optimal Assignments. First, consider the class $(m_1, \ldots, m_d)$. Obviously, for every instance of this class consisting of the set of jobs U, $opt(U) = 1$. Now change the jth entry to $r_j = am_j$. We show, that $opt(U) = a$ for every instance with the set of jobs U of the class $(m_1, \ldots, r_j, \ldots, m_d)$ for all $j \in [d]$. First, we give an assignment with makespan a. Then we show that it is impossible to obtain a makespan smaller than a.

The assignment is computed as follows. Assign all jobs of workload a^{d-l}, $l \in [d]$ to the machines of speed a^{d-l}. This results in a completion time of one on machines with speed a^{d-l}, $l \neq j$ and a completion time of a on machines with speed a^{d-j}.

Now we show a lower bound for the optimal assignment. Assume, that there exists an assignment with makespan smaller than a. Observe, that all jobs with workload larger than a^{d-j} have to be assigned to the machines with speed larger than a^{d-j}. Now look at the jobs with workload a^{d-j}. They can only be assigned to the machines with speed at least a^{d-j}. At most $(a - 1)m_j$ of them can be assigned to the machines with speed a^{d-j}. Now, all jobs of workload larger than a^{d-j} and the m_j remaining jobs of workload a^{d-j} have to be assigned to the machines with speed larger than a^{d-j}. The makespan cannot be smaller than a, because a lower bound for the optimal assignment for these jobs on these machines is given by a:

$$\frac{\sum_{l=1}^{j-1}(m_l a^{d-l}) + m_j a^{d-j}}{\sum_{l=1}^{j-1}(m_l a^{d-l})} = \frac{\sum_{l=1}^{j-1}(m_l a^{d-l}) + (a-1)\sum_{l=1}^{j-1}(m_l a^{d-l})}{\sum_{l=1}^{j-1}(m_l a^{d-l})} = a . \quad (1)$$

Cross-Monotonicity. In the following, we assume, that there is a cross-monotonic cost-sharing method $x(\cdot)$. Let $\Gamma(m_1, \ldots, m_d) = \prod_{l=1}^{d} \binom{n_l}{m_l}$. For all instances of the class $(m_1, \ldots, m_d)$, the average cost share of the agents with jobs in $N_k, k \in [d]$ is

$$\chi_k := \chi_k((m_1, \ldots, m_d)) := \Gamma(m_1, \ldots, m_d)^{-1} \sum_{\substack{U \subseteq N \\ \forall l: |U_l| = m_l}} \sum_{i \in U_k} x_i(U) . \quad (2)$$

Now change the jth profile entry to $r_j = am_j$. Then, the average cost-share for agents with jobs in N_k is $\chi_k((m_1, \ldots, r_j, \ldots, m_d))$. Define $\Gamma = \Gamma(m_1, \ldots, m_d)$ and also $\Gamma_j = \Gamma(m_1, \ldots, r_j, \ldots, m_d)$. We will utilize cross-monotonicity to bound it from above in terms of χ_k.

Consider the set $U_j \subseteq N_j$ with $|U_j| = r_j$ and $U_j \subseteq U \subseteq N$. First, let $k = j$. Every single cost-share of an agent $i \in U_j$ for the set U is not larger than his cost-share for the set $(U \backslash U_j) \cup \{i\} \cup \tilde{U}$, with $\tilde{U} \subset U_j \backslash \{i\}, |\tilde{U}| = m_j - 1$. Especially, it is not larger than the average value of the cost-shares for i for each of these $\binom{r_j - 1}{m_j - 1}$ sets. Therefore an upper bound of $\chi_j((m_1, \ldots, r_j, \ldots, m_d))$ is given by:

$$\chi_j^{(j)} := \Gamma_j^{-1} \sum_{\substack{U \subseteq N \\ \forall l \neq j: |\tilde{U}_l| = m_l \\ |U_j| = r_j}} \sum_{i \in U_j} \sum_{\substack{\tilde{U} \subset U_j \backslash \{i\} \\ |\tilde{U}| = m_j - 1}} \frac{x_i((U \backslash U_j) \cup \{i\} \cup \tilde{U})}{\binom{r_j - 1}{m_j - 1}} . \quad (3)$$

Now, let $k \neq j$. Every single cost-share of an agent $i \in U \backslash U_j$ for U is not larger than the cost-share for i for $(U \backslash U_j) \cup \tilde{U}, \tilde{U} \subset U_j, |\tilde{U}| = m_j$. With the same argument as above, the following upper bound of $\chi_k((m_1, \ldots, r_j, \ldots, m_d))$ results:

$$\chi_k^{(j)} := \Gamma_j^{-1} \sum_{\substack{U \subseteq N \\ \forall l \neq j: |\tilde{U}_l| = m_l \\ |U_j| = r_j}} \sum_{i \in U_k} \sum_{\substack{\tilde{U} \subset U_j \\ |\tilde{U}| = m_j}} \frac{x_i((U \backslash U_j) \cup \tilde{U})}{\binom{r_j}{m_j}} . \quad (4)$$

We now give a lemma on the relationship between the average cost-shares and their bounds.

Lemma 2. $a\chi_j = \chi_j^{(j)}$ and $\chi_k = \chi_k^{(j)}$ for $j \in [d]$ and $k \in [d], k \neq j$.

Proof. We first look at χ_j and $\chi_j^{(j)}, j \in [d]$. Observe, that both sums are over the same subsets of N_l with m_l elements for $l \in [d] \backslash \{j\}$. It therefore suffices to consider both sums for fixed subsets $U_l \subset N_l, |U_l| = m_l, l \in [d] \backslash \{j\}$ only. Let $U = \cup_{l \in [d] \backslash \{j\}} U_l$. Define:

$$\tilde{\chi}_j := \Gamma^{-1} \sum_{\substack{\tilde{U} \subset N_j \\ |\tilde{U}| = m_j}} \sum_{i \in \tilde{U}} x_i(U \cup \tilde{U}) \text{ and} \tag{5}$$

$$\tilde{\chi}_j^{(j)} := \Gamma_j^{-1} \binom{r_j - 1}{m_j - 1}^{-1} \sum_{\substack{U' \subseteq N_j \\ |U'| = r_j}} \sum_{i \in U'} \sum_{\substack{\tilde{U} \subset U' \setminus \{i\} \\ |\tilde{U}| = m_j - 1}} x_i(U \cup \{i\} \cup \tilde{U}). \tag{6}$$

$\tilde{\chi}_j$ and $\tilde{\chi}_j^{(j)}$ are related to each other the same way than χ_j and $\chi_j^{(j)}$. Now,

$$\tilde{\chi}_j^{(j)} = \Gamma_j^{-1} \binom{r_j - 1}{m_j - 1}^{-1} \sum_{\substack{U' \subset N_j \\ |U'| = r_j}} \sum_{\substack{\tilde{U} \subset U' \\ |\tilde{U}| = m_j}} \sum_{i \in \tilde{U}} x_i(U \cup \tilde{U}) \tag{7}$$

$$= \Gamma_j^{-1} \binom{r_j - 1}{m_j - 1}^{-1} \binom{n_j - m_j}{r_j - m_j} \sum_{\substack{\tilde{U} \subset N_j \\ |\tilde{U}| = m_j}} \sum_{i \in \tilde{U}} x_i(U \cup \tilde{U}). \tag{8}$$

Equation (7) is a simple combinatorial observation. To obtain (8), we investigate how often each subset of N_j with m_j elements occurs. For each subset $\tilde{U}$ with m_j elements, to determine a superset $U' \supset \tilde{U}$ with r_j elements, we have $\binom{n_j - m_j}{r_j - m_j}$ possibilities. Combining Equations (5) and (8) we get:

$$\tilde{\chi}_j^{(j)} = \Gamma_j^{-1} \Gamma \binom{r_j - 1}{m_j - 1}^{-1} \binom{n_j - m_j}{r_j - m_j} \tilde{\chi}_j = \frac{r_j}{m_j} \tilde{\chi}_j = a\tilde{\chi}_j. \tag{9}$$

Therefore, $a\chi_j = \chi_j^{(j)}$ for $j \in [d]$. With similar argumentation, we can conclude that $\chi_k = \chi_k^{(j)}$ for $j \in [d]$ and $k \in [d] \setminus \{j\}$.   □

Budget-Balance. Let us now assume, that $x(\cdot)$ is not only cross-monotonic but also α-budget-balanced. We have seen that the optimal cost for all instances of $(m_1, \ldots, m_d)$ and therefore the average optimal cost is one. With the same argument, the average optimal cost of all instances in class $(m_1, \ldots, r_j, \ldots, m_d)$ for $j \in [d]$ is a. Then we can conclude:

$$\sum_{k=1}^{d} \chi_k \leq 1 \text{ and } \sum_{k=1}^{j-1} \chi_k + a\chi_j + \sum_{k=j+1}^{d} \chi_k \geq a\alpha \ \forall j \in [d]. \tag{10}$$

Summation of these equations yields $\alpha \leq \frac{a-1+d}{da}$. For every $\epsilon > 0$ and a sufficient large a, this results in $\alpha \leq 1/(d + \epsilon)$. Note, that it suffices to consider the optimal cost instead of the LPT cost in Equation (10). If an α-fraction of the optimal cost cannot be recovered, in particular it cannot be recovered for a non-optimal cost.   □

4   The Core and Other Solution Concepts

Since an α-budget-balanced cross-monotonic cost-sharing method is in the α-core, Theorem 5 tells us that the cost-sharing methods defined in the proofs of Theorems 1 and 2 yield the best approximation factor possible. Theorem 6 provides us with a cost-sharing method in the $(m + 1)/(2m)$-core for the scheduling problem with arbitrary jobs and related machines.

Theorem 5. *For the scheduling problem with identical jobs and machines, there is no cost-sharing method in the α-core for $\alpha > (m+1)/(2m)$.*

Proof. We show that for $\alpha > (m+1)/(2m)$, there is no cost-allocation function in the α-core for the set U with $|U| = m+1$. Let $U' \subset U, |U'| = m$. Assume, there is a cost-allocation function $\xi(\cdot) : N \to \mathbb{R}$ for the set U in the α-core. Since we have that $\sum_{i \in U'} \xi(i) \leq opt(U') = 1$, there is an agent $k \in U'$ with $\xi(k) \leq 1/m$. Then,

$$\sum_{i \in U} \xi(i) = \xi(k) + \sum_{i \in U \setminus \{k\}} \xi(i) \leq 1/m + opt(U \setminus \{k\}) = 1/m + 1 . \quad (11)$$

From $2\alpha \leq \sum_{i \in U} \xi(i) \leq (1+m)/m$, we can conclude that $\alpha \leq (m+1)/(2m)$. $\square$

Theorem 6. *There is a cost-sharing method in the $(m+1)/(2m)$-core of the scheduling problem with arbitrary jobs and related machines computable in time $O(n \log n)$.*

Proof. Let $U \subseteq N$. Let $\hat{U} \subset U$ be the set of jobs, that LPT assigns until the makespan is reached and let $\tau = |m(\hat{U})|$. Furthermore, we denote by $m_{opt}(U)$ the machines that an optimal assignment uses to assign the set U.

To define the cost-sharing method, we look at two different cases. In the first case $\tau < |\hat{U}|$, i.e. if the makespan first occurs, there is at least one machine that is assigned more than one job. Then, define $x_i(U) = w_i/S(m(\hat{U}))$ for all $i \in \hat{U}$ and $x_i(U) = 0$ for all $i \notin \hat{U}$. In the second case $\tau = |\hat{U}|$, i.e. if the makespan first occurs, LPT has assigned at most one job to each machine. Let $\tau \geq 3$. We will omit the proof of the subcase $\tau \in \{1, 2\}$ due to space restrictions . We define $A(U) = S(m(\hat{U}))opt(U) - W(\hat{U})$. Let $x_i(U) = 0$ for $i \notin \hat{U}$. For $i \in \hat{U}$, let

$$x_i(U) = \begin{cases} \frac{w_i}{S(m(\hat{U}))} & \text{if } A(U) < \frac{\tau-1}{\tau+1}W(\hat{U}) \\ \frac{w_i}{S(m(\hat{U}))-s_\tau} & \text{otherwise .} \end{cases}$$

First observe, that LPT determines the running time. We have to show for the given cost-sharing method $x(\cdot)$ that $x(\cdot, U)$ is in the α-core for all $U \subseteq N$. We start with the first case in which $\tau < |\hat{U}|$. $x(U)$ is smaller than $opt(U)$, since by Lemma 1 it holds, that $x(U) = W(\hat{U})/S(m(\hat{U})) \leq opt(\hat{U}) \leq opt(U)$. Lemma 1 also provides the approximation factor. Next, we show the core condition. Let $U' \subseteq U$. From the proof of Lemma 1 we can conclude, that $m(\hat{U}) \supseteq m_{opt}(\hat{U})$. Therefore,

$$\sum_{i \in U'} x_i(U) = \frac{W(U' \cap \hat{U})}{S(m(\hat{U}))} \leq \frac{W(U' \cap \hat{U})}{S(m_{opt}(\hat{U}))} \leq \frac{W(U' \cap \hat{U})}{S(m_{opt}(U' \cap \hat{U}))} \leq opt(U') . \quad (12)$$

Consider the second case, $\tau = |\hat{U}|$. Then, $lpt(U) = opt(U) = w_\tau/s_\tau$. Due to space restrictions, we omit the case $A(U) < (\tau - 1)/(\tau + 1)W(\hat{U})$ and for the remaining case only show the core condition. If $A(U) \geq (\tau - 1)/(\tau + 1)W(\hat{U})$, then $x(U) = W(\hat{U})/(S(m(\hat{U})) - s_\tau)$.

Let $U' \subseteq U$. If $U' = \hat{U}$, then $\sum_{i \in U'} x_i(U) = x(U) \leq opt(U) = opt(U')$. Otherwise, $(U' \cap \hat{U}) \subset \hat{U}$. Since $lpt(U) = opt(U)$, it holds that $m(\hat{U}) = m_{opt}(\hat{U})$ and

$m_{opt}(U' \cap \hat{U}) \subseteq m_{opt}(\hat{U}) \setminus \{\tau\}$, since an optimal assignment for a proper subset of $\hat{U}$ does not use the machine τ anymore. Thus,

$$\sum_{i \in U'} x_i(U) = \frac{W(U' \cap \hat{U})}{S(m_{opt}(\hat{U})) - s_\tau} \leq \frac{W(U' \cap \hat{U})}{S(m_{opt}(U' \cap \hat{U}))} \leq opt(U') . \qquad (13)$$

$\square$

Finally, we state Theorem 7, whose proof is available in the full version.

Theorem 7. *There is a $3/5$-budget-balanced strategyproof cost-sharing mechanism for the scheduling game with arbitrary jobs and related machines. It is $3m/(4m - 1)$-budget-balanced for the scheduling game with identical jobs and related machines and 1-budget-balanced for identical jobs and identical machines. Its running time is the running time of LPT.*

Acknowledgements. We would like to thank Rainer Feldmann, Martin Gairing, and Karsten Tiemann for many helpful discussions.

References

1. A. Archer, J. Feigenbaum, A. Krishnamurthy, and R. Sami. Approximation and collusion in multicast cost sharing. *Games and Economic Behaviour*, 47:36–71, 2004.
2. A. Archer and E. Tardos. Truthful mechanisms for one-parameter agents. *Proceedings of the 42th IEEE Symposium on Foundations of Computer Science*, pages 482–491, 2001.
3. L. Beccetti, J. Könemann, S. Leonardi, and M. Pál. Sharing the cost more efficiently: improved approximation for multicommodity rent-or-buy. In *Proceedings of the 16th Annual ACM-SIAM Symposium on Discrete Algorithms*, pages 375–384, 2005.
4. A. Czumaj. Selfish Routing on the Internet. *Chapter 42 in Handbook of Scheduling: Algorithms, Models, and Performance Analysis*, 2004.
5. N. Devanur, M. Mihail, and V. Vazirani. Strategyproof cost sharing mechanisms for set cover and facility location problems. In *Proceedings of ACM Conference on Electronic Commerce*, pages 108–114, 2003.
6. J. Feigenbaum, A. Krishnamurthy, R. Sami, and S. Shenker. Hardness results for multicast cost sharing. *Theoretical Computer Science*, 304(1-3):215–236, 2003.
7. J. Feigenbaum, C. Papadimitriou, and S. Shenker. Sharing the cost of multicast transmissions. *Journal of Computer and System Sciences*, 63:21–41, 2001.
8. D. Friesen. Tighter bounds for the multifit processor scheduling algorithm. *SIAM Journal on Computing*, 13(1):170–181, 1984.
9. D. Friesen. Tighter bounds for lpt scheduling on uniform processors. *SIAM Journal on Computing*, 16(3):554–560, 1987.
10. D. Friesen and M. Langston. Bounds for multifit scheduling on uniform processors. *SIAM Journal on Computing*, 12(1):60–70, 1983.
11. M. Gairing, T. Lücking, B. Monien, and K. Tiemann. Nash Equilibria, the Price of Anarchy and the Fully Mixed Nash Equilibrium Conjecture. In *Proceedings of the 32nd International Colloquium on Automata, Languages and Programming*, volume 3580 of *LNCS*, pages 51–65, 2005.
12. R. Graham. Bounds on multiprocessing timing anomalies. *SIAM Journal of Applied Mathematics*, 17(2):416–429, 1969.

13. A. Gupta, A. Srinivasan, and E. Tardos. Cost-Sharing Mechanisms for Network Design. *Proceedings of the 7th International Workshop on Approximation Algorithms for Combinatorial Optimization Problems*, 3122:139–152, 2004.
14. D. Hochbaum and D. Shmoys. Using dual approximation algorithms for scheduling problems: theoretical and practical results. *Journal of the ACM*, 34(1):144–162, 1987.
15. D. Hochbaum and D. Shmoys. A polynomial approximation scheme for scheuduling on uniform processors: using the dual approximation approach. *SIAM Journal on Computing*, 17(3):539–551, 1988.
16. E. Horowitz and S. Sahni. Exact and approximate algorithms for scheduling nonidentical processors. *Journal of the Association for Computing Machinery*, 23(2):317–327, 1976.
17. N. Immorlica, M. Mahdian, and V. Mirrokni. Limitations of cross-monotonic cost sharing schemes. In *Proceedings of the 16th Annual ACM-SIAM Symposium on Discrete Algorithms*, pages 602–611, 2005.
18. K. Jain and V. Vazirani. Applications of approximate algorithms to cooperative games. In *Proceedings of the 33th Annual ACM Symposium on Theory of Computing*, pages 364–372, 2001.
19. K. Kent and D. Skorin-Kapov. Population monotonic cost allocation on msts. In *Operational Research Proceedings KOI*, pages 43–48, 1996.
20. J. Könemann, S. Leonardi, and G. Schäfer. A group-strategyproof mechanism for steiner forests. In *Proceedings of the 16th Annual ACM-SIAM Symposium on Discrete Algorithms*, pages 612–619, 2005.
21. J. Könemann, S. Leonardi, G. Schäfer, and S. van Zwam. From primal-dual to cost shares and back: a stronger LP relaxation for the steiner forest problem. In *Proceedings of the 32th Int. Colloquium on Automata, Languages, and Programming*, pages 930–942, 2005.
22. J. K. Lenstra, D. B. Shmoys, and E. Tardos. Approximation algorithms for scheduling unrelated parallel machines. In *Proceedings of the 28th Annual Symposium on Foundations of Computer Science (FOCS'87)*, pages 217–224, 1987.
23. S. Leonardi and G. Schäfer. Cross-monotonic cost-sharing methods for connected facility location games. In *ACM Conference on Electronic Commerce*, pages 224–243, 2004.
24. D. Mishra and B. Rangarajan. Cost sharing in a job scheduling problem using the shapley value. In *Proceedings of the 6th ACM Conference on Electronic Commerce*, pages 232–239, 2005.
25. H. Moulin and S. Shenker. Strategyproof sharing of submodular costs: budget balance versus efficiency. *Economic Theory*, 18:511–533, 2001.
26. N. Nisan and A. Ronen. Algorithmic Mechanism Design. *Games and Economic Behaviour*, 35:166–196, 2001. Extended abstract appeard at STOC'99.
27. M. Pál and E. Tardos. Group strategyproof mechanisms via primal-dual algorithms. In *Proceedings of the 44th Annual IEEE Symposium on Foundations of Computer Science*, pages 584–593, 2003.
28. P. Penna and C. Ventre. The Algorithmic Structure of Group Strategyproof Budget-Balanced Cost-Sharing Mechanisms. In *Proceedings of the 23rd International Symposium on Theoretical Aspects of Computer Science, to appear*, 2006.
29. L. S. Shapley. On balanced sets and cores. *Naval Research Logistics Quarterly*, 14:453–460, 1967.
30. E. V. Shchepin and N. Vakhania. An optimal rounding gives a better approximation for scheduling unrelated machines. *Operations Research Letters*, 33:127–133, 2005.

Tighter Approximation Bounds for LPT Scheduling in Two Special Cases

Annamária Kovács

Max-Planck Institut für Informatik,
Stuhlsatzenhausweg 85,
66123 Saarbrücken, Germany
Fax: +49-681-93-25-199
panni@mpi-inf.mpg.de

Abstract. $Q||C_{\max}$ denotes the problem of scheduling n jobs on m machines of different speeds such that the makespan is minimized. In the paper two special cases of $Q||C_{\max}$ are considered: Case I, when $m-1$ machine speeds are equal, and there is only one faster machine; and Case II, when machine speeds are all powers of 2. Case I has been widely studied in the literature, while Case II is significant in an approach to design so called *monotone* algorithms for the scheduling problem.

We deal with the worst case approximation ratio of the classic list scheduling algorithm 'Longest Processing Time (LPT)'. We provide an analysis of this ratio Lpt/Opt for both special cases: For one fast machine, a tight bound of $(\sqrt{3}+1)/2 \approx 1.366$ is given. When machine speeds are powers of 2 (2-divisible machines), we show that in the worst case $41/30 < Lpt/Opt < 42/30 = 1.4$.

To our knowledge, the best previous lower bound for both problems was $4/3 - \epsilon$, whereas the best known upper bounds were $3/2 - 1/2m$ for Case I [6] resp. $3/2$ for Case II [10]. For both the lower and the upper bound, the analysis of Case II is a refined version of that of Case I.

1   Introduction

We consider the offline task scheduling problem on related (uniform) machines $(Q||C_{\max})$. In the input of this problem we are given a *speed vector* $\langle s_1, s_2, \ldots, s_m \rangle$ representing the speeds of m machines, and a *job vector* $\langle t_1, t_2, \ldots, t_n \rangle$, where t_j is the *size* of the jth job, $1 \leq j \leq n$. In general, machine speeds and job sizes are arbitrary positive numbers. We assume that $s_i \leq s_{i+1}$ $(1 \leq i < m)$, and $t_j \geq t_{j+1}$ $(1 \leq j < n)$, i.e., machine speeds are non-decreasing and job sizes are non-increasing. The goal is to assign the jobs to the machines, so that the overall finish time is minimized: If jobs assigned to machine i are $\{t_\gamma^i\}_{\gamma=1}^{\Gamma}$ then the *work* assigned to i is $w_i := \sum_{\gamma=1}^{\Gamma} t_\gamma^i$ and the *finish time* of i is $f_i := w_i/s_i$. The *makespan* to be minimized is $\max_{i=1}^{m} f_i$. This problem is NP-hard even for 2 identical machines [14], but it has an approximation scheme; for constant m a FPTAS exists [7, 8].

A classic, simple approximation algorithm for $Q||C_{\max}$ is the so called 'Longest Processing Time first' algorithm, or LPT for short. This algorithm picks the jobs

T. Calamoneri, I. Finocchi, G.F. Italiano (Eds.): CIAC 2006, LNCS 3998, pp. 187–198, 2006.
© Springer-Verlag Berlin Heidelberg 2006

one by one in decreasing order, and always assigns the next job to the machine where it will have the smallest completion time. In this paper we analyse the worst case ratio of LPT in two special cases of $Q||C_{max}$:

- Case I (one fast machine): $s_1 = s_2 = \ldots = s_{m-1} = 1, s_m = s > 1$;
- Case II (2-divisible speeds): $s_i = 2^{l_i}, \quad l_i \in \mathbb{Z}$.

For a particular instance of the scheduling problem, let Lpt denote the makespan produced by the LPT schedule, and Opt denote the optimum makespan. We provide tight bounds for the worst case of Lpt/Opt in Case I and 'nearly' tight bounds in Case II. A complete version of this paper is available at [9].

Related work. The approximation ratio of LPT for arbitrary machine speeds was first considered by Gonzalez, Ibarra, and Sahni in [6], where the authors prove that $Lpt/Opt < 2$, whereas for any $\epsilon > 0$ an instance exists so that $Lpt/Opt > 3/2 - \epsilon$. These bounds were later improved to $(1.512, 19/12)$ by Dobson [4], respectively to $(1.52, 1.67)$ by Friesen [5].

Case I has been studied in a number of papers: Liu and Liu [12] give approximation bounds in terms of m and s for a variation of LPT, and for list schedules (the case when jobs are given in any fixed order). Gonzalez et al. [6] obtain the lower and upper bounds $4/3 - \epsilon < Lpt/Opt \le 3/2 - 1/2m$. For $m = 2$ they prove the tight bound of $\frac{1+\sqrt{17}}{4}$. Cho and Sahni [2] analyse general list schedules for both arbitrary machine speeds and for Case I. For the latter they obtain the tight bound $\frac{1+\sqrt{5}}{2}$ if $m = 2$, and $3 - 4/(m+1)$ if $m \ge 3$. Li and Shi [11] consider the same special case, and suggest better heuristics than list scheduling for the online problem. Finally for $m = 2$, Mireault, Orlin, and Vohra [13] provide a complete analysis of Lpt/Opt in terms of s_2/s_1.

Case II has been recently studied from a different point of view: A scheduling algorithm is *monotone*, if increasing the speed of any particular machine does not decrease the work assigned to that machine. The monotonicity of an algorithm gained relevance in the context of mechanism design. If each machine speed is only known to the machine itself, we need to motivate that machines declare their true speeds to the scheduling mechanism. As shown by Archer and Tardos [1], such motivation is possible only if the scheduling algorithm used by the mechanism is monotone. Auletta et al. [15] conjecture that LPT is monotone if machine speeds are 2-divisible (or divisible, in general). In [10] we prove this conjecture; moreover we show that in case of 2-divisible speeds, LPT is a 3/2-approximation algorithm. For arbitrary input speeds we obtain a monotone 3-approximation algorithm LPT* by running LPT with machine speeds rounded to powers of 2. An improved approximation bound for LPT improves the bound for LPT* at the same time.

Our result. We present an instance of the $Q||C_{max}$ problem with speed vector $\langle 1, 1, \ldots, 1, 2^r \rangle$ $(r \in \mathbb{N})$, so that for this instance $Lpt/Opt > \frac{\sqrt{3}+1}{2} - \epsilon$ for arbitrary $\epsilon > 0$, if r and m are large enough. (The instance is the same for arbitrary $s > 1$ instead of 2^r.) With this we improve the previously known lower bound $4/3 - \epsilon$ for the approximation ratio of LPT in Case I [6] as well as in Case II [10]. It is interesting to mention that in both previous papers the bound $4/3$ is conjectured to be tight in the respective case.

After that we show that the lower bound $\frac{\sqrt{3}+1}{2} - \epsilon$ is actually tight in Case I, i.e., for any instance with one fast machine $Lpt/Opt < \frac{\sqrt{3}+1}{2}$ holds. To our best knowledge, the previous upper bound was $Lpt/Opt \leq 3/2 - 1/2m$ [6].

On the other hand, for Case II we show that the lower bound $\frac{\sqrt{3}+1}{2}$ is not tight: we could construct an instance with 2-divisible machines such that $Lpt/Opt > \frac{(\sqrt{409}+29)}{36} - \epsilon$, where $\frac{(\sqrt{409}+29)}{36} \approx 1.3673... > \frac{41}{30} > \frac{\sqrt{3}+1}{2}$. However, this instance relies on calculation with exact job sizes, completion times etc., and is valid only if LPT favours faster machines in case of ties. If LPT breaks ties arbitrarily, another instance exists [9] with approximation ratio $Lpt/Opt > 955/699 - \epsilon$, where $955/699 > \frac{\sqrt{3}+1}{2}$.

Both of these instances are further developed variants of the first instance. On the one hand, this slight improvement over $\frac{\sqrt{3}+1}{2}$ is of theoretical interest; on the other hand the new instances give an impression about how troublesome it might be to provide a tight approximation bound for 2-divisible machines.

Instead, with hardly more effort than in Case I, and following the same lines, it is now natural to prove an upper bound of 1.4 for 2-divisible machines. This improves on our previous upper bound of 1.5, and automatically provides a better worst case ratio of 2.8 for the monotone algorithm LPT* of [10].

Overview. In the rest of Section 1 we introduce further notation and definitions and state some basic observations. Section 2 presents an instance that proves the lower bound $\frac{\sqrt{3}+1}{2} - \epsilon$ for arbitrary $\epsilon > 0$ in both special cases. In Section 3 we show that this bound is tight in Case I, whereas Section 4 gives an upper bound of 1.4 in Case II. Some intuition about both upper bound proofs can be found at the beginning of the respective sections. Finally, Section 5 provides an example with improved lower bound for Case II. The proofs omitted from this short version can be found at [9].

Notation and definitions. We use t_j to denote both the jth job, and the size of the jth job in formulas. Throughout the paper t denotes the size of t_n. We will use the short expressions 1-job, y-job, t-job for a job of size 1, y, t, etc. Similarly, a 1-machine or a 4-machine means a machine of speed 1 or 4, respectively.

The work and the finish time of machine i in LPT is denoted by w_i, resp. $f_i = w_i/s_i$. In the upper bound proofs these values will be defined disregarding the last job t_n. The *completion time* of a job t_j assigned to machine i is the finish time of i right after t_j was scheduled.

The formal definition of LPT is as follows:

LPT algorithm: Input: $\langle s_1, \ldots, s_m \rangle$ and $\langle t_1, \ldots, t_n \rangle$

At step j of LPT let w_i^j denote the work of machine i ($1 \leq i \leq m$). LPT assigns t_j to machine h if $(w_h^j + t_j)/s_h = \min_i(w_i^j + t_j)/s_i$, and h is the largest machine index with this property.

In the above definition, LPT decides for the faster (higher index) machine in case of ties. Nevertheless, all our upper bound results hold if LPT prefers lower index machines (for simplicity we did not consider other definitions). Furthermore, the lower bound example in Section 2 is valid if ties are broken arbitrarily.

Next, we present a frequently used simple tool, called *principle of domination* [5, 3]. In the proofs of Sections 3 and 4 we assume a *minimal* counter-example, meaning that it has the smallest number of machines, and for this number of machines the smallest number of jobs. Let an instance of the $Q||C_{\max}$ problem be a minimal counter-example for an approximation upper bound of LPT. Let OPT be a fixed optimal schedule of this instance.

Definition 1. [5] *We say that machine i dominates machine i^* if*
(i) $s_i \le s_{i^*}$ and
(ii) LPT assigns the jobs $\tau_1, \ldots, \tau_k$ to i (disregarding t_n); OPT assigns the jobs $\tau_1^*, \ldots, \tau_l^*$ to i^*, and there is a function $F : \{\tau_1^*, \ldots, \tau_l^*\} \to \{\tau_1, \ldots, \tau_k\}$ such that for each τ_j, $\sum_{f(\tau_v)=\tau_j} \tau_v \le \tau_j$.

Proposition 1. (principle of domination [5]) *In a minimal counter-example for an upper bound on Lpt/Opt, no machine i dominates a machine i^*.*

The proof uses the argument that in case i dominates i^*, deleting i and all jobs (but t_n) assigned to i would result in a smaller counter-example. Note that as a corollary, there are no empty machines in OPT.

We conclude the introduction with a simple observation:

Proposition 2. *Let $s_i = s_{i+1}$. If in LPT t_j is the first job assigned to $i + 1$, then t_{j+1} is the first job assigned to i.*   □

2   A Lower Bound: $\frac{\sqrt{3}+1}{2} - \epsilon$

In this section we present an instance of the $Q||C_{\max}$ problem with $m - 1$ machines of speed 1 and one machine of speed 2^r ($r \in \mathbb{N}$). The instance is valid if ties are broken arbitrarily; moreover the use of speed 2^r instead of some $s > 1$ is not essential. The approximation ratio of LPT on this instance can be arbitrarily close to $(\sqrt{3} + 1)/2 \approx 1.366$. In particular, $Lpt > \sqrt{3} + 1 - \epsilon'$ and $Opt < 2 + \epsilon'$, where $\epsilon' > 0$ is arbitrarily small if m and r are large enough.

We will call the machine of speed 2^r the *fast machine*.

Theorem 1. *For any $\epsilon > 0$ there is a speed vector $\langle s_1 = \ldots = s_{m-1} = 1, s_m = 2^r \rangle$ and a job vector $\langle t_1, \ldots, t_n \rangle$, s. t. for this instance $Lpt/Opt > (\sqrt{3}+1)/2 - \epsilon$.*

The proof is given by the following instance:

Instance A. Let $x = 3 - \sqrt{3} \approx 1.268$ and $y = \sqrt{3} - 1 \approx 0.732$. We start by describing the assignment of jobs to machines in LPT (see Figure 1): The fast machine first receives $2^r - 1$ jobs of size x; then it is filled with as many jobs of size 1 as fit below time 2; finally it gets $2^r - 1$ jobs of size y. At this point the number of jobs on the fast machine is $2 \cdot (2^r - 1) + \lfloor 2 \cdot 2^r - (2^r - 1) \cdot x \rfloor$, and the total work on the fast machine amounts to at least $(2^r - 1) \cdot x + 2 \cdot 2^r - (2^r - 1) \cdot x - 1 + (2^r - 1) \cdot y = 2^r(2 + y) - 1 - y$.

The set of 1-machines is divided into blocks. The number of 1-machines in one block is $(x - 1)/\delta$, where $\delta > 0$ is arbitrarily small and it divides $x - 1$ evenly. The LPT schedule on a block is as follows: Each 1-machine has a large and a small job. The large jobs range from $x - \delta$ down to 1 by steps of δ and the small jobs range from y up to $1 - \delta$ by steps of δ. Every 1-machine has total work $y + x - \delta = 2 - \delta$.

We claim that if $1/2^r < \delta$, then the above assignment is an LPT schedule: all x-jobs on the fast machine are completed by time $x - x/2^r$; after that 1-machines receive their first jobs, all of size less than x. These jobs would have higher completion time on m. Since an additional 1-job on a 1-machine would not be completed before time 2, the 1-jobs are all assigned to m. Now the 1-machines receive their second jobs with completion time $2 - \delta < 2 - 1/2^r$ where $2 - 1/2^r$ is a lower bound on the current completion time of m. Finally, after (at most) $2^r - 1$ y-jobs, a last job of size y is assigned to one of the 1-machines, yielding makespan $(y + 2 - \delta) = \sqrt{3} + 1 - \delta$. On the fast machine this last job would have been completed after $(2^r(2 + y) - 1)/2^r = \sqrt{3} + 1 - 1/2^r > \sqrt{3} + 1 - \delta$.

Now we rearrange the jobs on the machines in order to get the optimum schedule. We claim that a block of 1-machines can be used to exchange an x-job for a 1-job *or* to exchange a 1-job for a y-job. The first happens if we shift the large jobs within a block, insert a job of size x instead of $x - \delta$, and take out a job of size 1. The second happens, if we shift the small jobs within a block, insert a 1-job and take out a y-job. In either case the new finish time on 1-machines will be 2.

Let the number of blocks be $2 \cdot (2^r - 1) + \lfloor 2 \cdot 2^r - (2^r - 1) \cdot x \rfloor$, so that every job of size x or size 1 on the fast machine can be exchanged for a y-job. Moreover, we put the very last job of size y on the fast machine. Now the total work on the fast machine is at most $y \cdot (2(2^r - 1) + 2 \cdot 2^r - (2^r - 1) \cdot x) + y = y \cdot 4 \cdot 2^r - y \cdot 2^r \cdot x + y(x - 1) = 2^r \cdot y \cdot (4 - x) + y \cdot (x - 1) = 2^r \cdot 2 + y \cdot (x - 1)$. Thus, the optimum makespan is at most $2 + y(x - 1)/2^r$. Clearly, the desired bound is obtained if $\epsilon' > \delta > 1/2^r > y(x - 1)/2^r$ for some appropriate ϵ'.

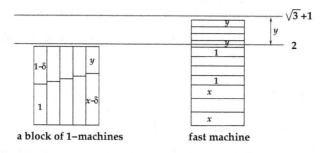

Fig. 1. Instance A: the assignment of jobs before the last job in LPT

3   Tight Bound

We consider the special case of $Q||C_{\max}$ when $s_1 = s_2 = \ldots = s_{m-1} = 1$ and $s_m = s > 1$. We show that in this case the bound given in Section 2 is tight:

Theorem 2. *For any instance of the $Q||C_{\max}$ problem for which $s_1 = s_2 = \ldots = s_{m-1} = 1$ and $s_m = s > 1$ holds, $Lpt/Opt < (\sqrt{3}+1)/2$.*

The proof is by contradiction: we regard an instance with minimum number of machines, for which $Lpt/Opt \geq (\sqrt{3}+1)/2$. We fix any optimal schedule of this instance and denote it by OPT.

This proof – and also the proof in Section 4 – is based on the following elementary technique: Our starting point is the LPT schedule. First we rearrange the jobs of LPT within 1-machines. Then we pick jobs $\{t_j^*\}$ of machine m and put them to 1-machines according to how they are scheduled in OPT. We will have to put other jobs from 1-machines back to machine m. This exchanging process will be carried out sometimes one by one, other times by moving sets of jobs. We will calculate the minimum possible ratio: (work moved to m)/(work moved from m). This ratio depends on which time period of machine m the jobs $\{t_j^*\}$ are taken from. Propositions 3 and 4 provide a technical tool for differentiating these time periods. Lemmas 1, 2 and 3 yield the proof of Theorem 2.

For sake of convenience, we assume w.l.o.g. that $Opt = 2$, and so $Lpt \geq \sqrt{3}+1$. Let $t = t_n$ be the size of the last job, and f_i denote the finish time of machine i *before the last job is scheduled*. $Lpt \geq \sqrt{3}+1$ implies $f_i \geq \sqrt{3}+1-t$ for $1 \leq i \leq m-1$ and $f_m \geq \sqrt{3}+1-t/s$.

Analogues to the following lemma can already be found in [6].

Lemma 1. *If $t \leq \sqrt{3}-1$, or $t > 1$, then $\frac{Lpt}{Opt} < \frac{\sqrt{3}+1}{2}$.*   □

In the rest of the proof we assume $\sqrt{3}-1 < t \leq 1$. Now in OPT there are at most 2 jobs on every 1-machine, since $3(\sqrt{3}-1) > 2$. Furthermore, in LPT every 1-machine has finish time $f_i \geq \sqrt{3}+1-1 = \sqrt{3}$. Thus, on a 1-machine in LPT there is either a job of size $\geq \sqrt{3}$, or at least two jobs. Let $t_a \geq t_{a+1} \geq \ldots \geq t_{a+m-2} = t_b$ be the first jobs assigned to the 1-machines in LPT as described by Proposition 2 (see Fig. 2). Let $t_a' \leq t_{a+1}' \leq \ldots \leq t_b'$ denote the second jobs on the respective machines if they exist (these are not consecutive jobs).[1] If $t_a', t_{a+1}', \ldots, t_{a+v}'$ do not exist, then let $t_a' = t_{a+1}' = \ldots = t_{a+v}' = 0$. The proofs of the following two propositions are based on the principle of domination:

Proposition 3. *Let $t_a > 2-t$. In OPT let t' be any job on a 1-machine and t'' be another job on the same machine if such a t'' exists. Now $t' > t_a'$ holds. Furthermore, if $t' \in \{t_a', t_{a+1}', \ldots, t_b'\}$, then $t'' \in \{t_a, t_{a+1}, \ldots, t_b\}$.*   □

Proposition 4. *Let t^* be a job assigned to m in LPT and to a 1-machine in OPT. Let T^* denote the completion time of t^* in LPT. If $t^* > t_a$, then $T^* \leq t^* \leq 2$. If $t^* > t_a'$, then $T^* \leq \max(2, t_b + t^*)$.*   □

---

[1] A different order, due to jobs of equal size would be easy to handle by reordering the 1-machines.

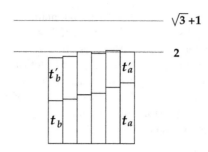

Fig. 2. The first two jobs on 1-machines in LPT

Corollary 1. *Let t^* and T^* be defined as in Proposition 4. If $t_a > 2 - t$, then $T^* \leq \max(2, t_b + t^*)$.*

Proof. Since $t_a > 2 - t$, by Proposition 3, $t^* > t'_a$. Now Proposition 4 implies $T^* \leq \max(2, t_b + t^*)$.   □

Lemma 2. *If $\sqrt{3} - 1 < t \leq 1$ and $t_b \leq 1$, then $\frac{Lpt}{Opt} < \frac{\sqrt{3}+1}{2}$.*

Proof. First of all, we put t_n on machine m, so that it has total work at least $(\sqrt{3} + 1)s$. Let $t^* \leq 2$ be a job that is on a 1-machine in OPT, but on machine m in LPT. Now either $t^* \geq t_a$ or $t^* \leq t_b \leq 1$.

We consider two cases. Suppose first, that $t_a > 2 - t$. By Proposition 4 and Corollary 1, the completion time of any t^* in LPT is at most $t^* \leq 2$, resp. at most $\max(2, t_b + t^*) = 2$. We start by rearranging the jobs *within* 1-machines in LPT: From 1-machines with at least two jobs, we match jobs that belong to the same 1-machine in OPT, and delete the matched jobs together with a 1-machine. As a consequence of Proposition 3, any job *not* in $\{t_a, t_{a+1}, \ldots, t_b\}$ that stays on a 1-machine is by now deleted. We can rearrange the jobs so, that on every remaining 1-machine there is either one job of size at least $\sqrt{3}$, or (at least) two jobs, so that at most one of these jobs remains on the 1-machine in OPT.

Now we put jobs from m to 1-machines. If there is no remaining job on the 1-machine, then we exchange total work of ≤ 2 for one job of size at least $\sqrt{3}$, or for two jobs of total size at least $2t$. Otherwise we exchange $t^* \leq t_b \leq 1$, for one job of size at least t. The size reduction cannot be smaller than $\min(\sqrt{3}/2, \ 2t/2, \ t/1) = \min(\sqrt{3}/2, \ t)$. The reduced work is at most $2s$, so we must have

$$\min(\frac{\sqrt{3}}{2}, t) \cdot 2s + (\sqrt{3} - 1)s \leq 2s$$

so that either $\sqrt{3}/2 \cdot 2 + \sqrt{3} - 1 \leq 2$, a contradiction; or $2t + \sqrt{3} - 1 \leq 2$, that is $t \leq (3 - \sqrt{3})/2$, contradicting to $\sqrt{3} - 1 < t$.

Second, suppose that $t_a \leq 2 - t < \sqrt{3} + 1 - t$. Now in LPT there are at least two jobs on each 1-machine. First we rearrange jobs within 1-machines, so that every job that is on a 1-machine in OPT, gets on its final place, and there are still at least two jobs of size $\geq t'_a$ on every 1-machine.

Now we put jobs $\{t^*\}$ from machine m to 1-machines. If $2 \geq t^* > 2 - t$, then we exchange it for two jobs of total size $\geq 2t$. If $1 \geq t^* > t_a'$, then we exchange it for one job of size at least t. In both cases the size reduction of the t^* is not less than $t/1$, and according to Proposition 4, completion time of t^* in LPT is at most $\max(2, t_b + t^*) = 2$. If $t^* \leq t_a'$, we exchange it for a larger job, so there is no size reduction. Finally, if $2 - t \geq t^* > 1$, then $t^* \geq t_a$, since t^* was on machine m. In this case t^* has completion time at most $t^* \leq 2 - t$. The size reduction can be $t/(2-t)$. We get the inequality:

$$\frac{t}{(2-t)} \cdot (2-t) + t \cdot t + (\sqrt{3} - 1) \leq 2$$

Solving the inequality yields $-\sqrt{3} \leq t \leq \sqrt{3} - 1$, contradicting to $t > \sqrt{3} - 1$. □

Observe, that the conditions in Instance A correspond to the second part of Lemma 2, therefore the obtained bounds for t were tight.

Lemma 3 is proved by a similar reasoning, the proof is omitted here.

Lemma 3. *If* $\sqrt{3} - 1 < t \leq 1$ *and* $t_b > 1$, *then* $\frac{Lpt}{Opt} < \frac{\sqrt{3}+1}{2}$. □

4   A 1.4 Upper Bound for 2-Divisible Machines

In this section we deal with 2-divisible machines. We start with the formal definition of 2-divisibility. Due to technical reasons, we allow fractional machine speeds (e.g., 1/2) in the definition. After that, we state the main result.

Definition 2. *The speed vector* $\langle s_1, s_2, \ldots, s_m \rangle$, *or the machines are called* 2-*divisible if* $s_i = 2^{l_i}$ $(l_i \in \mathbb{Z})$ *for all* i, *and* $s_i \leq s_{i+1}$ $(1 \leq i < m)$.

Theorem 3. *Let* $\langle s_1, \ldots, s_m \rangle$ *and* $\langle t_1, \ldots, t_n \rangle$ *be an instance of* $Q \| C_{\max}$. *If* $\langle s_1, \ldots, s_m \rangle$ *is 2-divisible, then* $\frac{Lpt}{Opt} < 1.4$

Just like in the previous section, we assume that the contrary holds, and we fix a minimal counter-example with 2-divisible machines. Let OPT be an arbitrary optimal schedule of this instance.

The proof technique is similar to that in the previous section: We start from the LPT schedule, then we rearrange jobs, so that more and more jobs get to their final place in OPT. We delete machines that received all their jobs according to OPT. We strive to get into a state, when the set of remaining machines has more total work than $Opt \cdot S$, where S denotes the sum of speeds of the remaining machines. Recall that $t = t_n$.

Since job sizes can be normalized, we assume w.l.o.g. that $Opt = 2$. Moreover, since machine sizes can be normalized too, we may assume that $1/2 < t \leq 1$. This implies that $1/2$ is the smallest possible size of a nonempty machine in OPT, and the instance is minimal, so $s_1 \geq 1/2$. We will call machines of speed at least 2 *fast machines*. Let f_i denote the finish time of machine i in LPT, *before* t_n is scheduled. We assume that $Lpt \geq 2.8$, and the instance was minimal. Consequently, $2.8 > f_i \geq 2.8 - t/s_i$ for $1 \leq i \leq m$.

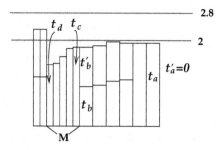

Fig. 3. Jobs on 1/2-machines and 1-machines in LPT

We will exchange the jobs in several rounds. In the first round, machines of speed 1/2 receive their final job, and can be deleted. After this we show, that we got into a similar situation as in Lemmas 2 and 3. Despite the similarity, we have to deal with two additional difficulties: On the one hand, the first round of exchanges has already resulted in some reduction of work by the time we want to apply the arguments of the lemmas. This is a minor problem, and in most cases it does not affect the original argument. On the other hand, we may have more than one fast machines, and therefore we cannot assume that at the beginning they have finish time $f_i \geq 2.8$ (recall, that in Section 3 we could assume $f_m \geq \sqrt{3}+1$, because at the beginning of the exchanges t_n was put on top of machine m). The second difficulty is more crucial, and this is the intrinsic reason why the $(\sqrt{3}+1)/2$ worst case ratio does not hold in case II.

As a first step, we delete the job t_n from LPT. Let M denote the (possibly empty) set of 1/2-machines that are assigned only 1 job in LPT, and let $t_c, t_{c+1}, \ldots, t_d$ be these jobs (see Fig 3). Obviously, t_d is the smallest among them, and $t_d \geq \max(t, 2.8 \cdot 1/2 - t) \geq 1.4/2 = 0.7$.

Now we do the first round of exchanges: In OPT there is one job of size at most 1 on every 1/2-machine. By the principle of domination all of these jobs precede t_c; in LPT they are assigned to machines of speed at least 1, and all of them have completion time at most 2, otherwise they would have been assigned to a 1/2-machine. In particular, t_n is not one of these jobs.

In LPT there is one job of size at least 0.7 on every machine in M. We exchange these jobs for the job in OPT, and then delete all machines of M together with their new job. The resulting schedule will be called LPT$_0$. Let f_i^0 be the finish time of machine i in LPT$_0$. A short case analysis yields Lemma 4 below. The lemma excludes that $t < 0.8$, by impliying that in LPT$_0$ we are left with total remaining work of at least $2S + t_n$, where S is the remaining total speed. On the other hand, Lemma 5 shows that if $t \geq 0.8$, then after the first round, our problem becomes analoguous to Case I in Section 3. The proof of this lemma is straightforward.

Lemma 4. *If* $t < 0.8$, *then* $f_i^0 \geq 2$ *for all* i.   □

Lemma 5. *Let* $0.8 \leq t$. *If* $f_i^0 < 2$, *then* $s_i = 1$. *Moreover, every 1/2-machine was deleted in the first round.*   □

The rest of the proof follows the same lines as the proof in Section 3. We assume $0.8 \leq t \leq 1$. *Instead of* LPT$_0$, *our starting schedule is* LPT: We delete all the 1/2-machines and their jobs. On the remaining machines we calculate *with the original sizes* of jobs, as they are in LPT. Nevertheless, we keep in mind, that every job of size at most 1, and of completion time at most 2 on a 1-machine or on a fast machine, can 'shrink' to size t_d before putting it to its machine in OPT. Such a shrinkage is equivalent to an exchange of 2 jobs in the first round.

We 'put back' the job t_n on top of an arbitrary fast machine. After that, we put jobs from fast machines to 1-machines and vice versa, and perform essentially the same case analysis as in Lemmas 2 and 3.

5   Improved Lower Bound for 2-Divisible Machines

We describe an instance on 2-divisible machines which has an approximation bound arbitrarily close to $(\sqrt{409} + 29)/36 \approx 1.3673... > 41/30 > (\sqrt{3} + 1)/2$. Instance B is a refined version of Instance A of Section 2. We are able to improve on the lower bound by exchanging jobs larger than x for jobs of size $t = t_n$. However, this instance is not suitable if in LPT ties are broken in favour of slow machines. Therefore, in [9] we also present another instance which is valid for any kind of tie-breaking and has approximation bound $955/699 - \delta/2 > (\sqrt{3} + 1)/2$. Besides providing slightly better lower bounds than $(\sqrt{3} + 1)/2$, these kinds of examples are of interest, because they also give an insight into the potential difficulties in determining the worst case bound of LPT on 2-divisible machines.

Theorem 4. *If we restrict the problem* $Q||C_{\max}$ *to 2-divisible speed vectors, then the approximation ratio of* LPT *in the worst case is* $Lpt/Opt > (\sqrt{3}+1)/2$. *Moreover, if in* LPT *ties are broken in favour of faster machines, then in the worst case* $Lpt/Opt > 41/30$.

The proof is given by instance B:

Instance B. In this case we assume that in LPT ties are always broken in favour of faster machines. Just like in Instance A, we have plenty of blocks of 1-machines and a fast machine of speed 2^r. Moreover, we have several 4-machines. First we describe the assigment of jobs in LPT (see Fig 4): Let $x = 1.25$ and $y = 0.75$. A block of 1-machines is scheduled like in Instance A: The large jobs in a block range from $x - \delta$ down to 1; the small jobs range from y up to $1 - \delta$. Every block will be later used for exchanging an x-job for a 1-job or exchanging a 1-job for a y-job. On any 4-machine in LPT there are 10 jobs: 4 jobs of size x; 3 jobs of size 1; 2 jobs of size y and 1 job of size $t \leq y$, where $t = t_n$. The total work on a 4-machine is $9.5 + t$. Let $z = 8 - 9t \geq 8 - 9y = x$. On the fast machine there are $2^r/4$ jobs of size z. After that, it is filled up with x-jobs until time x; with 1-jobs until time 2; with y-jobs until time 9.5/4 and with t-jobs until $2 + t - \delta$. The total work on the fast machine is at least $2^r(2 - \delta + t) - t$.

It is straightforward to check that this is an LPT schedule, either by setting $1/2^r < \delta$, like in Instance A, or by allowing one more job of size between 1 and

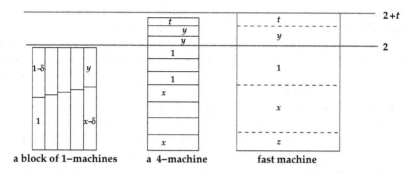

Fig. 4. LPT schedule modulo the last job in Instance B

y on the fast machine. Finally, a last job of size t is assigned to a 1-machine yielding makespan $2 + t - \delta$. On the fast machine this job would be completed after $2 + t - \delta$; on a 4-machine, the finish time would be $(9.5 + 2t)/4 \geq 2 + t$, where the last inequality holds because $t \leq y = 0.75$.

The goal is to determine a possibly large t value, so that the optimum makespan can be arbitrarily close to 2. In order to get an optimum makespan we rearrange the jobs as follows. As a first step, we put the very last job on the fast machine. Second, we exchange every job of size x or 1 for a job of size y using the blocks of 1-machines. At this point, on every 4-machine there are 9 jobs of size y and 1 job of size t. Since $(10 \cdot y)/4 < 2$, the job of size t can be exchanged for a y without violating the desired optimum makespan. Thus, we can use 4-machines to exchange all y-jobs on the fast machine for t-jobs. Moreover, on $2^r/4$ of the 4-machines we also exchange all the jobs for t-jobs. On these $2^r/4$ machines we will have 10 jobs of size t. Finally we use each of these $2^r/4$ machines for exchanging a z-job for a t-job. This is possible, since $(9 \cdot t + z)/4 = 2$. Now every job on the fast machine is exchanged for a t-job. If we calculate with fractional jobs on the fast machine, the following inequality models the desired shrinkage of work (we calculate with no shrinkage above time 2):

$$\frac{z}{4} \cdot \frac{t}{z} + \left(x - \frac{z}{4}\right) \cdot \frac{t}{x} + (2 - x) \cdot \frac{t}{1} + t \leq 2$$

$$\frac{t}{4} + t - \frac{(8 - 9t)t}{4 \cdot 1.25} + 0.75t + t \leq 2$$

By solving the inequality we get: $\frac{-\sqrt{409}-7}{18} \leq t \leq \frac{\sqrt{409}-7}{18} \approx 0.734$. Using $t = \frac{\sqrt{409}-7}{18}$ yields the approximation $(2 + t)/2 = (\sqrt{409} + 29)/36 > 41/30$. The surplus on the optimum makespan due to calculating with fractional jobs and due to the very last job is not more than $(x + 1 + t)/2^r < \epsilon$ if r is large enough.

6   Conclusions

For the classic LPT algorithm, we have shown a tight worst case approximation bound of $\frac{\sqrt{3}+1}{2}$ in case of one fast machine; and 'nearly' tight lower and upper bounds, (more than) $41/30$ and $42/30$, for the same problem on 2-divisible machines. In our instances providing approximation within ϵ distance to the lower bounds, the number of machines $m = \mathcal{O}(1/\epsilon^2)$, and the ratio of speeds $s_m/s_1 = \mathcal{O}(1/\epsilon)$. However, for relatively large ϵ (e.g., if we just want to demonstrate $Lpt/Opt > 4/3$), with modified x and y values, it suffices to take one 4-machine, and altogether 20 1-machines.

We do not exclude, that – if LPT prefers faster machines in case of ties –, Instance B of Section 5 actually yields the basic construction for a tight bound on 2-divisible machines (the bound itself can be a bit higher). However, proving such a tight bound seems to require a lengthy and technical elaboration.

Finally, we turn to the monotone algorithm LPT* of [10]. Based on Instance A, for any $\epsilon > 0$ it is possible to create an instance for arbitrary machines so, that LPT* performs as bad as $Lpt^*/Opt > \sqrt{3} + 1 - \epsilon$ on this instance [9]. On the other hand, Theorem 3 implies a worst case ratio $Lpt^*/Opt \leq 2.8$.

References

1. A. Archer and É. Tardos. Truthful mechanisms for one-parameter agents. In *Proc. 42nd IEEE Symp. on Found. of Comp. Sci. (FOCS)*, pages 482–491, 2001.
2. Y. Cho and S. Sahni. Bounds for list schedules on uniform processors. *SIAM Journal on Computing*, 9(1):91–103, 1980.
3. E.G. Coffman, M.R. Garey, and D.S. Johnson. An application of bin-packing to multiprocessor scheduling. *SIAM Journal on Computing*, 7(1):1–17, 1978.
4. G. Dobson. Scheduling independent tasks on uniform processors. *SIAM Journal on Computing*, 13(4):705–716, 1984.
5. D.K. Friesen. Tighter bounds for LPT scheduling on uniform processors. *SIAM Journal on Computing*, 16(3):554–560, 1987.
6. T. Gonzalez, O.H. Ibarra, and S. Sahni. Bounds for LPT schedules on uniform processors. *SIAM Journal on Computing*, 6(1):155–166, 1977.
7. D.S. Hochbaum and D.B. Shmoys. A polynomial approximation scheme for scheduling on uniform processors: Using the dual approximation approach. *SIAM J. Comp.*, 17(3):539–551, 1988.
8. E. Horowitz and S. Sahni. Exact and approximate algorithms for scheduling nonidentical processors. *Journal of the ACM*, 23:317–327, 1976.
9. A. Kovács. Tighter approximation bounds for LPT scheduling in two special cases. Extended version: http://www.mpi-inf.mpg.de/~panni/approx.ps.
10. A. Kovács. Fast monotone 3-approximation algorithm for scheduling related machines. In *Proc. 13th Ann. Europ. Symp. on Algo. (ESA)*, LNCS. Springer, 2005.
11. R. Li and L. Shi. An on-line algorithm for some uniform processor scheduling. *SIAM Journal on Computing*, 27(2):414–422, 1998.
12. J.W.S. Liu and C.L. Liu. Bounds on scheduling algorithms for heterogeneous computing systems. In *Proc. Intern. Feder. of Inf. Proc. Soc.*, p. 349–353, 1974.
13. P. Mireault, J.B. Orlin, and R.V. Vohra. A parametric worst case analysis of the LPT heuristic for two uniform machines. *Oper. Res.*, 45(1):116–125, 1997.
14. D.S. Johnson M.R. Garey. *Computers and Intractability; A Guide to the Theory of NP-completeness*. Freeman, San Francisco, 1979.
15. V. Auletta R. De Prisco P. Penna and G. Persiano. Deterministic truthful approximation mechanisms for scheduling related machines. In *Proc. 21st STACS*, volume 2996 of *LNCS*, p. 608–619. Springer, 2004.

Inapproximability Results for Orthogonal Rectangle Packing Problems with Rotations

Miroslav Chlebík[1] and Janka Chlebíková[2,*]

[1] MPI for Mathematics in the Sciences, D-04103 Leipzig, Germany
chlebik@mis.mpg.de
[2] Faculty of Mathematics, Physics and Informatics,
Mlynská dolina, 842 48 Bratislava, Slovakia
chlebikova@fmph.uniba.sk

Abstract. Recently Bansal and Sviridenko [4] proved that there is no asymptotic PTAS for 2-DIMENSIONAL ORTHOGONAL RECTANGLE BIN PACKING without rotations allowed, unless P = NP. We show that similar approximation hardness results hold for several rectangle packing problems even if rotations by ninety degrees around the axes are allowed. Moreover, for some of these problems we provide explicit lower bounds on asymptotic approximation ratio of any polynomial time approximation algorithm.

1 Introduction

We focus on orthogonal packing problems of rectangles into bins in 2 and 3-dimensions, where ninety-degree rotations of rectangles around any of the axes are allowed. These problems have many real-world applications in areas like job scheduling, container loading, and cutting objects out of a strip of material in such a way that the amount of material wasted is minimal.

Notation and terminology. Throughout this paper we only consider offline versions of the problems. In all 2-dimensional variants of the problems, the input consists of a list $\mathscr{L} = \{R^1, R^2, \ldots, R^n\}$ of 2-dimensional rectangles in the Euclidean space $\mathbb{R}^2$ and a 2-dimensional rectangular bin $\mathbb{B} = [0, b_1] \times [0, b_2]$ (for which the notation (b_1, b_2) is used as well). Each rectangle R^i is given with an (initial) *orientation* related to the coordinate axes and side-lengths denoted as $(w(R^i), h(R^i))$ and called width and height, respectively. The generalization to the higher dimensions is straightforward. In the 3-dimensional strip version of the problems we suppose that the last dimension of the bin $\mathbb{B}$ is unlimited and we call such bin $\mathbb{B} = (b_1, b_2, \infty)$ a *strip*. All rectangles of the list $\mathscr{L}$ need to be packed into bins without overlap. The most interesting and well-studied version of these problems is the so-called *orthogonal* version, where the edges of packed rectangles and bins are always parallel to the coordinate axes. In problems *without rotations* rectangles have to be placed into the bin with given orientation and a feasible solution is called *oriented packing*. In problems *with rotations allowed* rectangles to be placed may be *rotated around any of the axes by 90°*

---

* The author has been supported by VEGA grant no. 1/3106/06.

T. Calamoneri, I. Finocchi, G.F. Italiano (Eds.): CIAC 2006, LNCS 3998, pp. 199–210, 2006.
© Springer-Verlag Berlin Heidelberg 2006

and a feasible solution is referred to as *r-packing*. In the 3-dimensional case, if only rotations around the *z-axis* (the last one) are allowed, a packing is called *z-oriented*.

Given a list $\mathscr{L}$ of 2-dimensional rectangles and a 2-dimensional bin $\mathbb{B} = (b_1, b_2)$. The goal of 2-DIMENSIONAL BIN PACKING (2-BP) and 2-DIMENSIONAL BIN PACKING WITH ROTATIONS (2-BP^r) is to find an oriented packing and an *r*-packing, respectively, of all rectangles of $\mathscr{L}$ into the minimum number of copies of $\mathbb{B}$. In 3-dimensional strip versions of the problems, a list $\mathscr{L}$ of 3-dimensional rectangles and a 3-dimensional strip $\mathbb{B} = (b_1, b_2, \infty)$ are given. In the problems 3-DIMENSIONAL STRIP PACKING (3-SP) and 3-DIMENSIONAL STRIP PACKING WITH ROTATIONS (3-SP^r) we are looking for an oriented packing and an *r*-packing, respectively, that minimizes h such that all rectangles of $\mathscr{L}$ are packed into the bin (b_1, b_2, h). If only 90° rotations around the *z*-axis (the unlimited direction of the strip $\mathbb{B}$) are allowed, the problem is called *z*-ORIENTED 3-DIMENSIONAL STRIP PACKING.

The standard measure of algorithm quality for bin packing problems is the *asymptotic approximation ratio*. For a minimization problem it is defined as $\rho_{\mathcal{A}}^{\infty} = \lim_{n \to \infty} \sup_I \left\{ \frac{\mathcal{A}(I)}{\mathrm{OPT}(I)} : \mathrm{OPT}(I) \geq n \right\}$, where I ranges over the set of all problem instances, and $\mathcal{A}(I)$ (resp. $\mathrm{OPT}(I)$) denote the value of the solution returned by $\mathcal{A}$ (resp. the optimum value) for an input instance I. For a maximization problem, $\frac{\mathcal{A}(I)}{\mathrm{OPT}(I)}$ is replaced by $\frac{\mathrm{OPT}(I)}{\mathcal{A}(I)}$ so that always $\rho_{\mathcal{A}}^{\infty} \geq 1$. We say, that a problem admits an *asymptotic approximation scheme* (shortly, APTAS), if for any $\varepsilon > 0$ there is a polynomial time algorithm with an asymptotic approximation ratio less than $1 + \varepsilon$. For other optimization terminology we refer to Ausiello et al. [1].

Overview. For 1-BP, Fernandez de la Vega & Lueker [10] designed an APTAS. More precisely, for any positive integer k they provided a polynomial time algorithm $\mathcal{A}_k$ that uses at most $(1 + \frac{1}{k})\mathrm{OPT} + 1$ bins. Later, Karmarkar & Karp [15] gave a single algorithm with asymptotic approximation ratio 1 that uses $\mathrm{OPT} + O(1 + \log^2 \mathrm{OPT})$ bins. For the 2-BP problem Caprara [5] presented an algorithm with currently the best asymptotic approximation ratio 1.691. On the negative side, Bansal & Sviridenko [4] proved that there is no APTAS for 2-BP, unless P = NP. Interestingly, they provided an APTAS for a restricted version of *d*-BP in which the items and the bins are *d*-cubes; this result was independently obtained by Correa & Kenyon [8]. For 3-BP, Li & Cheng [17] and Csizik & van Vliet [9] designed algorithms with asymptotic ratio at most 4.84. This asymptotic ratio was later improved to $4 + \varepsilon$ by Jansen & Solis-Oba [11]. The algorithms from [17] and [9] generalize to the problem *d*-BP with asymptotic approximation ratio at most 1.691^d. For the problem 2-SP, the breakthrough result was obtained by Kenyon & Rémila [16] who gave an APTAS. For 3-SP, Miyazawa & Wakabayashi [19] presented an algorithm with asymptotic approximation ratio at most 2.64, which was improved to $2 + \varepsilon$ by Jansen & Solis-Oba [11]. On the other hand, it is easy to see that approximation hardness result for

2-BP implies that no APTAS for 3-SP can exist, unless P = NP (see Section 2.1 for more details).

When ninety-degree rotations are allowed, only weaker results are known. Some algorithms for the versions without rotations provide upper bounds on asymptotic approximation ratio for versions with rotations allowed as well. The results by Miyazawa & Wakabayashi [18] were the first ones where rotations are exploited in non-trivial way. Currently the best upper bounds on asymptotic approximation ratio for the problems 2-BP^r, 3-BP^r, 3-SP^r, and 3-SP^z, are $2 + \varepsilon$, 4.89, 2.76, and 2.64, respectively, see [19] and [12]. Moreover, Jansen & Stee provided an APTAS for 2-SP^r ([12]).

Rectangle Packing without and with Rotations. When dealing with packing problems *without rotation*, one can always assume that a bin $\mathbb{B}$ is a unit cube (resp., a base of a strip $\mathbb{B}$ is a unit cube), as the problems are invariant under *heterogeneous scaling*, i.e., the one which scales by different factors in different coordinate directions. However, this is not true for problems with rotations allowed. It is unclear if the problems with rotations allowed, where the bin $\mathbb{B}$ is a unit cube, are easier to approximate than the general one. For some problems, algorithms with better asymptotic approximation ratio were suggested in such restricted case. For example, when a base of the strip in the problem 3-SP^z is a unit square, an algorithm with asymptotic approximation ratio at most 2.528 is known [18].

Using heterogeneous scaling one can show that 2-BP can be viewed as a particular case of general 2-BP^r with highly excentric instances. Let a list $\mathcal{L} = (R^1, R^2, \ldots, R^n)$ of rectangles with dimensions $R^i = (r_1^i, r_2^i)$, $i = 1, 2, \ldots, n$, and a bin $\mathbb{B} = (b_1, b_2)$ be an instance of 2-BP. One can find positive scaling factors λ_1, λ_2, and use scaling $(x_1, x_2) \mapsto (\lambda_1 x_1, \lambda_2 x_2)$ to map any R^i to $\widetilde{R}^i = (\widetilde{r}_1^i, \widetilde{r}_2^i)$, and the bin $\mathbb{B}$ to $\widetilde{\mathbb{B}} = (\widetilde{b}_1, \widetilde{b}_2)$, so that it holds that $\min\{\widetilde{r}_1^i : 1 \leq i \leq n\} > \widetilde{b}_2$. It is easy to see that the only way a rectangle $\widetilde{R}^i$ can fit into the bin $\widetilde{\mathbb{B}}$, even if ninety-degree rotations are allowed, is that $\widetilde{R}^i$ is not rotated. Similarly, 3-SP can be handled as a particular case of 3-SP^r or 3-SP^z. Thus, for problems 2-BP^r, 3-SP^r, and 3-SP^z without any restriction on the bin $\mathbb{B}$, non-existence of an APTAS easily follows from results by Bansal & Sviridenko [4] for 2-BP (see Sections 2 and 2.1 for more details). However, for the most interesting case of a unit *square* bin $\mathbb{B}$, one can hardly obtain hardness results in such a way.

Main results. In this paper we prove non-existence of an APTAS (unless P = NP) for 2-DIMENSIONAL BIN PACKING WITH ROTATIONS into unit square bins (Section 2), 3-DIMENSIONAL STRIP PACKING WITH ROTATIONS and z-ORIENTED 3-DIMENSIONAL STRIP PACKING (Section 2.1) into a strip with unit square base. The methods allow to give explicit lower bounds on asymptotic approximation ratio of any polynomial time approximation algorithm (unless P = NP). For example, we provide a lower bound $1 + \frac{1}{3792}$ for 2-DIMENSIONAL BIN PACKING WITH ROTATIONS, and $1 + \frac{1}{2196}$ for the same problem without rotations.

We prove also non-existence of an APTAS for a related 3-dimensional packing problem where the goal is to pack the maximum number of rectangles from a given collection into a *single* cube bin (Section 3).

General technique. Recall, that for pairwise disjoint sets X, Y, Z, and a set of ordered triples $T \subseteq X \times Y \times Z$, a *matching in* T is a subset $M \subseteq T$ in which no two ordered triples in M agree in any coordinate. The goal of the MAXIMUM 3-DIMENSIONAL MATCHING problem (shortly, MAX-3DM) is to find a matching in T of maximum cardinality. A k-bounded MAX-3DM is restricted to instances, in which each element of $X \cup Y \cup Z$ occurs at most k times in T.

Kann [14] showed that the 3-bounded MAX-3DM problem is Max SNP-complete (hence also APX-complete). Thus, using PCP-theorem, the existence of a PTAS for it would imply that P = NP. Petrank [20] proved a refined approximation hardness result that an NP-hard gap occurs also on instances with perfect matching. Unfortunately, the estimates that are implicit in his proof provide lower bound $1 + \varepsilon$ with extremely small $\varepsilon > 0$. To achieve explicit inapproximability results it is more convenient to use the following NP-hard gap type result for 2-bounded instances of MAX-3DM.

Theorem A. [7] *There are instances* $T \subseteq X \times Y \times Z$ *of 2-bounded* MAX-3DM *with* $|X| = |Y| = |Z|(:= q)$ *and every element of* $X \cup Y \cup Z$ *occurring in exactly 2 triples in* T *such that it is NP-hard to distinguish between instances with* $\mathrm{OPT}(T) > 0.979338843q$ *and* $\mathrm{OPT}(T) < 0.9690082645q$.

Both mentioned approximation hardness results for bounded MAX-3DM suit well as a starting point to inapproximability results for various (multidimensional) packing, covering, and scheduling problems, see e.g., [21], [6], and [4].

2   2-Dimensional Bin Packing with Rotations

In this section we build on ideas from [4] and introduce a general parametrised version of a gap preserving reduction from bounded MAX-3DM to 2-DIMENSIONAL BIN PACKING. We show that with properly chosen parameters this reduction can be used to obtain approximation hardness results for 2-DIMENSIONAL BIN PACKING WITH ROTATIONS into unit square bin.

The Bin Packing reduction. Let $\mathcal{T}$ be an infinite set of instances (ordered triples) T of MAX-3DM with the optimum value $\mathrm{OPT}(T)$, with the property that for some efficiently computable function $\alpha(T) < \beta(T)$ it is NP-hard to decide of whether $\mathrm{OPT}(T) \geq \beta(T)$, or $\mathrm{OPT}(T) < \alpha(T)$. For a fixed instance $T \in \mathcal{T}$ let $X := \Pi_1(T)$, $Y := \Pi_2(T)$, and $Z := \Pi_3(T)$, where $\Pi_i(T) = \{p_i : (p_1, p_2, p_3) \in T\}$ for $i = 1, 2, 3$, and X, Y, Z are pairwise disjoint sets. The objects in X, Y, Z, and T will be denoted as $\{x_i : 1 \leq i \leq |X|\}$, $\{y_j : 1 \leq j \leq |Y|\}$, $\{z_k : 1 \leq k \leq |Z|\}$, and $\{t_l : 1 \leq l \leq |T|\}$, respectively. (In fact, we will use this general reduction for instances from Theorem A, where $|X| = |Y| = |Z|$ holds.) Of course, any $t_l \in T$ is of the form $t_l = (x_i, y_j, z_k) \in X \times Y \times Z$. Let

$n = |X| + |Y| + |Z|$, $q = \max\{|X|, |Y|, |Z|\}$, and $r = 32q$. The reduction has several parameters: a gap location $\beta(T)$, $\delta \in \left(0, \frac{1}{500}\right]$, and $p \in \left[\frac{1}{4} + 9\delta, \frac{1}{2} - 20\delta\right]$.

We first define an integer for each object in X, Y, Z, and T as follows: $x'_i = ir^3 + i^2 r + 1$, for $1 \le i \le |X|$, $y'_j = jr^6 + j^2 r^4 + 2$, for $1 \le j \le |Y|$, $z'_k = kr^9 + k^2 r^7 + 4$, for $1 \le k \le |Z|$. For each triple $t_l = (x_i, y_j, z_k) \in T$ we define an integer $t'_l = r^{10} - x'_i - y'_j - z'_k + 15$. Put $c = \frac{r^{10}+15}{\delta}$ and observe that $0 < x'_i, y'_j, z'_k < \frac{\delta c}{10}$ for all i, j, k, and $t'_l + x'_i + y'_j + z'_k = c\delta$ whenever $t_l = (x_i, y_j, z_k) \in T$.

For each $x_i \in X$ (resp., $y_j \in Y$ and $z_k \in Z$) we define a pair of rectangles $A_{X,i}$, $A'_{X,i}$ (resp., $A_{Y,j}$, $A'_{Y,j}$ and $A_{Z,k}$, $A'_{Z,k}$) with width about $\frac{1}{4}$ and with heights about $\frac{1}{2} + p$ and $\frac{1}{2} - p$ as follows:

$$A_{X,i} = \left(\frac{1}{4} - 4\delta + \frac{x'_i}{c}, \frac{1}{2} + p + 4\delta - \frac{x'_i}{c}\right), \quad A'_{X,i} = \left(\frac{1}{4} + 4\delta - \frac{x'_i}{c}, \frac{1}{2} - p - 4\delta + \frac{x'_i}{c}\right),$$

$$A_{Y,j} = \left(\frac{1}{4} - 3\delta + \frac{y'_j}{c}, \frac{1}{2} + p + 3\delta - \frac{y'_j}{c}\right), \quad A'_{Y,j} = \left(\frac{1}{4} + 3\delta - \frac{y'_j}{c}, \frac{1}{2} - p - 3\delta + \frac{y'_j}{c}\right),$$

$$A_{Z,k} = \left(\frac{1}{4} - 2\delta + \frac{z'_k}{c}, \frac{1}{2} + p + 2\delta - \frac{z'_k}{c}\right), \quad A'_{Z,k} = \left(\frac{1}{4} + 2\delta - \frac{z'_k}{c}, \frac{1}{2} - p - 2\delta + \frac{z'_k}{c}\right).$$

For each $t_l \in T$ we define two rectangles B_l and B'_l such that

$$B_l = \left(\frac{1}{4} + 8\delta + \frac{t'_l}{c}, \frac{1}{2} + p + \delta - \frac{t'_l}{c}\right) \quad \text{and} \quad B'_l = \left(\frac{1}{4} - 8\delta - \frac{t'_l}{c}, \frac{1}{2} - p - \delta + \frac{t'_l}{c}\right).$$

Let $\mathscr{A}_X = \{A_{X,1}, A_{X,2}, \ldots, A_{X,|X|}\}$, $\mathscr{A}'_X = \{A'_{X,1}, A'_{X,2}, \ldots, A'_{X,|X|}\}$ and define sets of rectangles $\mathscr{A}_Y$, $\mathscr{A}'_Y$, $\mathscr{A}_Z$, and $\mathscr{A}'_Z$ analogously. Put $\mathscr{A} = \mathscr{A}_X \cup \mathscr{A}_Y \cup \mathscr{A}_Z$ and $\mathscr{A}' = \mathscr{A}'_X \cup \mathscr{A}'_Y \cup \mathscr{A}'_Z$. Similarly, let $\mathscr{B} = \{B_1, B_2, \ldots, B_{|T|}\}$ and $\mathscr{B}' = \{B'_1, B'_2, \ldots, B'_{|T|}\}$. We define also $\mathscr{D}$ to be a collection of $|T| + n - 4\beta(T)$ dummy rectangles, each of the size $\left(\frac{3}{4} - 10\delta, 1\right)$.

The collection of rectangles $\mathscr{A} \cup \mathscr{A}' \cup \mathscr{B} \cup \mathscr{B}' \cup \mathscr{D}$, together with a unit square bin is now viewed as an instance of the 2-BP^r problem and denoted by $f(T)$. Our aim is to relate the optimum value $\mathrm{OPT}'(f(T))$ of 2-BP^r for an instance $f(T)$ to $\mathrm{OPT}(T)$. Informally, the dimensions of rectangles and dummy rectangles are chosen such that if $\mathrm{OPT}(T) \ge \beta(T)$, the rectangles can be packed into bins in such a way that their number is within a factor $(1 + O(\delta))$ of the total area of rectangles. On the other hand, if $\mathrm{OPT}(T) < \frac{\beta(T)}{\gamma}$ for a constant $\gamma > 1$, then the number of bins needed to pack all rectangles of $f(T)$ is larger than the total area of rectangles by a constant factor $\gamma' > 1$ independent of δ for $\delta > 0$ small enough.

The reduction given by Bansal & Sviridenko ([4]) can be viewed as a particular case of the Bin Packing reduction with $\delta = \frac{1}{500}$, a set T of instances $T \subseteq X \times Y \times Z$ of 3-bounded MAX-3DM with $|X| = |Y| = |Z| = q$, and a gap location $\beta(T) = q$ (as it follows from the Petrank's result [20]). The parameter p is an important novelty of this paper. The crucial point is that for the proper choice of the parameter p we can prove that even if rotations are allowed it is not advantageous to use them.

Remark 1. Bansal & Sviridenko [4] (see also [3]) claim to prove not only non-existence of APTAS (unless P = NP) for 2-DIMENSIONAL BIN PACKING (without rotations), but also APX-hardness for it. However, such result does not follow from their proof. The given reduction from 3-bounded MAX-3DM to 2-BP is not an *L*-reduction (or an approximation preserving reduction), but it is rather a *gap preserving reduction* that preserves one but not all gaps.

We start with the following simple lemma valid for the choice of $p \in [\frac{1}{4} + 9\delta, \frac{1}{2} - 20\delta]$.

Lemma 1. *(i) For every r-packing of $f(T)$ all rectangles from $\mathscr{A} \cup \mathscr{B}$ contained in the same bin are either in their initial orientations or all are rotated by ninety degrees.*

(ii) For every r-packing of $f(T)$ if a bin contains exactly 4 rectangles from $\mathscr{A} \cup \mathscr{B}$, then all rectangles from $\mathscr{A} \cup \mathscr{B} \cup \mathscr{A}' \cup \mathscr{B}'$ packed in this bin are either in their initial orientations or all are rotated by ninety degrees.

For oriented packings some properties of the Bansal's and Sviridenko's reduction [4] (that corresponds to $p = 0$) are preserved to our general situation with the parameter p introduced. The proofs of Lemmas 3 and 4 given in [4] work in this case as well, as widths of rectangles are the same in both reductions.

Definition 1. *([4]) We say that two rectangles A and A' from $\mathscr{A} \cup \mathscr{A}' \cup \mathscr{B} \cup \mathscr{B}'$ are buddies if $\{A, A'\}$ corresponds to a pair of rectangles for a single element from X, Y, Z or T, e.g., $\{A, A'\} = \{A_{X,i}, A'_{X,i}\}$ for some $x_i \in X$ and similarly for the other sets Y, Z, and T.*

Observation 1. *For any rectangle, $A \in \mathscr{A}$ implies $w(A) + h(A) = \frac{3}{4} + p$, $A' \in \mathscr{A}'$ implies $w(A') + h(A') = \frac{3}{4} - p$, $B \in \mathscr{B}$ implies $w(B) + h(B) = \frac{3}{4} + p + 9\delta$, and $B' \in \mathscr{B}'$ implies $w(B') + h(B') = \frac{3}{4} - p - 9\delta$.*

Observation 2. *For any two rectangles A, A' in $\mathscr{A} \cup \mathscr{A}' \cup \mathscr{B} \cup \mathscr{B}'$, $h(A) + h(A') = 1$ if and only if A and A' are buddies.*

In the following lemma we observe some basic properties for oriented packing of rectangles from $\mathscr{A} \cup \mathscr{B} \cup \mathscr{A}' \cup \mathscr{B}'$ into unit square bin.

Lemma 2. *Consider a unit square bin containing exactly 4 rectangles from $\mathscr{A} \cup \mathscr{B}$ for an oriented packing of $f(T)$. Then the bin contains at most 8 rectangles from $\mathscr{A} \cup \mathscr{B} \cup \mathscr{A}' \cup \mathscr{B}'$ and if it contains exactly 8 rectangles then, for any $h \in [4\delta, \frac{1}{2} - p - 4\delta]$, each rectangle intersects exactly one of the lines $L_1 = \{(x, y) : y = h\}$ and $L_2 = \{(x, y) : y = 1 - h\}$.*

Lemma 3. *For any rectangles A_1, A_2, $A_3 \in \mathscr{A}$ and $B \in \mathscr{B}$, $w(A_1) + w(A_2) + w(A_3) + w(B) = 1$ if and only if $\{A_1, A_2, A_3, B\} = \{A_{X,i}, A_{Y,j}, A_{Z,k}, B_l\}$ for some integers i, j, k, and l such that $t_l = (x_i, y_j, z_k) \in T$. A similar statement holds also for rectangles A'_1, A'_2, $A'_3 \in \mathscr{A}'$, $B' \in \mathscr{B}'$.*

Lemma 4. *Let A_1, A_2, A_3, $A_4 \in \mathscr{A} \cup \mathscr{A}'$ be such that no two of them are buddies. Then $\sum_{i=1}^{4} w(A_i) \neq 1$.*

Definition 2. *Given an r-packing of a bin by some rectangles from $f(T)$. The bin is called* well-packed, *if it contains exactly 4 rectangles from $\mathscr{A} \cup \mathscr{B}$ and 4 rectangles from $\mathscr{A}' \cup \mathscr{B}'$.*

Now the crucial fact is, that for any choice of the parameter p from the interval $\left[\frac{1}{4} + 9\delta, \frac{1}{2} - 20\delta\right]$, we can characterize the structure of well-packed bins similarly as it has been done in [4] for oriented packings.

Lemma 5. *A bin is well-packed if and only if it contains the rectangles $A_{X,i}$, $A_{Y,j}$, $A_{Z,k}$, B_l, $A'_{X,i}$, $A'_{Y,j}$, $A'_{Z,k}$, B'_l, for some $t_l = (x_i, y_j, z_k) \in T$.*

Proof. The 8-tuple of rectangles corresponding to a triple as above can be packed in a square bin $\mathbb{B} = [0,1]^2$ even without using rotations. Starting from the bottom left corner of the bin $\mathbb{B}$ and moving to the right, each of rectangles $A_{X,i}$, $A_{Y,j}$, $A_{Z,k}$, and B_l is placed such that it touches the bottom of the bin $\mathbb{B}$. As $w(A_{X,i}) + w(A_{Y,j}) + w(A_{Z,k}) + w(B_l) = 1$ (Lemma 3), the rectangles can be packed in this way. The rectangles $A'_{X,i}$, $A'_{Y,j}$, $A'_{Z,k}$, and B'_l can be placed in the remaining gaps starting from the top left corner of the bin $\mathbb{B}$ and moving towards the right touching the top of the bin. Clearly, such packing is possible due to the size properties of rectangles.

Now we show that any well-packed bin contains rectangles that correspond to a triple in T. Due to Lemma 1(ii), all rectangles are either in their initial orientations or all are rotated by ninety degrees. We can assume that they are all in the initial orientation in a well-packed bin; the case when all are rotated by 90° can be discussed similarly. Fix $h \in \left[4\delta, \frac{1}{2} - p - 4\delta\right]$ and consider the lines $L_1 = \{(x,y) : y = h\}$ and $L_2 = \{(x,y) : y = 1 - h\}$. Due to Lemma 2, each rectangle must intersect exactly one of the lines L_1 and L_2. Moreover, as any rectangle has width larger than $\frac{1}{5}$, each of lines L_1 and L_2 intersects exactly 4 rectangles. Let $\{A_1, A_2, A_3, A_4\}$ denote the rectangles that intersect L_1 such that A_i is to the left of A_j for $i < j$. Similarly, let $\{A_5, A_6, A_7, A_8\}$ denote the rectangles that intersect L_2 in the left to right order. Thus, we have that

$$\sum_{i=1}^{4} w(A_i) \leq 1, \tag{1}$$

$$\sum_{i=1}^{4} w(A_{i+4}) \leq 1. \tag{2}$$

Observe that for each $i = 1, 2, 3, 4$ the rectangle A_i must overlap with A_{i+4} in the x-coordinate. Thus, we have that

$$h(A_i) + h(A_{i+4}) \leq 1 \quad \text{for } i = 1, 2, 3, 4. \tag{3}$$

From (3) it follows that, for each $i = 1, 2, 3, 4$, at most one of A_i, A_{i+4} belongs to $\mathscr{A} \cup \mathscr{B}$. Consequently, for each $i = 1, 2, 3, 4$ exactly one of A_i, A_{i+4} is from $\mathscr{A} \cup \mathscr{B}$ and another one is from $\mathscr{A}' \cup \mathscr{B}'$. Using these facts, we can use the same arguments as in [4]:

(i) First observe that *at most 1 from rectangles* $\{A_1, \ldots, A_8\}$ *belongs to* $\mathscr{B}$. Indeed, if $k \geq 2$ of them belong to $\mathscr{B}$ and $4 - k$ belong to $\mathscr{A}$, then the sum of widths of these rectangles from $\mathscr{A} \cup \mathscr{B}$ would be > 1, a contradiction with the fact that any line in y-direction intersects at most 1 rectangle from $\mathscr{A} \cup \mathscr{B}$.

(ii) *If no rectangle from* $\{A_1, \ldots, A_8\}$ *belongs to* $\mathscr{B}$, *than the same is true for* $\mathscr{B}'$. The height of any rectangle in $\mathscr{B}'$ is larger then $\frac{1}{2} - p - \delta$ so such a rectangle cannot form a pair $\{A_i, A_{i+4}\}$ with a rectangle from $\mathscr{A}$. Thus, in this case four rectangles belong to $\mathscr{A}$ and four to $\mathscr{A}'$. Using Observation 1 we get $\sum_{i=1}^{8}(w(A_i) + h(A_i)) = 6$, thus it must be the case that each of (1), (2) and (3) must hold with equality. By Observation 2, A_i and A_{i+4} are buddies for each $i = 1, 2, 3, 4$. In particular, no two rectangles among A_1, A_2, A_3, and A_4 are buddies. Now Lemma 4 contradicts with $\sum_{i=1}^{4} w(A_i) = 1$ that has been observed earlier. Thus this case is impossible.

So, necessarily *exactly one of rectangles* $\{A_1, A_2, \ldots, A_8\}$ *belongs to* $\mathscr{B}$, say B_l.

(iii) As, due to (3), no pair $\{A_i, A_{i+4}\}$ can contain a rectangle from $\mathscr{B}'$ and a rectangle from $\mathscr{A}$, there can be *at most one rectangle from* $\mathscr{B}'$. But if there are no rectangles from $\mathscr{B}'$, then the sum of widths of all 8 rectangles would be > 2, a contradiction.

Consequently, there is *exactly 1 rectangle from* $\mathscr{B}'$, *1 from* $\mathscr{B}$, *3 from* $\mathscr{A}$, *and 3 from* $\mathscr{A}'$. Using Observation 1 we get $\sum_{i=1}^{8}(w(A_i) + h(A_i)) = 6$, thus each of (1), (2), and (3) holds with equality. In particular, for each $i = 1, 2, 3, 4$, A_i and A_{i+4} are buddies due to Observation 2. Let $m \in \{1, 2\}$ be such that B_l intersects the line L_m. Let $A_{m_1}, A_{m_2}, A_{m_3}$ denote the other three rectangles (from $\mathscr{A} \cup \mathscr{A}'$) which are also intersected by L_m. Thus we have that $w(A_{m_1}) + w(A_{m_2}) + w(A_{m_3}) + w(B_l) = 1$. None of $A_{m_1}, A_{m_2}, A_{m_3}$ can lie in $\mathscr{A}'$ because otherwise $w(A_{m_1}) + w(A_{m_2}) + w(A_{m_3}) + w(B_l) > (\frac{1}{4} + 8\delta) + (\frac{1}{4} + \delta) + 2(\frac{1}{4} - 4\delta) = 1 + \delta$, a contradiction. Hence $\{A_{m_1}, A_{m_2}, A_{m_3}\} \subseteq \mathscr{A}$, and using Lemma 3 we get that $\{A_{m_1}, A_{m_2}, A_{m_3}\} = \{A_{X,i}, A_{Y,j}, A_{Z,k}\}$ for integers i, j, k such that $t_l = (x_i, y_j, z_k)$, where t_l is the corresponding triple for the rectangle B_l. This completes the proof.

Now we can prove the main theorem of this section

Theorem 1. *There is a constant* $\rho > 1$ *such that it is* NP-*hard to approximate* 2-DIMENSIONAL BIN PACKING WITH ROTATIONS *into unit square bins with an asymptotic approximation ratio less than* ρ.

Proof. Recall that the Bin Packing reduction f started from a set $\mathscr{T}$ of instances of MAX-3DM such that for $T \in \mathscr{T}$ it is NP-hard to decide of whether $\mathrm{OPT}(T) \geq \beta(T)$, or $\mathrm{OPT}(T) < \alpha(T)$.

(a) Assume first that $T \in \mathscr{T}$ is such that $\mathrm{OPT}(T) \geq \beta(T)$. We will show that the corresponding instance $f(T)$ of the 2-BP^r problem has its optimum $\mathrm{OPT}'(f(T))$ of size at most $|T| + n - 3\beta(T)$. Consider a matching M in T consisting of $\beta(T)$ triples. For each triple $t_l = (x_i, y_j, z_k) \in M$ we create a well-packed bin with rectangles $\{A_{X,i}, A_{Y,j}, A_{Z,k}, B_l, A'_{X,i}, A'_{Y,j}, A'_{Z,k}, B'_l\}$ packed.

For each $t_l \in T \setminus M$ we can put B_l and B'_l along with a dummy rectangle into a bin; in this way we use $|T| - \beta(T)$ dummy rectangles.

For each of $n - 3\beta(T)$ elements in $X \cup Y \cup Z$ that are not covered by M, we put in a bin the corresponding buddies A and A' along with one dummy rectangle. The rest of the dummy rectangles is used in this way and all rectangles from $f(T)$ are packed into $|T| + n - 3\beta(T)$ bins.

(b) Assume now that $T \in \mathcal{T}$ satisfies $\mathrm{OPT}(T) < \alpha(T)$. Our aim is to estimate $\mathrm{OPT}'(f(T))$ from below. Consider for an instance $f(T)$ any feasible solution of 2-BP^r. There will be exactly $N_d = |T| + n - 4\beta(T)$ bins with dummy rectangles, each of them can contain at most one rectangle from $\mathscr{A} \cup \mathscr{B}$. Let us consider now bins without dummy rectangles. If such bin is not well-packed then it either contains at most 3 rectangles from $\mathscr{A} \cup \mathscr{B}$ or else it contains at most 3 rectangles from $\mathscr{A}' \cup \mathscr{B}'$. Let N_g denote the number of well-packed bins. Among the bins without dummy rectangles which are not well-packed, let N_{b_2} denote the number of bins with at most 3 rectangles from $\mathscr{A} \cup \mathscr{B}$, and let N_{b_1} denote the number of the rest rectangles (i.e., N_{b_1} is the number of bins with 4 rectangles from $\mathscr{A} \cup \mathscr{B}$, but with at most 3 rectangles from $\mathscr{A}' \cup \mathscr{B}'$).

Since all $|T| + n$ rectangles from $\mathscr{A} \cup \mathscr{B}$ have to be packed, we have the constraint that
$$4N_g + 4N_{b_1} + 3N_{b_2} + N_d \geq |T| + n,$$
or equivalently
$$4N_g + 4N_{b_1} + 3N_{b_2} \geq 4\beta(T). \tag{4}$$

With the choice of parameter $p = \frac{1}{4} + 9\delta$ and assuming $\delta \in (0, \frac{1}{500}]$ as small as we need, rectangles from $\mathscr{A} \cup \mathscr{B}$ are roughly $(\frac{1}{4}, \frac{3}{4})$ each, and those from $\mathscr{A}' \cup \mathscr{B}'$ are roughly $(\frac{1}{4}, \frac{1}{4})$ each. In what follows we will count rectangles from $\mathscr{A} \cup \mathscr{B}$ with weight 3, and those from $\mathscr{A}' \cup \mathscr{B}'$ with weight 1 each. Easy area's estimate shows that the total weight of rectangles packed to a unit square bin cannot exceed 16. Further, any bin containing a dummy rectangle can contain rectangles from $\mathscr{A} \cup \mathscr{B} \cup \mathscr{A}' \cup \mathscr{B}'$ of weight at most 4. Observe that each of N_{b_1} bins contains rectangles of weight at most 15. Hence the second constraint derived from the fact that all rectangles have to be packed reads as follows:
$$16N_g + 15N_{b_1} + 16N_{b_2} + 4N_d \geq 4(|T| + n).$$

Using $N_d = |T| + n - 4\beta(T)$ and adding the constraint (4) to the last one we get
$$20N_g + 19N_{b_1} + 19N_{b_2} \geq 20\beta(T).$$

Since the set of well-packed bins corresponds to a feasible solution for a matching (by Lemma 5), $N_g < \alpha(T)$. Thus, assuming $\mathrm{OPT}(T) < \alpha(T)$ we get
$$\mathrm{OPT}'(f(T)) > N_g + N_{b_1} + N_{b_2} + N_d \geq \frac{20}{19}\beta(T) - \frac{1}{19}N_g + N_d$$
$$> |T| + n - 3\beta(T) + \frac{1}{19}(\beta(T) - \alpha(T)).$$

It easily follows that our reduction f is a gap preserving reduction assuming that we started from $(\alpha(T), \beta(T))$-gap version of the bounded MAX-3DM problem for which $\frac{\beta(T) - \alpha(T)}{|T| + n - 3\beta(T)}$ is bounded below by a positive constant.

Now suppose that for a fixed constant ρ, $1 < \rho < 1 + \frac{1}{19}\frac{\beta(T)-\alpha(T)}{|T|+n-3\beta(T)}$, there exists a polynomial time algorithm $\mathcal{A}_\rho$ and a constant C such that for instances $f(T)$ if $\mathrm{OPT}'(f(T)) > C$, then $\mathcal{A}_\rho \leq \rho\mathrm{OPT}'(f(T))$. Thus, for any corresponding instance T of MAX-3DM we could distinguish whether $\mathrm{OPT}(T) \geq \beta(T)$, or $\mathrm{OPT}(T) < \alpha(T)$, which is an NP-hard problem. Hence, it is NP-hard to achieve an asymptotic approximation ratio $\leq \rho$ for the problem 2-DIMENSIONAL BIN PACKING WITH ROTATIONS into unit square bins.

Using the NP-hard gap result from Theorem A we can obtain an explicit lower bound $1 + \frac{1}{3792}$ on asymptotic approximation ratio of any polynomial time approximation algorithm (unless P = NP) for 2-DIMENSIONAL BIN PACKING WITH ROTATIONS into unit square bins. For the same problem *without* rotations our method provides a lower bound $1 + \frac{1}{2196}$.

2.1   3-Dimensional Strip Packing Problems

Let a list of 2-dimensional rectangles $\mathscr{L} = \{(r_1^1, r_2^1), (r_1^2, r_2^2), \ldots, (r_1^n, r_2^n)\}$ with a bin $\mathbb{B} = (b_1, b_2)$ be an instance of the 2-DIMENSIONAL BIN PACKING problem. For a fixed $t > 0$ we define an instance of the 3-DIMENSIONAL STRIP PACKING problem as a list of 3-dimensional rectangles $\mathscr{L}_t = \{(r_1^i, r_2^i, t) : 1 \leq i \leq n\}$ with a strip (b_1, b_2, ∞). It is easy to prove, that if $\mathrm{OPT}(\mathscr{L})$ denote the optimum of an instance $\mathscr{L}$ for 2-BP (resp., 2-BP^r) then $t \cdot \mathrm{OPT}(\mathscr{L})$ denote the optimum of the corresponding 3-dimensional instance $\mathscr{L}_t$ for 3-SP (resp., for 3-SP^z and 3-SP^r provided $t > \max\{b_1, b_2\}$).

Hence, non-existence of APTAS for 2-BP ([4]) implies non-existence of APTAS for the 3-SP problem, unless P = NP. Moreover, using a heterogeneous scaling one can obtain some inapproximability results also for 3-SP^z and 3-SP^r already from hardness results for 2-BP, e.g., for instances of 3-SP^z and 3-SP^r with a strip $(b, 1, \infty)$ for any fixed $b \in (0, \frac{1}{2})$. However, for a strip with square base we have to use Theorem 1 instead.

Theorem 2. *There is no APTAS for any of 3-dimensional strip packing problems 3-SP, 3-SP^z, and 3-SP^r on instances with a strip $(1, 1, \infty)$, unless* P = NP.

3   Maximum Rectangle Packing Problem

Another rectangle bin packing problem well studied in the literature (e.g., [13], [2]) is the following:

Definition 3. *Given a collection of d-dimensional rectangles together with a d-dimensional rectangular bin $\mathbb{B}$. The goal of the MAXIMUM d-DIMENSIONAL RECTANGLE PACKING problem is to pack the maximum number of rectangles from the collection into a single bin $\mathbb{B}$.*

Other variants of this problem are studied as well, e.g., each of the rectangles can be associated with weight, and the goal is to maximize the total weight of packed rectangles. In some variants ninety-degree rotations of rectangles are allowed. But even in the simplest case, namely the 2-dimensional unweighted case without rotations, only a $(2+\varepsilon)$-approximation algorithm is known [13]. The question of whether there is an APTAS is open. However, in the 3-dimensional case the problem can be settled in the negative.

Theorem 3. *Unless* $P = NP$, *there is no* APTAS *for the* MAXIMUM 3-DIMENSIONAL RECTANGLE PACKING *problem with unit cube bin. The same result holds also for z-oriented packings and for r-packings, in both cases with a bin* $(1, 1, b)$, *where* $b \in \left(0, \frac{1}{4}\right)$.

Proof. We can use the hardness result of Theorem 2 for 3-SP with the strip $(1, 1, \infty)$. Namely, there is a constant $\rho > 1$ and an infinite family $\mathcal{F}$ of instances of the 3-SP problem with the strip $(1, 1, \infty)$, such that for a certain computable function $\alpha : \mathcal{F} \to \mathbb{N}$ it is NP-hard to distinguish for $\mathscr{L} \in \mathcal{F}$ whether $\text{OPT}(\mathscr{L}) \leq \alpha(\mathscr{L})$, or $\text{OPT}(\mathscr{L}) > \rho \cdot \alpha(\mathscr{L})$. Moreover, each rectangle in $\mathscr{L}$ is a small perturbation of either $\left(\frac{1}{4}, \frac{1}{2}, 1\right)$ or $\left(\frac{3}{4}, 1, 1\right)$.

For oriented packings (i.e., without rotations) and for any $\mathscr{L} \in \mathcal{F}$ denote by $\mathscr{L}'$ a rescaled copy of $\mathscr{L}$ by a factor $1/\alpha(\mathscr{L})$ in the direction of the z-axis. Then clearly, it is NP-hard to decide whether $\text{OPT}(\mathscr{L}') \leq 1$, or $\text{OPT}(\mathscr{L}') > \rho$ for an instance $\mathscr{L}$ of the 3-SP problem with the strip $(1, 1, \infty)$. In the former case all rectangles of $\mathscr{L}'$ can be packed into the unit cube bin. In the latter one we easily obtain that less than $|\mathscr{L}'| - \lfloor (\rho - 1)\alpha(\mathscr{L}) \rfloor$ can be packed into this bin.

For z-oriented packings we can use the same arguments starting instead from the NP-hard gap derived for the problem 2-BP^r with unit square bin $[0, 1]^2$ (the proof of Theorem 1).

For r-packings we rescale $\mathscr{L}$ by a factor $b/\alpha(\mathscr{L})$, $b \in (0, \frac{1}{4})$, in the direction of z-axis. Then it is NP-hard to decide whether all, or only a fraction strictly less than 1 of the rectangles of $\mathscr{L}$ can be packed into the bin $(1, 1, b)$. The special uniform structure of instances in our hardness result for 2-BP^r implies that all r-packings for such rescaled instances are, in fact, z-oriented packings. Thus the results follow as above.

References

1. G. Ausiello, P. Crescenzi, G. Gambosi, V. Kann, A. Marchetti-Spaccamela, and M. Protasi, *Complexity and approximation*, Springer, 1999.
2. B. S. Baker, A. R. Calderbank, E. G. Coffman, and J. C. Lagarias, *Approximation algorithms for maximizing the number of squares packed into a rectangle*, SIAM J. on Algebraic and Discrete Methods **4** (1983), 383–397.
3. N. Bansal, J. R. Correa, C. Kenyon, and M. Sviridenko, *Bin packing in multiple dimensions: inapproximability results and approximation schemes*, Manuscript, February 2004.
4. N. Bansal and M. Sviridenko, *New approximability and inapproximability results for 2-dimensional bin packing*, Proceedings of the 15th Annual ACM-SIAM Symposium on Discrete Algorithms, SODA, 2004, pp. 189–196.

5. A. Caprara, *Packing 2-dimensional bins in harmony*, Proceedings of the 43th Annual IEEE Symposium on Foundations of Computer Science, FOCS, 2002, pp. 490–499.
6. C. Chekuri and S. Khanna, *On multi-dimensional packing problems*, Proc. of the 10th ACM-SIAM Symposium on Discrete Algorithms, SODA, 1999, pp. 185–194.
7. M. Chlebík and J. Chlebíková, *Complexity of approximating bounded variants of optimization problems*, Theoretical Computer Science 354(2006), 320–338.
8. J. R. Correa and C. Kenyon, *Approximation schemes for multidimensional packing*, Proceedings of the 15th ACM-SIAM Symposium on Discrete Algorithms, SODA, 2004, pp. 179–188.
9. J. Csirik and A. van Vliet, *An on-line algorithm for multidimensional bin packing*, Operation Research Letters **13** (1993), 149–158.
10. W. Fernandez de la Vega and G. S. Lueker, *Bin packing can be solved within* $(1+\varepsilon)$ *in linear time*, Combinatorica **1** (1981), 349–355.
11. K. Jansen and R. Solis-Oba, *An asymptotic approximation algorithm for 3d-strip packing*, to appear in Proceedings of the 17th Annual ACM-SIAM Symposium on Discrete Algorithms, SODA, 2006.
12. K. Jansen and R. Stee, *On strip packing with rotations*, Proceedings of the 37th ACM Symposium on Theory of Computing, STOC, 2005, pp. 755–761.
13. K. Jansen and G. Zhang, *On rectangle packing: maximizing benefits*, Proceedings of the 15th Annual ACM-SIAM Symposium on Discrete Algorithms, SODA, 2004, pp. 197–206.
14. V. Kann, *Maximum bounded 3-dimensional matching is MAX SNP complete*, Information Processing Letters **37** (1991), 27–35.
15. N. Karmarkar and R. M. Karp, *An efficient approximation scheme for the one-dimensional bin-packing problem*, Proceedings of the 23rd IEEE Symposium on Foundations of Computer Science, FOCS, 1982, pp. 312–320.
16. C. Kenyon and E. Rémila, *A near optimal solution to a two-dimensional cutting stock problem*, Mathematics of Operations Research **25** (2000), 645–656, Preliminary version in *Proceedings of the 37th Annual IEEE Symposium on Foundations of Computer Science*, FOCS, 1996, pp. 31–36.
17. K. Li and K. H. Cheng, *On three-dimensional packing*, SIAM J. Comput. **19** (1990), 847–867.
18. F. K. Miyazawa and Y. Wakabayashi, *Approximation algorithms for the orthogonal z-oriented three-dimensional packing problems*, SIAM J. Comput. **29** (2000), 1008–1029.
19. _____, *Packing problems with orthogonal rotations*, Proceedings of the 6th Latin American Symposium on Theoretical Informatics, LATIN, Buenos Aires, Argentina, LNCS 2976, Springer-Verlag, 2004, pp. 359–368.
20. E. Petrank, *The hardness of approximation: Gap location*, Computational Complexity **4** (1994), 133–157.
21. G. J. Woeginger, *There is no asymptotic PTAS for two-dimensional vector packing*, Information Processing Letters **64(6)** (1997), 293–297.

Approximate Hierarchical Facility Location and Applications to the Shallow Steiner Tree and Range Assignment Problems*
(Extended Abstract)

Erez Kantor and David Peleg

Department of Computer Science and Applied Mathematics,
The Weizmann Institute of Science, Rehovot 76100, Israel
{erez.kantor, david.peleg}@weizmann.ac.il

Abstract. The paper concerns a new variant of the *hierarchical facility location problem* on metric powers ($\mathrm{HFL}_\beta[h]$), which is a multi-level uncapacitated facility location problem defined as follows. The input consists of a set F of locations that may open a facility, subsets $D_1, D_2, \ldots, D_{h-1}$ of locations that may open an intermediate transmission station and a set D_h of locations of clients. Each client in D_h must be serviced by an open transmission station in D_{h-1} and every open transmission station in D_l must be serviced by an open transmission station on the next lower level, D_{l-1}. An open transmission station on the first level, D_1 must be serviced by an open facility. The cost of assigning a station j on level $l \geq 1$ to a station i on level $l - 1$ is c_{ij}. For $i \in F$, the cost of opening a facility at location i is $f_i \geq 0$. It is required to find a feasible assignment that minimizes the total cost. A constant ratio approximation algorithm is established for this problem. This algorithm is then used to develop constant ratio approximation algorithms for the *bounded depth steiner tree* and the *bounded hop strong-connectivity range assignment* problems.

1 Introduction

The $\mathrm{HFL}_\beta[h]$ problem. The paper concerns a new variant of the *hierarchical facility location problem* on metric powers ($\mathrm{HFL}_\beta[h]$), which is a multi-level uncapacitated facility location (UFL) problem defined as follows. The input consists of a tuple $\langle h, S, F = D_0, D_1, \ldots, D_h, c, f \rangle$, where $S = \{1, 2, \ldots, n\}$ is a set of locations, $F \subseteq S$ contains locations at which one may open a facility, $D_1, D_2, \ldots, D_{h-1} \subseteq S$ are subsets of locations which may open an intermediate transmission station and $D_h \subseteq S$ is a set of locations of clients.

A *feasible assignment* is an assignment of stations to stations one level down, which satisfies the following requirements.
(1) A station i at level $0 \leq l \leq h - 1$ must be *open* if there exists a station $j \in D_{l+1}$ assigned to it. (Here we refer to a facility as a station on level 0.)
(2) Every client $j \in D_h$ must be assigned to some station $i \in D_{h-1}$.

---

* Supported in part by a grant from the Israel Ministry of Science and Technology.

T. Calamoneri, I. Finocchi, G.F. Italiano (Eds.): CIAC 2006, LNCS 3998, pp. 211–222, 2006.
© Springer-Verlag Berlin Heidelberg 2006

(3) Every open station $j \in D_l$ on level $1 \leq l \leq h - 1$ must be assigned to some station $i \in D_{l-1}$.

The cost $c_{ij} \geq 0$ of assigning a station j on level $l \geq 1$ to a station i on level $l - 1$ is specified via a metric. Specifically, the input for the HFL$_\beta[h]$ problem includes a positive real *power* parameter $\beta \geq 1$ and a distance metric between the locations in S, denoted by $dist(i, j)$ for every $i, j \in S$. The cost is defined as $c_{ij} = dist^\beta(i, j)$. We refer to this cost type as a *metric power*. For $i \in F$, the cost of opening a facility at location i is $f_i \geq 0$. It is required to find a feasible assignment that minimizes the total cost. Observe that UFL of [15] and HFL$_\beta[1]$ are equivalent problems when $\beta = 1$, but for $h \geq 2$, HFL$_\beta[h]$ is different from h-UFL of [9] even when $\beta = 1$.

The bounded depth steiner tree problem. The problem H-STEINER(h, s) is defined as follows. Given an undirected complete graph $G(V, E)$ with nonnegative edge weights w, representing a general metric space, a subset $\mathcal{V} \subseteq V$, a source node $s \in V$ and a positive constant integer h, a *h-steiner tree* $T(h, s)$ is a tree of depth at most h rooted at s that spans $\mathcal{V}$, i.e., such that for any node $v \in \mathcal{V}$ there is a path of at most h hops (edges) from v to s. The cost of a tree $t(h, s)$, denoted by COST$(t(h, s))$, is the sum of its edge weights. The goal is to find a minimum weight h-steiner tree rooted at s. The related *bounded diameter minimum Steiner tree* problem *(BDST)* is defined similarly, except that the parameters s and h are replaced by a constant integer parameter d, and the Steiner tree is required to be of diameter at most d. These problems find applications in communication network design. The BDST problem is NP-hard (see problem ND4 in [8]). It is shown in [2] based on the result of [13, 7] that the problem on general graphs has no better than $\ln n$ approximation.

The bounded depth range assignment problem. The formal definition of range assignment problems is as follows. Given a set of locations $S = \{1, 2, \ldots, n\}$, represented as points in 2 or 3-dimensional Euclidean space, a *range assignment* for S is a function $r : S \to \mathbb{R}^+$ assigning a nonnegative real $r(i)$ to every location $i \in S$. The *cost* of a range assignment r is defined as its total power consumption, that is, $cost(r) = \sum_{i \in S} r(i)^\beta$ for some fixed positive real $\beta > 1$. A range assignment r for a set S of stations determines a logical directed communication graph $G_r = (S, E)$ such that for every $i, j \in S$, the directed edge (j, i) occurs in E if and only if $r(j) \geq dist(i, j)$. In this case, we say that j *covers* i, or that j can transmit to i.

Depending on the particular application, the communication graph is required to satisfy a certain property Π. For any desired graph property Π, the MIN-RANGE(Π) problem is defined as follows: Given a set of points S, find a range assignment r for S such that G_r satisfies Π and $cost(r)$ is minimized. The *bounded depth* or *bounded-hop strong connectivity* *(hSC)* problem is when G_r must contain a directed path of at most h hops from every station $j \in S$ to every other station $i \in S$, for some integer $h \geq 1$.

Another property that interests us in the current paper is the following. A *central base* under the assignment r is a node of outdegree $n - 1$ in G_r. Equivalently, a node i is a central base if $r(i) \geq \rho_i$, where $\rho_i = \max\{dist(i,j) \mid j \in S\}$ is the *radius* of i (namely, the distance from i to the node farthest from it). Denote by $CB_r(S)$ the set of central base nodes in S. The property is now stated as:

Bounded-hop center forest (hCF): For every $i \in S$, G_r must contain a directed path of at most h hops from i to some central base node $cb_r(i)$ in $CB_r(S)$.

It is obvious that the $(h-1)CF$ property implies the hSC property, i.e., if G_r enjoys the $(h-1)CF$ property then for every $i,j \in S$ there is path of at most h hops from i to j (composed of a subpath of at most $h-1$ hops from i to its central base $cb_r(i)$, and then to j), hence G_r is h-strongly connected. This means that every feasible solution to the MIN-RANGE($(h-1)CF$) problem is also a feasible solution to the MIN-RANGE(hSC) problem.

Related work. The classical *uncapacitated facility location problem* UFL is define as follows: Given set F of locations at which facilities can be built. A fixed cost f_i is incurred if a facility is opened at location i. In addition, the input contains a set D of clients to be serviced by the opened facilities, and if the client j is assigned to a facility at location i, there is an associated service cost of the form $c_{ij} = dist(i,j)$, where $dist(i,j)$ is symmetric and satisfies the triangle inequality. The goal is to find an assignment from clients to facilities that minimizes the total cost.

Several papers deal with the UFL problem [14, 15, 4]. The first constant factor approximation algorithm is given by Shmoys et al [15]. Currently, the best result for this problem is a 1.52-approximation algorithm of [14].

Another known extension of the UFL problem is the *h-level uncapacitated facility location problem*, denoted h-UFL, where each client must be serviced by a sequence of different facilities. These sequences are defined by a hierarchy of production and distribution systems and can be presented as facility paths. The set of admissible facility paths is given. Each facility has a fixed cost. Each client incurs a transportation cost for being served. The input specifies the transportation cost for each client and each facility path. A 3-approximation algorithm for this problem is presented in [9].

The bounded depth and bounded diameter steiner tree problems were also studies previously. An $O(\log n)$ approximation algorithm for the bounded diameter steiner tree problem on general graphs when the diameter is constant is shown in [11, 3]. An algorithm for the bounded depth minimum spanning tree of total expected cost $O(\log n)$ times the optimal minimum cost h-hop spanning tree is presented in [1].

Several variations of the range assignment problem have been studied in the past. The following results were obtained.

Strong connectivity (SC): In the one-dimensional case, i.e., when the stations are located on the real line, the problem has an $O(n^4)$-time algorithm [10]. When the stations are spread in d-dimensional space ($d > 1$), finding an optimal solution for MIN-RANGE(SC) is NP-hard [6, 10], and moreover, it is APX-hard for $d \geq 3$ [10]. On the positive side, the problem has a 2-approximation algorithm based on constructing a minimum spanning tree [10].

Bounded-hop strong connectivity (hSC): For the case where the locations of S are on the line, an $O(hn^3)$-time 2-approximation algorithm for $\beta = 2$ and any $h > 0$ is described in [5]. Lower and upper bounds are shown in [6] on the optimal cost for any 2-dimensional instances where h is an arbitrary constant. It is also shown therein that when S is a family of *well-spread* instances (namely, the locations in S are suitably distributed), the MIN-RANGE(hSC) problem on S admits a polynomial time approximation algorithm with constant ratio, i.e., MIN-RANGE(hSC) is in APX. Additionally, it is shown that the MIN-RANGE(hSC) problem with a uniform instance probability is in Av-APX.

Our results. The paper presents constant factor approximation algorithms for the above problems. We first show an approximation algorithm of ratio $\left(1 + 3^\beta\right) \cdot \left(3^{\beta+1}\right)^{h-1}$ for the HFL$_\beta[h]$ problem. We then use this algorithm to derive constant approximations for the other problems. Specifically, we show that for complete graphs whose weight function is a metric, there exists a polynomial time approximation algorithm for the bounded depth steiner tree problem with ratio $1.52 \cdot 9^{h-2}$ for constant h, and similarly, a polynomial time approximation algorithm for the bounded diameter steiner tree problem with ratio $1.52 \cdot 9^{\left(\lfloor \frac{d}{2} \rfloor - 2\right)}$ for constant d. Finally, we present a polynomial time approximation algorithm with ratio $\left(1 / \left(\sqrt[h]{2} - 1\right)\right)^\beta \left(1 + 3^\beta\right) \left(3^{\beta+1}\right)^{h-2}$ for MIN-RANGE(hSC) on general metrics for constant h and β. These are the first constant approximation guarantees known for those problems.

As may be expected, the MIN-RANGE(hSC) problem is NP-hard on general metric spaces for constant h. Proof details are deferred to the full version.

2   Approximating the Hierarchical Facility Location

In this section we develop a constant ratio approximation algorithm for HFL$_\beta[h]$.

2.1   An ILP Representation for HFL$_\beta[h]$

The HFL$_\beta[h]$ problem can be represented in a straightforward manner as an integer linear program where for every $i \in F$, the boolean variable y_i indicates whether i is an open facility, and for $1 \leq l \leq h$, the boolean variable x_{ij}^l indicates that the open transmission station $j \in D_l$ on level l is assigned to the station at location $i \in D_{l-1}$ on level $l - 1$. The resulting program is defined as follows.

Program HFL$_\beta^-[h, I]$:   Minimize
$$cost(x, y) = \sum_{i \in F} f_i y_i + \sum_{l=1}^{h} \sum_{i \in D_{l-1}} \sum_{j \in D_l} c_{ij} x_{ij}^l,$$
subject to
(C1.1) $\sum_{i \in D_{h-1}} x_{ij}^h = 1$ for every $j \in D_h$.
(C1.2) $x_{ij}^l \leq \sum_{k \in D_{l-2}} x_{ki}^{l-1}$, for every $j \in D_l$, $i \in D_{l-1}$, and $l \in \{2, 3, \ldots, h\}$.
(C1.3) $x_{ij}^1 \leq y_i$, for every $i \in F$ and $j \in D_1$.
(C1.4) $x_{ij}^l \in \{0, 1\}$, for every $j \in D_l$, $i \in D_{l-1}$ and $l \in \{1, 2, \ldots, h\}$.
(C1.5) $y_i \in \{0, 1\}$, for every $i \in F$.

Constraint (C1.1) ensures that every client $j \in D_h$ is assigned to some open transmission station $i \in D_{h-1}$ in level $h - 1$. Constraint (C1.2) ensures that for every level l ($2 \leq l \leq h$), every open transmission station $j \in D_l$ on level l is assigned to an open transmission station $i \in D_{l-1}$ on level $l - 1$. Constraint (C1.3) ensures that every open transmission station $j \in D_1$ on the first level is assigned to an open facility $i \in F$. Note that that if location j occurs on two consecutive levels, i.e., $j \in D_l \cap D_{l-1}$ and j is an open transmission station on level l, then j can be assigned to itself, i.e., $x_{jj}^l = 1$ satisfying constraint (C1.1) or constraint (C1.2) respectively at cost $c_{jj} = 0$.

Unfortunately, while the integer linear program $\mathrm{HFL}_\beta^-[h, I]$ ensures feasible assignment for integer numbers, its fractional relaxation $\mathrm{HFL}_\beta^-[h, \mathbb{R}]$ fails to approximate the problem. In particular, to obtain $\mathrm{HFL}_\beta^-[h, \mathbb{R}]$, the boolean constraints (C1.4) and (C1.5) are replaced with

(C1.4^f) $x_{ij}^l \geq 0$, for every $i \in D_{l-1}$, $j \in D_l$ and $l \in \{1, 2, \ldots, h\}$.
(C1.5^f) $y_i \geq 0$, for every $i \in F$.

However, the solution for the fractional linear program $\mathrm{HFL}_\beta^-[h, \mathbb{R}]$ can be far away from the integer solution to $\mathrm{HFL}_\beta^-[h, I]$, particularly for the constraints (C1.1-C1.3).

Therefore, we consider an alternative integer linear program $\mathrm{HFL}_\beta[h, I]$ making use of *flow* variables z, where the optimal solution of the linear relaxation $\mathrm{HFL}_\beta[h, \mathbb{R}]$ is close to the optimal solution of $\mathrm{HFL}_\beta[h, I]$. Consider fixed $0 \leq \ell_1 < \ell_2 < \ldots < \ell_k < h$. For every $i_{\ell_1} \in D_{\ell_1}, \ldots, i_{\ell_k} \in D_{\ell_k}$, let $\mathcal{D}[i_{\ell_1}, \ldots, i_{\ell_k}]$ denote the collection of all possible choices of nodes $i_0 \in D_0, \ldots, i_h \in D_h$, except that the choices from $D_{\ell_1}, \ldots, D_{\ell_k}$ are fixed to be $i_{\ell_1}, \ldots, i_{\ell_k}$, respectively. For example, if $h = 4, \ell_1 = 2$ and $\ell_2 = 3$, then for every $i_2 \in D_2$ and $i_3 \in D_3$,

$$\mathcal{D}[i_2, i_3] = D_0 \times D_1 \times \{i_2\} \times \{i_3\} \times D_4 = \{(i_0, i_1, i_2, i_3, i_4) \mid i_0 \in D_0, i_1 \in D_1, i_4 \in D_4\}.$$

In particular, $\mathcal{D} = \mathcal{D}[\ ] = D_0 \times D_1 \times \ldots \times D_h$.

For $\bar{i} = (i_0, i_1, \ldots, i_h) \in \mathcal{D}$, the variable $z_{\bar{i}}$ represents the amount of flow going from i_h via $i_{h-1}, \ldots, i_1, i_0$. We refer to the edge (i_{l-1}, i_l) for $i_{l-1} \in D_{l-1}$ and $i_l \in D_l$ as a *channel*, and the variable $x_{i_{l-1} i_l}^l$ represents its capacity.

The resulting program is defined as follows.

Program $\mathrm{HFL}_\beta[h, I]$:   Minimize
$$cost(x, y) = \sum_{i \in F} f_i y_i + \sum_{l=1}^{h} \sum_{i \in D_{l-1}} \sum_{j \in D_l} c_{ij} x_{ij}^l,$$
subject to the following constraints.
(C2.1) $\sum_{\bar{i} \in \mathcal{D}[i_h]} z_{\bar{i}} = 1$, for every $i_h \in D_h$.
(C2.2.l) $x_{i_{l-1} i_l}^l \geq \sum_{\bar{i} \in \mathcal{D}[i_{l-1}, i_l, i_h]} z_{\bar{i}}$, for every $1 \leq l \leq h - 1$ and every $i_{l-1} \in D_{l-1}, i_l \in D_l$ and $i_h \in D_h$.
(C2.2.h) $x_{i_{h-1} i_h}^h \geq \sum_{\bar{i} \in \mathcal{D}[i_{h-1}, i_h]} z_{\bar{i}}$, for every $i_{h-1} \in D_{h-1}$ and $i_h \in D_h$.
(C2.3) $y_{i_0} \geq \sum_{\bar{i} \in \mathcal{D}[i_0, i_h]} z_{\bar{i}}$, for every $i_0 \in F$ and $i_h \in D_h$.
(C2.4) $x_{ij}^l \in \{0, 1\}$, for every $i \in D_{l-1}$, $j \in D_l$ and $1 \leq l \leq h$.
(C2.5) $y_i \in \{0, 1\}$, for every $i \in F$.
(C2.6) $z_{\bar{i}} \in \{0, 1\}$, for every $\bar{i} \in \mathcal{D}$.

Constraint (C2.1), coupled with (C2.6), ensures that from every client $i_h \in D_h$ there is a unit flow to some facility $i_0 \in F$. Constraint (C2.2) ensures that if there is flow in $z_{\bar{i}}$ for some $\bar{i} = (i_0, i_1, \ldots, i_h)$ then the related channels from i_l to i_{l-1} (for $l = h, h-1, \ldots, 1$) have sufficient capacity, i.e., are open. Specifically, if $z_{\bar{i}} = 1$ then constraint (C2.2) implies that $x^l_{i_{l-1} i_l} = 1$, for every $1 \le l \le h$. Constraint (C2.3) ensures that if there exists some flow in $z_{\bar{i}}$ for a chain $\bar{i}$ that ends at facility $i_0 \in F$ then the facility i_0 is open, i.e., $z_{\bar{i}} = 1$ implies that $y_{i_0} \ge 1$.

Note that the optimal assignment for $\mathrm{HFL}_\beta[h, I]$ is the same as that of $\mathrm{HFL}_\beta^-[h, I]$. However, when we relax this linear program and look at $\mathrm{HFL}_\beta[h, \mathbb{R}]$, constraint (C2.2) ensures that the sum of flows from each client $i_h \in D_h$ that cross a channel (i_{l-1}, i_l) does not exceed the (fractional) capacity of this channel, $x^l_{i-1,l}$, and (C2.3) ensures that the sum of flows from each client $i_h \in D_h$ to some facility i_0 in F does not exceed the fraction y_{i_0} to which this facility is open. As proved later on, this guarantees that the optimum fractional solution for $\mathrm{HFL}_\beta[h, \mathbb{R}]$ is close to the optimal solution for the integer linear program $\mathrm{HFL}_\beta[h, I]$, as illustrated in Figure 1.

Also note that in the h-UFL variant of [9], the target function charges the z entries instead of the x entries as in our variant.

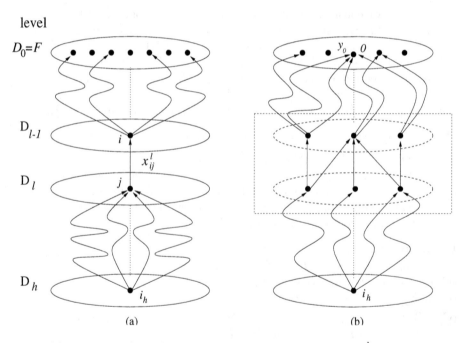

(a)   (b)

Fig. 1. (a) Constraint (C2.2.l) ensures that the fractional value of x^l_{ij} is greater than the sum of flows from the client i_h that crosses the channel between j and i. (b) Constraint (C2.3) ensures that for every client i_h, the fractional value of y_i is greater than the sum of flows from i_h to the facility at location i. The dashed box illustrates how the flow can split and join in its way up to the facilities.

2.2 Approximation Algorithm for $\mathrm{HFL}_\beta[h]$

We derive a constant approximation algorithm for the $\mathrm{HFL}_\beta[h]$ problem where h is constant. The algorithm consists of h rounds. In each round l we assign values only to the variables $x^l_{i_{l-1}i_l}$ of level l. Specifically, in the first round, decide on the set T of transmission stations in D_{h-1} to be opened (for $h = 1$ we decide on set T of facilities) and assign every client $j \in D_h$ to the closest open transmission station $\hat{i}$ in T, i.e., such that $c_{\hat{i}j} = \min\{c_{ij} \mid i \in T\}$, by setting $\hat{x}^h_{\hat{i}j}$ to 1. For $h \geq 2$, in the next round, define an instance $\langle h - 1, S, F, D_1, D_2, \ldots, D_{h-2}, T, c, f \rangle$ of the problem $\mathrm{HFL}_\beta[h - 1]$ (i.e., where the clients belong to T instead of D_{h-1}), and apply the same procedure for $\mathrm{HFL}_\beta[h - 1]$.

In each round l, consider instance of the current $\mathrm{HFL}_\beta[h - l + 1]$, and decide the assignment of the x^l variables on the last level and the transmission stations that will be open on level $l - 1$. This is done by solving the linear relaxation of the integer linear program $\mathrm{HFL}_\beta[h, I]$ and then rounding the vector x^l of the optimal fractional solution to a boolean vector $\hat{x}^l$. Finally, when left with an instance of $\mathrm{HFL}_\beta[h]$ for $h = 1$, we decide which facilities to be open and assign each station to his closest facility, i.e. decide the assignment of $\hat{x}^1$ and $\hat{y}$.

The rounding procedure consists of two phases using the filtering and rounding technique of [12] for solving the $k\text{-median}$ problem. (For $h = 1$ the procedure resemble the approximation algorithm of [15] for solving the UFL problem.) The first phase modifies the fractional solution (x, y, z) obtained for the linear program into a new fractional solution $(\bar{x}, \bar{y}, \bar{z})$ that enjoys the *closeness* property, namely, that whenever a node $j \in D_h$ is fractionally assigned to a fractionally open transmission station $i \in D_{h-1}$, the cost c_{ij} of this assignment is not too large. (In this section for simplicity we denote sometimes $i \in D_{h-1}$ and $j \in D_h$ instead of $i_{h-1} \in D_{h-1}$ and $i_h \in D_h$. When we denote x^l_{ij} the i, j's entries refers to locations $i \in D_{l-1}$ and $j \in D_l$.) The second phase rounds the last level (the vector $\bar{x}^h$) of the new fractional solution $(\bar{x}, \bar{y}, \bar{z})$ to a boolean vector $\hat{x}^h$. We then show (for $h \geq 2$) that there exists a mixed (integral-fractional) solution $(\hat{x}^h \bar{x}^{h-1} \bar{x}^{h-2} \ldots \bar{x}^1, \bar{y}, \hat{z})$, (namely, integral on the last level $\hat{x}^h$ and fractional on the rest of the levels) that is close to the optimal solution. In particular, this mixed assignment increases the total cost by at most a constant factor.

We now define for the next round an instance of the linear program for $\mathrm{HFL}_\beta[h-1, \mathbb{R}]$, where the clients are the open stations of T, and solve it recursively. We then continue to assign the clients of T to lower levels (with the same procedure). At the end of the algorithm, $(\hat{x}, \hat{y}, \hat{z})$ becomes an integral feasible assignment that approximates the optimal solution of $\mathrm{HFL}_\beta[h, I]$ by a constant factor.

Note that in $\mathrm{HFL}_\beta[h, \mathbb{R}]$ the minimization of the target function $cost(x, y)$ implies that in constraint (C2.2.h), the inequality can be change into an equality, requiring

$$x^h_{ij} = \sum_{\bar{i} \in \mathcal{D}[i,j]} z_{\bar{i}}, \quad \text{for every } i \in D_{h-1} \text{ and } j \in D_h. \qquad (1)$$

(Note that this is not the case for constraint (C2.2.l) for $1 \leq l < h$.) Therefore, by constraint (C2.1),

$$\sum_{i \in D_{h-1}} x_{ij}^h = \sum_{\bar{i} \in \mathcal{D}[j]} z_{\bar{i}} = 1, \quad \text{for every } j \in D_h. \tag{2}$$

Given a real parameter $g_j > 0$ for every $j \in D_h$, a feasible solution (x, y, z) to the linear program $\text{HFL}_\beta[h, \mathbb{R}]$ is *g-close* if it satisfies the following closeness property:

(C2.7) For every $i \in D_{h-1}$ and $j \in D_h$, if $x_{ij}^h > 0$ then $c_{ij} \le g_j$.

Note that this property implies that if $z_{i_0 i_1 \ldots i_{h-2} ij} > 0$, i.e., there is flow on channel (i, j), then $c_{ij} \le g_j$.

We now describe Procedure GEN_CLOSE implementing the first phase of the rounding procedure. Given a feasible fractional solution (x, y, z) and a fixed fraction $0 < \alpha < 1$, define for every client $j \in D_h$ an (α, j)-*count* ℓ_j, an (α, j)-*weight* $c_j(\alpha)$ and an (α, j)-*fraction* α_j as follows. Given a client $j \in D_h$, consider the permutation π of the transmission stations D_{h-1} ordered by distance from location j, i.e., such that $c_{\pi(1)j} \le c_{\pi(2)j} \le \ldots \le c_{\pi(|D_{h-1}|)j}$. For $k \ge 1$, the k transmission stations closest to j, $\{\pi(1), \pi(2), \ldots, \pi(k)\}$, contribute a fraction $\alpha_j(k) = \sum_{\ell=1}^k x_{\pi(\ell)j}^h \le 1$ of j's assignment, and each of them does it at a cost not exceeding $c_{\pi(k)j}$. Let ℓ_j be the smallest k such that the fraction $\alpha_j(k)$ exceeds α, i.e., $\ell_j = \min\{k \mid \sum_{\ell=1}^k x_{\pi(\ell)j}^h \ge \alpha\}$. (Such ℓ_j exists by Equality (2).) Denote the set of transmission stations on level $h - 1$ *farther* away from j than $\pi(\ell_j)$ by $F_j = \{\pi(\ell) \mid \ell \ge \ell_j\}$ and denote the set of transmission stations on level $h - 1$ *closer* to j than $\pi(\ell_j)$ by $C_j = \{\pi(\ell) \mid \ell \le \ell_j\}$. Note that $\pi(\ell_j)$ belongs to both sets, i.e., $\pi(\ell_j) \in F_j \cap C_j$. Define the (α, j)-*weight* as $c_j(\alpha) = c_{\pi(\ell_j)j}$ and the (α, j)-*fraction* as $\alpha_j = \alpha_j(\ell_j) = \sum_{i \in C_j} x_{ij}^h$. Note that by definition, $\alpha_j \ge \alpha$.

Procedure GEN_CLOSE operates as follows. For every $\bar{i} = (i_0, i_1, \ldots, i_{h-2}, i, j) \in \mathcal{D}$, allow flow through $\bar{i}$ only if i is one of the transmission stations close to j, setting $\bar{z}_{i_0 i_1 \ldots i_{h-2} ij} = z_{i_0 i_1 \ldots i_{h-2} ij}/\alpha_j$ if $i \in C_j$ and $\bar{z}_{i_0 i_1 \ldots i_{h-2} ij} = 0$ otherwise. To accommodate that flow, for every $i \in D_{h-1}$ and $j \in D_h$, set the capacity of the channel (i, j) as $\bar{x}_{ij}^h = x_{ij}^h/\alpha_j$ if $i \in C_j$ and $\bar{x}_{ij}^h = 0$ otherwise. For the intermediate levels, we must use α as a lower bound on all possible values of α_j, so for every $i_l \in D_l$, $i_{l-1} \in D_{l-1}$ and for every $l \in \{1, 2, \ldots, h - 1\}$, set $\bar{x}_{i_{l-1} i_l}^l = \min\{1, x_{i_{l-1} i_l}^l/\alpha\}$. Finally, for every $i \in F$, set $\bar{y}_i = \min\{1, y_i/\alpha\}$.

Lemma 1. *Given an instance of* $\text{HFL}_\beta[h, \mathbb{R}]$, *a feasible fractional solution* (x, y, z) *and a fixed* $0 < \alpha < 1$, *define the function* $g : D_h \to \mathbb{R}^+$ *as* $g_j = c_j(\alpha)$, *for every* $j \in D_h$. *Then the fractional solution* $(\bar{x}, \bar{y}, \bar{z})$ *constructed by Procedure* GEN_CLOSE *is g-close and feasible. Moreover ignoring the costs associated with the last level, the new solution is at most* $1/\alpha$ *times more expensive than the original one, namely,*

$$\sum_{i \in F} f_i \bar{y}_i + \sum_{l=1}^{h-1} \sum_{i \in D_{l-1}} \sum_{j \in D_l} c_{ij} \bar{x}_{ij}^l \le \frac{1}{\alpha} \left(\sum_{i \in F} f_i y_i + \sum_{l=1}^{h-1} \sum_{i \in D_{l-1}} \sum_{j \in D_l} c_{ij} x_{ij}^l \right).$$

(Throughout, proofs are omitted.)

We now show how to exploit this closeness property to find a subset of transmission stations on the first level, $T \subseteq D_{h-1}$, and show that there exists a $3^\beta g$-close integral assignment such that these transmission stations are opened and the cost is close to the optimal solution.

For a set of vertices U, denote $dist(j, U) = \min\{dist(i, j) \mid i \in U\}$ and let $d_j = \sqrt[\beta]{g_j}$. We present a rounding procedure named h-ROUND, that given an instance of $\mathrm{HFL}_\beta[h, I]$ and a g-close feasible fractional solution $(\bar{x}, \bar{y}, \bar{z})$ for the relaxed $\mathrm{HFL}_\beta[h, \mathbb{R}]$ instance, picks a subset of transmission stations $T \subseteq D_{h-1}$ satisfying the property that $dist(j, T) \leq 3d_j$ for every $j \in D_h$ and show that for $h = 1$ there exists solution close to the optimal using only facilities from T.

For $h \geq 2$ we show that there exists a mixed solution close to the optimal and using only the transmission stations of T on level $h - 1$.

We then approximate the optimal solution for $\mathrm{HFL}_\beta[h - 1, I]$ for the tuple $\langle h - 1, S, F, D_1, D_2, \ldots, D_{h-2}, T, c, f \rangle$, namely, with T as the set of clients, and get a cost that is close to the optimal solution for the original problem.

The input is $\langle D_{h-1}, D_h, c, f, \bar{x}^h, g \rangle$, where $\langle h, S, F, D_1, D_2, \ldots, D_h, c, f \rangle$ is the input for $\mathrm{HFL}_\beta[h, I]$ and $(\bar{x}, \bar{y}, \bar{z})$ is a feasible fractional solution generated by Procedure GEN_CLOSE(x, y, z) for some $0 < \alpha < 1$, where (x, y, z) is the optimal solution for $\mathrm{HFL}_\beta[h, \mathbb{R}]$ that satisfying the g-closeness property for the given function $g : D_h \to \mathbb{R}^+$ where $g_j = c_j(\alpha)$. Initially the procedure finds the set T of transmission stations i such that there exists a client j that integrally supplies i in $\hat{x}^h$, i.e., $T = \{i \in D_{h-1} \mid \exists j \in D_h \text{ such that } \bar{x}_{ij}^h = 1\}$. Next, the procedure finds the set B' of clients that are fractionally or integrally assigned to transmission stations in T, $B' = \{j \in D_h \mid \exists i \in T \text{ s.t. } \bar{x}_{ij}^h > 0\}$. Throughout the execution of the procedure, $\hat{D}_h$ denotes the set of clients in D_h that are still unassigned under the current solution T, i.e., $\hat{D}_h = \{j \in D_h \mid \forall i \in T, \ \bar{x}_{ij}^h = 0\}$. The procedure iteratively picks new transmission stations in D_{h-1} that will be opened in addition to those selected in previous iterations. In each iteration it first finds the client $j^* \in \hat{D}_h$ that has the smallest g_j, i.e., $g_{j^*} = \min\{g_j \mid j \in \hat{D}_h\}$. Let A be the set of transmission stations that supply to j^* in $\bar{x}^h$, i.e., $A = \{i \in D_{h-1} \mid \bar{x}_{ij^*}^h > 0\}$. Let B be the subset of unassigned clients that are supplied in $\bar{x}^h$ by some transmission station from A, i.e., $B = \{j \in \hat{D}_h \mid \exists i \in A \text{ s.t. } \bar{x}_{ij}^h > 0\}$. We pick an arbitrary transmission station $i^* \in A$ to be opened, (for $h = 1$ we pick facility $i^* \in A$ which f_{i^*} is the smallest in A), and add i^* to T. We then update the set of unassigned clients (deleting from $\hat{D}_h$ the clients that belong to B). These iterations are repeated until $\hat{D}_h$ becomes empty. The resulting set T is returned as the set of transmission stations on level D_{h-1} that will be opened. In the full paper we show that there exists mixed assignment close to the optimal solution of $\mathrm{HFL}_\beta[h]$, where the transmission stations/facilities, on level $h - 1$ is the set T the output of Procedure h-ROUND and that there exists assignment close to the optimal solution of $\mathrm{HFL}_\beta[h]$ using only facilities from T.

We now describe how to exploit the properties of Procedure h-ROUND and provide a constant approximation algorithm h-AlgHFL for $\mathrm{HFL}_\beta[h]$. Given the input tuple $\langle h, S, F, D_1, D_2, \ldots, D_h, c, f \rangle$, we solve the linear program $\mathrm{HFL}_\beta[h, \mathbb{R}]$ and get a fractional solution (x, y, z). Next we filter this solution to a g-close solution

$(\bar{x}, \bar{y}, \bar{z})$ using Procedure GEN_CLOSE. We find a subset $T \subseteq D_{h-1}$ of transmission stations or a subset $T \subseteq F$ of facilities (for $h = 1$) that will be opened (by executing Procedure h-ROUND), and assign each client $j \in D_h$ to the closest transmission station/facility in T respectively, i.e., for every $j \in D_h$ we find $i^* = argmin\{c_{ij} \mid i \in T\}$ and set $\hat{x}^h_{i^*j} \leftarrow 1$. For $h = 1$ in addition we set $y_i = 1$ for every $i \in T$, and for $h \geq 2$, we recursively call Algorithm h-AlgHFL for $\text{HFL}_\beta[h-1, I]$ where the clients in the input are the opened transmission stations of the set T, i.e., the input is the tuple $\langle h-1, S, F, D_1, D_2, \ldots, D_{h-2}, T, c, f \rangle$. At the end the algorithm returns $(\hat{x}, \hat{y})$ as the output.

The analysis is defered to the full paper, where the following is proved.

Theorem 1. *Algorithm h-AlgHFL is a polynomial time approximation algorithm with ratio $\left(1 + 3^\beta\right) \cdot \left(3^{\beta+1}\right)^{h-1}$ for the $\text{HFL}_\beta[h]$ problem.*

When $\beta = 1$ combining with the ratio 1.52 of [14] for the UFL implies the following.

Theorem 2. *Algorithm h-AlgHFL is a polynomial time approximation algorithm with ratio $1.52 \cdot 9^{h-1}$ for the $\text{HFL}[h]$ problem.*

3   Approximating the Shallow Minimum Steiner Tree Problems

The H-STEINER(h, s) problem is a special case of the $\text{HFL}[h]$ problem, (or $\text{HFL}_\beta[h]$, where $\beta = 1$). A given instance $I = \langle h, s, V, \mathcal{V}, \omega \rangle$ for the H-STEINER(h, s) problem can be transformed into an instance I' of $\text{HFL}[h-1]$ by setting $D_{h-1} = \mathcal{V} \backslash \{s\}$, $D_{h-2} = D_{h-3} = \cdots = D_1 = V \backslash \{s\}$, $F = V \backslash \{s\}$, $c_{uv} = \omega(u, v)$ and $f_v = \omega(v, s)$ for every $u, v \in V \backslash \{s\}$. The relations between the two instances are stated in the following lemma (whose proof is omitted).

Lemma 2. *(1) Let (x, y) be a feasible solution for the defined $\text{HFL}[h-1]$ instance I'. Then there exists an H-STEINER(h, s) tree $t(h, s)$ for the original instance I such that $\text{COST}(t(h, s)) \leq \text{COST}(x, y)$.*
(2) Let $t(h, s)$ be an H-STEINER(h, s) tree for the given instance. Then there exists a feasible solution (x, y) for the corresponding $\text{HFL}[h-1]$ problem such that $\text{COST}(x, y) = \text{COST}(t(h, s))$.

By Lemma 2 we get that an optimal solution for $\text{HFL}[h-1]$ can be transformed to a minimal H-STEINER(h, s) tree. Combining with Theorem 2 we get the following.

Theorem 3. *There is a polynomial time approximation algorithm with ratio $1.52 \cdot 9^{h-2}$ for the H-STEINER(h, s) problem.* ∎

Using the above result, we present a constant ratio approximation algorithm for the *BDST* problem for a positive constant d. The reduction from BDST to H-STEINER is done slightly differently for the cases of even and odd d. The (standard) reduction details are deferred to the full paper.

Theorem 4. *There is a polynomial time approximation algorithm with ratio* $1.52 \cdot 9^{(\lfloor \frac{d}{2} \rfloor - 2)}$ *for BDST problem with constant positive integer d.*

4   Approximation Algorithm for MIN-RANGE(hSC)

Relating strong connectivity to center forest. We first show that the optimal solution to MIN-RANGE($(h-1)CF$) is a $\left(1/(\sqrt[h]{2}-1)\right)^{\beta}$ approximation for MIN-RANGE(hSC).

Let $r : S \to \mathbb{R}^+$ be a range assignment, $\varphi > 0$ a constant and ℓ an integer. Consider some $i \in S$. We say that r is a (φ, ℓ)-*assignment* for i if there exists some node $j \in S$ such that j is farthest from i, i.e., $\rho_i = dist(i,j)$, G_r contains a directed ℓ-hop path from i to j, $i = i_1 \to i_2 \to \ldots \to i_\ell \to i_{\ell+1} = j$ and

$$r(i_k) \le \frac{\rho_{i_k}}{\varphi} \quad \text{for every } 1 \le k \le \ell. \tag{3}$$

As this path belongs to G_r, $r(i_k) \ge dist(i_k, i_{k+1})$ for every $1 \le k \le \ell$.

Hereafter, fix $\varphi_\ell = 1/(\sqrt[\ell]{2} - 1)$.

Lemma 3. *If r is a (φ_ℓ, ℓ)-assignment for i, then $\varphi_\ell \cdot r(i_\ell) \ge \rho_{i_\ell}$.*

Lemma 4. *An optimal solution for MIN-RANGE($(h-1)CF$) is a $(\varphi_h)^{\beta}$ approximation for MIN-RANGE(hSC).*

An approximation for MIN-RANGE($(h - 1)CF$). The MIN-RANGE(hCF) problem is a special case of the HFL$_\beta[h]$ problem. A given instance $S = \{1, 2, \ldots, n\}$ for MIN-RANGE(hCF) is transformed into an instance of HFL$_\beta[h]$ by setting $D_1 = D_2 = \ldots = D_h = S$, $F = S$, $c_{ij} = dist^{\beta}(i,j)$ and $f_i = \rho_i^{\beta} = \max\{c_{ij} \mid j \in S\}$ for every $i, j \in S$. The solution for the instance $\langle S, F, D_1, \ldots, D_h, c, f \rangle$ of the HFL$_\beta[h]$ problem is the solution for the hCF problem for S, β. By Theorem 1 we get the following lemma, which together with Lemma 4 yields our final result.

Lemma 5. *There is a polynomial time approximation algorithm with ratio* $\left(1 + 3^{\beta}\right) \cdot \left(3^{\beta+1}\right)^{h-2}$ *for MIN-RANGE($(h - 1)CF$).*

Lemma 4 now yields our final result.

Theorem 5. *There is a polynomial time approximation algorithm with ratio* $\left(1/\left(\sqrt[h]{2} - 1\right)\right)^{\beta} \left(1 + 3^{\beta}\right) \left(3^{\beta+1}\right)^{h-2}$ *for MIN-RANGE(hSC).*

In contrast, we have the following.

Theorem 6. *The MIN-RANGE(hSC) problem is NP-hard on general metric spaces for constant h.*

References

1. E. Althaus, S. Funke, S. Har-Peled, J. Koenemann, E. A. Ramos, and M. Skutella. Approximation k-hop minimum-spanning trees. *Operations Research Letters, 33*, pages 115–120, 2005.
2. J. Bar-ilan, G. Kortsarz, and D. Peleg. Generalized aubmodular cover problems and applications. In *Proc. 4th Israel Symp. on Theory of Computing and Systems*, pages 110–118, 1996.
3. M. Charikar, C. Chekuri, T. Cheung, Z. Dai, A. Goel, S. Guha, and M. Li. Approximation algorithms for directed Steiner problems. In *Proc. 9th ACM-SIAM Symp. on Discrete Algorithms*, pages 192–200, 1998.
4. F.A. Chudak. Improved approximation algorithm for uncapacitated facility location problem. In *Proc. 6th Conf. on Integer Programing and Combinatorial Optimization*, pages 180–194, 1998.
5. A. E. F. Clementi, P. Penna, A. Ferreira, S. Perennes, and R. Silvestri. The minimum range assignment problem on linear radio networks. *Algorithmica*, 35(2):95–110, 2003.
6. A. E. F. Clementi, P. Penna, and R. Silvestri. On the power assignment problem in radio networks. *Mobile Network Applic.*, 9(2):125–140, 2004.
7. U. Feige. A threshold of ln n for approximating set cover. In *Proc. 28th ACM Symp. on Theory of Computing*, pages 314–318, 1996.
8. M.R. Garey and D.S. Johnson. Computers and intractability: A guide to the theory of NP-completeness. In *W. H. Freeman and Company*, 1979.
9. Aardal K., A.F. Chudak, and B.D. Shmoys. A 3-approximation algorithm for the K-level uncapacitated facility location problem. *Information Processing Letters*, 72:161–167, 1999.
10. L. M. Kirousis, E. Kranakis, D. Kriznac, and A. Pelc. Power consumption in packet radio networks. In *Proc. 14th Symp. on Theoretical Aspects of Computer Science*, pages 363–374, 1997.
11. G. Kortsarz and D. Peleg. Approximating the weight of shallow steiner trees. *Discrete Applied Math.*, pages 265–285, 1999.
12. J.H. Lin and J.S. Vitter. $\varepsilon-$approximations with small packing constraint violation. In *Proc. 24th ACM Symp. on Theory of Computing*, pages 771–782, 1992.
13. C. Lund and M. Yannakakis. On the hardness of approximating minimization problems. *J. ACM*, 41:960–981, 1994.
14. M. Mahdian, Y. Ye, and J. Zhang. A 1.52-approximation algorithm for the uncapacitated facility location problem. In *Proc. 5th Workshop on Approximation Algorithms for Combinatorial Optimization Problems*, pages 229–242, 2002.
15. B.D. Shmoys, E. Tardos, and Aardal K. Approximation algorithms for facility location problems. In *Proc. 29th ACM Symp. on Theory of Computing*, pages 265–274, 1997.

An Approximation Algorithm for a Bottleneck Traveling Salesman Problem*

Ming-Yang Kao and Manan Sanghi

Department of Electrical Engineering and Computer Science,
Northwestern University,
Evanston, IL 60208, USA
{kao, manan}@cs.northwestern.edu

Abstract. Consider a truck running along a road. It picks up a load L_i at point β_i and delivers it at α_i, carrying at most one load at a time. The speed on the various parts of the road in one direction is given by $f(x)$ and that in the other direction is given by $g(x)$. Minimizing the total time spent to deliver loads $L_1, \ldots, L_n$ is equivalent to solving the Traveling Salesman Problem (TSP) where the cities correspond to the loads L_i with coordinates (α_i, β_i) and the distance from L_i to L_j is given by $\int_{\alpha_i}^{\beta_j} f(x)dx$ if $\beta_j \geq \alpha_i$ and by $\int_{\beta_j}^{\alpha_i} g(x)dx$ if $\beta_j < \alpha_i$. This case of TSP is polynomially solvable with significant real-world applications.

Gilmore and Gomory obtained a polynomial time solution for this TSP [6]. However, the bottleneck version of the problem (BTSP) was left open. Recently, Vairaktarakis showed that BTSP with this distance metric is NP-complete [10].

We provide an approximation algorithm for this BTSP by exploiting the underlying geometry in a novel fashion. This also allows for an alternate analysis of Gilmore and Gomory's polynomial time algorithm for the TSP. We achieve an approximation ratio of $(2 + \gamma)$ where $\gamma \geq \frac{f(x)}{g(x)} \geq \frac{1}{\gamma}$ $\forall x$. Note that when $f(x) = g(x)$, the approximation ratio is 3.

1 Introduction

Consider n cities $C_1, C_2, \ldots, C_n$. Let c_{ij} be the distance from C_i to C_j. The problem of finding a tour that visits each city exactly once and minimizes the total travel distance is known as the traveling salesman problem (TSP). The bottleneck traveling salesman problem (BTSP) is to find a tour that visits each city exactly once and minimizes the maximum distance traveled between any two adjacent cities on the tour. Both the TSP and the BTSP are NP-hard in general [5]. We consider the distance metric first proposed by Gilmore and Gomory in [6] which has widespread practical applications [9, 2, 11, 7] and for which the TSP is polynomial time solvable. Unfortunately, the BTSP remains NP-hard for this distance metric [10].

---

* Supported in part by NSF Grant EIA-0112934.

T. Calamoneri, I. Finocchi, G.F. Italiano (Eds.): CIAC 2006, LNCS 3998, pp. 223–235, 2006.
© Springer-Verlag Berlin Heidelberg 2006

Let each city C_i be specified by the coordinates (α_i, β_i) for $i = 1, 2, \ldots, n$. The distance metric considered in this paper is given by:

$$d(C_i, C_j) = c(\alpha_i, \beta_j) = \begin{cases} \int_{\alpha_i}^{\beta_j} f(x)dx & \text{if } \beta_j \geq \alpha_i \\ \int_{\beta_j}^{\alpha_i} g(x)dx & \text{if } \beta_j < \alpha_i \end{cases}$$

where $f(\cdot)$ and $g(\cdot)$ are integrable and $f(x), g(x) \geq 0$. Note that if $f(x) = g(x) = 1$, then $c(\alpha_i, \beta_j) = |\alpha_i - \beta_j|$. Also note that in [6], Gilmore and Gomory solve the TSP with a less restrictive condition viz. $f(x) + g(x) \geq 0$ for all x.

Problem 1 (Gilmore-Gomory's Traveling Salesman Problem (GG-TSP)).
INPUT: n pairs of numbers $(\alpha_0, \beta_0), (\alpha_1, \beta_1), \ldots, (\alpha_{n-1}, \beta_{n-1})$.
OUTPUT: A permutation $\pi : \{0, \ldots, n-1\} \to \{0, \ldots, n-1\}$ such that $\sum_{i=0}^{n-1} c(\alpha_{\pi(i+1 \bmod n)}, \beta_{\pi(i)})$ is minimized.

Problem 2 (Gilmore-Gomory's Bottleneck TSP (GG-BTSP)).
INPUT: n pairs of numbers $(\alpha_0, \beta_0), (\alpha_1, \beta_1), \ldots, (\alpha_{n-1}, \beta_{n-1})$.
OUTPUT: A permutation $\pi : \{0, \ldots, n-1\} \to \{0, \ldots, n-1\}$ such that $\max_{i=0,\ldots,n-1} c(\alpha_{\pi(i+1 \bmod n)}, \beta_{\pi(i)})$ is minimized.

Results. GG-TSP can be solved in $O(n \log n)$ time [6, 11]. GG-BTSP can also be solved in $O(n \log n)$ time if either $f(x) = 0$ or $g(x) = 0$ [6, 11]. However, in general GG-BTSP is NP-hard [10]. In fact, the reduction used in [10] proves NP-hardness for the special case when $f(x) = g(x) = 1$.

In this paper, we give a $(2+\gamma)$-approximation algorithm for GG-BTSP where $\gamma \geq \frac{f(x)}{g(x)} \geq \frac{1}{\gamma} \forall x$. Note that this result immediately implies the following:

1. a 3-approximation algorithm when $c(\alpha_i, \beta_j) = |\alpha_i - \beta_j|$.
2. a $(2 + \max\{\frac{b}{a}, \frac{a}{b}\})$-approximation algorithm when $f(x) = a$ and $g(x) = b$.
3. a 3-approximation algorithm when $f(x) = g(x)$.
4. a $(2 + \frac{b}{a})$-approximation algorithm when $a \leq f(x), g(x) \leq b$.

Further, we uncover some interesting properties of the underlying geometry of the problem that shed new light on the structure of an optimal solution and hence allows for an alternate analysis of the polynomial time TSP algorithm presented in [6, 11].

Paper Layout. Section 2 discusses some applications of GG-TSP and GG-BTSP. Section 3 formulates an equivalent problem of GG-BTSP, called BBCA, on bipartite graphs and defines some concepts and notations used in the paper. Section 4 derives a lower bound on the optimum bottleneck cost. Finally, Section 5 presents the approximation algorithm for GG-BTSP.

2   Applications

The original motivation for the formulation of GG-TSP and GG-BTSP was job sequencing on a single state variable machine [6]. Consider a furnace and let temperature be its state variable. A number of jobs are to be given a heat treatment in the furnace. The ith job will be started at temperature β_i and taken out of the furnace at temperature α_i. The temperature is then changed for the next job. Heating the furnace requires $f(x)$ amount of energy while cooling requires $g(x)$ when the temperature is x. The furnace is at temperature α_0 to start with and is required to be in state β_0 at the end. Sequencing the jobs to minimize the total energy is equivalent to GG-TSP. Sequencing to minimize the maximum energy required for changing between two jobs is equivalent to GG-BTSP.

Another application of this problem formulation is in reconstructing sequential order from inaccurate adjacency information. Consider n women standing in a circle with each facing the clockwise direction. Each woman reports her own height α_i, and the height β_i of the one in front of her. Given this information, we want to reconstruct the order of the women in the circle. When the women make some errors in estimating heights, we may want to construct an ordering which minimizes the maximum of the differences between the height α_j reported by the jth woman from the height β_i reported by the ith woman for each pair of women i and j such that j is in front of i in the ordering. This problem is equivalent to GG-BTSP with $f(x) = g(x) = 1$.

One practical field where the problem of reconstructing such sequential order arises naturally is in interpreting nuclear magnetic resonance (NMR) spectroscopy data for solving a NMR protein structure. In NMR spectroscopy experiments [14, 3], the individual nuclei in a protein sample respond at specific resonance frequencies when exposed to an oscillating radio frequency field. These resonant frequencies are called *chemical shifts* and they serve as identifiers of the corresponding atoms. The data from NMR experiments consist of spectral peaks where a peak can correspond to a pair of chemical shifts of atoms in adjacent amino acids on the protein backbone. The goal is to determine the correct order of these chemical shifts from such adjacency information provided in NMR spectral data. For NMR data interpretation, corresponding to the women in the aforementioned example we have spectral peaks, and corresponding to the pairs of reported heights we have pairs of chemical shifts associated with each peak. Some good references for extracting the adjacency information from NMR experiments can be found in [12, 13]. There is also extensive work in automatic resonance frequency assignment algorithms [15, 13, 12, 8, 4, 1].

Note that though we discuss the problem in terms of reconstructing a circular order, the transformation to reconstructing linear order, as is required for NMR data interpretation, is achieved in polynomial time. If the first and the last element in the linear order are known, the linear order problem can be reduced to GG-BTSP in linear time by assigning α_i of the first element and β_j of the last element the identifier ∞, i.e., if f is the index for the first element and ℓ is the index for the last one, then $\alpha_f = \beta_\ell = \infty$. This forces a minimum cost circular order to place the last element before the first one. If the first and the last element are not known, then there are $2\binom{n}{2}$ options for them and hence if the time complexity of GG-BTSP is T, we can solve this linear order problem in $O(n^2 T)$ time. Similarly, if either the first element or the last element is given, then we can solve the linear order problem in $O(nT)$ time. In this paper, we provide an approximation algorithm for GG-BTSP with a runtime of $O(n \log n)$; i.e., $T = O(n \log n)$.

3   Preliminaries

We first define an equivalent problem of GG-BTSP on bipartite graphs in Section 3.1. The rest of the paper focusses on solving this equivalent problem. Then we define some notations in Section 3.2, discuss some concepts in Section 3.3 and present basic lemmas in Section 3.4.

3.1   Problem Definition

Problem 3 (Bottleneck Bipartite Cyclic Augmentation (BBCA)).
INPUT: A bipartite graph $G = (U, V, H)$ where H is a perfect matching, and a function $\phi : U \cup V \rightarrow \mathbb{R}$.
OUTPUT: A set of edges M such that the bipartite graph $G' = (U, V, H \cup M)$ is a hamiltonian cycle and $\max_{(u,v) \in M} c(\phi(u), \phi(v))$ is minimized.

For $w \in U \cup V$, $\phi(w)$ is called the *potential* of w. The *cost* of an edge (u, v) where $u \in U$ and $v \in V$ is given by $c((u, v)) = c(\phi(u), \phi(v))$. The *cost* of a matching M is given by $c_M = \max_{e \in M} c(e)$. A set of edges M such that $G' = (U, V, H \cup M)$ is a cycle is called a *cyclic augmentation* of $G = (U, V, H)$.

Lemma 1. *GG-BTSP and BBCA can be reduced to each other in linear time.*

3.2   Notations

For the remainder of the paper, let $u_0, u_1, \ldots, u_{n-1}$ be the n vertices in U such that $\phi(u_0) \leq \phi(u_1) \leq \cdots \leq \phi(u_{n-1})$. Similarly, let $v_0, v_1, \ldots, v_{n-1}$ be the n vertices in V such that $\phi(v_0) \leq \cdots \leq \phi(v_{n-1})$.

If M is a matching between U and V, then let G_M denote the graph $G(U, V, H \cup M)$. Note that for any matching M, G_M is a set of simple cycles. If G_M contains exactly one cycle, then M is a cyclic augmentation. For $g \in U \cup V$, let e_g^M denote the edge adjacent to vertex g in M.

3.3   Concepts

It is useful to visualize the vertices in U as being arranged on the horizontal axis with their abscissa being the potential $\phi(u_i)$. Similarly the vertices in V can be visualized as being at a higher ordinate and with their abscissa being their corresponding potential. An edge (u_i, v_j) is a straight line connecting $\phi(u_i)$ and $\phi(v_j)$. See Figure 1 for an example visualization.

In our figures, we will represent the edges in H by dashed lines and the edges in M by solid lines.

Left, Right and In-between Edges. For any three edges $e_1 = (u_a, v_b)$, $e_2 = (u_c, v_d)$ and $e_3 = (u_p, v_q)$. If $a < c$, then e_1 is said to be on the *left* and e_2 is said to be on the *right*. The edge e_2 is said to be *in-between* e_1 and e_3 if $a < c < p$ and $b < d < q$. Let η_{e_1, e_2} be the number of edges in-between e_1 and e_2.

$$\eta_{e_1, e_2} = |\{ (u_r, v_s) \mid a < r < c, b < s < d\}|.$$

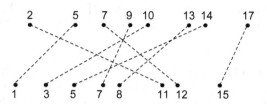

Fig. 1. A visual representation of $G(U, V, H)$ where the potentials of the vertices in U are $1, 3, 5, 7, 8, 11, 12, 15$, the potentials of the vertices in V are $2, 5, 7, 9, 10, 13, 14, 17$, and H consists of 8 edges connecting the potential pairs $(1, 5), (3, 10), (5, 14), (7, 9), (8, 13), (11, 2), (12, 7)$ and $(15, 17)$

Cross State and Straight State. Given two edges (u_a, v_b) and (u_c, v_d) such that $a < c$, the edges are said to be in a *straight state* if $b < d$ and in a *cross state* if $b > d$. In our visualization, the edges in a cross state will intersect while those in a straight state will not (see Figure 2). An edge e_1 is said to *cross* e_2 if e_1 and e_2 are in a cross state. Note that if e_1 crosses e_2, then $\eta_{e_1, e_2} = 0$.

Cross Number. The *cross number* of a matching M, denoted by Γ_M, is the number of pairs of edges which are in a cross state in M. Observe that

$$\Gamma_M = \frac{|\{(g, h) \mid g, h \in U \text{ and } e_g^M \text{ crosses } e_h^M\}|}{2}.$$

Note that the cross number of M is the number of intersections in its visualization (see Figure 2 for an example).

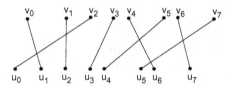

Fig. 2. The cross number of this matching M is 5. The edges $e_{u_2}^M$ and $e_{u_3}^M$ are in a straight state. The edges $e_{u_5}^M$ and $e_{u_7}^M$ are in a cross state.

Exchange. Given a matching M with two edges $e_1 = (u_a, v_b)$ and $e_2 = (u_c, v_d)$, an *exchange* on e_1 and e_2 returns a matching M' such that $M' = M \otimes (e_1, e_2) = (M \setminus \{e_1, e_2\}) \cup \{(u_a, v_d), (u_c, v_b)\}$. Note that if e_1 and e_2 are in a straight state in M, then their replacement edges are in a cross state in M'. Such exchanges are called *straight-to-cross* exchanges. Similarly, *cross-to-straight* exchanges are the ones on two edges in a cross state to result in two in a straight state. A *null exchange* on M is defined to be the operation which returns matching M.

Direct Pair. The set of vertices $\{u_i, v_i\}$ is called the ith *direct pair.* For any i $(0 \le i \le n - 2)$, the ith and the $(i + 1)^{th}$ direct pairs are said to be *consecutive.*

Let $M_D = \{(u_0, v_0), (u_1, v_1), ..., (u_{n-1}, v_{n-1})\}$. Note that M_D is a matching with the minimum cost over all possible matchings. However, M_D may not be a cyclic augmentation.

Cluster. A *cluster* is the union of consecutive direct pairs which belong to the same cycle in G_{M_D}. The ith cluster from the left is denoted by ψ_i. Therefore, ψ_1 is the cluster containing the leftmost direct pair and ψ_j is the cluster containing the leftmost direct pair in $U \cup V \setminus (\psi_1 \cup \cdots \cup \psi_{j-1})$.

Note that the clusters define a partition of $U \cup V$. All the vertices in a cluster belong to the same cycle in G_{M_D} but all the vertices in the same cycle in G_{M_D} need not be in the same cluster. See Figure 3 for an illustration of clusters.

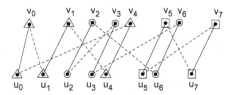

Fig. 3. For this graph, G_{M_D} has three cycles $C_1 = u_0 v_0 u_1 v_1 u_4 v_4$, $C_2 = u_2 v_2 u_6 v_6 u_3 v_3$ and $C_3 = u_5 v_5 u_7 v_7$. The vertices in C_1 are marked by a triangle, the ones in C_2 by a circle and those in C_3 by a square. Therefore, the clusters are $\psi_1 = \{u_0, v_0, u_1, v_1\}$, $\psi_2 = \{u_2, v_2, u_3, v_3\}$, $\psi_3 = \{u_4, v_4\}$, $\psi_4 = \{u_5, v_5\}$, $\psi_5 = \{u_6, v_6\}$, and $\psi_6 = \{u_7, v_7\}$.

Exchange Graph. The *exchange graph* $\mathcal{X}$ for $G = (U, V, H)$ is a multigraph whose vertices correspond to the cycles in G_{M_D}. There is an edge between two cycles C and C' for every pair of consecutive clusters ψ_i and ψ_{i+1} such that ψ_i has vertices in C and ψ_{i+1} has vertices in C'. The *weight* of this edge is $\max\{c(\phi(u_{\text{right}}^{\psi_i}), \phi(v_{\text{left}}^{\psi_{i+1}})), c(\phi(u_{\text{left}}^{\psi_{i+1}}), \phi(v_{\text{right}}^{\psi_i}))\}$, where $(u_{\text{right}}^{\psi_i}, v_{\text{right}}^{\psi_i})$ are the rightmost direct pair in ψ_i and $(u_{\text{left}}^{\psi_{i+1}}, v_{\text{left}}^{\psi_{i+1}})$ are the leftmost direct pair in ψ_{i+1}. If $(u_{\text{right}}^{\psi_i}, v_{\text{right}}^{\psi_i}) = (u_k, v_k)$, then the *label* of the corresponding edge is $(k, k+1)$.

3.4   Lemmas

Lemma 2. *Given matchings M and M' between U and V, there exists a sequence of exchanges $x_0, x_1, \ldots, x_m$ for $m < n$ which transforms M to M'.*

In Lemmas 3 through 6 below, for any two edges $e_1 = (u_a, v_b)$ and $e_2 = (u_c, v_d)$ such that $e_1, e_2 \in M$ and $a < c$, let $M' = M \otimes (e_1, e_2)$.

Lemma 3. *If e_1 and e_2 are in a straight state, then $c_{M'} \geq c_M$.*

Lemma 4.

1. *If e_1 and e_2 are in the same cycle in G_M, then their replacement edges are in different cycles in $G_{M'}$.*
2. *If e_1 and e_2 are in different cycles in G_M, then their replacement edges are in a same cycle in $G_{M'}$.*

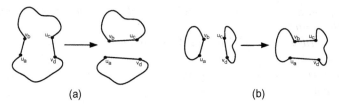

Fig. 4. (a) An exchange between edges in a same cycle splits the cycle in two. (b) An exchange between edges in two different cycles joins the cycles.

Lemma 5. *If e_1 crosses e_2 and for some $u_p \in U$, $e_{u_p}^{M'}$ crosses $e_{u_a}^{M'}$, then $e_{u_p}^{M}$ crosses at least one of e_1 and e_2. By symmetry, if e_1 crosses e_2 and for some $u_p \in U$, $e_{u_p}^{M'}$ crosses $e_{u_c}^{M'}$, then $e_{u_p}^{M}$ crosses at least one of e_1 and e_2.*

Lemma 6.

1. *If e_1 and e_2 are in a straight state, then $\Gamma_{M'} = \Gamma_M + 1 + 2\eta_{e_1,e_2}$.*
2. *If e_1 and e_2 are in a cross state, then $\Gamma_{M'} = \Gamma_M - 1 - 2\eta_{e_{u_a}^{M'},e_{u_c}^{M'}}$.*

4 Lower Bound on the Optimum Bottleneck Cyclic Augmentation

As observed in Section 3.2, for any perfect matching M between U and V, the graph G_M is a collection of simple cycles. Note that M_D is the minimum cost matching of $G(U, V, H)$. However, M_D may not be a cyclic augmentation i.e., $H \cup M_D$ may not be a hamiltonian cycle. Our strategy for solving BBCA is to begin with G_{M_D} and transform M_D into a cyclic augmentation by means of exchanges.

Recall from Lemma 4(2) that an exchange between two edges in different cycles, say C_1 and C_2, yields a graph in which all the vertices in C_1 and C_2 are in one cycle (see Figure 4). Alternately, from Lemma 4(1), an exchange between two edges in the same cycle yields a graph in which the vertices in that cycle are split into two distinct cycles. Furthermore, from Lemma 2, we know that for any two matchings M and M', M can be converted to M' by a sequence of exchanges. In this section we present Lemma 7 which identifies some useful properties of a minimum cost cyclic augmentation which allows us to restrict the search space for suitable exchanges to convert M_D to an approximately optimal cyclic augmentation. Then, using Lemma 8, we reduce our search space to exchanges corresponding to the edges in the exchange graph $\mathcal{X}$. As will be shown in Lemma 9, this allows us to derive a good lower bound on the cost of the optimal cyclic augmentation.

Lemma 7. *There exists a minimum bottleneck cost cyclic augmentation M^* for $G = (U, V, H)$ such that the following properties hold true:*
(P_1) Any edge $e \in M^*$ crosses either some edges on its left or some on its right but not both.

(P_2) *For* $e_1, e_2, e_3 \in M^*$, *if* e_1 *crosses* e_2 *and* e_3, *then no other edge in* M^*
crosses both e_2 *and* e_3.

(P_3) *If two vertices* $u_p, u_q \in U$ *are in the same cycle in* G_{M_D}, *then* $e_{u_p}^{M^*}$ *and* $e_{u_q}^{M^*}$
do not cross.

(P_4) *If two vertices* $u_p, u_q \in U$ *are in the same cycle in* G_{M_D} *and* u_p *is on the*
left of u_q, *then*

1. $e_{u_p}^{M^*}$ *cannot cross any edge to the right of* $e_{u_q}^{M^*}$; *and*
2. $e_{u_q}^{M^*}$ *cannot cross any edge to the left of* $e_{u_p}^{M^*}$.

Proof. The proof is by construction. Given a minimum cost cyclic augmentation
M', we show that it can be transformed to a minimum cost cyclic augmentation
M^* which satisfies the above 4 properties.

For each property P_i, given the smallest set of vertices $W \subseteq U \cup V$ for which
P_i does not hold in M', we give a transformation T_i for constructing a new
matching of cost no more than that of M' and a cross number smaller than
that of M'. The algorithm for the construction begins with any minimum cost
cyclic augmentation and repeatedly finds the smallest i such that P_i does not
hold. Use T_i to correct this violation till a matching for which all the properties
hold true is obtained. For the correctness and termination of this algorithm, we
ensure that each of the transformations T_i satisfies the following two conditions.
Assuming that P_j holds for all $j < i$, given any cyclic augmentation M' and the
smallest set of vertices $W \in M'$ such that the edges incident to W do not satisfy
P_i, $T_i(M', W)$ returns a matching M'' such that

1. M'' is a cyclic augmentation of cost no more than that of M'; and
2. $\Gamma_{M''} < \Gamma_{M'}$.

The first condition above ensures that after every transformation, we get a cyclic
augmentation of the minimum cost. The second condition ensures that the total
number of crosses decreases monotonically. Hence, we terminate with a minimum
cost cyclic augmentation which either satisfies all the properties or has no crosses.
Since the only matching with no crosses is M_D and all the 4 properties do hold
for M_D, in either case we are guaranteed to construct M^*.

To give a flavor of the transformations, we present only T_1 here.

Transformation T_1. Let $W = \{u_a, u_b, u_c\} \subseteq U$ and $e_1 = (u_a, v_q)$, $e_2 = (u_b, v_p)$,
$e_3 = (u_c, v_d)$ such that e_2 crosses e_1 and e_3; e_1 is to the left of e_2 and e_3 is to
the right of e_2. This implies that $a < b < c$ and $d < p < q$.

Let $M'' = M' \otimes \{e_1, e_3\}$. Since, M' was a cyclic augmentation, M'' will contain
two cycles with $e_1' = (u_a, v_d)$ and $e_3' = (u_c, v_q)$ in different cycles. Therefore, e_2
will be in the same cycle as either e_1' or e_3'. Suppose e_2 is in the same cycle as
e_1'. Let $M''' = M'' \otimes \{e_{u_b}^{M''}, e_3'\}$. Now M''' is a cyclic augmentation and the edges
$e_{u_a}^{M'''}$, $e_{u_b}^{M'''}$ and $e_{u_c}^{M'''}$ do not violate P_1. The other case when e_2 is in the same
cycle as e_3' is symmetric.

We have transformed M' to M''' using one cross-to-straight exchange and
one straight-to-cross exchange. However, M' can be transformed to M''' using

only cross-to-straight exchanges (see Figure 5). Therefore, the cost of M''' is no more than that of M' (using Lemma 3). Further, using Lemma 6(2) we have $\Gamma_{M'''} < \Gamma_{M'}$.

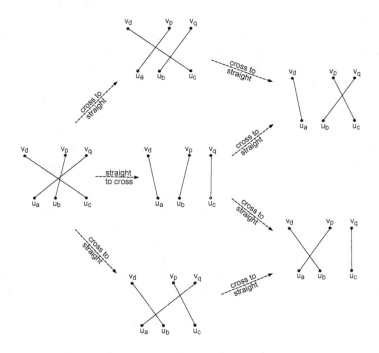

Fig. 5. Illustration of transformation T_1

The next lemma uses the properties established in Lemma 7 to restrict the space of exchanges required for transforming M_D to an optimum bottleneck cyclic augmentation M^* to the exchanges corresponding to edges in $\mathcal{X}$.

Lemma 8. M^* can be constructed by performing a series of exchanges on G_{M_D} where each exchange corresponds to a unique edge in the exchange graph $\mathcal{X}$.

Proof. We will construct a sequence of exchanges $x_0, \ldots, x_{n-1}$ where each x_i is either a null exchange or corresponds to a unique edge in the exchange graph $\mathcal{X}$. Let $M_k = M_D \otimes x_0 \otimes \cdots \otimes x_k$. It suffices to show that $M_{n-1} = M^*$. A vertex in M_k is said to be *satisfied* if its adjacent vertex in M_k is the same as its adjacent vertex in M^*.

Let $H(k)$ denote the statement that in M_k either all the vertices in the first $k + 1$ direct pairs are satisfied or exactly two are not, and that at least one of those two, h, is in the $(k + 1)^{\text{th}}$ direct pair and the other is the one adjacent to it, g, in M_k such that $\phi(g) \leq \phi(h)$.

The proof of this lemma is by induction on $H(k)$ as follows. Note that if $H(n - 1)$ is true, then all the vertices in M_{n-1} must be satisfied, i.e. M_{n-1} is the same as M^*.

Base Case: $k = 0$. Let x_0 be a null exchange. Either the vertices of first direct pair are satisfied or they are not. In either case, $H(0)$ is true.

Induction Step: $H(k-1)$ holds for some k, where $0 \leq k-1 \leq n-2$. Then, if all of the first k direct pairs are satisfied, let x_k be a null exchange and $H(k)$ will be true.

However, if two vertices of the first k direct pairs are not satisfied, let the two vertices be g and h such that $\phi(g) \leq \phi(h)$ and h belongs to the kth direct pair. Note that in this case M_{k-1} must contain the edge (g, h). Without loss of generality, let $g \in U$ and $h \in V$. Since all the first k direct pairs except g and h are satisfied, the vertices adjacent to g and h in M^* must be to the right of the kth direct pair.

Now let x_k be the exchange between (g, h) and (u_{k+1}, v_{k+1}). Using P_3 and P_4 we can conclude that h and u_{k+1} (or v_{k+1}) cannot belong to same clusters. So the exchange x_k corresponds to an edge labeled $(k, k+1)$ in the exchange graph $\mathcal{X}$.

We need to show that either $(g, v_{k+1}) \in M^*$ or $(h, u_{k+1}) \in M^*$. To prove this by contradiction, suppose this is not so. That is $(g, v_{k+1}) \notin M^*$ and $(h, u_{k+1}) \notin M^*$. Then let the vertex adjacent to g in M^* be g' and that adjacent to v_{k+1} be v'_{k+1}. Similarly, let the vertex adjacent to h in M^* be h' and that adjacent to u_{k+1} be u'_{k+1}. We know now that g' should be to the right of v_{k+1} and h' should be right of u_{k+1}. By P_1, v'_{k+1} should be to the right of h', and u'_{k+1} should be to the right of g'. But by P_2 this is not possible. Hence, we have reached a contradiction.

Therefore, by induction we can conclude that $H(n-1)$ is true.

Let c_{MST} be the weight of the heaviest edge in a minimum spanning tree over $\mathcal{X}$ and let $c_{LB} = \max\{c_{M_D}, c_{MST}\}$. Let the cost of the optimal bottleneck cyclic augmentation M^* be c_{OPT}.

Lemma 9. $c_{OPT} \geq c_{LB}$.

5   Approximation Algorithm for GG-BTSP

For finding the minimum bottleneck cost augmentation, we first construct a minimum spanning tree of the exchange graph and then perform exchanges corresponding to the edges of the spanning tree such that no exchange exceeds the weight of the heaviest edge in the spanning tree by a factor of $(2 + \gamma)$.

From Lemma 8, there exists a set of exchanges corresponding to the edges of exchange graph such that the resulting augmentation is of optimum cost. Furthermore, as observed in proof to Lemma 9, these edges form a spanning tree. However, note that the spanning tree corresponding to the optimal augmentation need not be the minimum spanning tree over the exchange graph. Furthermore, the cost of the augmentation is only lower bounded by the heaviest weighted edge in the corresponding spanning tree. Hence, there is scope for improving the analysis by tightening the lower bound.

Algorithm 1. Approx-BTSP

1. Let $M \leftarrow M_D$.
2. Construct the exchange graph $\mathcal{X}$.
3. Find a minimum spanning tree T of $\mathcal{X}$.
4. Sort the edges in T in the increasing order of their label. Let the ordered edges be
   $e_1, \ldots, e_m$.
5. For $i = 1$ to m,
   (a) Let the label of e_i be $(a_i, a_i + 1)$.
   (b) **if** $c(e^M_{u_{a_i}}) \le c_{\text{LB}}$,
          then $M \leftarrow M \otimes (e^M_{u_{a_i}}, e^M_{u_{(a_i+1)}})$;
          else $M \leftarrow M \otimes (e^M_{v_{a_i}}, e^M_{u_{(a_i+1)}})$.
6. Output $M_{\textbf{OUT}} = M$.

Lemma 10. *At Step 5b of Approx-BTSP, either $c(e^M_{u_{a_i}}) \le c_{\text{LB}}$ or $c(e^M_{v_{a_i}}) \le c_{\text{LB}}$.*

Proof. This is proven by induction on $H(k)$ where $H(k)$ denotes the statement that after k iterations of the algorithm

1. Either $c(e^M_{u_{a_k}}) \le c_{\text{LB}}$ or $c(e^M_{v_{a_k}}) \le c_{\text{LB}}$; and
2. For all $x > a_k$, $e^M_{u_x} = e^M_{v_x} = (u_x, v_x)$.

Base Case: $k = 1$, $M = M_D$ and hence all the edges in M have cost at most c_{LB}. Therefore, $H(1)$ is true.

Induction Step: $H(k-1)$ holds for some k, where $0 \le k-1 \le m-1$. Note that, $a_k \ge a_{k-1} + 1$ since the edges were sorted according to their labels. Without loss of generality, assume that at $(k-1)^{\text{th}}$ iteration $c(e^M_{u_{a_{k-1}}}) \le c_{\text{LB}}$. Therefore after the $(k-1)^{\text{th}}$ iteration, $e^M_{u_{a_{k-1}}} = (u_{a_{k-1}}, v_{a_{k-1}+1})$ and for all $x > a_{k-1} + 1$, $e^M_{u_x} = e^M_{v_x} = (u_x, v_x)$. Note that weight of the edge labeled $(a_{k-1}, a_{k-1} + 1)$ is at least $c(u_{a_{k-1}}, v_{a_{k-1}+1})$ and since $e_{k-1} \in T$, $c(e^M_{v_{a_{k-1}+1}}) \le c_{\text{LB}}$.

Therefore at the kth iteration, $c(e^M_{v_{a_{k-1}+1}}) \le c_{\text{LB}}$ and for all $x > a_{k-1} + 1$, $e^M_{u_x} = e^M_{v_x} = (u_x, v_x)$. Since $a_k \ge a_{k-1} + 1$, $H(k)$ holds.

Lemma 11. *If $\gamma \ge \frac{f(x)}{g(x)} \ge \frac{1}{\gamma} \; \forall x$, then $c(b, a) \le \gamma \cdot c(a, b)$.*

Theorem 1. *Running time of Algorithm Approx-BTSP is $O(n \log n)$.*

Theorem 2. *If $\gamma \ge \frac{f(x)}{g(x)} \ge \frac{1}{\gamma} \; \forall x$, then $M_{\textbf{OUT}}$ is a cyclic augmentation of cost no more than $(2 + \gamma) \cdot c_{\text{OPT}}$.*

Proof. We need to prove the following two parts:

1. $M_{\textbf{OUT}}$ is a cyclic augmentation.
2. $c_{M_{\textbf{OUT}}} \le (2 + \gamma) \cdot c_{\text{OPT}}$.

Note that every exchange performed in the algorithm is between edges belonging to two different cycles. Therefore, using Lemma 4, the number of cycles in M decreases with every iteration. If M_D has $m + 1$ cycles, the minimum spanning tree T contains m edges and hence after m iterations M consists of just one cycle and is hence a cyclic augmentation. This completes the proof of Part 1.

For Part 2, we will show that the following invariant holds true for the algorithm: $c_M \leq (2 + \gamma) \cdot c_{\text{LB}}$. Consider the ith iteration of the algorithm. Let the matching before the ith iteration be M' and the one after be M''. Assuming $c_{M'} \leq (2 + \gamma) \cdot c_{\text{LB}}$, we need to show that $c_{M''} \leq (2 + \gamma) \cdot c_{\text{LB}}$. From Lemma 10, either $c(e_{u_{a_i}}^{M'}) \leq c_{\text{LB}}$ or $c(e_{v_{a_i}}^{M'}) \leq c_{\text{LB}}$. Without loss of generality, assume $c(e_{u_{a_i}}^{M'}) \leq c_{\text{LB}}$. Let $e_{u_{a_i}}^{M'} = (u_{a_i}, h)$. $M'' = (M' \setminus \{(u_{a_i}, h), (u_{a_i+1}, v_{a_i+1})\}) \cup \{(u_{a_i}, v_{a_i+1}), (u_{a_i+1}, h)\}$. Clearly $c((u_{a_i}, v_{a_i+1})) \leq c_{\text{LB}}$ because weight of edge labeled $(a_i, a_i + 1)$ is at least $c((u_{a_i}, v_{a_i+1}))$.

So all we need to show is that $c((u_{a_i+1}, h)) \leq (2 + \gamma) \cdot c_{\text{LB}}$.

$$c(\phi(u_{a_i+1}), \phi(v_{a_i})), \ c(\phi(u_{a_i}), \phi(v_{a_i})), \ c(\phi(u_{a_i}), \phi(h)) \leq c_{\text{LB}}$$

$$
\begin{aligned}
c(\phi(u_{a_i+1}), \phi(u_{a_i})) &\leq c(\phi(u_{a_i+1}), \phi(v_{a_i})) + c(\phi(v_{a_i}), \phi(u_{a_i})) \\
&\leq c(\phi(u_{a_i+1}), \phi(v_{a_i})) + \gamma \cdot c(\phi(u_{a_i}), \phi(v_{a_i})) \ \text{(using Lemma 11)} \\
&\leq (1 + \gamma) \cdot c_{\text{LB}} \\
c((u_{a_i+1}, h)) &= c(\phi(u_{a_i+1}), \phi(h)) \\
&\leq c(\phi(u_{a_i+1}), \phi(u_{a_i})) + c(\phi(u_{a_i}), \phi(h)) \\
&\leq (1 + \gamma) \cdot c_{\text{LB}} + c_{\text{LB}} \\
&= (2 + \gamma) \cdot c_{\text{LB}}
\end{aligned}
$$

References

[1] C. BAILEY-KELLOGG, S. CHAINRAJ, AND G. PANDURANGAN, *A Random Graph Approach to NMR Sequential Assignment*, in Proceedings of the 8th Annual International Conference on Computational Molecular Biology, 2004, pp. 58–67.

[2] M. O. BALL AND M. J. MAGAZINE, *Sequencing of Insertions in Printed Circuit Board Assembly*, Operations Research, 36 (1988), pp. 192–201.

[3] J. CAVANAGH, W. J. FAIRBROTHER, A. G. PALMER III, AND N. J. SKELTON, *Protein NMR Spectroscopy: Principles and Practice*, Academic Press, New York, NY, 1996.

[4] Z.-Z. CHEN, T. JIANG, G. LIN, J. WEN, D. XU, J. XU, AND Y. XU, *Approximation Algorithms for NMR Spectral Peak Assignment*, Theoretical Computer Science, 299 (2003), pp. 211–229.

[5] M. R. GAREY AND D. S. JOHNSON, *Computers and Intractability: A Guide to the Theory of NP-Completeness*, W. H. Freeman & Co., New York, NY, USA, 1979.

[6] P. C. GILMORE AND R. E. GOMORY, *Sequencing a One State-Variable Machine: A Solvable Case of the Traveling Salesman Problem*, Operations Research, 12 (1964), pp. 655–679.

[7] G. GUTIN AND A. P. PUNNEN, *The Traveling Salesman Problem and Its Variations*, Kluwer Academic Publishers, Dordrecht, The Netherlands, 2002.

[8] T. K. HITCHENS, J. A. LUKIN, Y. ZHAN, S. A. MCCALLUM, AND G. S. RULE, *MONTE: An Automated Monte Carlo Based Approach to Nuclear Magnetic Resonance Assignment of Proteins*, Journal of Biomolecular NMR, 25 (2003), pp. 1–9.

[9] S. S. REDDI AND C. V. RAMAMOORTHY, *On the Flow-Shop Sequencing Problem with No Wait in Process*, Operational Research Quarterly, 23 (1972), pp. 323–331.

[10] G. L. VAIRAKTARAKIS, *On Gilmore-Gomory's open question for the bottleneck TSP*, Operations Research Letters, 31 (2003), pp. 483–491.

[11] ———, *Simple Algorithms for Gilmore-Gomory's Traveling Salesman and Related Problems*, Journal of Scheduling, 6 (2003), pp. 499–520.

[12] O. VITEK, J. VITEK, B. CRAIG, AND C. BAILEY-KELLOGG, *Model-Based Assignment and Inference of Protein Backbone Nuclear Magnetic Resonances*, Statistical Applications in Genetics and Molecular Biology, 3 (2004), pp. 1–22.

[13] X. WAN, D. XU, C. M. SLUPSKY, AND G. LIN, *Automated Protein NMR Resonance Assignments*, in Proceedings of the 2nd IEEE Computer Society Conference on Bioinformatics, 2003, pp. 197–208.

[14] K. WÜTHRICH, *NMR of Proteins and Nucleic Acids*, John Wiley & Sons, New York, NY, 1986.

[15] Y. XU, D. XU, D. KIM, V. OLMAN, J. RAZUMOVSKAYA, AND T. JIANG, *Automated Assignment of Backbone NMR Peaks Using Constrained Bipartite Matching*, Computing in Science and Engineering, 4 (2002), pp. 50–62.

On the Minimum Common Integer
Partition Problem

Xin Chen[1], Lan Liu[2], Zheng Liu[2], and Tao Jiang[2,3]

[1] School of Physical and Mathematical Sciences, Nanyang Tech. Univ., Singapore
`ChenXin@ntu.edu.sg`
[2] Department of Computer Science, Univ. of California at Riverside, USA
`lliu, zliu, jiang@cs.ucr.edu`
[3] Currently visiting at Tsinghua University, Beijing, China

Abstract. We introduce a new combinatorial optimization problem in this paper, called the *Minimum Common Integer Partition* (MCIP) problem, which was inspired by computational biology applications including ortholog assignment and DNA fingerprint assembly. A *partition* of a positive integer n is a multiset of positive integers that add up to exactly n, and an *integer partition* of a multiset S of integers is defined as the multiset union of partitions of integers in S. Given a sequence of multisets $S_1, \cdots, S_k$ of integers, where $k \geq 2$, we say that a multiset is a *common integer partition* if it is an integer partition of every multiset S_i, $1 \leq i \leq k$. The MCIP problem is thus defined as to find a common integer partition of $S_1, \cdots, S_k$ with the minimum cardinality. It is easy to see that the MCIP problem is NP-hard since it generalizes the well-known Set Partition problem. We can in fact show that it is APX-hard. We will also present a $\frac{5}{4}$-approximation algorithm for the MCIP problem when $k = 2$, and a $\frac{3k(k-1)}{3k-2}$-approximation algorithm for $k \geq 3$.

1 Introduction

Computational molecular biology has emerged as one of the most exciting interdisciplinary fields in the past two decades, in part because various biological applications have spawned a large number of interesting combinatorial problems such as multiple sequence alignment [12], sorting by reversals [20], and recently the minimum common partition problem [10]. These problems have attracted considerable attention from computer scientists who took the challenge to design efficient and effective algorithms for solving them [5, 14, 13]. In this paper, we introduce a new combinatorial optimization problem, called the *Minimum Common Integer Partition* problem (MCIP), which was inspired by our recent work on ortholog assignment and DNA fingerprint assembly.

By a *partition* of a positive integer n we mean a multiset $\{n_1, n_2, \cdots, n_r\}$ of positive integers that add up to exactly n, *i.e.* $\sum_{i=1}^{r} n_i = n$, where n_i is called a *part* of n [2, 4]. Given a multiset $S = \{x_1, x_2, \cdots, x_m\}$ of integers with a partition for each integer x_i, $1 \leq i \leq m$, we can define an *integer partition* of S as the multiset union of these partitions, that is $\biguplus_{i=1}^{m} P(x_i)$. By definition, S is an integer partition of itself. A multiset is said to be a *common integer partition* of

T. Calamoneri, I. Finocchi, G.F. Italiano (Eds.): CIAC 2006, LNCS 3998, pp. 236–247, 2006.

a sequence of multisets $S_1, S_2, \ldots, S_k (k \geq 2)$ if it is an integer partition of every multiset S_i, $1 \leq i \leq k$. The minimum common integer partition problem is thus defined as follows: given a sequence of multisets $S_1, S_2, \cdots, S_k$ of integers, find a common integer partition of them with the minimum cardinality. We denote the minimum common integer partition by $\text{MCIP}(S_1, S_2, \cdots, S_k)$ (or simply MCIP when the input multisets are clear from the context). Note that, now MCIP denotes both the MCIP problem and also its solution on a particular instance, but this overloading is a common pratice and should not cause any confusion given the context. For simplicity, we also denote by $\text{MCIP}(S_1, S_2, \cdots, S_k)$ (or simply k-MCIP) the restricted version of the MCIP problem when the number of input multisets is fixed to be k throughout the paper.

For example, the integer 3 has only three partitions, i.e., $\{3\}, \{2, 1\}$, and $\{1, 1, 1\}$, while the integer 10 has 190569292 partitions [2]. We can see that the number of partitions increases quite rapidly with the integer n. For multiset $S = \{3, 3, 4\}$, $\{2, 2, 3, 3\}$ is an integer partition of S and $\{1, 1, 2, 2, 4\}$ is another one. For a pair of multisets $S = \{3, 3, 4\}$ and $T = \{2, 2, 6\}$, both $\{2, 2, 3, 3\}$ and $\{1, 1, 2, 2, 4\}$ are common integer partitions of S and T, while the first one gives the minimum cardinality, i.e., $\text{MCIP}(S, T) = \{2, 2, 3, 3\}$. Note that the minimum common integer partition is not necessarily unique. So, the notation $\text{MCIP}(S_1, S_2, \cdots, S_k)$ is not really a function, strictly speaking. But we will use it as a function throughout the paper for simplicity.

The necessary and sufficient condition for a sequence of multisets $S_1, S_2, \ldots, S_k$ to have a common integer partition is that they have the same summation over their integer elements. Multisets with this property are called *related*. Verifying whether a sequence of multisets of integers are related can be done easily in linear time, and thus for the rest of the paper we will assume, without loss of generality, that the input multisets are all related.

Clearly, the MCIP problem is NP-hard since it generalizes the well-known Set Partition problem [7]. In this paper, we show that the MCIP problem is APX-hard and hence has no polynomial-time approximation algorithm (PTAS) unless $P = NP$. We also present a $\frac{5}{4}$-approximation algorithm for the 2-MCIP using a heuristic for the *Maximum Set Packing* problem, and a $\frac{3k(k-1)}{3k-2}$-approximation algorithm for the general k-MCIP problem, where $k \geq 3$.

1.1   Biological Background

Although the MCIP problem is quite a natural extension of the Set Partition problem, its formulation was mainly motivated by our recent work on ortholog assignment and DNA fingerprint assembly in computational molecular biology. The following gives a brief account of the background. Since it contains discussions that involve the knowledge of some biological experiments, the reader who is not interested in the biological relevance may feel free to skip some (or all) of the paragraphs in this subsection.

Ortholog assignment. Orthologous genes are typically the evolutionary and functional counterparts in different species, and therefore the prediction (or assignment) of orthologs is a common task in computational biology. While it is usually done using sequence homology search [19], we have recently proposed an alternative and promising approach to assign orthologs via genome rearrangement [9, 10]. This new approach has inspired us to formulate several interesting combinatorial optimization problems, *e.g.*, Signed Reversal Distance with Duplicates (SRDD), Minimum Common String Partition (MCSP), and Maximum Cycle Decomposition (MCD), which have attracted increasing attention from the algorithms community [6, 13, 11, 16]. In particular, the MCSP problem, which is the most related to MCIP, is defined as follows: Given two input strings, partition them into the same collection of substrings so that the number of resultant substrings is minimized. For example, the MCSP for $\{aaabbbccc, bbbaaaccc\}$ is $\{aaa, bbb, ccc\}$. The restricted version of MCSP where the number of symbols that occur in an input string multiple times (called duplicated symbols; the other symbols are called singletons) is no more than l in each input string, is denoted by MCSP-l. It is known that the MCSP-l problem is NP-hard [8], when $l \geq 1$. In other words, even when there is only one symbol with multiple copies in input strings, we still cannot find the MCSP in polynomial time unless P=NP.

It is easy to transform an instance of MCSP-1 into an instance of 2-MCIP where each integer represents the size of a block consisting of only the duplicated symbol so that an optimal solution to the 2-MCIP problem would in most cases give an optimal solution to the MCSP-1 problem with the same cardinality [8]. Therefore, we hope that the study of MCIP will help the design of good approximation algorithms for MCSP-1 and MCSP in general.

DNA fingerprint assembly. In the ongoing *Oligonucleotide Fingerprinting Ribosomal Genes* (OFRG) project [21], we collaborate with microbiologists and statisticians to provide a high-throughput method for identifying different microbial organisms. Briefly, the microbiologists build an rDNA clone library after DNA extraction and *Polymerase Chain Reaction* (PCR) amplification. The rDNA clones are assigned fingerprints (binary strings where 0 indicate nonbinding between a clone and a probe, and 1 otherwise) through a series of hybridization experiments, each using a single 10-nucleotide DNA probe. These 10-nucleotide DNA probes comprise a probe set and the size of the probe set determines the length of a fingerprint. Then, clones are identified by clustering their fingerprints with those of known sequences. By mapping sequence data to hybridization patterns, clones can be identified (or at least differentiated). Compared with direct sequencing, the method saves significant cost without sacrificing too much discriminating ability.

Although OFRG is a cost-effective approach, we are trying to scale it up in order to process a large number of samples from applications such as identifying microorganisms involved in the development of the mucosal and systemic immune system. One possible way of enhancing OFRG is inspired by new (but proven) technologies such as microbead clone libraries and multiplex flow cytometry. By producing clone libraries on microbeads, we are able to simultaneously hybridize a set of probes to thousands of clones in seconds, which is a significant improvement over the current array platform. However, we will still need multiple hybridizations, each using a different probe (sub)set, as the size of the desired probe set in OFRG exceeds the maximum discriminating size of the cytometry technology. Thus we obtain a *partial fingerprint* from each run of hybridization because only a subset of the probes are used in each hybridization.

The *DNA fingerprint assembly problem* aims at inferring a *complete fingerprint* (with respect to the overall probe set) for each clone from partial fingerprints by minimizing the total number of distinct complete fingerprints. We assume that all the probe subsets share a small number of common probes which are called the *linking probes*. That is, these linking probes will be used for each run of hybridization. A complete fingerprint can thus be obtained from partial fingerprints that share the same bits on the linking probes. More specifically, after each run of the hybridization, we assign a *weight* to each distinct partial fingerprint as the number of clones that produced this partial fingerprint in the hybridization. Then we divide all partial fingerprints into groups based on their bits on the positions of linking probes. The partial fingerprints in a group are compatible with each other and may correspond to the same complete fingerprint. For each group, the fingerprint assembly problem can be viewed as $MCIP(S_1, S_2, \cdots, S_k)$, with k being the number of the probe subsets (*i.e.* the number of hybridizations) and S_i containing the weight of each partial fingerprint in this particular group from the ith hybridization. Hence, complete fingerprints for each group can be obtained by combining their respective partial fingerprints via the minimum common integer partition of the weights. Such a solution would represent the minimum number of *distinct* complete fingerprints (or clones) that have produced the group of partial fingerprints.

2   Some Basic Facts

Throughout the paper, we assume that the multisets given as input to MCIP are related as mentioned before. Due to page constraint, we omit the proofs of all the lemmas and Theorem 4 (See [22] for the details of the proofs).

We denote the size of the minimum common integer partition by $|MCIP(S_1, S_2, \cdots, S_k)|$ (or simply $|k\text{-}MCIP|$ if the input multisets are clear from the context). Because every integer in any input multiset will be partitioned into one or more integers in the minimum common integer partition, the following lemma gives a trivial, but useful lower bound.

Lemma 1. $|MCIP(S_1, S_2, \cdots, S_k)| \geq max(|S_1|, |S_2|, \cdots, |S_k|)$, *where* $|\cdot|$ *is the size of a multiset.*

In the case of 2-MCIP, we use $\langle S, T \rangle$ to denote the two input multisets, where $S = \{x_1, x_2, \cdots, x_m\}$ and $T = \{y_1, y_2, \cdots, y_n\}$ such that $\sum_{i=1}^{m} x_i = \sum_{i=1}^{n} y_i$. A greedy algorithm that constructs a common integer partition of $\langle S, T \rangle$ is to iteratively add the smaller one of two integers randomly selected from the two input multisets. More precisely, the algorithm can be described in pseudo-code as in Figure 1, and runs in time linear in n. The following lemma gives an upper bound for 2-MCIP, which is very useful in the subsequent discussion.

Algorithm 2-APPROX-MCIP(S, T)
input Two related multisets S and T
output A common integer partition CIP
of S and T
begin
$CIP := \emptyset;$
while $S \neq \emptyset$ **do**
arbitrarily pick $x_i \in S$ and $y_j \in T$;
$S := S \setminus \{x_i\};$ $T := T \setminus \{y_j\};$
$z := \min(x_i, y_j);$ $CIP := CIP \uplus \{z\};$
if $x_i \neq z$ $S := S \uplus \{x_i - z\};$
if $y_i \neq z$ $T := T \uplus \{y_i - z\};$
end.

Fig. 1. A 2-approximation algorithm for 2-MCIP

Algorithm $\frac{5}{4}$-Approx-MCIP(S, T)
input Two related multisets S and T
output A common integer partition CIP
of S and T
begin
remove_common_integer(S,T);
approximate_set_packing(S,T);
$CIP := CIP(S_1, T_1) \uplus CIP(S_2, T_2);$
$CIP := CIP \uplus$ 2-APPROX-MCIP$(S_3, T_3);$
return CIP;
end.

Fig. 2. A $\frac{5}{4}$-approximation algorithm for 2-MCIP

Lemma 2. $|MCIP(S,T)| \leq |S| + |T| - 1.$

As its name suggests, 2-APPROX-MCIP(S,T) is a 2-approximation algorithm for the problem of 2-MCIP, which is implied by Lemma 1 and Lemma 2.

Lemma 3. *The algorithm* 2-APPROX-MCIP*(S,T) achieves an approximation ratio of 2.*

Given a common integer partition $CIP(S,T)$ of $\langle S, T \rangle$, we say that x_i is *mapped* to y_j if there exists an element in $CIP(S,T)$ such that it is a part of x_i as well as a part of y_j. Notice that an integer in S (or T) can be mapped to two or more integers in T (or S). Two integers a_1 and a_h in $\langle S, T \rangle$ (*i.e.*, $a_1, a_h \in S \uplus T$) are said to be *connected* if there exist a sequence of integers $a_2, \cdots, a_{h-1}$ in $\langle S, T \rangle$ such that a_i is mapped to a_{i+1}, for each $i \in [1, h-1]$. Thus, all the integers that are connected to each other in S and T will constitute a *connected component* (or simply *component*) of $\langle S, T \rangle$. We say that these connected components are *induced* by the given common integer partition $CIP(S,T)$.

Lemma 4. *Suppose that* $CIP(S,T)$ *denotes a common integer partition of S and T. Then*

1. *every connected component* $\langle S_1, T_1 \rangle$ *induced by $CIP(S,T)$ is a pair of related multisets;*
2. *for every connected component* $\langle S_1, T_1 \rangle$*, all the integers in $CIP(S,T)$ that are parts of integers in S_1 or T_1 constitute a common integer partition $CIP(S_1, T_1)$ of S_1 and T_1 such that $|CIP(S_1, T_1)| \geq |S_1| + |T_1| - 1$.*

2.1   The Maximum Related Multiset Partition

In this subsection, we define a new combinatorial optimization problem, *maximum related multiset partition (MRMP)*, to assist solving the MCIP problem.

S_1 and T_1 are said to be a pair of *related submultisets* of two related multisets S and T if S_1 is a (nonempty) submultiset of S, T_1 is a (nonempty) submultiset of T, and they are related. We write $\langle S_1, T_1 \rangle \subseteq \langle S, T \rangle$ to denote the related submultisets. Obviously,$\langle S, T \rangle \subseteq \langle S, T \rangle$. Furthermore, S and T are said to be *basic* if they have one and only one pair of related submultisets, namely $\langle S, T \rangle$. For example, consider $S = \{3, 3, 4\}$ and $T = \{2, 2, 6\}$. They have three pairs of related submultisets: $\langle \{3, 3\}, \{6\} \rangle$, $\langle \{4\}, \{2, 2\} \rangle$, and $\langle S, T \rangle$. Therefore, S and T are not a pair of basic related multisets. An example of two basic related multisets is $\langle \{1, 4\}, \{2, 3\} \rangle$.

A *multiset partition* (or simply *partition*) of a multiset S is a sequence of disjoint submultisets $S_1, S_2, \cdots, S_l$ of S whose union is S, i.e. $S = \biguplus_{i=1}^{l} S_i$. By definition, S is a multiset partition of itself. It is important to remember that multiset partition and the integer partition are two different concepts in this paper. Given two multisets S and T of integers, a sequence of multiset pairs $\langle S_1, T_1 \rangle, \langle S_2, T_2 \rangle, \cdots, \langle S_l, T_l \rangle$ is called a *related multiset partition* if $\{S_1, S_2, \cdots, S_l\}$ is a multiset partition of S, $\{T_1, T_2, \cdots, T_l\}$ is a multiset partition of T, and, moreover, for each $i \in [1, l]$, S_i and T_i are a pair of related multisets. The maximum related multiset partition problem is then defined as to find a related multiset partition of two given multisets S and T, maximizing the number of related multiset pairs in the partition. We denote by $MRMP(S, T)$ (or 2-$MRMP$) the maximum related multiset partition of S and T, and by $|MRMP(S, T)|$ (or $|2$-$MRMP|$) the size of the partition, i.e., the number of related multiset pairs in the partition.

Lemma 5. *Given a common integer partition $CIP(S, T)$, we can transform it into a related multiset partition of S and T, denoted as $RMP(S, T)$, such that $|RMP(S, T)| \geq |S| + |T| - |CIP(S, T)|$.*

The following lemma establishes the relationship between MCIP and MRMP, showing their (complementary) equivalence.

Lemma 6. *If S and T are related multisets, then $|MCIP(S, T)| + |MRMP(S, T)| = |S| + |T|$.*

Since a pair of basic related multisets S and T cannot be partitioned further into related submultisets, i.e., $|MRMP(S, T)| = 1$, the following lemma is trivially implied by Lemma 6.

Lemma 7. *If S and T are a pair of basic related multisets, then $|MCIP(S, T)| = |S| + |T| - 1$.*

The following lemmas will be crucial to the approximation algorithms. We define the size of a pair of related multisets S and T as the sum of the size of S and the size of T, i.e., $|\langle S, T \rangle| = |S| + |T|$.

Lemma 8. *If the minimum size of any related submultiset of S and T is c, then $|MCIP(S, T)| \geq \frac{c-1}{c}(|S| + |T|)$.*

Lemma 9. *Given two related multisets, $S = \{x_1, x_2, \cdots, x_m\}$ and $T = \{y_1, y_2, \cdots, y_n\}$. If x_i and y_j are identical, then $\{x_i\} \uplus MCIP(S \backslash \{x_i\}, T \backslash \{y_j\})$ is a minimum common integer partition of S and T, i.e., $|MCIP(S,T)| = |MCIP(S \backslash \{x_i\}, T \backslash \{y_j\})| + 1$.*

Unfortunately, the result in Lemma 9 cannot be extended to the case of k multisets when $k \geq 3$. An interesting counterexample is $\{6, 5, 1, 4, 2\}$, $\{6, 5, 1, 3, 3\}$, $\{6, 4, 2, 3, 3\}$. Their minimum common integer partition is of size 6, but any common integer partition including 6 as an element is of size at least 7. In the following, we will use a procedure **remove_common_integer**$(S_1, S_2, \cdots, S_k)$ to remove all common integer elements existing in every multiset of $\{S_1, S_2, \cdots, S_k\}$ (and add them into the solution). The optimality of this operation is guaranteed only when $k = 2$, as shown in Lemma 9.

3   Hardness of Approximation

It is easy to see that MCIP is NP-hard because there is a straightforward reduction from the *Set Partition* problem. This section is devoted to proving that MCIP is APX-hard.

In the sequel, we prove the APX-completeness of 2-MCIP by an L-reduction from the *Maximum Bounded 3-Dimensional Matching* problem (denoted as MAX 3DM-3). The MAX 3DM-3 problem is defined as follows: Given a set $D \subseteq X \times Y \times Z$, where X, Y and Z are disjoint sets and moreover, each element in X, Y and Z occurs in at least one and at most three triples in D [17], the goal is to find a matching $M \subseteq D$ for D of the maximum cardinality, *i.e.*, a largest set $M \subseteq D$ such that no two elements in M agree in any coordinate. In this problem, without loss of generality, we can assume that $n = |X| \leq |Y| \leq |Z|$. Since each element in X occurs at least once and at most three times in D, the number of triples is at least n and at most $3n$, *i.e.*, $n \leq |D| \leq 3n$. It also implies that $|Y| \leq 3n$ and $|Z| \leq 3n$. Further observe that each triple can intersect at most six other triples, which implies that the maximum matching contains at least $|D|/7$ triples. Let $|MAX$ $3DM$-$3|$ denote the size of maximum matching of $|D|$. It is easy to see that $\lceil \frac{n}{7} \rceil \leq |MAX$ $3DM$-$3| \leq n$.

Let $X = \{x_1, x_2, \cdots, x_{|X|}\}$, $Y = \{y_1, y_2, \cdots, y_{|Y|}\}$, $Z = \{z_1, z_2, \cdots, z_{|Z|}\}$, and $D = \{d_1, d_2, \cdots, d_{|D|}\}$ where $d_i = (x_{i^X}, y_{i^Y}, z_{i^Z})$ for each $i \in [1, |D|]$ and i^X (i^Y or i^Z, respectively) is the corresponding index of the integer x_{i^X} (y_{i^Y} or z_{i^Z}, respectively) in X (Y or Z, respectively). We can define a function f to construct an instance of 2-MCIP as follows:

- A multiset $\tilde{X} = \{\tilde{x}_i | \tilde{x}_i = 4^i, \forall x_i \in X\}$;
- A multiset $\tilde{Y} = \{\tilde{y}_i | \tilde{y}_i = 4^{|X|+i}, \forall y_i \in Y\}$;
- A multiset $\tilde{Z} = \{\tilde{z}_i | \tilde{z}_i = 4^{|X|+|Y|+i}, \forall z_i \in Z\}$;
- A multiset $\tilde{D} = \{\tilde{d}_i | \tilde{d}_i = \tilde{x}_{i^X} + \tilde{y}_{i^Y} + \tilde{z}_{i^Z}, \forall d_i \in D\}$;
- An integer $e = \sum_{i=1}^{|D|} \tilde{d}_i - \sum_{i=1}^{|X|} \tilde{x}_i - \sum_{i=1}^{|Y|} \tilde{y}_i - \sum_{i=1}^{|Z|} \tilde{z}_i$.
- Two multisets $S = \tilde{D}$ and $T = \tilde{X} \cup \tilde{Y} \cup \tilde{Z} \cup \{e\}$.

Since each element in X, Y and Z is assumed to occur at least once in D while some elements occur more than once, it always holds that $e > 0$. Obviously, $\sum S = \sum T$. Therefore, $\langle S, T \rangle$ is an instance of 2-MCIP that we can obtain in time linear in n.

Let $|2\text{-}MCIP|$ denote the size of the minimum common integer partition of $\langle S, T \rangle$. Then, we have the following lemma.

Lemma 10. *For any instance of MAX 3DM-3, $|2\text{-}MCIP| \leq 70 \cdot |MAX\ 3DM\text{-}3|$.*

Given a common integer partition 2-CIP of $\langle S, T \rangle$, we define a function g to construct a subset (denoted as $3DM$-3) of D by including all the triples $d_i = (x_{i^X}, y_{i^Y}, z_{i^Z})$ $(1 \leq i \leq |D|)$ whose corresponding integers $\tilde{d}_i = \tilde{x}_{i^X} + \tilde{y}_{i^Y} + \tilde{z}_{i^Z}$ are not connected to the integer e in the common integer partition 2-CIP.

Lemma 11. *For any instance D of MAX 3DM-3, the subset $3DM$-3 constructed by the function g is a matching of D.*

Let $|2\text{-}MRMP|$ be the size of the maximum related multiset partition of S and T. Let $|2\text{-}RMP|$ be the size of a related multiset partition of S and T, induced by a given common partition 2-CIP.

Lemma 12. $|2\text{-}MRMP| = |MAX\ 3DM\text{-}3| + 1$.

Lemma 13. $|MAX\ 3DM\text{-}3| - |3DM\text{-}3| \leq |2\text{-}CIP| - |2\text{-}MCIP|$.

Lemma 14. $MAX\ 3DM\text{-}3 \leq_L 2\text{-}MCIP$.

Theorem 1. *The k-MCIP problem is APX-complete, for any $k \geq 2$.*

Proof. Since the MAX 3DM-3 problem is APX-complete [17] and MAX 3DM-3 $\leq_L$ 2-MCIP by Lemma 14, 2-MCIP is APX-hard. In addition, by Lemma 3, there exists a polynomial-time 2-approximation algorithm for 2-MCIP, which implies that 2-MCIP is APX-complete. In Section 5, we will present a k-approximation algorithm for k-MCIP, which implies that k-MCIP is APX-complete, for any $k \geq 2$. □

4   Approximation of 2-MCIP Via Maximum Set Packing

In this section, we will give a $\frac{5}{4}$-approximation algorithm for the 2-MCIP problem by considering basic related submultisets of sizes three and four between S and T. As mentioned earlier, we assume that there are no common integer elements between the two input multisets S and T, without loss of generality.

We can construct an instance of the *Maximum Set Packing* problem [1], in which the collection C consists of all the basic related submultisets of sizes three and four between S and T. Since the cardinality of each multiset in C is bounded from the above by a constant, it is actually an instance of the *Maximum k-Set Packing* problem where $k = 4$. Hurkens and Schrijver [15] show that the Maximum k-Set Packing problem is approximable within ratio $k/2 + \epsilon$ for any

$\epsilon > 0$. For the weighted version of the Maximum k-Set Packing problem, where each set is given a non-negative weight, Arkin and Hassin [3] show that it is approximable within ratio $k - 1 + \epsilon$ for any $\epsilon > 0$.

In the following, we consider a special weighted Maximum k-Set Packing problem on C, where the weight for each basic related multiset of size three is 2 and the weight for a multiset of size four is 1, and the goal is to find a collection of disjoint multisets of maximum total weight. Call any collection of pairwise disjoint multisets a *packing*. We design a heuristic algorithm, which is implemented in the procedure **approximate_set_packing**(S,T), to find a packing as follows: first find a *maximal* set packing, and then recursively replace a multiset of size four in the packing by a multiset of size three, or replace a multiset of size three by two multisets of size three, or add some multiset into the packing so that the resultant collection is still a packing (but with one more multiset of size three after a replacement or with one more multiset after an addition), until no such replacement or addition could be made further.

The above heuristic algorithm can be made to run in $O(|U| \cdot |C|^2)$ time. Due to the space limitation, the running time analysis is omitted here, which can be found in [22] .

Let q_3 and q_4 denote the numbers of basic related multisets of sizes three and four in the packing found by our heuristic algorithm, and q_3^* and q_4^* the numbers of basic related multisets of sizes three and four in an optimal weighted set packing, respectively. It is obvious that $2q_3 + q_4 \leq 2q_3^* + q_4^*$. Moreover, we can obtain the following relationship. [1]

Lemma 15. $2q_3^* + q_4^* \leq 4(q_3 + q_4)$.

Let q_3' and q_4' be the numbers of basic related submultisets of sizes three and four in the related multiset partition induced by a given minimum common partition $MCIP(S, T)$. It is obvious that $2q_3' + q_4' \leq 2q_3^* + q_4^*$. The following is a tighter lower bound for 2-MCIP.

Lemma 16. $|MCIP(S,T)| \geq \frac{4}{5}(m + n) - \frac{1}{5}(2q_3^* + q_4^*)$, where $m = |S|$ and $n = |T|$.

The following lemma gives a tighter upper bound for 2-MCIP.

Lemma 17. $|MCIP(S,T)| \leq m + n - q_3 - q_4 - 1$.

As mentioned earlier, we run the procedure **approximate_set_packing**(S,T) to find the three disjoint submultisets $\langle S_1, T_1 \rangle$, $\langle S_2, T_2 \rangle$ and $\langle S_3, T_3 \rangle$. A $\frac{5}{4}$- approximation algorithm for 2-MCIP can then be obtained, as illustrated in Figure 2. The algorithm runs in time $O((m + n)^9)$, which is dominated by the running time of the procedure **approximate_set_packing**(S,T), as there are $m + n$ elements in the universe and the size of the collection C could reach

---

[1] The $(k/2 + \epsilon)$-approximation algorithm given by Hurkens and Schrijver [15] can also find a packing of C satisfying the inequality in Lemma 15, but only in quasi-polynomial time.

```
Algorithm k-APPROX-MCIP(S₁, ···, Sₖ)
  input Related multisets S₁, ···, Sₖ
  output A common integer partition CIP
         of S₁, ···, Sₖ
  begin
    CIP := 2-APPROX-MCIP(S₁, S₂);
    for i = 3 to k do
      CIP := 2-APPROX-MCIP(CIP, Sᵢ);
    return CIP;
  end.
```

Fig. 3. A k-approximation algorithm for k-MCIP

```
Algorithm
  (3k(k-1))/(3k-2)-APPROX-MCIP(S₁, ···, Sₖ)
  input Related multisets S₁, ···, Sₖ
  output A common integer partition CIP
         of S₁, ···, Sₖ
  begin
    remove_common_integer(S₁, ···, Sₖ);
    CIP := k-APPROX-MCIP(S₁, ···, Sₖ);
    return CIP;
  end.
```

Fig. 4. A $\frac{3k(k-1)}{3k-2}$-approximation algorithm for k-MCIP

$\Theta((m + n)^4)$ in the worst case. We believe that the running time can be further reduced by a more careful implementation and analysis of the procedure **approximate_set_packing**(S,T).

Theorem 2. *The algorithm $\frac{5}{4}$-APPROX-MCIP is a $\frac{5}{4}$-approximation algorithm for 2-MCIP.*

Proof. By Lemmas 16 and 17, the approximation ratio α given by algorithm $\frac{5}{4}$-APPROX-MCIP is

$$\alpha \leq \frac{m + n - q_3 - q_4 - 1}{\frac{4}{5}(m + n) - \frac{1}{5}(2q_3^* + q_4^*)} = \frac{5}{4} \cdot \frac{m + n - q_3 - q_4 - 1}{m + n - \frac{1}{4}(2q_3^* + q_4^*)}$$

It suffices to show that $m + n - q_3 - q_4 - 1 \leq m + n - \frac{1}{4}(2q_3^* + q_4^*)$, which is equivalent to showing $2q_3^* + q_4^* \leq 4(q_3 + q_4 + 1)$. By lemma 15, we know that $2q_3^* + q_4^* \leq 4(q_3 + q_4)$. Therefore, $\alpha \leq \frac{5}{4}$. □

5   Approximation of k-MCIP

In this section, we will discuss how to approximate the general k-MCIP $(k \geq 3)$ problem.

Using the algorithm 2-Approx-MCIP(S,T) in the previous section, we give an approximation algorithm to solve the k-MCIP $(k \geq 3)$ problem, as described in Figure 3. First, we give an upper bound on the performance of this algorithm.

Lemma 18. $|MCIP(S_1, S_2, \cdots, S_k)| \leq \sum_{i=1}^{k} |S_i| - k + 1$.

Theorem 3. *The algorithm k-APPROX-MCIP is a k-approximation algorithm for the k-MCIP $(k \geq 2)$ problem.*

Proof. By Lemma 1 and Lemma 18, the size of the common integer partition CIP returned from k-APPROX-MCIP$(S_1, S_2, \cdots, S_k)$ is such that $max\{|S_1|, |S_2|, \cdots, |S_k|\} \leq |MCIP(S_1, S_2, \cdots, S_k)| \leq |CIP(S_1, S_2, \cdots, S_k)| \leq \sum_{i=1}^{k} |S_i| - k + 1$, from which the theorem follows. □

As described in Figure 4, the algorithm k-APPROX-MCIP can be slightly improved by employing the procedure **remove_common_integer**$(S_1, S_2, \cdots, S_k)$. To show that this improved algorithm achieves an approximation ratio less than k, we need the following lemma.

Lemma 19. *If there is no integer element common to all the multisets in $\{S_1, S_2, \cdots, S_k\}$, then it holds that $|MCIP(S_1, S_2, \cdots, S_k)| \geq \frac{3k-2}{3k(k-1)} \sum_{i=1}^{k} |S_i|$.*

Theorem 4. *The algorithm $\frac{3k(k-1)}{3k-2}$-APPROX-MCIP is a $\frac{3k(k-1)}{3k-2}$-approximation algorithm for the k-MCIP ($k \geq 2$) problem.*

Clearly, the algorithm $\frac{3k(k-1)}{3k-2}$-APPROX-MCIP$(S_1, \cdots, S_k)$ runs in $O(\sum_i |S_i| \cdot log(\sum_i |S_i|))$ time. Let us compare Theorem 4 with Theorem 3. Clearly, $\frac{3k(k-1)}{3k-2}$ is always smaller than k, for any $k \geq 2$. For example, when $k = 2$, the above algorithm gives approximation ratio 1.5, and when $k = 3$, its approximation ratio is $\frac{18}{7}$, which is much better than the ratio 3 in Theorem 3. However, when k becomes large, $\frac{3k(k-1)}{3k-2}$ is only slightly smaller than k, since $\frac{3k(k-1)}{3k-2} = \Theta(k)$. It is an interesting open question whether k-MCIP has an approximation algorithm with a ratio that is asymptotically better than k.

6   Concluding Remarks

It is interesting to observe that although 2-MCIP is in some sense similar to other integer partition/summation problems such as Knapsack and Bin Packing, it is much more difficult to approximate. For example, Knapsack and Bin Packing all have an FPTAS (fully polynomial-time approximation scheme) or asymptotic PTAS, but Theorem 1 implies that it is unlikely for 2-MCIP to have a PTAS.

Acknowledgments

We would like to thank David P. Woodruff for several useful discussions. This project is supported in part by NSF grants CCR-0309902 and DBI-0133265, NSFC grant 60528001, National Key Project for Basic Research (973) grant 2002CB512801, and a fellowship from the Center for Advanced Study, Tsinghua University.

References

1. G. Ausiello, P. Crescenzi, G. Gambosi, V. Kann, A. Marchetti-Spaccamela, and M. Protasi. *Complexity and Approximation*, Springer, 1999.
2. G.E. Andrews. *The Theory of Partitions*, Addison-Wesley, 1976.
3. E.M. Arkin and R. Hassin. On local search for weighted packing problems. *Math. Oper. Res.* 23, pp. 640-648, 1998.
4. G.E. Andrews and K. Eriksson. *The Integer Partitions*, Cambridge, 2004.
5. S. Altschul and D. Lipman. Trees, stars, and multiple sequence alignment. *SIAM Journal on Applied Math.* 49(1), pp. 197-209, 1989.

6. M. Chrobak, P. Lolman, and J. Sgall. The greedy algorithm for the minimum common string partition problem. *Proc. of 7th International Workshop on Approximation Algorithms for Combinatiorial Optimization Problems (APPROX)*, pp. 84-95, 2004.

7. T.H. Cormen, C.E. Leiserson, R.L. Rivest, C. Stein. *Introduction to algorithms*, The MIT Press, 2nd edition, p. 1017, 2001.

8. X. Chen. The minimum common partition problem revisited. *manuscript*, 2005.

9. X. Chen, J. Zheng, Z. Fu, P. Nan, Y. Zhong, S. Lonardi, and T. Jiang. Computing the assignment of orthologous genes via genome rearrangement. *Proc. of 3rd Asia Pacific Bioinformatics Conference (APBC'05)*, pp. 363-378, 2005.

10. X. Chen, J. Zheng, Z. Fu, P. Nan, Y. Zhong, S. Lonardi, and T. Jiang. The assignment of orthologous genes via genome rearrangement. *IEEE/ACM Transactions on Computational Biology and Bioinformatics*, 2(4), pp. 302-315, 2005.

11. Z. Fu. Assignment of orthologous genes for multichromosomal genomes using genome rearrangement. *UCR CS Technical report*, 2004.

12. D. Gusfield. *Algorithms on Strings, Tree, and Sequences: Computer Science and Computational Biology*, Cambridge University Press, 1997.

13. A. Goldstein, P. Kolman, and J. Zheng. Minimum common string partition problem: hardness and approximations. *Proc. of 15th International Symposium on Algorithms and Computation* (ISAAC), LNCS 3341, pp. 473-484, 2004.

14. S. Hannenhalli and P.A. Pevzner. Transforming cabbage into turnip (polynomial algorithm for sorting signed permutations by reversals). *Proc. 27th Ann. ACM Symp. Theory of Comput.* (STOC'95), pp. 178-189, 1995.

15. C. Hurkens and A. Schrijver. On the size of systems of sets every t of which have an SDR, with an application to the worst-case ratio of heuristics for packing problems. *SIAM J. Discrete Mathematics*, 2, pp. 68-72, 1989.

16. P. Kolman. Approximating reversal distance for strings with bounded number of duplicates in linear time. *Proc. of 30 International Symposium on Mathematical Foundations of Computer Science (MFCS)*, pp. 580-590, 2005.

17. V. Kann. Maximum bounded 3-dimensional matching is MAX SNP-complete. *Information Processing Letters*, 37: 27-35, 1991.

18. C.H. Papadimitriou and M. Yannakakis. Optimization, approximation, and complexity classes. *J. Computer and System Sciences*, 43: 425-440, 1991.

19. M. Remm, C. Storm, and E. Sonnhammer. Automatic clustering of orthologs and in-paralogs from pairwise species comparisons. *J. Mol. Biol.*, 314, pp. 1041-1052, 2001.

20. D. Sankoff. Mechanisms of genome evolution: models and inference. *Bull. Int. Stat. Instit.* 47, pp. 461-475, 1989.

21. L. Valinsky, A. Scupham, G.D. Vedova, Z. Liu, A. Figueroa, K. Jampachaisri, B. Yin, E. Bent, R. Mancini-Jones, J. Press, T. Jiang, and J. Borneman. Oligonucleotide Fingerprinting of Ribosomal RNA Genes (OFRG), pp. 569-585. In G. A. Kowalchuk, F. J. de Bruijn, I. M. Head, A. D. L. Akkermans, J. D.van Elsas (eds.) *Molecular Microbial Ecology Manual* (2nd ed). Kluwer Academic Publishers, Dordrecht, The Netherlands, 2004.

22. Available at http://www.cs.ucr.edu/~lliu/paper/mcip_ciac_full.pdf.

Matching Subsequences in Trees

Philip Bille[1,*] and Inge Li Gørtz[2,**]

[1] IT University of Copenhagen
beetle@itu.dk
[2] Technical University of Denmark
ilg@imm.dtu.dk

Abstract. Given two rooted, labeled trees P and T the tree path subsequence problem is to determine which paths in P are subsequences of which paths in T. Here a path begins at the root and ends at a leaf. In this paper we propose this problem as a useful query primitive for XML data, and provide new algorithms improving the previously best known time and space bounds.

1 Introduction

We say that a tree is *labeled* if each node is assigned a character from an alphabet Σ. Given two sequences of labeled nodes p and t, we say that p is a *subsequence* of t, denoted $p \sqsubseteq t$, if p can be obtained by removing nodes from t. Given two rooted, labeled trees P and T the *tree path subsequence problem* (TPS) is to determine which paths in P are subsequences of which paths in T. Here a path begins at the root and ends at a leaf. That is, for each path p in P we must report all paths t in T such that $p \sqsubseteq t$.

This problem was introduced by Chen [3] who gave an algorithm using $O(\min(l_P n_T, n_P l_T + n_T))$ time and $O(l_P d_T)$ space. Here, n_S, l_S, and d_S denotes the number of nodes, number of leaves, and depth, respectively, of a tree S. Note that in the worst-case this is quadratic time and space. In this paper we show the following result:

Theorem 1. *For trees P and T the tree path subsequence problem can be solved in $O(\min\left(l_P n_T, n_P l_T + n_T, \frac{n_P n_T}{\log n_T} + n_P \log n_P\right))$ time and $O(n_P + n_T)$ space.*

Hence, if one of the trees has few leaves we match the previous time bounds, while improving the space to linear. The latter bound improves the worst-case time by a logarithmic factor whenever $\log n_P = O(n_T / \log n_T)$. Note that – in the worst-case – the number of pairs consisting of a path from P and a path T is $\Omega(n_P n_T)$, and therefore we need at least as many bits to report the solution to TPS. Hence, on a RAM with logarithmic word size our worst-case bound is optimal.

---

* This work is part of the DSSCV project supported by the IST Programme of the European Union (IST-2001-35443).

** This work was performed while the author was a PhD student at the IT University of Copenhagen.

T. Calamoneri, I. Finocchi, G.F. Italiano (Eds.): CIAC 2006, LNCS 3998, pp. 248–259, 2006.

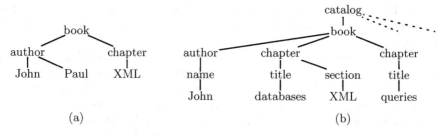

Fig. 1. (a) The trie of queries 1,2,3, or the tree for query 4. (b) A fragment of a catalog of books.

More importantly, all our algorithms use linear space, whereas the previous ones used quadratic space in the worst-case. For practical applications this makes it possible to solve TPS on larger trees and speed up the running time since more of the computation can be kept in main memory.

The first two time bounds are useful when the number of leaves in one of the trees has few leaves. In this case our contribution is the reduction to linear space. If, on the other hand, the number of leaves in both trees are proportional to the number of nodes in the tree the last time bound is the best. In this paper we present the first algorithm with subquadratic worst-case time and space bound.

Applications. We propose TPS as a useful query primitive for XML data. The key idea is that an XML document D may be viewed as a rooted, labeled tree. For example, suppose that we want to maintain a catalog of books for a bookstore. A fragment of a possible XML tree, denoted D, corresponding to the catalog is shown in Fig. 1(b). In addition to supporting full-text queries, such as find all documents containing the word "John", we can also use the tree structure of the catalog to ask more specific queries, such as the following examples:

1. Find all books written by John,
2. find all books written by Paul,
3. find all books with a chapter that has something to do with XML, or
4. find all books written by John and Paul with a chapter that has something to do with XML.

The queries 1,2, and 3 correspond to a *path query* on D, that is, compute which paths in D that contains a specific path as a subsequence. For instance, computing the paths in D that contain the path of three nodes labeled "book", "chapter", and "XML", respectively, effectively answers query 3. Most XML-query languages, such as XPath [4], support such queries.

Using a simple algorithm (a describtion of which we omit due to lack of space) a path query can be solved in linear time. Specifically, if q is a path consisting of n_q nodes, answering the path query on D takes $O(n_q + n_D)$ time. Hence, if we are given path queries $q_1, \ldots, q_k$ we can answer them in $O((n_{q_1} + \cdots + n_{q_k})n_D)$ time. If, however, the paths overlap we can do better by constructing the *trie*, Q, of $q_1, \ldots, q_k$. Answering all paths queries now correspond to solving TPS on

Q and D. As example the queries 1,2, and 3 form the trie shown in Fig. 1(a). Depending on the overlap between $q_1, \ldots, q_k$, n_Q is up to a linear factor smaller than $n_{q_1} + \cdots + n_{q_k}$.

Next consider query 4. This query cannot be answered by solving a TPS problem but is an instance of the *tree inclusion problem* (TI). Here we want to decide if P is *included* in T, that is, if P can be obtained from T by *deleting* nodes of T. Deleting a node y in T means making the children of y children of the parent of y and then removing y. It is straightforward to check that we can answer query 4 by deciding if the tree in Fig. 1(a) can be included in the tree in Fig. 1(b).

Recently, TI has been recognized as an important XML query primitive and has recieved considerable attention, see e.g., [9, 13, 12, 14, 10, 11]. Unfortunately, TI is NP-complete in general [8] and therefore the existing algorithms are based on heuristics. Observe that a necessary condition for P to included in T is that all paths in P are subsequences of paths in T. Hence, we can use TPS to quickly rule out trees that cannot be included T. We believe that in this way TPS can be used as an effective "filter" for many tree inclusion problems that occur in practice.

Technical Overview. Given two strings (or labeled paths) a and b, it is straightforward to determine if a is a subsequence of b by scanning the character from left to right in b. This uses $O(|a| + |b|)$ time. We can solve TPS by applying this algorithm to each of the pair of paths in P and T, however, this may use as much as $O(n_P n_T (n_P + n_T))$ time. Alternatively, Baeza-Yates [2] showed how to preprocess b in $O(|b| \log |b|)$ time such that testing whether a is a subsequence of b can be done in $O(|a| \log |b|)$ time. Using this data structure on each path in T we can solve the TPS problem, however, this may take as much as $O(n_T \log n_T + n_P^2 \log n_T)$. Hence, the availiable subsequence algorithms on strings does not provide an immediate solution.

Inspired by the work of Chen [3] we take another approach. We provide a framework for solving TPS. The main idea is to traverse T while maintaining a subset of nodes in P, called the *state*. When reaching a leaf z in T the state represents the paths in P that are a subsequences of the path from the root to z. At each step the state is updated using a simple procedure defined on subset of nodes. The result of Theorem 1 is obtained by taking the best of two algorithms based on our framework: The first one uses a simple data structure to maintain the state. This leads to an algorithm using $O(\min(l_P n_T, n_P l_T + n_T))$ time. At a high level this algorithm resembles the algorithm of Chen [3] and achieves the same running time. However, we improve the analysis of the algorithm and show a space bound of $O(n_P + n_T)$. This should be compared to the worst-case quadratic space bound of $O(l_P d_T)$ given by Chen [3]. Our second algorithm takes a different approach combining several techniques. Starting with a simple quadratic time and space algorithm, we show how to reduce the space to $O(n_P \log n_T)$ using a decomposition of T into disjoint paths. We then divide P into small subtrees of logarithmic size called *micro trees*. The micro trees are then preprocessed such that subsets of nodes in a micro tree can be maintained in constant time and space. Intuitively, this leads to a logarithmic improvement of the time and space bound.

Notation and Definitions. In this section we define the notation and definitions we will use throughout the paper. For a graph G we denote the set of nodes and edges by $V(G)$ and $E(G)$, respectively. Let T be a rooted tree. The root of T is denoted by root(T). The *size* of T, denoted by n_T, is $|V(T)|$. The *depth* of a node $y \in V(T)$, depth(y), is the number of edges on the path from y to root(T) and the depth of T, denoted d_T, is the maximum depth of any node in T. The parent of y is denoted parent(y). A node with no children is a leaf and otherwise it is an internal node. The number of leaves in T is denoted l_T. Let $T(y)$ denote the subtree of T rooted at a node $y \in V(T)$. If $z \in V(T(y))$ then y is an ancestor of z and if $z \in V(T(y))\backslash\{y\}$ then y is a proper ancestor of z. If y is a (proper) ancestor of z then z is a (proper) descendant of y. We say that T is *labeled* if each node y is assigned a character, denoted label(y), from an alphabet Σ. The path from y to root(T), of nodes root$(T) = y_1, \ldots, y_k = y$ is denoted path(y). Hence, we can formally state TPS as follows: Given two rooted tree P and T with leaves $x_1, \ldots, x_r$ and $y_1, \ldots, y_s$, respectively, determine all pairs (i, j) such that path$(x_i) \sqsubseteq$ path(y_j). For simplicity we will assume that leaves in P and T are always numbered as above and we identify each of the paths by the number of the corresponding leaf.

Throughout the paper we assume a standard RAM model of computation with logarithmic word size. We use a standard instruction set including bitwise boolean operations, shifts, addition, multiplication, etc.

2   A Framework for Solving TPS

In this section we present a simple general algorithm for the tree path subsequence problem. The key ingredient in our algorithm is the following procedure. For any $X \subseteq V(P)$ and $y \in V(T)$ define:

DOWN(X, y): Return the set CHILD$(\{x \in X \mid$ label$(x) =$ label$(y)\}) \cup \{x \in X \mid$ label$(x) \neq$ label$(y)\}$.

The notation CHILD(X) denotes the set of children of X. Hence, DOWN(X, y) is the set consisting of nodes in X with a different label than y and the children of the nodes X with the same label as y. We will now show how to solve TPS using this procedure.

First assign a unique number in the range $\{1, \ldots, l_P\}$ to each leaf in P. Then, for each i, $1 \leq i \leq l_P$, add a *pseudo-leaf* $\perp_i$ as the single child of the ith leaf. All pseudo-leaves are assigned a special label $\beta \notin \Sigma$. The algorithm traverses T in a depth first order and computes at each node y the set X_y. We call this set the *state* at y. Initially, the state consists of $\{\text{root}(P)\}$. For $z \in$ child(y), the state X_z can be computed from state X_y as follows: $X_z = \text{DOWN}(X_y, z)$.

If z is a leaf we report the number of each pseudo-leaf in X_z as the paths in P that are subsequences of path(z). See Fig. 2 for an example. To show the correctness of this approach we need the following lemma.

Lemma 1. *For any node $y \in V(T)$ the state X_y satisfies the following property: If $x \in X_y$ then* path$($parent$(x)) \sqsubseteq$ path(y).

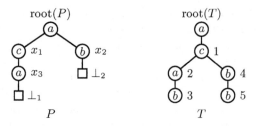

Fig. 2. Letters inside nodes are labels, and the identifier of each node is written outside the node. Initially we have $X = \{\text{root}(P)\}$. Since $\text{label}(\text{root}(P)) = a = \text{label}(\text{root}(T))$ we replace $\text{root}(P)$ with is children and get $X_{\text{root}(T)} = \{x_1, x_2\}$. Since $\text{label}(1) = \text{label}(x_1) \neq \text{label}(x_2)$ we get $X_1 = \{x_3, x_2\}$. Continuing this way we get $X_2 = \{\perp_1, x_2\}$, $X_3 = \{\perp_1, \perp_2\}$, $X_4 = \{x_3, \perp_2\}$, and $X_5 = \{x_3, \perp_2\}$. The nodes 3 and 5 are leaves of T and we thus report paths 1 and 2 after computing X_3 and path 2 after computing X_5.

Proof. By induction on the number of iterations of the procedure. Initially, $X = \{\text{root}(P)\}$ satisfies the property since $\text{root}(P)$ has no parent. Suppose that X_y is the current state and $z \in \text{child}(y)$ is the next node in the depth first traversal of T. By the induction hypothesis X_y satisfies the property, that is, for any $x \in X_y$, $\text{path}(\text{parent}(x)) \sqsubseteq \text{path}(y))$. Then, $X_z = \text{DOWN}(X_y, z) = \text{CHILD}(\{x \in X_y \mid \text{label}(x) = \text{label}(z)\}) \cup \{x \in X_y \mid \text{label}(x) \neq \text{label}(z)\}$.

Let x be a node in X_y. There are two cases. If $\text{label}(x) = \text{label}(z)$ then $\text{path}(x) \sqsubseteq \text{path}(z)$ since $\text{path}(\text{parent}(x)) \sqsubseteq \text{path}(y)$. Hence, for any child x' of x we have $\text{path}(\text{parent}(x')) \sqsubseteq \text{path}(z)$. On the other hand, if $\text{label}(x) \neq \text{label}(z)$ then $x \in X_z$. Since $y = \text{parent}(z)$ we have $\text{path}(y) \sqsubseteq \text{path}(z)$, and hence $\text{path}(\text{parent}(x)) \sqsubseteq \text{path}(y) \sqsubseteq \text{path}(z)$. □

By the above lemma all paths reported at a leaf $z \in V(T)$ are subsequences of $\text{path}(z)$. The following lemma shows that the paths reported at a leaf $z \in V(T)$ are *exactly* the paths in P that are subsequences of $\text{path}(z)$.

Lemma 2. *Let z be a leaf in T and let $\perp_i$ be a pseudo-leaf in P. Then, $\perp_i \in X_z \Leftrightarrow \text{path}(\text{parent}(\perp_i)) \sqsubseteq \text{path}(z)$.*

Proof. It follows from Lemma 1 that $\perp_i \in X_z \Rightarrow \text{path}(\text{parent}(\perp_i)) \sqsubseteq \text{path}(z)$. It remains to show that $\text{path}(\text{parent}(\perp_i)) \sqsubseteq \text{path}(z) \Rightarrow \perp_i \in X_z$. Let $\text{path}(z) = z_1, \ldots, z_k$, where $z_1 = \text{root}(T)$ and $z_k = z$, and let $\text{path}(\text{parent}(\perp_i)) = y_1, \ldots, y_\ell$, where $y_1 = \text{root}(P)$ and $y_\ell = \text{parent}(\perp_i)$. Since $\text{path}(\text{parent}(\perp_i)) \sqsubseteq \text{path}(z)$ there are nodes $z_{j_i} = y_i$ for $1 \leq i \leq k$, such that (i) $j_i < j_{i+1}$ and (ii) there exists no node z_j with $\text{label}(z_j) = \text{label}(y_i)$, where $j_{i-1} < j < j_i$. Initially, $X = \{\text{root}(P)\}$. We have $\text{root}(P) \in X_{z_j}$ for all $j < j_1$, since z_{j_1} is the first node on $\text{path}(z)$ with label $\text{label}(\text{root}(P))$. When we get to z_{j_1}, $\text{root}(P)$ is removed from the state and y_2 is inserted. Similarly, y_i is in all states X_{z_j} for $j_{i-1} \leq j < j_i$. It follows that $\perp_i$ is in all states X_{z_j} where $j \geq j_\ell$ and thus $\perp_i \in X_{z_k} = X_z$. □

The next lemma can be used to give an upper bound on the number of nodes in a state. The proof is omitted due to lack of space.

Lemma 3. *For any* $y \in V(T)$ *the state* X_y *has the following property: Let* $x \in X_y$. *Then no ancestor of* x *is in* X_y.

It follows from Lemma 3 that $|X_y| \leq l_P$ for any $y \in V(T)$. If we store the state in an unordered linked list each step of the depth-first traversal takes time $O(l_P)$ giving a total $O(l_P n_T)$ time algorithm. Since each state is of size at most l_P the space used is $O(n_P + l_P n_T)$. In the following sections we show how to improve these bounds.

3   A Simple Algorithm

In this section we consider a simple implementation of the above algorithm, which has running time $O(\min(l_P n_T, n_P l_T))$ and uses $O(n_P + n_T)$ space. We assume that the size of the alphabet is $n_T + n_P$ and each character in Σ is represented by an integer in the range $\{1, \ldots, n_T + n_P\}$. If this is not the case we can sort all characters in $V(P) \cup V(T)$ and replace each label by its rank in the sorted order. This does not change the solution to the problem, and assuming at least a logarithmic number of leaves in both trees it does not affect the running time. To get the space usage down to linear we will avoid saving all states. For this purpose we introduce the procedure UP, which reconstructs the state X_z from the state X_y, where $z = \text{parent}(y)$. We can thus save space as we only need to save the current state.

We use the following data structure to represent the current state X_y: A *node dictionary* consists of two dictionaries denoted X^c and X^p. The dictionary X^c represents the node set corresponding to X_y, and the dictionary X^p represents the node set corresponding to the set $\{x \in X_z \mid x \notin X_y \text{ and } z \text{ is an ancestor of } y\}$. That is, X^c represents the nodes in the current state, and X^p represents the nodes that is in a state X_z, where z is an ancestor of y in T, but not in X_y. We will use X^p to reconstruct previous states. The dictionary X^c is indexed by Σ and X^p is indexed by $V(T)$. The subsets stored at each entry are represented by doubly-linked lists. Furthermore, each node in X^c maintains a pointer to its parent in X^p and each node x' in X^p stores a linked list of pointers to its children in X^p. With this representation the total size of the node dictionary is $O(n_P + n_T)$.

Next we show how to solve the tree path subsequence problem in our framework using the node dictionary representation. For simplicity, we add a node $\top$ to P as a the parent of $\text{root}(P)$. Initially, the X^p represents $\top$ and X^c represents $\text{root}(P)$. The DOWN and UP procedures are implemented as follows:

DOWN$((X^p, X^c), y)$:   1.   Set $X := X^c[\text{label}(y)]$ and $X^c[\text{label}(y)] := \emptyset$.
                                   2.   For each $x \in X$ do:
                                        (a) Set $X^p[y] := X^p[y] \cup \{x\}$.
                                        (b) For each $x' \in \text{child}(x)$ do:
                                              i.   Set $X^c[\text{label}(x')] := X^c[\text{label}(x')] \cup \{x\}$.
                                                ii. Create pointers between x' and x.
                                   3.   Return (X^p, X^c).

UP$((X^p, X^c), y)$:   1.   Set $X := X^p[y]$ and $X^p[y] := \emptyset$.
                  2.   For each $x \in X$ do:
                        (a)   Set $X^c[\mathrm{label}(x)] := X^c[\mathrm{label}(x)] \cup \{x\}$.
                        (b)   For each $x' \in \mathrm{child}(x)$ do:
                              i.   Remove pointers between x' and x.
                              ii.   Set $X^c[\mathrm{label}(x')] := X^c[\mathrm{label}(x')] \setminus \{x'\}$.
                  3.   Return (X^p, X^c).

The next lemma shows that UP correctly reconstructs the former state.

Lemma 4. *Let $X_z = (X^c, X^p)$ be a state computed at a node $z \in V(T)$, and let y be a child of z. Then, $X_z = \mathrm{UP}(\mathrm{DOWN}(X_z, y), y)$.*

Proof. Let $(X_1^c, X_1^p) = \mathrm{DOWN}(X_z, y)$ and $(X_2^c, X_2^p) = \mathrm{UP}((X_1^c, X_1^p), y)$. We will first show that $x \in X_z \Rightarrow x \in \mathrm{UP}(\mathrm{DOWN}(X_z, y), y)$.

Let x be a node in X^c. There are two cases. If $x \in X^c[\mathrm{label}(y)]$, then it follows from the implementation of DOWN that $x \in X_1^p[y]$. By the implementation of UP, $x \in X_1^p[y]$ implies $x \in X_2^c$. If $x \notin X^c[\mathrm{label}(y)]$ then $x \in X_1^c$. We need to show $\mathrm{parent}(x) \notin X_1^p[y]$. This will imply $x \in X_2^c$, since the only nodes removed from X_1^c when computing X_2^c are the nodes with a parent in $X_1^p[y]$. Since y is unique it follows from the implementation of DOWN that $\mathrm{parent}(x) \in X_1^p$ implies $x \in X^c[\mathrm{label}(y)]$.

Let x be a node in X^p. Since y is unique we have $x \in X^p[y']$ for some $y' \neq y$. It follows immediately from the implementation of UP and DOWN that $X^p[y'] = X_1^p[y'] = X_2^p[y']$, when $y' \neq y$, and thus $X^p = X_2^p$.

We will now show $x \in \mathrm{UP}(\mathrm{DOWN}(X_z, y), y) \Rightarrow x \in X_z$. Let x be a node in X_2^c. There are two cases. If $x \notin X_1^c$ then it follows from the implementation of UP that $x \in X_1^p[y]$. By the implementation of DOWN, $x \in X_1^p[y]$ implies $x \in X^c[\mathrm{label}(y)]$, i.e., $x \in X^c$. If $x \in X_1^c$ then by the implementation of UP, $x \in X_2^c$ implies $\mathrm{parent}(x) \notin x_1^p[y]$. It follows from the implementation of DOWN that $x \in X^c$. Finally, let x be a node in X_2^p. As argued above $X^p = X_2^p$, and thus $x \in X^p$.   □

From the current state $X_y = (X^c, X^p)$ the next state X_z is computed as follows:

$$X_z = \begin{cases} \mathrm{DOWN}(X_y, z) & \text{if } y = \mathrm{parent}(z), \\ \mathrm{UP}(X_y, y) & \text{if } z = \mathrm{parent}(y). \end{cases}$$

The correctness of the algorithm follows from Lemma 2 and Lemma 4. We will now analyze the running time of the algorithm. The procedures DOWN and UP uses time linear in the size of the current state and the state computed. By Lemma 3 the size of each state is $O(l_P)$. Each step in the depth-first traversal thus takes time $O(l_P)$, which gives a total running time of $O(l_P n_T)$. On the other hand consider a path t in T. We will argue that the computation of all the states along the path takes total time $O(n_P + n_t)$, where n_T is the number of nodes in t. To show this we need the following lemma.

Lemma 5. *Let t be a path in T. During the computation of the states along the path t, any node $x \in V(P)$ is inserted into X^c at most once.*

Proof. Since t is a path we only need to consider the DOWN computations. The only way a node $x \in V(P)$ can be inserted into X^c is if parent$(x) \in X^c$. It thus follows from Lemma 3 that x can be inserted into X^c at most once. ☐

It follows from Lemma 5 that the computations of the all states when T is a path takes time $O(n_P + n_T)$. Consider a path-decomposition of T. A path-decomposition of T is a decomposition of T into disjoint paths. We can make such a path-decomposition of the tree T consisting of l_T paths. Since the running time of UP and DOWN both are linear in the size of the current and computed state it follows from Lemma 4 that we only need to consider the total cost of the DOWN computations on the paths in the path-decompostion. Thus, the algorithm uses time at most $\sum_{t \in T} O(n_p + n_t) = O(n_P l_T + n_T)$.

Next we consider the space used by the algorithm. Lemma 3 implies that $|X^c| \le l_P$. Now consider the size of X^p. A node is inserted into X^p when it is removed from X^c. It is removed again when inserted into X^c again. Thus Lemma 5 implies $|X^p| \le n_P$ at any time. The total space usage is thus $O(n_P + n_T)$. To summarize we have shown,

Theorem 2. *For trees P and T the tree path subsequence problem can be solved in $O(\min(l_P n_T, n_P l_T + n_T))$ time and $O(n_P + n_T)$ space.*

4   A Worst-Case Efficient Algorithm

In this section we consider the worst-case complexity of TPS and present an algorithm using subquadratic running time and linear space. The new algorithm works within our framework but does not use the UP procedure or the node dictionaries from the previous section.

Recall that using a simple linked list to represent the states we immediately get an algorithm using $O(n_P n_T)$ time and space. We first show how to modify the traversal of T and discard states along the way such that at most $O(\log n_T)$ states are stored at any step in the traversal. This improves the space to $O(n_P \log n_T)$. Secondly, we decompose P into small subtrees, called *micro trees*, of size $O(\log n_T)$. Each micro tree can be represented in a single word of memory and this way we can represent a state using only $O(\frac{n_P}{\log n_T})$ space. In total the space used to represent the $O(\log n_T)$ states is $O(\frac{n_P}{\log n_T} \cdot \log n_T) = O(n_P)$. Finally, we show how to preprocess P in linear time and space such that computing the new state can be done in constant time per micro tree. Intuitively, this achieves the $O(\log n_T)$ speedup.

Heavy Path Traversal. In this section we present the modified traversal of T. We first partition T into disjoint paths as follows. For each node $y \in V(T)$ let size$(y) = |V(T(y))|$. We classify each node as either *heavy* or *light* as follows. The root is light. For each internal node y we pick a child z of y of maximum size among the children of y and classify z as heavy. The remaining children are light. An edge to a light child is a *light edge*, and an edge to a heavy child is a *heavy edge*. The heavy child of a node y is denoted heavy(y). Let lightdepth(y) denote the number of light edges on the path from y to root(T).

Lemma 6 (Harel and Tarjan [7]). *For any tree T and node $y \in V(T)$,* lightdepth(y) $\leq \log n_T + O(1)$.

Removing the light edges, T is partitioned into *heavy paths*. We traverse T according to the heavy paths using the following procedure. For node $y \in V(T)$ define:

VISIT(y):   1. If y is a leaf report all leaves in X_y and return.
              2. Else let $y_1, \ldots, y_k$ be the light children of y and let $z =$ heavy(y).
              3. For $i := 1$ to k do:
                 (a) Compute $X_{y_i} :=$ DOWN(X_y, y_i)
                 (b) Compute VISIT(y_i).
              4. Compute $X_z :=$ DOWN(X_y, z).
              5. Discard X_y and compute VISIT(z).

The procedure is called on the root node of T with the initial state $\{\text{root}(P)\}$. The traversal resembles a depth first traversal, however, at each step the light children are visited before the heavy child. We therefore call this a *heavy path traversal*. Furthermore, after the heavy child (and therefore all children) has been visited we discard X_y. At any step we have that before calling VISIT(y) the state X_y is availiable, and therefore the procedure is correct. We have the following property:

Lemma 7. *For any tree T the heavy path traversal stores at most $\log n_T + O(1)$ states.*

Proof. At any node $y \in V(T)$ we store at most one state for each of the light nodes on the path from y to root(T). Hence, by Lemma 6 the result follows.   □

Using the heavy-path traversal immediately gives an $O(n_P \log n_T)$ space and $O(n_P n_T)$ time algorithm. In the following section we improve the time and space by an additional $O(\log n_T)$ factor.

Micro Tree Decomposition. A *micro tree* is a connected subgraph of P. A set of micro trees MS is a *micro tree decomposition* iff $V(P) = \cup_{M \in MS} V(M)$ and for any $M, M' \in MS$, $(V(M) \backslash \{\text{root}(M)\}) \cap (V(M') \backslash \{\text{root}(M')\}) = \emptyset$. Hence, two micro trees in a decomposition share at most one node and this node must be the root in at least one of the micro trees. If root(M') $\in V(M)$ then M is the *parent* of M' and M' is the *child* of M. A micro tree with no children is a *leaf* and a micro tree with no parent is a *root*. Note that we may have several root micro trees since they can overlap at the node root(P). We decompose P according to the following classic result:

Lemma 8 (Gabow and Tarjan [5]). *For any tree P and parameter $s > 1$, it is possible to build a micro tree decomposition MS of P in linear time such that $|MS| = O(n_P/s)$ and $|V(M)| \leq s$ for any $M \in MS$.*

Implementing the Algorithm. First decompose P according to Lemma 8 for a parameter s to be chosen later. Hence, each micro tree has at most s nodes and $|MS| = O(n_P/s)$. We represent the state X compactly using a bit vector for each micro tree. Specifically, for any micro tree M we store a bit vector $X_M = [b_1, \ldots, b_s]$, such that $X_M[i] = 1$ iff the ith node in a preorder traversal of M is in X. If $|V(M)| < s$ we leave the remaining values undefined. Later we choose $s = \Theta(\log n_T)$ such that each bit vector can be represented in a single word and the space used by the array is $O(n_P/\log n_T)$.

Next we define a DOWN_M procedure on each micro tree $M \in MS$. Due to the overlap between micro trees the DOWN_M procedure takes a bit b which will be used to propagate information between micro trees. For each micro tree $M \in MS$, bit vector X_M, bit b, and $y \in V(T)$ define:

$\text{DOWN}_M(X_M, b, y)$: Compute the state $X'_M := \text{CHILD}(\{x \in X_M \mid \text{label}(x) = \text{label}(y)\}) \cup \{x \in X_M \mid \text{label}(x) \neq \text{label}(y)\}$. If $b = 0$, return X'_M, else return $X'_M \cup \{\text{root}(M)\}$.

Later we will show how to implement DOWN_M in constant time for $s = \Theta(\log n_T)$. First we show how to use DOWN_M to simulate DOWN on P. We define a recursive procedure DOWN which traverse the hiearchy of micro trees. For micro tree M, state X, bit b, and $y \in V(T)$ define:

$\text{DOWN}(X, M, b, y)$: Let $M_1, \ldots, M_k$ be the children of M.
        1. Compute $X_M := \text{DOWN}_M(X_M, b, y)$.
        2. For $i := 1$ to k do:
            (a) Compute $\text{DOWN}(X, M_i, b_i, y)$, where $b_i = 1$ iff $\text{root}(M_i) \in X_M$.

Intuitively, the DOWN procedure works in a top-down fashion using the b bit to propagate the new state of the root of micro tree. To solve the problem within our framework we initially construct the state representing $\{\text{root}(P)\}$. Then, at each step we call $\text{DOWN}(R_j, 0, y)$ on each root micro tree R_j.

Lemma 9. *The above algorithm correctly simulates the DOWN procedure on P.*

Proof. Let X be the state and let $X' := \text{DOWN}(X, y)$. For simplicity, assume that there is only one root micro tree R. Since the root micro trees can only overlap at $\text{root}(P)$ it is straightforward to generalize the result to any number of roots. We show that if X is represented by bit vectors at each micro tree then calling $\text{DOWN}(R, 0, y)$ correctly produces the new state X'.

If R is the only micro tree then only line 1 is executed. Since $b = 0$ this produces the correct state by definition of DOWN_M. Otherwise, consider a micro tree M with children $M_1, \ldots, M_k$ and assume that $b = 1$ iff $\text{root}(M) \in X'$. Line 1 computes and stores the new state returned by DOWN_M. If $b = 0$ the correctness follows immediately. If $b = 1$ observe that DOWN_M first computes the new state and then adds $\text{root}(M)$. Hence, in both cases the state of M is correctly computed. Line 2 recursively computes the new state of the children of M. $\qquad\square$

If each micro tree has size at most s and DOWN_M can be computed in constant time it follows that the above algorithm solves TPS in $O(n_P/s)$ time. In the following section we show how to do this for $s = \Theta(\log n_T)$, while maintaining linear space.

Representing Micro Trees. In this section we show how to preprocess all micro trees $M \in MS$ such that DOWN_M can be computed in constant time. This preprocessing may be viewed as a "four russian trick" [1]. To achieve this in linear space we need the following auxiliary procedures on micro trees. For each micro tree M, bit vector X_M, and $\alpha \in \Sigma$ define:

$\text{CHILD}_M(X_M)$: Return the bit vector of nodes in M that are children of nodes in X_M.

$\text{EQ}_M(\alpha)$:   Return the bit vector of nodes in M labeled α.

By definition it follows that:

$$\text{DOWN}_M(X_M, b, y) = \begin{cases} \text{CHILD}_M(X_M \cap \text{EQ}_M(\text{label}(y))) \cup \\ \quad (X_M \backslash (X_M \cap \text{EQ}_M(\text{label}(y)))) & \text{if } b = 0, \\ \text{CHILD}_M(X_M \cap \text{EQ}_M(\text{label}(y))) \cup \\ \quad (X_M \backslash (X_M \cap \text{EQ}_M(\text{label}(y)))) \cup \{\text{root}(M)\} & \text{if } b = 1. \end{cases}$$

Recall that the bit vectors are represented in a single word. Hence, given CHILD_M and EQ_M we can compute DOWN_M using standard bit-operations in constant time.

Next we show how to efficiently implement the operations. For each micro tree $M \in MS$ we store the value $\text{EQ}_M(\alpha)$ in a hash table indexed by α. Since the total number of different characters in any $M \in MS$ is at most s, the hash table EQ_M contains at most s entries. Hence, the total number of entries in all hash tables is $O(n_P)$. Using perfect hashing we can thus represent EQ_M for all micro trees, $M \in MS$, in $O(n_P/s \cdot s) = O(n_P)$ space and $O(1)$ worst-case lookup time. The preprocessing time is expected $O(n_P)$ w.h.p.. To get a worst-case bound we use the deterministic dictionary of Hagerup et. al. [6] with $O(n_P \log n_P)$ worst-case preprocessing time.

Next consider implementing CHILD_M. Since this procedure is independent of the labeling of M it suffices to precompute it for all *structurally* different rooted trees of size at most s. The total number of such trees is less than 2^{2s} and the number of different states in each tree is at most 2^s. Therefore CHILD_M has to be computed for a total of $2^{2s} \cdot 2^s = 2^{3s}$ different inputs. For any given tree and any given state, the value of CHILD_M can be computed and encoded in $O(s)$ time. In total we can precompute all values of CHILD_M in $O(s2^{3s})$ time. Choosing the largest s such that $3s + \log s \leq n_T$ (hence $s = \Theta(\log n_T)$) this uses $O(n_T)$ time and space. Each of the inputs to CHILD_M are encoded in a single word such that we can look them up in constant time.

Finally, note that we also need to report the leaves of a state efficiently since this is needed in line 1 in the VISIT-procedure. To do this compute the state L

corresponding to all leaves in P. Clearly, the leaves of a state X can be computed by performing a bitwise AND of each pair of bit vectors in L and X. Computing L uses $O(n_P)$ time and the bitwise AND operation uses $O(n_T/s)$ time.

Combining the results, we decompose P, for s as described above, and compute all values of EQ_M and CHILD_M.

Then, we solve TPS using the heavy-path traversal. Since $s = \Theta(\log n_T)$ and from Lemmas 7 and 8 we have the following theorem:

Theorem 3. *For trees P and T the tree path subsequence problem can be solved in $O(\frac{n_P n_T}{\log n_T} + n_P \log n_P)$ time and $O(n_P + n_T)$ space.*

Combining the results of Theorems 2 and 3 this proves Theorem 1.

References

1. V. L. Arlazarov, E. A. Dinic, M. A. Kronrod, and I. A. Faradzev. On economic construction of the transitive closure of a directed graph (in russian). english translation in soviet math. dokl. 11, 1209-1210, 1975. *Dokl. Acad. Nauk.*, 194:487–488, 1970.
2. R. A. Baeza-Yates. Searching subsequences. *Theor. Comput. Sci.*, 78(2):363–376, 1991.
3. W. Chen. Multi-subsequence searching. *Inf. Process. Lett.*, 74(5-6):229–233, 2000.
4. J. Clark and S. DeRose. XML path language (XPath), avialiable as http://www.w3.org/TR/xpath, 1999.
5. H. N. Gabow and R. E. Tarjan. A linear-time algorithm for a special case of disjoint set union. In *Proc. of ACM Symp. on Theory of Computing*, pages 246–251, 1983.
6. T. Hagerup, P. B. Miltersen, and R. Pagh. Deterministic dictionaries. *J. Algorithms*, 41(1):69–85, 2001.
7. D. Harel and R. E. Tarjan. Fast algorithms for finding nearest common ancestors. *SIAM Journal of Computing*, 13(2):338–355, 1984.
8. P. Kilpeläinen and H. Mannila. Ordered and unordered tree inclusion. *SIAM Journal of Computing*, 24:340–356, 1995.
9. T. Schlieder and H. Meuss. Querying and ranking XML documents. *J. Am. Soc. Inf. Sci. Technol.*, 53(6):489–503, 2002.
10. T. Schlieder and F. Naumann. Approximate tree embedding for querying XML data. In *ACM SIGIR Workshop On XML and Information Retrieval*, 2000.
11. A. Termier, M. Rousset, and M. Sebag. Treefinder: a first step towards XML data mining. In *IEEE International Conference on Data Mining (ICDM)*, 2002.
12. H. Yang, L. Lee, and W. Hsu. Finding hot query patterns over an xquery stream. *The VLDB Journal*, 13(4):318–332, 2004.
13. L. H. Yang, M. L. Lee, and W. Hsu. Efficient mining of XML query patterns for caching. In *Proceedings of the 29th VLDB Conference*, pages 69–80, 2003.
14. P. Zezula, G. Amato, F. Debole, and F. Rabitti. Tree signatures for XML querying and navigation. In *LNCS 2824*, pages 149–163, 2003.

Distance Approximating Trees: Complexity and Algorithms

Feodor F. Dragan and Chenyu Yan

Department of Computer Science, Kent State University, Kent, OH 44242
dragan@cs.kent.edu, cyan@cs.kent.edu

Abstract. Let $\Delta \geq 1$ and $\delta \geq 0$ be real numbers. A tree $T = (V, E')$ is a *distance (Δ, δ)–approximating* tree of a graph $G = (V, E)$ if $d_H(u, v) \leq \Delta \, d_G(u, v) + \delta$ and $d_G(u, v) \leq \Delta \, d_H(u, v) + \delta$ hold for every $u, v \in V$. The *distance (Δ, δ)-approximating tree problem* asks for a given graph G to decide whether G has a distance (Δ, δ)-approximating tree. In this paper, we consider unweighted graphs and show that the distance $(\Delta, 0)$-approximating tree problem is NP-complete for any $\Delta \geq 5$ and the distance $(1, 1)$-approximating tree problem is polynomial time solvable.

1 Introduction

Many combinatorial and algorithmic problems are concerned with distances in a finite metric space induced by an undirected graph (possible weighted). An arbitrary metric space (in particular a finite metric defined by a general graph) might not have enough structure to exploit algorithmically. A powerful technique that has been successfully used recently in this context is to embed the given metric space in a simpler metric space such that the distances are approximately preserved in the embedding. New and improved algorithms have resulted from this idea for several important problems [1, 2, 7, 11, 12, 20]. Tree metrics are a very natural class of simple metric spaces since many algorithmic problems become tractable on them. If we approximate the graph by a tree such that the distance between a pair of vertices in the tree is at most some small factor of their distance in the graph, we can solve the problem on the tree and the solution interpret on the original graph.

Approximating general graph–distance d_G by a simpler distance (in particular, by tree–distance d_T) is useful also in such areas as communication networks, data analysis, motion planning, image processing, network design, and phylogenetic analysis. The goal is, for a given graph $G = (V, E)$, to find a sparse graph $H = (V, E')$ with the same vertex set, such that the distance $d_H(u, v)$ in H between two vertices $u, v \in V$ is reasonably close to the corresponding distance $d_G(u, v)$ in the original graph G. There are several ways to measure the quality of this approximation, two of them leading to the notion of a spanner. For $t \geq 1$ a spanning subgraph H of G is called a *multiplicative t–spanner* of G [9, 23, 24] if $d_H(u, v) \leq t \, d_G(u, v)$ for all $u, v \in V$. If $r \geq 0$ and $d_H(u, v) \leq d_G(u, v) + r$ for all $u, v \in V$, then H is called an *additive r–spanner* [19].

When H is a tree, one gets the notions of *multiplicative tree t–spanner* and *additive tree r–spanner*, respectively. Tree spanners of graphs were considered

---

T. Calamoneri, I. Finocchi, G.F. Italiano (Eds.): CIAC 2006, LNCS 3998, pp. 260–271, 2006.

in [6, 10, 25]. It was shown in [6] that for a given graph G and integer t, the problem to decide whether G has a multiplicative tree t–spanner is NP–complete for $t \geq 4$ and is linearly solvable for $t = 1, 2$. The status of the case $t = 3$ is open.

For many applications (e.g. in numerical taxonomy or in phylogeny reconstruction) the condition that H must be a spanning subgraph of G can be dropped (see [3, 26, 27]). In this case there is a striking way to measure how sharp d_H approximates d_G, based on the notion of a pseudoisometry between two metric spaces [20, 4]. Let $\Delta \geq 1$ and $\delta \geq 0$ be real numbers. Two graphs $G = (V, E)$ and $H = (V, E')$ are said to be (Δ, δ)–pseudoisometric [4] if for all $u, v \in V$, $d_H(u, v) \leq \Delta\, d_G(u, v) + \delta$ and $d_G(u, v) \leq \Delta\, d_H(u, v) + \delta$ hold. H is then said to be a distance (Δ, δ)–approximating graph for G (and vice-versa, G is a distance (Δ, δ)–approximating graph for H).

In this paper, continuing the line of research started in [4, 8], we will be interested in two special cases, when H is a tree and either $\Delta = 1$ or $\delta = 0$. A tree $T = (V, E')$ is a distance $(\Delta, 0)$–approximating tree of $G = (V, E)$ if $\frac{1}{\Delta} d_G(u, v) \leq d_T(u, v) \leq \Delta\, d_G(u, v)$ for all $u, v \in V$. A tree $T = (V, E')$ is a distance $(1, \delta)$–approximating tree of $G = (V, E)$ (or, simply, a distance δ-approximating tree of G) if $|d_G(u, v) - d_T(u, v)| \leq \delta$ for all $u, v \in V$. The distance (Δ, δ)–approximating tree problem asks for a given graph G to decide whether G has a distance (Δ, δ)-approximating tree.

In this paper, we consider unweighted graphs and show that the distance $(\Delta, 0)$-approximating tree problem is NP-complete for any $\Delta \geq 5$ and the distance $(1, 1)$-approximating tree problem is polynomial time solvable. The latter solves (algorithmically) the problem posed in [8] which asked to characterize/recognize the graphs admitting distance $(1, 1)$-approximating trees.

1.1 Previous Results and Their Implications

Let $G = (V, E)$ be a connected, undirected, loopless, and without multiple edges graph. The length of a path from a vertex u to a vertex v is the number of edges in this path. The distance $d_G(u, v)$ between the vertices u and v in G is the length of a shortest (u, v)-path.

A graph G is called chordal if no induced cycle of G has four or more edges. It is known that the class of chordal graphs does not admit any good tree spanners. Independently McKee [21] and Kratsch et al. [16] showed that, for every fixed integer t, there is a chordal graph without tree t–spanners (additive as well as multiplicative). Furthermore, recently Brandstädt et al. [5] have shown that, for any $t \geq 4$, the problem to decide whether a given chordal graph G admits a multiplicative tree t-spanner is NP-complete.

In contrast, in [4], Brandstädt et al. proved that every chordal graph G admits a tree $T(G)$ (constructable in linear time) which is both a $(3, 0)-$ and a $(1, 2)-$approximating tree of G. So, from the metric point of view chordal graphs do look like trees, but the notion of tree spanners failed to capture this. Note that the result is optimal in the sense that there are chordal graphs which do not admit any distance $(1, 1)-$approximating trees [8].

The result was used in [4, 8, 13] to provide efficient approximate solutions for several problems on chordal graphs. It is known that the (exact) distance matrix $D(G)$ of a chordal graph $G = (V, E)$ cannot be computed in less than "matrix-multiplication" time. Using a distance $(1, 2)$-approximating tree $T(G)$ of G, after a linear time preprocessing of G (and then of $T(G)$), in only $O(1)$ time, one can compute $d_G(x, y)$ with an error of at most 2 for any $x, y \in V$ (see [4] for further details). As another application, consider the p-center problem: given a graph G (or, more generally, a metric space) and an integer $p > 0$, we are searching for smallest radius r^* and a subset of vertices X of G with $|X| \leq p$ such that $d_G(v, X) \leq r^*$ for every vertex v of G. The problem is NP-hard even for chordal graphs. Solving the p-center problem on a distance $(1, 2)$-approximating tree $T(G)$ of G (on trees this problem is polynomial time solvable [15]), we will find an optimal covering radius r of $T(G)$ and a set of centers Y with $|Y| \leq p$. Then, Y can be taken as an approximate solution for G since $d_G(v, Y) \leq r + 2 \leq r^* + 4$ for all $v \in V$ (see [8] for further details). Clearly, similar results can be obtained for any graph admitting a good distance approximating tree.

The result was also used by Gupta in [13] for bandwidth approximation in chordal graphs. If a graph G has a distance (Δ, δ)–approximating tree $T(G)$ for some constants Δ and δ, then the bandwidth of a linear arrangements of G will be within some constant of the bandwidth of the same arrangement for $T(G)$. Gupta developed in [13] a simple randomized $O(log^{2.5} n)$-approximation algorithm for bandwidth minimization on trees and used it to get an approximation algorithm with a similar performance guarantee for chordal graphs (see [13] for further details). In [18], Krauthgamer et al. used the existence of good distance approximating trees for chordal graphs to obtain an embedding of any chordal graph into l_2 with a small r-dimensional volume distortion.

Later, in [8], Chepoi and Dragan extended the method of [4] from chordal graphs to all k-chordal graphs. A graph G is said to be k-chordal if no induced cycle of G has more than k edges. It was proven that, for every k-chordal graph $G = (V, E)$, there exists a tree $T = (V, F)$ (constructable in linear time) such that $|d_G(u, v) - d_T(u, v)| \leq \lfloor \frac{k}{2} \rfloor + \alpha$ for all vertices $u, v \in V$, where $\alpha = 1$ if $k \neq 4, 5$ and $\alpha = 2$ otherwise. Clearly, this result can be used to provide efficient approximate solutions for several problems on k-chordal graphs. Here, we will mention only one implication provided in [17]. Krauthgamer and Lee, in [17], proved first that the *Levin's conjecture on intrinsic dimensionality of graphs* holds for trees. Then, relying on low-distortion embeddings of k-chordal graphs into trees, due to [8], they extended that result to all k-chordal graphs: the *Levin's conjecture on intrinsic dimensionality of graphs* holds for all k-chordal graphs with bounded k (see [17] for further details).

Motivated by those applications of distance approximating trees, in this paper, we investigate the question how hard for a given graph G to find a good distance (Δ, δ)-approximating tree (for small Δ and δ). We prove that the distance $(\Delta, 0)$-approximating tree problem is NP-complete for any $\Delta \geq 5$ and the distance $(1, 1)$-approximating tree problem is polynomial time solvable. Due to space limitation, in this conference version, we present only the second result. The

NP-completeness proof will be given in the journal version. We reduce 3SAT to our problem. The reduction is too technical, involves complicated gadgets for the Boolean variables and hence omitted in this version.

1.2 Basic Notions, Notation and Facts

Let $G = (V, E)$ be a graph endowed with the shortest path metric $d_G(u, v)$. The *eccentricity* $ecc_G(v)$ of a vertex v is the maximum distance from v to any vertex in G. The *radius* $rad(G)$ of a graph G is the minimum eccentricity of a vertex in G and the *diameter* $diam(G)$ of G is the maximum eccentricity of a vertex.

For a subset $S \subseteq V$ of vertices of a graph G, by $G(S)$ we denote the subgraph of G induced by S. Let, for simplicity, $G - v := G(V \setminus \{v\})$ and $G - v - u := G(V \setminus \{v, u\})$, where v and u are vertices of G. Let also $G - uv$ denote the graph obtained from G by removing edge uv of G, i.e., $G - uv := (V, E \setminus \{uv\})$. A graph G is said to be *3-connected* if $G - u - v$ is connected for any pair of vertices $u, v \in V$. A graph G is said to be *2-connected* if $G - u$ is connected for any vertex $u \in V$. In a 2-connected graph G, if for some pair of vertices $x, y \in V$ the graph $G - x - y$ is disconnected, then we say that $\{x, y\}$ is a *2-cut* of G. In a connected graph G, if for some vertex $x \in V$ the graph $G - x$ is disconnected, then we say that x is a *1-cut vertex* (or, simply, *1-cut*) of G.

It is easy to see from the definitions of distance approximating trees that the following holds.

- A tree $T = (V, F)$ is a distance $(\Delta, 0)$-approximating tree of a graph $G = (V, E)$ if and only if $d_T(x, y) \leq \Delta$ holds for each edge $xy \in E$ and $d_G(u, v) \leq \Delta$ holds for each edge $uv \in F$.
- If T is a distance $(1, \delta)$-approximating tree for G, then T is a distance $(\delta + 1, 0)$-approximating tree for G.

2 Distance $(1, 1)$-Approximating Trees

In this section, we show that the distance $(1, 1)$-approximating tree problem is polynomial time solvable. For simplicity, in what follows, we will use the notion "distance 1-approximating tree" as a synonym to "distance $(1, 1)$-approximating tree".

2.1 3-Connected Graphs

A *star* is a tree with a vertex adjacent to all other vertices. We call that vertex *the center of the star*. Equivalently, a *star* is a tree of diameter at most 2.

Lemma 1. *For a 3-connected graph G, the following statements are equivalent.*

1. *G has a distance 1-approximating tree.*
2. *G has a distance 1-approximating tree which is a star.*
3. *$diam(G) \leq 3$ and $rad(G) \leq 2$.*

Proof. (1⟺2) Let T be a distance 1-approximating tree of G. If T is not a star, then there exists a path in T with length 3. Let (x', x, y, y') be such a path. Consider subtrees T_x and T_y obtained from T by removing edge xy, and assume that x belongs to T_x and y belongs to T_y. Since for any $u \in V(T_x) \setminus \{x\}$ and $v \in V(T_y) \setminus \{y\}$, $d_T(u, v) \geq 3$, we have $uv \notin E(G)$. This implies that $\{x, y\}$ is a 2-cut of G, contradicting with the 3-connectedness of G. Hence, T must be a star.

(2⟹3) Let T be a distance 1-approximating tree of G which is a star. Then, for any $x, y \in V$, we have $d_T(x, y) \leq 2$ and, therefore, $d_G(x, y) \leq 3$. Hence, $diam(G) \leq 3$. Let now u be the center of T. Then, for each $x \in V$, $d_T(x, u) \leq 1$, and therefore $d_G(x, u) \leq 2$. The latter implies $rad(G) \leq 2$.

(3⟹2) If $rad(G) \leq 2$, then, by definition, there exists a vertex $u \in V$ such that $d_G(x, u) \leq 2$, for any $x \in V$. Pick such a vertex u and construct a tree $T = (V, E')$ where each vertex $v \in V \setminus \{u\}$ is adjacent to u, i.e., construct a star on vertices V with the center u. Obviously, $0 \leq d_G(x, y) - d_T(x, y) \leq 1$, for any $x \in V \setminus \{u\}$. Moreover, since $diam(G) \leq 3$, we have $d_G(x, y) \leq 3$ for any $x, y \in V \setminus \{u\}$. As, for those vertices x and y, $d_T(x, y) = 2$, we conclude $d_G(x, y) - d_T(x, y) \leq 3 - 2 = 1$ and $d_G(x, y) - d_T(x, y) \geq 1 - 2 = -1$. Hence, T is a distance 1-approximating tree of G. □

Corollary 1. *Let G be an arbitrary (not necessarily 3-connected) graph. Then, G has a distance 1-approximating tree which is a star if and only if $diam(G) \leq 3$ and $rad(G) \leq 2$.*

2.2   2-Connected Graphs

A vertex of a tree is *inner* if it is not a leaf. An edge of a tree is an *inner edge* if it is not incident to a leaf.

Lemma 2. *If T is a distance 1-approximating tree of a connected graph G, then any inner edge of T is a 2-cut of G.*

Proof. For any inner edge xy of T, let T_x and T_y be the two subtrees of T obtained from T by removing edge xy. Let also x belong to T_x and y belong to T_y. Then, since T is a distance 1-approximating tree of G, for all $u \in V(T_x) \setminus \{x\}$ and $v \in V(T_y) \setminus \{y\}$, $uv \notin E(G)$. This implies that $\{x, y\}$ is a 2-cut of G separating $V(T_x) \setminus \{x\}$ from $V(T_y) \setminus \{y\}$. □

A *bistar* is a tree with only one inner edge. Equivalently, a *bistar* is a tree of diameter 3. The proof of the following lemma is omitted.

Lemma 3. *If T is a distance 1-approximating tree of a 2-connected graph G, then $diam(T) \leq 3$, i.e., T is a star or a bistar.*

To characterize 2-connected graphs admitting distance 1-approximating trees, we will need also the following easy observations (proofs are omitted).

Lemma 4. *Assume a graph G has a distance 1-approximating bistar T with the inner edge $c_1 c_2$. Then, the following properties hold:*

1. *$diam(G) \leq 4$ and $rad(G) \leq 3$;*
2. *for any $j = 1, 2$ and $x, y \in V(T_{c_j}) \cup \{c_1, c_2\}$, $d_G(x, y) \leq 3$ and $d_G(x, c_j) \leq 2$;*

3. *if $A_1, \ldots, A_k$ are the connected components of the graph $G - c_1 - c_2$ and T_{c_1}, T_{c_2} are the connected components of $T - c_1 c_2$, then, for any $i = 1, \ldots, k$, $V(A_i)$ is entirely contained either in $V(T_{c_1})$ or in $V(T_{c_2})$.*

Let now G be a graph with a 2-cut $\{a, b\}$ and $A_1, \ldots, A_k$ be the connected components of the graph $G - a - b$. For given 2-cut $\{a, b\}$ of G we can construct a new graph $H_{a,b}$ as follows. The vertex set of $H_{a,b}$ is $\{a, b, a_1, \ldots, a_k\}$. Edge aa_i ($i = 1, \ldots, k$) exists in $H_{a,b}$ if and only if for each $x, y \in V(A_i) \cup \{b\}$, $d_G(x, y) \leq 3$ and $d_G(x, a) \leq 2$ hold. Edge ba_i ($i = 1, \ldots, k$) exists in $H_{a,b}$ if and only if for each $x, y \in V(A_i) \cup \{a\}$, $d_G(x, y) \leq 3$ and $d_G(x, b) \leq 2$ hold. Edge $a_i a_j$ ($i, j = 1, \ldots, k$, $i \neq j$) exists in $H_{a,b}$ if and only if for each vertex $x \in V(A_i)$ and each vertex $y \in V(A_j)$, $d_G(x, y) \leq 3$ holds. No other edges exist in $H_{a,b}$.

The following lemma gives a characterization of those 2-connected graphs that admit distance 1-approximating trees. Denote the complement of a graph H by $\overline{H}$.

Lemma 5. *For a 2-connected graph G, the following statements are equivalent.*

1. *G has a distance 1-approximating tree.*
2. *G has a distance 1-approximating tree which is a star or a bistar.*
3. *$diam(G) \leq 3$ and $rad(G) \leq 2$ or $diam(G) \leq 4$ and there exists a 2-cut $\{a, b\}$ in G such that the graph $\overline{H_{a,b}}$ is bipartite.*

Proof. $(1 \Longleftrightarrow 2)$ is given by Lemma 3.

$(2 \Rightarrow 3)$ If G has a distance 1-approximating tree which is a star, then, by Corollary 1, $diam(G) \leq 3$ and $rad(G) \leq 2$. Assume now that a distance 1-approximating tree T of G is a bistar. Then, by Lemma 4, $diam(G) \leq 4$. Lemma 4 (together with Lemma 2) implies also that G has a 2-cut $\{a, b\}$ (which is the inner edge of T) such that for any connected component A_i ($i \in \{1, \ldots, k\}$) of $G - a - b$, either $V(A_i) \subset V(T_a)$ or $V(A_i) \subset V(T_b)$ holds. Since vertices $V(T_a) \cup \{b\}$ form a star in T with the center a, we have $d_G(x, y) \leq 3$ and $d_G(x, a) \leq 2$ for any $x, y \in V(T_a) \cup \{b\}$. By construction of $H_{a,b}$, vertices $\{a\} \cup \{a_i : V(A_i) \subset V(T_a)\}$ of $H_{a,b}$ will form a clique. Analogously, vertices $\{b\} \cup \{a_i : V(A_i) \subset V(T_b)\}$ form a clique in $H_{a,b}$. Since these two cliques cover all vertices of $H_{a,b}$, the complement $\overline{H_{a,b}}$ of $H_{a,b}$ is bipartite.

$(3 \Rightarrow 2)$ Clearly, if $diam(G) \leq 3$ and $rad(G) \leq 2$ then, by Corollary 1, G has a distance 1-approximating star. Assume now that $diam(G) \leq 4$ and there exists a 2-cut $\{a, b\}$ in G such that the graph $\overline{H_{a,b}}$ is bipartite. Let $A_1, \ldots, A_k$ be the connected components of the graph $G - a - b$. Vertices of $H_{a,b}$ can be partitioned into two cliques C_1 and C_2. Since a and b are not adjacent in $H_{a,b}$, they must be in different cliques. Assume, $a \in C_1$ and $b \in C_2$. By construction of $H_{a,b}$, for all $x, y \in \cup \{V(A_i) : a_i \in C_1\} \cup \{b\}$, $d_G(x, y) \leq 3$ and $d_G(x, a) \leq 2$ holds. Similarly, for all $x, y \in \cup \{V(A_i) : a_i \in C_2\} \cup \{a\}$, $d_G(x, y) \leq 3$ and $d_G(x, b) \leq 2$ holds. Hence, we can construct a bistar T of G as follows. Vertices a and b will form the inner edge of T. Vertices of A_i with $a_i \in C_1$ will be attached (i.e., made adjacent in T) to a. Vertices of A_i with $a_i \in C_2$ will be attached to b. It is easy to see that T is a distance 1-approximating tree of G. The only interesting case to mention here is when $x \in V(A_i)$, where $a_i \in C_1$, and $y \in V(A_j)$, where $a_j \in C_2$. For those x and y, we have $d_T(x, y) = 3$ and $2 \le d_G(x, y) \le 4$ (since $diam(G) \le 4$ and x and y are separated by $\{a, b\}$ in G). Thus, $-1 \le c_T(x, y) \le 1$ holds.   □

Corollary 2. *Let G be an arbitrary (not necessarily 2-connected) graph. Then, G has a distance 1-approximating tree which is a star or a bistar if and only if $diam(G) \le 3$ and $rad(G) \le 2$ or $diam(G) \le 4$ and there exists a 2-cut $\{a, b\}$ in G such that the graph $\overline{H_{a,b}}$ is bipartite.*

Lemma 5 implies also that the problem of checking whether a given 2-connected graph G has a distance 1-approximating tree is polynomial time solvable. More specifically, we have

Corollary 3. *It is possible, for a given 2-connected graph $G = (V, E)$, to check in $O(|V|^4)$ time whether G has a distance 1-approximating tree and, if such a tree exists, construct one within the same time bound.*

Proof. We can find in $O(|V||E|)$ time the distance matrix of G and all 2-cuts [14, 22] of G. Then, to check whether $diam(G) \le 3$ and $rad(G) \le 2$ and, if so, to construct a distance 1-approximating star of G as described in the proof of Lemma 1, one needs at most $O(|V|^2)$ time in total. To check if $diam(G) \le 4$ and whether there exists a 2-cut $\{a, b\}$ of G with $\overline{H_{a,b}}$ bipartite, one needs $O(|V|^4)$ total time. We just need, for each 2-cut $\{a, b\}$, to construct the graph $\overline{H_{a,b}}$ and check if it is bipartite. Construction of $\overline{H_{a,b}}$ for a given 2-cut $\{a, b\}$ and checking whether it is bipartite will take no more than $O(|V|^2)$ time (given the distance matrix of G). Since any graph G has at most $O(|V|^2)$ 2-cuts, to check if G has a distance 1-approximating bistar, one needs at most $O(|V|^4)$ time. If G admits such a bistar, then we can find one in linear time as described in the proof of Lemma 5.   □

2.3   Connected Graphs

In this subsection, we assume that G is a connected graph but not 2-connected. Therefore, there exists a vertex $v \in V(G)$, such that $G - v$ contains at least two connected components.

From Lemma 3 and its proof, the following lemma is obvious.

Lemma 6. *Let T be a distance 1-approximating tree of a connected graph G and (a, b, c) be a path in T. If both a and c are inner vertices of T, then at least one of these vertices is a 1-cut of G. Moreover, assuming c is a 1-cut, c separates vertices $V(T_c) \setminus \{c\}$ from other vertices of G, where T_c is the subtree of $T - bc$ containing c.*

A *2-connected component* of a graph G is a maximal by inclusion 2-connected subgraph of G or an edge uv of G such that both u and v are 1-cuts of G (such an edge is called a *bridge* of G). Two 2-connected components of G are neighbors if they share a common vertex (a 1-cut) of G.

Lemma 7. *Let G be a connected graph admitting a distance 1-approximating tree T and A be a 2-connected component of G. Then, for any two vertices $x, y \in V(A)$, $d_T(x, y) \le 3$. Moreover, if there exist vertices $x, y \in V(A)$, such that $d_T(x, y) = 3$, then $T(V(A))$ is a bistar.*

Proof. Assume that, for some vertices $x, y \in V(A)$, $d_T(x, y) \geq 4$ holds. Then, one can connect x and y in T with a path $P_T(x, y)$ of length at least 4. Pick three consecutive inner vertices a, b, c of path $P_T(x, y)$, they necessarily exist. According to Lemma 6, a or c is a 1-cut of G separating x from y in G. The latter is in contradiction with the assumption that $x, y \in V(A)$ and A is a 2-connected component of G. Hence, $d_T(x, y) \leq 3$, for any $x, y \in V(A)$, is proven.

Assume now that there exist vertices $x, y \in V(A)$, such that $d_T(x, y) = 3$. Then, one can find two vertices $\{c_1, c_2\}$ in G such that $T(V(A) \cup \{c_1, c_2\})$ is a bistar with the inner edge $c_1 c_2$. Let $x c_1, y c_2 \in E(T)$. We will show that both c_1 and c_2 are in A.

Suppose, neither c_1 nor c_2 is in A. Assume $c_1 \in V(B)$, $c_2 \in V(C)$, where B and C are 2-connected components of G. Let $V(B) \cap V(A) = \{v\}$ and $V(C) \cap V(A) = \{u\}$. We claim that $B = C$ or at least $v = u$. Suppose $B \neq C$ and $v \neq u$. Then, since $V(B) \cap V(C) = \emptyset$ (otherwise, A, B and C will be parts of one 2-connected component of G), $d_G(c_1, c_2) \geq 3$. As $d_T(c_1, c_2) = 1$, a contradiction with T being a distance 1-approximating tree of G arises. So, c_1, c_2 must be either in one 2-connected component of G or in two 2-connected components B and C such that $V(B) \cap V(A) = V(C) \cap V(A)$.

Without loss of generality, assume v is attached (i.e., adjacent in T) to c_1. Since $d_T(y, c_2) = 1$, we have $d_G(y, c_2) \leq 2$ and, hence, $y v \in E(G)$. On the other hand, $d_T(y, v) = 3$, contradicting the assumption that T is a distance 1-approximating tree of G.

Assume now that $c_1 \in V(A)$ and $c_2 \in V(B) \setminus \{v\}$. For any vertex $x' \in V(A)$ which is attached to c_1 and any vertex $y' \in V(A) \setminus \{c_1\}$ which is attached to c_2, $x' y' \notin E(G)$ must hold. Moreover, since $V(A) \cap V(B) = \{v\}$, one concludes that for all $x' \in V(A) \setminus \{v\}$, $x' c_2 \notin E(G)$. Hence, any path of A connecting a vertex attached to c_1 with a vertex attached to c_2 must use vertex c_1. Since there exist vertices $x, y \in V(A)$ such that $x c_1, y c_2 \in E(T)$, this is in contradiction with the assumption that A is 2-connected.

Thus, we conclude that $T(V(A))$ is a bistar. □

Corollary 4. *Let G be a connected graph admitting a distance 1-approximating tree T and A be a 2-connected component of G. Then, either $T(V(A))$ is a bistar or $T(V(A) \cup \{c\})$ is a star centered at some vertex c of G.*

In what follows, we will show that among all possible distance 1-approximating trees of G there is a tree T such that, for any 2-connected component A of G, $T(V(A))$ is connected, i.e., if $T(V(A) \cup \{c\})$ is a star for some vertex c of G, then c must be in A. To show that, we will need two lemmata (proofs can be found in the journal version).

A sequence $(B_0 := B, B_1, \ldots, B_{k-1}, B_k := A)$ is called *the chain of 2-connected components of G between A and B* if each B_i is a 2-connected component of G, B_i and B_j are different for $j \neq i$, B_{i-1}, B_i are neighbors sharing a 1-cut $v_i := V(B_{i-1}) \cap V(B_i)$ of G for any $i \in \{1, \ldots, k\}$, and $v_i \neq v_j$ for any $i \neq j$. Clearly, this chain is unique for any A and B.

Lemma 8. *Let G be a connected graph admitting a distance 1-approximating tree T, A and B be 2-connected components of G and $(B_0 := B, B_1, \ldots, B_{k-1}, B_k := A, Z)$ be the chain of 2-connected components of G between Z and B. If $T(V(A) \cup \{c\})$ is a star with the center c belonging to $V(Z) \setminus V(A)$, then for any $i \in \{0, \ldots, k-1\}$, $T(V(B_i))$ is a star centered at a 1-cut $v_{i+1} := V(B_{i+1}) \cap V(B_i)$ of G. Moreover, for any $i \in \{0, \ldots, k-1\}$ and any $x \in V(B_i)$, $xv_{i+1} \in E(G)$ must hold.*

Lemma 9. *Let G be a connected graph admitting a distance 1-approximating tree T and let A, Z be 2-connected components of G such that $V(A) \cap V(Z) = \{v\}$. Let also A' be that connected component of the graph $G - v$ which contains $A - v$. If $T(V(A) \cup \{c\})$ is a star centered at $c \in V(Z) \setminus \{v\}$, then for any vertices $x \in V(A')$, $y \in (V(G) \setminus V(A')) \setminus \{c, v\}$, $xy \notin E(T)$ holds. In particular, for any two vertices $y, z \in V(G) \setminus V(A')$, the path $P_T(x, y)$ between x and y in T does not contain any vertices of A'.*

In what follows, let G be a connected graph admitting a distance 1-approximating tree and let T denote a distance 1-approximating tree of G with minimum $|E(T) \setminus E(G)|$, i.e., with minimum number of non-graph edges. We will show that this tree T has a number of nice properties.

Theorem 1. *If T is a distance 1-approximating tree of G with minimum $|E(T) \setminus E(G)|$, then for any 2-connected component A of G, $T(V(A))$ is a star or a bistar.*

Proof. Since A is a 2-connected component of G, by Corollary 4, either $T(V(A))$ is a bistar or $T(V(A) \cup \{c\})$ is a star centered at some vertex c of G. By way of contradiction, assume that for A, $T(V(A) \cup \{c\})$ is a star centered at a vertex c of G not belonging to A. Let c belong to some 2-connected component Z of G. Necessarily, A and Z are neighbor (2-connected) components. Let $v := V(A) \cap V(Z)$ and A' be a connected component of $G - v$ containing $V(A) \setminus \{v\}$. By Lemma 8, for any 2-connected component B of G, which is different from A and belongs to A', $T(V(B))$ is a star centered at a 1-cut of G lying in B and closest to A. Moreover, if v' is that 1-cut, then for any $x \in V(B)$, $xv' \in E(G)$ holds (see Fig. 1). We have also that v is adjacent in G to c and to any vertex a $(a \neq v)$ of A (see Lemma 8).

We can transform tree T into a new tree T' as follows. Set $E(T') := E(T)$ and $V(T') := V(T)$. For each vertex $a \in V(A) \setminus \{v\}$, let $E(T') := (E(T') \setminus \{ac\}) \cup \{av\}$ (i.e., replace edge ac with edge av). We claim that T' is a distance 1-approximating tree of G, too. We need to show that $|d_{T'}(x, y) - d_G(x, y)| \leq 1$ holds for any two vertices $x, y \in V(G)$.

If $x, y \in V(A')$ then, by Lemma 8 and the way we transformed T into T', $d_{T'}(x, y) = d_T(x, y)$. If $x, y \in V(G) \setminus V(A')$ then, by Lemma 9 and the way T was transformed into T', $d_{T'}(x, y) = d_T(x, y)$. Hence, in these cases, $|d_{T'}(x, y) - d_G(x, y)| = |d_T(x, y) - d_G(x, y)| \leq 1$.

Consider now the case when $x \in V(A')$ and $y \in V(G) \setminus V(A')$. By Lemma 8, $d_{T'}(x, v) = d_G(x, v)$. Since v is a 1-cut of G, $d_G(x, y) = d_G(x, v) + d_G(v, y)$. By Lemma 9 and the way we transformed T into T', one concludes that $d_{T'}(x, y) =$

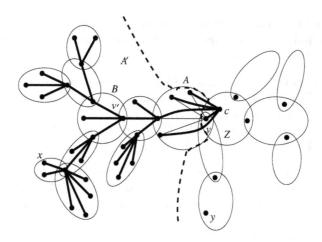

Fig. 1. Illustration to the proof of Theorem 1. A part of the tree T is shown using thick edges. Thin edges show some graph edges.

$d_{T'}(x, v) + d_{T'}(v, y)$. Combining these equalities, we get $|d_{T'}(x, y) - d_G(x, y)| = |d_{T'}(x, v) + d_{T'}(v, y) - (d_G(x, v) + d_G(v, y))| = |d_{T'}(v, y) - d_G(v, y)|$. But, by Lemma 9, $d_{T'}(v, y) = d_T(v, y)$. Hence, we get $|d_{T'}(x, y) - d_G(x, y)| = |d_T(v, y) - d_G(v, y)| \leq 1$.

Thus, T' is a distance 1-approximating tree of G. Since T' has original graph edges more than T has ($|E(T') \setminus E(G)| < |E(T) \setminus E(G)|$), a contradiction with the choice of T arises. Hence, the center c of star $T(V(A) \cup \{c\})$ must be in A. $\square$

Lemma 10. *Let T be a distance 1-approximating tree of G with minimum $|E(T) \setminus E(G)|$ and A be a 2-connected component of G such that $T(V(A))$ is a bistar. Then, for any other 2-connected component B of G, $T(V(B))$ is a star centered at a 1-cut of G which is closest to A (among all 1-cuts of G located in B).*

Corollary 5. *If T is a distance 1-approximating tree of G with minimum $|E(T) \setminus E(G)|$, then there is at most one 2-connected component A in G such that $T(V(A))$ is a bistar.*

The following lemma and its corollaries show that a distance 1-approximating tree T of G with $T(V(A))$ being a star for any 2-connected component A of G has also a very deterministic structure.

Lemma 11. *Let T be a distance 1-approximating tree of G with minimum $|E(T) \setminus E(G)|$ and A and B be two neighbor 2-connected components of G with $v := V(A) \cap V(B)$. If $T(V(A))$ is a star centered not at v, then $T(V(B))$ is a star centered at v.*

Proof. Since $T(V(A))$ is a star centered at some vertex $c \in V(A) \setminus \{v\}$, there must exist a vertex a in A such that $av \in E(G) \setminus E(T)$. By Lemma 10, $T(V(B))$ cannot be a bistar. If $T(V(B))$ is a star centered at some vertex $c' \in V(B) \setminus \{v\}$, then there must exist a vertex b in B such that $bv \in E(G) \setminus E(T)$. For these vertices a and b, $d_G(a,b) = 2$ and $d_T(a,b) = d_T(a,v) + d_T(v,b) = 2 + 2 = 4$ hold, contradicting with T being a distance 1-approximating tree of G. Hence, the center of $T(V(B))$ must be v. □

Corollary 6. *Let T be a distance 1-approximating tree of G with minimum $|E(T) \setminus E(G)|$ and A be a 2-connected component of G such that $T(V(A))$ is a star. If the center of this star $T(V(A))$ is not a 1-cut of G, then for any other 2-connected component B of G, $T(V(B))$ is a star centered at a 1-cut of G which is closest to A (among all 1-cuts of G located in B).*

Corollary 7. *Let T be a distance 1-approximating tree of G with minimum $|E(T) \setminus E(G)|$. If for every 2-connected component A of G, $T(V(A))$ is a star centered at a 1-cut of G, then there exists a 1-cut v in G such that*

a) *for any 2-connected component A of G containing v, $T(V(A))$ is a star centered at v,*
b) *for any 2-connected component B of G not containing v, $T(V(B))$ is a star centered at a 1-cut of G which is closest to v (among all 1-cuts of G located in B).*

Clearly, if $T(V(A))$ is a star for a 2-connected component A of G, then $diam(A) \leq 3$ and $rad(A) \leq 2$. And, if $T(V(B))$ is a bistar for a 2-connected component B of G, then $diam(B) \leq 4$ and $rad(B) \leq 3$.

Using all these auxiliary results, one can prove the following theorem (its proof is omitted in this conference version).

Theorem 2. *It is possible, for a given connected graph $G = (V, E)$, to check in $O(|V|^4)$ time whether G has a distance 1-approximating tree and, if such a tree exists, construct one within the same time bound.*

3   Conclusion

In this paper, we proved that the distance $(\Delta, 0)$-approximating tree problem is NP-complete for any $\Delta \geq 5$ and the distance $(1, 1)$-approximating tree problem is polynomial time solvable.

It remains an interesting open question to characterize/recognize the graphs admitting distance (Δ, δ)–approximating trees for $\Delta = 2, 3, 4$ and $\delta = 2, 3, 4$, or to prove that the problem remains NP-hard even for some of these small Δs and δs.

References

1. Y. Bartal, Probabalistic approximation of metric spaces and its algorithmic applications, *FOCS* 1996, pp. 184-193.
2. Y. Bartal, A. Blum, C. Burch, and A. Tomkins, A polylog(n)competitive algorithm for metrical task systems, *STOC* 1997, pp 711–719.
3. J.-P. Barthélemy and A. Guénoche, Trees and Proximity Representations, *Wiley*, New York, 1991.

4. A. Brandstädt, V. Chepoi, and F.F. Dragan, Distance Approximating Trees for Chordal and Dually Chordal Graphs, *Journal of Algorithms* 30 (1999), 166–184.
5. A. Brandstädt, F. Dragan, H.-O. Le, and V.B. Le, Tree Spanners on Chordal Graphs: Complexity and Algorithms, *Theor. Comput. Science* 310 (2004), 329-354.
6. L. Cai and D.G. Corneil, Tree spanners, *SIAM J. Disc. Math.* 8 (1995), 359–387.
7. M. Charikar, C. Chekuri, A. Goel, S. Guha, and S. Plotkin, Approximating a Finite Metric by a Small Number of Tree Metrics, *FOCS* 1998, pp. 379–388.
8. V. Chepoi and F.F. Dragan, A note on distance approximating trees in graphs, *European Journal of Combinatorics* 21 (2000), 761–766.
9. L.P. Chew, There are planar graphs almost as good as the complete graph, *J. of Computer and System Sciences,* 39 (1989), 205–219.
10. Y. Emek and D. Peleg, Approximating Minimum Max-Stretch Spanning Trees on Unweighted Graphs, *SODA* 2004, pp. 261-270.
11. J. Fakcharoenphol, S. Rao, and K. Talwar, A tight bound on approximating arbitrary metrics by tree metrics, *STOC* 2003, pp. 448-455.
12. U. Feige, Approximating the Bandwidth via Volume Respecting Embeddings, *J. Comput. System Sci.* 60 (2000), 510–539.
13. A. Gupta, Improved bandwidth approximation for trees and chordal graphs, *Journal of Algorithms* 40 (2001), 24–36.
14. J.E. Hopcroft and R.E. Tarjan, Dividing a graph into triconnected components, *SIAM J. Comput.* 2 (1973), 135–158.
15. O. Kariv and S.L. Hakimi, An algorithmic approach to network location problems, I: the p-centers, *SIAM J. Appl. Math.* 37 (1979), 513–538.
16. D. Kratsch, H.-O. Le, H. Müller, E. Prisner, and D. Wagner, Additive tree spanners, *SIAM J. Discrete Math.* 17 (2003), 332-340.
17. R. Krauthgamer and J.R. Lee, The intrinsic dimensionality of graphs, *STOC* 2003, pp. 438–447.
18. R. Krauthgamer, N. Linial, and A. Magen, Metric Embedding – Beyond one-dimensional distortion, *Discrete and Computational Geometry* 31 (2004), 339–356.
19. A.L. Liestman and T. Shermer, Additive graph spanners, *Networks,* 23 (1993), 343-364.
20. N. Linial, E. London, and Y. Rabinovich, The geometry of graphs and some its algorithmic applications, *Combinatorica* 15 (1995), 215–245.
21. T.A. McKee, personal communication to E. Prisner, 1995.
22. G.L. Miller and V. Ramachandran, A new graph triconnectivity algorithm and its parallelization, *Combinatorica* 12 (1992), 53–76.
23. D. Peleg and A.A. Schäffer, Graph Spanners, *J. Graph Theory,* 13(1989), 99-116.
24. D. Peleg and J.D. Ullman, An optimal synchronizer for the hypercube, *PODC* 1987, 77–85.
25. E. Prisner, Distance approximating spanning trees, *STACS'97,* LNCS 1200, 1997, pp. 499–510.
26. P.H.A. Sneath and R.R. Sokal, Numerical Taxonomy, *W.H. Freeman,* San Francisco, California, 1973.
27. D.L. Swofford and G.J. Olsen, Phylogeny reconstruction, In *Molecular Systematics (D.M. Hillis and C. Moritz, editors),* Sinauer Associates Inc., Sunderland, MA., 1990, 411–501.

How to Pack Directed Acyclic Graphs into Small Blocks[*]

Yuichi Asahiro[1], Tetsuya Furukawa[2], Keiichi Ikegami[3], and Eiji Miyano[4]

[1] Department of Social Information Systems, Kyushu Sangyo University,
Fukuoka 813-8503, Japan
asahiro@is.kyusan-u.ac.jp
[2] Department of Economic Engineering, Kyushu University,
Fukuoka 812-8581, Japan
furukawa@en.kyushu-u.ac.jp
[3] Department of Systems Innovation and Informatics,
Kyushu Institute of Technology, Fukuoka 820-8502, Japan
{ikegami@theory, miyano@}ces.kyutech.ac.jp

Abstract. The paper studies the following variant of clustering or laying out problems of graphs: Given a directed acyclic graph (DAG for short), the objective is to find a mapping of its nodes into blocks of size at most B that minimizes the maximum number of external arcs during traversals of the acyclic structure by following paths from the roots to the leaves. An external arc is defined as an arc connecting two distinct blocks. The problem can be shown to be NP-hard generally, and to remain intractable even if $B = 2$ and the height of DAGs is three. In this paper we provide a $\frac{3}{2}$ factor linear time approximation algorithm for $B = 2$, and prove that the $\frac{3}{2}$ ratio is optimal in terms of approximation guarantee. In the case of $B \geq 3$, we also show that there is no $\frac{3}{2} - \varepsilon$ factor approximation algorithm assuming P $\neq$ NP, where ε is arbitrarily small positive. Furthermore, we give a 2 factor approximation algorithm for $B = 3$ if the input is restricted to a set of layered graphs.

1 Introduction

In recent years, massive data sets with gigabytes or terabytes have emerged in a growing number of wide-range applications. Since the amount of such massive data sets is often too large to store in a fast main memory and they necessarily reside on a slower disk, the input/output (I/O) communication between the main memory and the disk can be a major performance bottleneck. Thus one of the crucial issues arising when computing with massive data sets is to develop external memory data structures and I/O-efficient algorithms, and recently, this area has received much attention, e.g., [1, 13].

External Memory Models and Previous Results. In this paper we consider the *two-level I/O model* introduced in [2], in which the memory hierarchy

---

[*] Supported in part by the Grant-in-Aid for Scientific Research on Priority Areas 16092223, for Scientific Research (C) 15500072, and for Young Scientists 15700021 and 17700022 from the Japanese Ministry of Education, Science, Sports and Culture.

T. Calamoneri, I. Finocchi, G.F. Italiano (Eds.): CIAC 2006, LNCS 3998, pp. 272–283, 2006.

consists of an internal memory of limited space, and an arbitrarily large external memory divided into fixed contiguous *blocks* of size B. We assume that an external memory query or modification, called a *block transfer*, transfers one block of B objects from the external memory to the internal one. While internal-memory access structures are typically constructed as a set of dynamically allocated objects linked by pointers and so-called *tall-and-skinny* structures are acceptable, external-memory data structures must be *short-and-fat* in order to decrease the number of the block transfers. If we want to use a search structure designed for the internal memory, for example, a binary tree, on the external one, then the following laying out problem naturally arises: Which nodes of the tree are mapped in which disk blocks such that only a few disk blocks are transfered to the internal memory although we may access a lot of nodes?

The above question was first proposed by Gil and Itai in [9], in which the inputs are limited to trees, and the goal is to minimize the expected number of block transfers over all queries. They presented an algorithm based on a dynamic programming method that can optimize the partition of n nodes into blocks of size B in $O(nB^2 \log \Delta)$ time, where Δ is the maximum degree of the nodes, and uses $O(B \log n)$ space. Subsequently, Alstrup, et al. decreased the time bound to $O(nB^2)$ by a tight analysis, and also presented a faster but approximate algorithm in the same setting in [3]. Also, Gil and Itai [9] proved that the problem of finding an optimal *compact* packing of trees is NP-hard, where the compact means that the total number of blocks must be minimized. Diwan, et al. [7] and independently Clark and Munro [6] considered the laying out problem of trees for the worst-case I/O communication, i.e., the goal is to minimize the maximum number of block transfers, and showed that the problem can be solved in polynomial time, and the problem subject to the compact constraint is NP-hard. In [3] Alstrup, et al. proposed an efficient cache oblivious layout of trees.

Our Problems and Contributions. In this paper we assume that the input is a *directed acyclic graph* (DAG). The DAG is a large class of the important data structures including persistent (or multiversion) B-trees and ordered binary-decision diagrams (OBDDs), and hence it is an ubiquitous data representation [4, 5, 10, 11]. More strictly, our problem of this paper, called the *Minimum Block Transfer Problem* (MBT for short), is formulated as follows: Given a DAG $G = (V, A)$ and a block size B, our goal is to find a mapping of its nodes into blocks of size at most B that *minimizes the maximum number of external arcs* when we traverse the acyclic structure by following paths from the roots to the leaves. The number of external arcs is defined as the number of arcs connecting two distinct blocks, that is, it denotes the number of the block transfers.

This problem was previously considered by Diwan, et al. [7], who pointed out the NP-hardness of the general MBT. However, as far as the authors know, its proof has not been published yet. In [7], they also provided a naive bottom-up-packing algorithm, which can optimize the partition of $|V|$ nodes into blocks and run in $O(|V|)$ time only if the input is restricted to trees as mentioned above. Additionally, for DAG layouts, a heuristic algorithm based on the same bottom-up idea was presented in the same paper, but its theoretical approximation guarantee was not shown.

In this paper we show the following results:

- We explicitly present the proof of the NP-hardness of the MBT. Furthermore, we prove that unfortunately it remains NP-hard even if each block size B is equal to two and the height of DAGs is three.
- Our proof method of the NP-hardness would be worthwhile in itself if we are looking for a stricter computational hierarchy; by the *gap-introducing reduction* [12] we show that assuming P $\neq$ NP there is no $\frac{3}{2} - \varepsilon$ factor approximation algorithm for any $B \geq 2$ and any positive ε.
- On the other hand, fortunately, we can obtain an optimal algorithm which runs in linear time if the height of DAGs is bounded above by two.
- A consequence of this algorithm is an approximation algorithm with approximation ratio $\frac{3}{2}$ for a DAG of arbitrary height if $B = 2$. Namely, our approximation algorithm is *optimal in terms of approximation guarantee*.
- In addition, as the first main step in obtaining a good approximation algorithm for the general problem, we provide a 2 factor approximation algorithm for $B = 3$ if the input is restricted to a set of layered DAGs.

2 Preliminaries

Let $G = (V, A)$ be a simple directed acyclic graph, i.e., G does not include any cycle, any multiple arcs, or any self-loop. V and A denote the sets of nodes and arcs, respectively. $d^-(v)$ and $d^+(v)$ represent the *indegree* and the *outdegree* of a node v, respectively. A node v is called a *source* or a *sink* if $d^-(v) = 0$ or $d^+(v) = 0$, respectively. A *path* of *length* k from a node u to a node u' in G is a sequence $\langle v_0, v_1, \cdots, v_k \rangle$ of nodes such that $u = v_0$, $u' = v_k$, and $(v_{i-1}, v_i) \in A$ for $i = 1, 2, \cdots, k$. The length of the longest path from a source to a node v is the *depth* of v in G, denoted by $\ell(v)$. The *height* of a node is the length of the longest path from the node to a sink, and the *height of a DAG G* is the largest height among all heights of its sources, denoted by $h(G)$. A directed graph is called a *layered graph* if its nodes are partitioned into a sequence of *layers*, and arcs are only permitted to connect nodes between successive layers. For a graph $G = (V, A)$ and a subset $V' \subseteq V$ of nodes and a subset $A' \subseteq A$ of arcs, $G[V'] = (V', A')$ denotes the *induced subgraph* of G such that its arc set A' consists of all arcs of G whose both endpoints belong to V'. For a node v, let $N^-(v) = \{u \mid (u, v) \in A\}$, $N^+(v) = \{u \mid (v, u) \in A\}$, $A^-(v) = \{(u, v) \mid u \in N^-(v)\}$, and $A^+(v) = \{(v, u) \mid u \in N^+(v)\}$.

If the fixed DAG structure G is too large to fit in main memory, then G is partitioned into small-sized *blocks* of size at most B, $\mathcal{P} = \{P_1, P_2, \cdots, P_k\}$ where $P_i \cap P_j = \emptyset$ for $i \neq j$, $\bigcup_{i=1}^k P_i = V$, $|P_i| \leq B$, and thus $k \geq \lceil |V|/B \rceil$. We call $\mathcal{P}$ is a *packing* if $|P_i| \leq B$ for all i's. An arc (u, v) is said to be *packed* (under a packing $\mathcal{P}$) if $\{u, v\} \subseteq P$ for some $P \in \mathcal{P}$; otherwise, an arc is called an *external arc*. Let q be a path from a source to a sink. Given a packing $\mathcal{P}$, the *block transfer $bt_{\mathcal{P}}(q)$* of q under $\mathcal{P}$ is the number of external arcs (u, v)'s on q, and furthermore, the *block transfer* $bt_{\mathcal{P}}(G)$ of a DAG G under $\mathcal{P}$ is $\max_{q \in Q} bt_{\mathcal{P}}(q)$, where Q is a set of paths from sources to sinks. Now our problem is formulated as follows:

Minimum Block Transfers with B (MBT(B)):
INSTANCE: A directed acyclic graph $G = (V, A)$ and an integer $B \in [1..|V|]$.
GOAL: Find a packing $\mathcal{P}$ that minimizes the block transfer $bt_{\mathcal{P}}(G)$ (this packing is denoted by $\mathcal{OPT}$, and is termed an *optimal* packing).

We assume that a set V of nodes of an input DAG $G = (V, A)$ is partitioned into layers $V_0, \ldots, V_{h(G)}$, i.e., $V = V_0 \cup \cdots \cup V_{h(G)}$ and $V_i \cap V_j = \emptyset$ for $i \neq j$ where V_0 contains all the sources in G and $v \in V_{\ell(v)}$. By this partition, there always exists an arc (u, v) for each node $v \in V_i$ ($1 \leq i \leq h(G)$) such that $u \in V_{i-1}$. This partition of V can be obtained in linear time $O(|V| + |A|)$ based on the topological sort. Note that this definition of layering does not indicate the input DAG is a layered graph, since some arc may connect non-consecutive layers.

The block transfer under $\mathcal{OPT}$ has the following trivial lower bound caused by $h(G)$ and B.

Proposition 1. *For DAG G and the block size B, $bt_{\mathcal{OPT}}(G) \geq \lfloor h(G)/B \rfloor$.*

For MBT(B), a greedy heuristic, called GREEDY, has been proposed [7], which basically tries to pack nodes in a bottom-up manner as mentioned in Sect. 1. Although details are omitted here, it is not a good approximation algorithm:

Proposition 2. GREEDY *is a B factor approximation algorithm for the MBT(B).*

3 Optimal Algorithms for Flat DAGs

In this section we provide optimal algorithms for DAGs of height at most two, which are used in the approximation algorithms presented in the next section. We begin with a quite simple proposition.

Proposition 3. $bt_{\mathcal{OPT}}(G) = 0$ *for the MBT(B) if and only if every connected component of G has at most B nodes.*

By the above proposition, we obtain a polynomial time algorithm for the case $h(G) = 1$: All we have to do is to calculate the number of nodes in each connected component in $O(|V| + |A|)$ time by the breadth-first search. If every connected component has at most B nodes, then we can make the block transfer zero by packing all the nodes into a single block. Otherwise, the block transfer is one. This linear time algorithm will be referred to as HeightOne in the following.

Lemma 1. HeightOne *is an optimal algorithm which runs in $O(|V| + |A|)$ time for the MBT(B) if the height of DAGs is one (i.e., $h(G) = 1$).*

Now we turn our attention to the case $B = 2$ and show that we can obtain an optimal solution in $O(|V| + |A|)$ time for the height $h(G) = 2$. Recall that the node set V of G is partitioned into three layers V_0, V_1, and V_2 as mentioned above. The main idea of the optimal algorithm is first reducing the MBT(2) to the 2-SAT problem [8], and then obtaining an optimal packing $\mathcal{OPT}$ by finding a truth assignment of the reduced 2-CNF predicate f. If f is satisfiable, then $bt_{\mathcal{OPT}}(G) = 1$; otherwise, $bt_{\mathcal{OPT}}(G) = 2$. The key point is that the predicate f represents conditions on paths of length 2. Let v be a node in V_1 associated with a variable x. In order to decrease the block transfer to one, all the incoming arcs to v (i.e., $A^-(v)$) or all the outgoing arcs from v (i.e., $A^+(v)$) have to be packed, which corresponds to assigning *true* or *false* to the variable x, respectively.

Algorithm HeightTwo
>   (**Input:** DAG G of height two, **Output:** Packing $\mathcal{P}$)
>
> **Step 1.** Transform G to a 2-CNF predicate f by the following manner:
>   **Step 1-1:** Assign one variable x_i to each node $v_i \in V_1$.
>   **Step 1-2:** Make the following clauses associated with each $v_i \in V_1$:
>   **Rule (i):** $(\overline{x_i})$ if the indegree of v_i satisfies that $d^-(v_i) \geq 2$, and
>   **Rule (ii):** (x_i) if the outdegree of v_i satisfies that $d^+(v_i) \geq 2$.
>   **Step 1-3:** For each pair of v_i and $v_j \in V_1$, make the following clauses
>   **Rule (iii):** $(\overline{x_i} \vee \overline{x_j})$ if $d^-(v_i) = d^-(v_j) = 1$ and two arcs (u, v_i) and (u, v_j) exist for some $u \in V_0$, and
>   **Rule (iv):** $(x_i \vee x_j)$ if $d^+(v_i) = d^+(v_j) = 1$ and two arcs (v_i, u) and (v_j, u) exist for some $u \in V_2$.
>   **Step 1-4:** Construct f by adding the clauses made in Steps 1-2 and 1-3 conjunctively.
> **Step 2.** Solve f by a polynomial-time algorithm for the 2-SAT [8].
> **Step 3.** If f is unsatisfiable, then output $\{\{v\} \mid v \in V\}$ as a packing $\mathcal{P}$. If f is satisfiable, then output a packing $\mathcal{P}$ according to the satisfying truth assignment obtained in Step 2 as follows: For each variable x_i if $x_i = true$ then add $\{v_i\} \cup N^-(v)$ to $\mathcal{P}$, otherwise add $\{v_i\} \cup N^+(v)$ to $\mathcal{P}$. For each node v not contained in such blocks, add $\{v\}$ to $\mathcal{P}$.

For expositions of the behavior of HeightTwo, consider a DAG in Fig. 1-(a). We construct clauses $(\overline{x_1})$ by the rule (i), $(\overline{x_2} \vee \overline{x_3})$ by the rule (iii), and $(x_1 \vee x_2)$ by the rule (iv). Then the final predicate f is $(\overline{x_1}) \wedge (\overline{x_2} \vee \overline{x_3}) \wedge (x_1 \vee x_2)$. A truth assignment that makes f *true* is $(x_1, x_2, x_3) = (false, true, false)$ and it corresponds to the packing $\{\{v_1, s\}, \{q, v_2\}, \{v_3, t\}, \{p\}\}$ that makes the block transfer one as shown in Fig. 1-(b).

Lemma 2. HeightTwo *is an optimal algorithm which runs in* $O(|V| + |A|)$ *time for the MBT(2) if the height of DAGs is two.*

Proof. **Running Time.** There is an algorithm for the 2-SAT that runs in linear time of the number of clauses, i.e., $O(|V| + |A|)$ time [8]. It follows that the whole running time is also $O(|V| + |A|)$.

Correctness of Reduction. We show that $bt_{\mathcal{OPT}}(G) = 1$ iff f is satisfiable. ($\Rightarrow$) Consider a node $v_i \in V_1$. One can see that at most one arc in the paths passing through v_i can be packed. However, in the case $d^-(v_i) = |N^-(v_i)| \geq 2$, e.g., $N^-(v_1) = \{p, q\}$ in Fig. 1, packing either of the arcs (p, v_1) and (q, v_1), say,

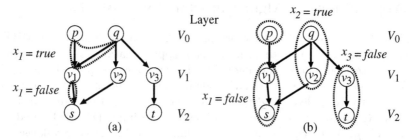

Fig. 1. (a) Assigning variables to nodes. (b) Packing according to a truth assignment.

(p, v_1), does not reduce the block transfer to one, because no arc in the path $\langle q, v_1, s \rangle$ is packed. Therefore in such a case, no arc in $A^-(v_i)$ is packed and all arcs in $A^+(v_i)$ have to be packed in $\mathcal{OPT}$ in order to achieve $bt_{\mathcal{OPT}}(G) = 1$. This situation is expressed by the clause $(\overline{x_i})$ in the rule (i): Packing all arcs in $A^-(v_i)$ corresponds to assigning $true$ to x_i, which makes the clause $(\overline{x_i})$ $false$. As for the clause (x_i) in the rule (ii), the discussion is similar. By the assumption that $bt_{\mathcal{OPT}}(G) = 1$, only one or none of (x_i) and $(\overline{x_i})$ is included in f and hence we can choose $true/false$-value of those variables according to $\mathcal{OPT}$.

Next we consider the clause $(\overline{x_i} \vee \overline{x_j})$ constructed by the rule (iii). Suppose two arcs (u, v_i) and (u, v_j) exist for $u \in V_0$ and $v_i, v_j \in V_1$. Since both of the two arcs can not be packed simultaneously, one of them, say, (u, v_i), is not packed under $\mathcal{OPT}$, that corresponds to assigning $false$ to x_i and makes the clause $(\overline{x_i} \vee \overline{x_j})$ $true$. As for the clause $(x_i \vee x_j)$ by the rule (iv), a similar argument can be applied to v_i and v_j. In summary, there exists a satisfying truth assignment for the variable x_i's in f according to $\mathcal{OPT}$.

($\Leftarrow$) We construct a packing according to a satisfying truth assignment. Assume that in the truth assignment, $true$ is assigned to a variable x_i that implies $A^-(v_i)$ contains at most one arc, e.g., (q, v_2) for v_2 in Fig. 1. If no other arcs that emanate from q satisfy the condition of the rule (iii), then (q, v_2) can be packed. If such an arc exists, e.g., (q, v_3) in Fig. 1, f includes the clause $(\overline{x_2} \vee \overline{x_3})$. Since f is satisfiable, x_3 must be assigned $false$ in the truth assignment that means (q, v_3) cannot be packed. A similar argument can be applied to any variable which is assigned $false$. Therefore we can choose a packing according to the truth assignment for f. Finally, the block transfer must be one by such a packing because for each variable x_i, either of $true$ or $false$ is assigned, which implies that all arcs in $A^-(v_i)$ or $A^+(v_i)$ are packed for each $v_i \in V_1$. This completes the proof. $\square$

From Lemmas 1 and 2 we obtain Theorem 1:

Theorem 1. *There is an optimal algorithm which runs in $O(|V|+|A|)$ time for the MBT(2) if the height of DAGs is at most two.*

A very similar strategy to the above gives us the following theorem:

Theorem 2. *There is an optimal algorithm which runs in polynomial time for the MBT(3) if the height of DAGs is at most two.*

4   Approximation Algorithms for Small Blocks

In this section we present a $\frac{3}{2}$ factor approximation algorithm for the MBT(2), called DAGPack, which is optimal in terms of approximation guarantee as shown in the next section. It uses the algorithms in Sect. 3 as sub-procedures.

We first present two approximation algorithms for DAGs of height three and four, respectively, called HeightThree and HeightFour, that are also used in DAGPack. Recall that we assume a node set V of an input DAG is layered as V_0 through $V_{h(G)}$. A description of the algorithm HeightThree is as follows:

> **Algorithm HeightThree**
>   (**Input:** DAG G of height three, **Output:** Packing $\mathcal{P}$)
> **Step 1:** Apply HeightOne to $G[V_0 \cup V_1]$ and also to $G[V_2 \cup V_3]$. Let $\mathcal{P}_1$ and $\mathcal{P}_2$ be the obtained packings, respectively.
> **Step 2:** Do the following according to $\mathcal{P}_1$ and $\mathcal{P}_2$:
>   – If $bt_{\mathcal{P}_1}(G[V_0 \cup V_1]) = bt_{\mathcal{P}_2}(G[V_2 \cup V_3]) = 0$, then output $\mathcal{P} = \mathcal{P}_1 \cup \mathcal{P}_2$ and halt.
>   – If $bt_{\mathcal{P}_2}(G[V_2 \cup V_3]) = 1$, then output $\mathcal{P} = \{\{v\} \mid v \in V\}$ and halt.
>   – Otherwise, go to Step 3.
> **Step 3:** Let R be a set of nodes, each of which is reachable to some node in V_3. If $R = V$, output $\mathcal{P} = \{\{v\} \mid v \in V\}$ and halt.
> **Step 4:** Apply HeightTwo to $G[V - R]$ and then HeightOne to both $G[R \cap (V_0 \cup V_1)]$ and $G[R \cap (V_2 \cup V_3)]$. Let $\mathcal{P}_3$, $\mathcal{P}_4$ and $\mathcal{P}_5$ be the obtained packings, respectively. Output $\mathcal{P} = \mathcal{P}_3 \cup \mathcal{P}_4 \cup \mathcal{P}_5$ and halt.

Lemma 3. HeightThree *is a $\frac{3}{2}$ factor approximation algorithm for DAGs of height three that runs in $O(|V| + |A|)$ time.*

Proof. The total running time is linear since both HeightOne and HeightTwo run in linear time from Lemmas 1 and 2. Let us proceed to show the approximation ratio. We have to verify the following four cases to prove the $\frac{3}{2}$ ratio. Note that it is enough to show $bt_{\mathcal{OPT}}(G) \geq 2$ for each case, because the height of the input DAG is now three and the block transfer under any packing is at most three.

Case (i). In Step 2, $bt_{\mathcal{P}_1}(G[V_0 \cup V_1]) = bt_{\mathcal{P}_2}(G[V_2 \cup V_3]) = 0$: HeightThree outputs $\mathcal{P} = \mathcal{P}_1 \cup \mathcal{P}_2$, and then $bt_{\mathcal{P}}(G) = 1$. From Proposition 1, $bt_{\mathcal{OPT}}(G) \geq 1$ that implies $\mathcal{P}$ is optimal.
Case (ii). In Step 2, $bt_{\mathcal{P}_2}(G[V_2 \cup V_3]) = 1$: HeightThree outputs $\mathcal{P} = \{\{v\} \mid v \in V\}$ and then $bt_{\mathcal{P}}(G) = 3$. Since $\mathcal{P}_2$ is optimal for $G[V_2 \cup V_3]$ from Lemma 1, $bt_{\mathcal{OPT}}(G[V_2 \cup V_3]) = 1$. In addition to that, $bt_{\mathcal{OPT}}(G[V_0 \cup V_1 \cup V_2]) \geq 1$ from Proposition 1, and hence $bt_{\mathcal{OPT}}(G) \geq 2$.
Case (iii). HeightTree halts in Step 3: The discussion is similar to Case (ii).
Case (iv). In Step 3, $R \neq V$ and then Step 4 is processed: We further divide the case into two subcases. Note that the height of $G[V - R]$ is at most two and there is no arc (u, v) for $u \in V - R$ and $v \in R$ by the definition of R.
   (a).   No arc (u, v) for $u \in R$ and $v \in V - R$ exists: $G[V - R]$ and $G[R]$ are disjoint. $\mathcal{P}_3$ is optimal for $G[V - R]$ from Lemma 2. As for $G[R]$, a similar argument to the above cases (i), (ii), and (iii) can be also applied.

(b).   An arc (u, v) for $u \in R$ and $v \in V - R$ exists: If (u, v) is packed under $\mathcal{OPT}$, $bt_{\mathcal{OPT}}(G) \geq 2$ holds, because u cannot be packed with nodes in R and there exists a path of length three in $G[R]$, which goes through u. Otherwise, i.e., if (u, v) is not packed under $\mathcal{OPT}$, then the discussion is similar to Case (iv)-(a). $\square$

Next we present a description of `HeightFour` for DAGs of height four, which plays the most important role in `DAGPack`.

Algorithm HeightFour
    (**Input:** DAG G of height four, **Output:** Packing $\mathcal{P}$)
Step 1: Apply `HeightOne` to $G[V_0 \cup V_1]$ and let $\mathcal{P}_1$ be the obtained packing. If $bt_{\mathcal{P}_1}(G[V_0 \cup V_1]) = 0$, then output $\mathcal{P} = \mathcal{P}_1 \cup \{\{v\} \mid v \in V_2 \cup V_3 \cup V_4\}$ and halt.
Step 2: Apply `HeightTwo` to $G[V_0 \cup V_1 \cup V_2]$ and let $\mathcal{P}_2$ be the obtained packing. If $bt_{\mathcal{P}_2}(G[V_0 \cup V_1 \cup V_2]) = 1$, then output $\mathcal{P} = \mathcal{P}_2 \cup \{\{v\} \mid v \in V_3 \cup V_4\}$ and halt.
Step 3: Output $\mathcal{P} = \{\{v\} \mid v \in V\}$ and halt.

Lemma 4. `HeightFour` *is a* $\frac{3}{2}$ *factor approximation algorithm for DAGs of height four that runs in* $O(|V| + |A|)$ *time.*

Proof. First, the running time is clearly $O(|V|+|A|)$ from Lemmas 1 and 2. Then, we show $bt_{\mathcal{P}}(G) \leq \frac{3}{2} bt_{\mathcal{OPT}}(G)$ for all the cases that `HeightFour` terminates in (i) Step 1, (ii) Step 2, and (iii) Step 3.

Case (i): Since $bt_{\mathcal{P}_1}(G[V_0 \cup V_1]) = 0$, it holds that $bt_{\mathcal{P}}(G) = 3$. From Proposition 1, $bt_{\mathcal{OPT}}(G) \geq 2$ and hence $bt_{\mathcal{P}}(G) \leq \frac{3}{2} bt_{\mathcal{OPT}}(G)$.
Case (ii): Since $bt_{\mathcal{P}_2}(G[V_0 \cup V_1 \cup V_2]) = 1$, it holds that $bt_{\mathcal{P}}(G) = 3$ and again $bt_{\mathcal{P}}(G) \leq \frac{3}{2} bt_{\mathcal{OPT}}(G)$ as in Case (i).
Case (iii): No arc is packed under $\mathcal{P}$, and then $bt_{\mathcal{P}}(G) = 4$ holds. Since $\mathcal{P}_2$ is optimal for $G[V_0 \cup V_1 \cup V_2]$ from Lemma 2 and Step 3 is processed, it holds that $bt_{\mathcal{OPT}}(G[V_0 \cup V_1 \cup V_2]) = 2$. From Proposition 1, $bt_{\mathcal{OPT}}(G[V_2 \cup V_3 \cup V_4]) \geq 1$ and hence $bt_{\mathcal{OPT}}(G) \geq 3$. This implies that $bt_{\mathcal{P}}(G) \leq \frac{4}{3} bt_{\mathcal{OPT}}(G)$. $\square$

Now we can present a $\frac{3}{2}$ factor approximation algorithm for DAGs of general height. Its basic idea is as follows: (1) We divide the whole graph into subgraphs of height four, and then (2) iteratively apply `HeightFour` to those subgraphs:

Algorithm DAGPack
    (**Input:** DAG G, **Output:** Packing $\mathcal{P}$)
Step 1: Let $W_0 = V_0 \cup \cdots \cup V_4$, $W_1 = V_4 \cup \cdots \cup V_8$, $\ldots$, $W_k = V_{4k} \cup \cdots \cup V_{h(G)}$, where $k = \lfloor h(G)/4 \rfloor$.
Step 2: Apply `HeightFour` to each subgraph $G[W_i]$ for $0 \leq i \leq k - 1$. Let $\mathcal{P}_i$ be the obtained packing for $G[W_i]$.
Step 3: Let $h' = h(G) - 4k$ $(1 \leq h' \leq 3)$, which is the height of $G[W_k]$. Apply `HeightOne`, `HeightTwo` and `HeightThree` to $G[W_k]$ if $h' = 1$, 2 and 3, respectively. Let $\mathcal{P}_k$ be the obtained packing.

Step 4: Output $\mathcal{P} = (\mathcal{P}_1 - \{\{v\} \mid v \in V_4\}) \cup (\mathcal{P}_2 - \{\{v\} \mid v \in V_8\}) \cup \cdots \cup \mathcal{P}_k$.

See the description of `HeightFour` again and recall that when we apply `HeightFour` to each subgraph $G[W_i]$ in Step 2, each node v in the last layer V_{4i} of the subgraph $G[W_i]$, e.g., V_4 of $G[W_0]$, is not packed with other nodes within $G[W_i]$. Namely, $\mathcal{P}_i$ always contains $\{\{v\} \mid v \in V_{4i}\}$ for $0 \le i \le k-1$. The reason why we remove this part from each $\mathcal{P}_i$, e.g., $\{\{v\} \mid v \in V_4\}$ from the packing $\mathcal{P}_1$ in Step 4 is that it may conflict to the packing determined by applying `HeightFour` to the next subgraph $G[W_{i+1}]$. Since the packing $\{\{v\} \mid v \in V_{4i}\}$ does not reduce the block transfer at all, we choose to determine blocks that the nodes in V_{4i} belong to, as a part of processing $G[W_{i+1}]$ instead of $G[W_i]$.

Let G_r be a subgraph of G constructed by removing arcs that do not appear in any subgraphs $G[W_i]$'s. The following lemma holds.

Lemma 5. *For any packing $\mathcal{Q}$, $bt_{\mathcal{Q}}(G_r) = bt_{\mathcal{Q}}(G)$.*

Proof. The layer V_{4i} of nodes is shared by $G[W_{i-1}]$ and $G[W_i]$. That is, arcs from V_{4i-1} to V_{4i}, and from V_{4i} to V_{4i+1} are included inside $G[W_{i-1}]$ and $G[W_i]$, respectively, and hence appear in G_r. However, arcs connected $V_{4i-4} \cup V_{4i-3} \cup V_{4i-2} \cup V_{4i-1}$ with $V_{4i+1} \cup V_{4i+2} \cup V_{4i+3} \cup V_{4i+4}$ do not appear in G_r. Let (u, v) denote such an arc, where $u \in V_{4i-1}$ and $v \in V_{4i+1}$. Consider a path $\langle x, y, v \rangle$ of length 2 for some $x \in V_{4i-1}$ and $y \in V_{4i}$, where the arcs (x, y) and (y, v) exist inside $G[W_{i-1}]$ and $G[W_i]$, respectively. Recall that such a path surely exists by the definition of the layers. Under any packing $\mathcal{Q}$ (including $\mathcal{OPT}$), $bt_{\mathcal{Q}}(\langle x, y, v \rangle) \ge 1 \ge bt_{\mathcal{Q}}(\langle u, v \rangle)$ from Proposition 1, and then $bt_{\mathcal{Q}}(\langle x, y, T \rangle) \ge bt_{\mathcal{Q}}(\langle u, T \rangle)$, where T is an arbitrary path that starts from v and is included in $G[W_i]$. Therefore the maximum block transfer of paths through the nodes in $G[V_{4i-1} \cup W_i]$ does not depend on (u, v) (except ties). The same arguments can be applied for the cases $u \in V_{4i-2}, V_{4i-3}$, or V_{4i-4} and $v \in V_{4i+2}, V_{4i+3}$ or V_{4i+4}, and moreover, for any pair of (non-consecutive) subgraphs $G[W_i]$'s. $\qquad \square$

Theorem 3. `DAGPack` *is a $\frac{3}{2}$ factor approximation algorithm for the MBT(2), that runs in linear time.*

Proof. We can conclude that `DAGPack` is a $\frac{3}{2}$ factor approximation algorithm because $bt_{\mathcal{P}}(G) = bt_{\mathcal{P}}(G_r) = \sum_i bt_{\mathcal{P}}(G[W_i]) \le \frac{3}{2} \sum_i bt_{\mathcal{OPT}}(G[W_i]) = \frac{3}{2} bt_{\mathcal{OPT}}(G_r) = \frac{3}{2} bt_{\mathcal{OPT}}(G)$, where the first and the last equalities hold from Lemma 5, and the inequality is based on Theorem 1, Lemmas 3 and 4.

We briefly analyze the running time: Steps 1 and 4 can be achieved in linear time. In Steps 2 and 3, the linear-time procedures `HeightOne`, `HeightTwo`, `HeightThree`, and `HeightFour` are executed for (independent) induced subgraphs, therefore, the total running time is also linear. $\qquad \square$

The same strategies give us the following theorem (the proof is omitted).

Theorem 4. *There is a 2 factor polynomial time approximation algorithm for the MBT(3) if the input is restricted to a set of layered graphs.*

5   Hardness of Approximation

We show that, given a DAG G and an integer B, the problem of finding an optimal packing for G is NP-hard even if $B = 2$ and the height of G is three. The 3-SAT problem is reduced to this problem as follows. Given a 3-CNF predicate f, we construct DAG G of height three satisfying conditions (i) and (ii): (i) There exists an optimal packing $\mathcal{OPT}$ such that $bt_{\mathcal{OPT}}(G) = 2$ if f is satisfiable. (ii) If there exists a packing $\mathcal{OPT}$ such that $bt_{\mathcal{OPT}}(G) = 2$, then f is satisfiable.

Theorem 5. *(I) The MBT(2) and hence the general MBT(B) are NP-hard. (II) The MBT(2) remains NP-hard even if the height of DAGs is three.*

Proof. Suppose that the predicate f uses n variables, $U = \{u_1, u_2, \cdots, u_n\}$, and contains m clauses, $C = \{c_1, c_2, \cdots, c_m\}$, where the i-th clause includes exactly three variables $x_{i,1}$, $x_{i,2}$ and $x_{i,3}$. The reduced DAG G consists of two subgraphs, SG_1 and SG_2. SG_1 is the *variable gadget* associated with the variable set U, and SG_2 is the *clause gadget* associated with the clause set C.

The variable gadget SG_1 is divided into n components $G_{u_i} = (V_{u_i}, A_{u_i})$ for $i = 1, 2, \cdots, n$, corresponding to the variable u_i of f. Since G_{u_1}, G_{u_2}, $\cdots$, G_{u_n} are in the same form, Fig. 2-(a) illustrates only G_{u_i}, where, only for clarity of exposition, we assume that the positive literal u_i and its negation $\overline{u_i}$ appear three times and twice in f, respectively. In this case $G_{u_i} = (V_{u_i}, A_{u_i})$ has 16 nodes and 16 arcs as shown in Fig. 2-(a). V_{u_i} consists of three node sets, $U_i = \{t_i, f_i, u_{i,0}, \overline{u_{i,0}}, v_{i,1}, v_{i,2}\}$, $T_i = \{u_{i,1}, u_{i,2}, u_{i,3}, u'_{i,1}, u'_{i,2}, u'_{i,3}\}$, and $F_i = \{\overline{u_{i,1}}, \overline{u_{i,2}}, u'_{i,1}, u'_{i,2}\}$. The first node set U_i plays a key role in the remaining of the proof. T_i and F_i include 2×3 and 2×2 nodes, corresponding to the numbers of occurrences of u_i and its negation $\overline{u_i}$, respectively. If, for example, there are four positive literals u_i's in f, then two nodes $\{u_{i,4}, u'_{i,4}\}$ and two arcs $\{(u_{i,0}, u_{i,4}), (u_{i,4}, u'_{i,4})\}$ are further added to T_i ($\subset V_{u_i}$) and A_{u_i}, respectively.

The clause gadget SG_2 consists of m components since f includes m clauses. Fig. 2-(b) illustrates one of them. For each clause $c_j = x_{j,1} \vee x_{j,2} \vee x_{j,3}$ for $j = 1, 2, \cdots, m$ of f, we introduce a component $G_{c_j} = (V_{c_j}, A_{c_j})$ which has six nodes and six arcs as follows: $V_{c_j} = \{x_{j,1}, x_{j,2}, x_{j,3}, c_{j,0}, c_{j,1}, c_{j,2}\}$ and $A_{c_j} = \{(c_{j,0}, c_{j,1}), (c_{j,0}, c_{j,2}), (c_{j,1}, x_{j,1}), (c_{j,1}, x_{j,2}), (c_{j,2}, x_{j,2}), (c_{j,2}, x_{j,3})\}$.

There is also a set of arcs which connects SG_1 with SG_2. For example, we assume that the clause c_j is the conjunction of the α-th occurrence of positive literal u_p, the β-th occurrence of negative literal $\overline{u_q}$, and the γ-th occurrence of negative literal $\overline{u_r}$. Then, we connect $x_{j,1}$ with $u'_{p,\alpha} \in V_{u_p}$, $x_{j,2}$ with $\overline{u'_{q,\beta}} \in V_{u_q}$, and $x_{j,3}$ with $\overline{u'_{r,\gamma}} \in V_{u_r}$. The correctness of our reduction is based on the following key lemmas, Lemmas 6 and 7, but their proofs are omitted:

Lemma 6. *For each component G_{u_i} of SG_1, the block transfer $bt_{\mathcal{P}}(G_{u_i})$ is at least two under any packing $\mathcal{P}$. Furthermore, every packing $\mathcal{P}_2$ with $bt_{\mathcal{P}_2}(G_{u_i}) = 2$ has to satisfy either of the following two necessary conditions: (C1) $\{\{t_i, u_{i,0}\}, \{\overline{u_{i,0}}, f_i\}, \{\overline{u_{i,1}}, u'_{i,1}\}, \cdots, \{\overline{u_{i,j_1}}, u'_{i,j_1}\}\} \subseteq \mathcal{P}_2$, and (C2) $\{\{t_i, \overline{u_{i,0}}\}, \{u_{i,0}, f_i\}, \{u_{i,1}, u'_{i,1}\}, \cdots, \{u_{i,j_2}, u'_{i,j_2}\}\} \subseteq \mathcal{P}_2$, where j_1 and j_2 mean the numbers of occurrences of $\overline{u_i}$ and u_i, respectively.*

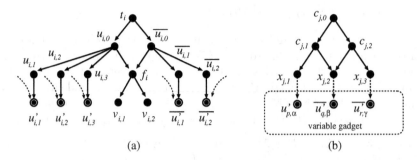

Fig. 2. (a) Variable gadget G_{u_i}. (b) Clause gadget G_{c_j}.

Lemma 7. *For each component G_{c_j} of SG_2, the block transfer $bt_\mathcal{P}(G_{c_j})$ is at least two under any packing $\mathcal{P}$. Furthermore, every packing $\mathcal{P}_2$ with $bt_{\mathcal{P}_2}(G_{c_j}) = 2$ has to satisfy at least one of the following three conditions: (C3) $\{x_{j,1}, u'_{p,\alpha}\} \in \mathcal{P}_2$, (C4) $\{x_{j,2}, \overline{u'_{q,\beta}}\} \in \mathcal{P}_2$, and (C5) $\{x_{j,3}, \overline{u'_{r,\gamma}}\} \in \mathcal{P}_2$.*

Now we show the two conditions (i) and (ii) at the beginning of this section are satisfied for the reduced DAG G.

(i) Suppose that there is a satisfying truth assignment for the 3-CNF predicate f. Then we can obtain the following optimal packing $\mathcal{OPT}$ under which the block transfer $bt_{\mathcal{OPT}}(G)$ is two:

1. Depending on the truth assignment that makes f *true*, for example, if $u_i = true$, then $\mathcal{OPT} \supseteq \{\{t_i, u_{i,0}\}, \{\overline{u_{i,0}}, f_i\}, \{\overline{u_{i,1}}, \overline{u'_{i,1}}\}, \{\overline{u_{i,2}}, \overline{u'_{i,2}}\}, \{u_{i,1}\}, \{u_{i,2}\}, \{u_{i,3}\}, \{v_{i,1}\}, \{v_{i,2}\}\}$; otherwise, if $u_i = false$, then $\mathcal{OPT} \supseteq \{\{t_i, \overline{u_{i,0}}\}, \{u_{i,0}, f_i\}, \{u_{i,1}, u'_{i,1}\}, \{u_{i,2}, u'_{i,2}\}, \{u_{i,3}, u'_{i,3}\}, \{\overline{u_{i,1}}\}, \{\overline{u_{i,2}}\}, \{v_{i,1}\}, \{v_{i,2}\}\}$. The mapping for $u'_{i,1}, u'_{i,2}$, and $u'_{i,3}$ in the former case (or for $\overline{u'_{i,1}}$ and $\overline{u'_{i,2}}$ in the latter case) is determined in the next.

2. As for each component G_{c_j} of SG_2 for $j = 1, \cdots, m$, $\mathcal{OPT}$ depends again on the truth assignment: (1) If all the three literals $u_p, \overline{u_q}$, and $\overline{u_r}$ are assigned to be *true*, $\mathcal{OPT} \supseteq \{\{x_{j,1}, u'_{p,\alpha}\}, \{x_{j,2}, \overline{u'_{q,\beta}}\}, \{x_{j,3}, \overline{u'_{r,\gamma}}\}\}$, and the other nodes $c_{j,i}$'s remain single, i.e., $\{c_{j,i}\} \in \mathcal{OPT}$ for $i = 0, 1, 2$. (2) If two literals, say, u_p and $\overline{u_q}$, are assigned to be *true*, $\mathcal{OPT} \supseteq \{\{x_{j,1}, u'_{p,\alpha}\}, \{x_{j,2}, \overline{u'_{q,\beta}}\}, \{c_{j,2}, x_{j,3}\}\}$ and the other nodes remain single. As for another example, in case u_p and $\overline{u_r}$ are *true*, $\mathcal{OPT} \supseteq \{\{x_{j,1}, u'_{p,\alpha}\}, \{x_{j,3}, \overline{u'_{r,\gamma}}\}, \{c_{j,0}, c_{j,1}\}, \{c_{j,2}, x_{j,2}\}\}$. (3) If only one literal, say, u_p is assigned to be *true*, $\mathcal{OPT} \supseteq \{\{x_{j,1}, u'_{p,\alpha}\}, \{c_{j,0}, c_{j,2}\}, \{c_{j,1}, x_{j,2}\}\}$ and $x_{j,3}$ remains single. In the case only u_q is assigned to be *true*, $\mathcal{OPT} \supseteq \{\{x_{j,2}, \overline{u'_{q,\beta}}\}, \{c_{j,1}, x_{j,1}\}, \{c_{j,2}, x_{j,3}\}, \{c_{j,0}\}\}$.

(ii) Suppose that there exists an optimal packing $\mathcal{OPT}$ such that $bt_{\mathcal{OPT}}(G) = 2$. Then the predicate f is satisfiable by constructing the satisfying truth assignment. From Lemma 6, $\mathcal{OPT}$ has to satisfy either (C1) or (C2) for every variable gadget G_{u_i}. If (C1) is satisfied for G_{u_i}, then we assign $u_i = true$ and call three nodes $u'_{i,1}, u'_{i,2}$, and $u'_{i,3}$ *free*; otherwise, if (C2) is satisfied, then $\overline{u_i} = true$ is assigned and two nodes $\overline{u'_{i,1}}$ and $\overline{u'_{i,2}}$ become *free*. From Lemma 7, $\mathcal{OPT}$ must satisfy at least one of the three conditions (C3), (C4), and (C5). It follows that at least one of the three nodes $x_{j,1}$, $x_{j,2}$, and $x_{j,1}$ in the j-th clause gadget is surely connected with a free node in the variable gadgets. This means that the above truth assignment satisfies all the clauses in the predicate f. □

Theorem 6. *If, for some $\varepsilon > 0$, there is a $\frac{3}{2} - \varepsilon$ factor polynomial time approximation algorithm for the MBT(2), then P = NP.*

Proof. Suppose that there is a $\frac{3}{2} - \varepsilon$ factor polynomial time approximation algorithm for some $\varepsilon > 0$. If a predicate f of the 3-SAT is satisfiable, the algorithm can find a packing $\mathcal{P}$ with $bt_{\mathcal{P}}(G) = 2 \times (\frac{3}{2} - \varepsilon) < 3$ for G reduced from f. However, if f is unsatisfiable, $bt_{\mathcal{Q}}(G) = 3$ under any packing $\mathcal{Q}$ from the proof of Theorem 5. Thus, this approximation algorithm can be used for deciding the 3-SAT in polynomial time, which implies P = NP. □

Corollary 1. Algorithm DAGPack *is optimal in terms of approximation guarantee for the MBT(2).*

Theorem 7. *For general $B \geq 3$, (I) the MBT(B) is NP-hard even if the height of DAGs is three, and moreover, (II) if, for some $\varepsilon > 0$, there is a $\frac{3}{2} - \varepsilon$ factor polynomial time approximation algorithm for the MBT(B), then P = NP.*

References

1. *Handbook of massive data sets*, J. Abello, P.M. Pardalos, M.G.C. Resende Eds., Kluwer Academic Pub., 2002.
2. A. Aggarwal and J.S. Vitter. The input/output complexity of sorting and related problems. *Commun. ACM*, 31 (9), pp.1116–1127, 1988.
3. S. Alstrup, M.A. Bender, E.D. Demaine, M. Farach-Colton, T. Rauhe, M. Thorup. Efficient tree layout in a multilevel memory hierarchy. *CoRR cs.DS/0211010*, 2002.
4. S. Amer-Yahia, N. Koudas, A. Marian, D. Srivastava, D. Toman. Structure and content scoring for XML. In *Proc. 31st VLDB*, pp.361–372, 2005.
5. L. Arge, A. Danner, S.-M. Teh. I/O-efficient point location using persistent B-trees. In *Proc. 5th ALENEX*, pp.82–92, 2003.
6. D. Clark and J. Munro. Efficient suffix trees on secondary storage. In *Proc. 7th Annual ACM-SIAM Symposium on Discrete Algorithms*, pp.383–391, 1996.
7. A.A. Diwan, S. Rane, S. Seshadri, S. Sudarshan. Clustering techniques for minimizing external path length. In *Proc. 22nd VLDB*, pp.432–353, 1996.
8. S. Even, A. Itai, and A. Shamir. On the complexity of timetable and multicommodity flow problems. *SIAM J. Comput.*, 5, pp.691–703, 1976.
9. J. Gil and A. Itai. Packing trees. In *Proc. 3rd Annual European Symposium on Algorithms*, pp.113–127, 1995 (full version: *J. Algorithms*, 32 (2), pp.108–132, 1999).
10. D.G. Kirkpatrick. Optimal search in planar subdivisions. *SIAM J. Comput.*, 12 (28), pp.28–35, 1983.
11. P.J. Varman and R.M. Verma. An efficient multiversion access structure. *IEEE Trans. on Knowledge and Data Engineering*, 9 (3), 1997.
12. V.V. Vazirani. *Approximation Algorithms*, Springer, 2001.
13. J.S. Vitter. External memory algorithms and data structures: Dealing with massive data. *ACM Comput. Surveys*, 33 (2), pp.209–271, 2001.

On-Line Coloring of H-Free Bipartite Graphs

H.J. Broersma[1], A. Capponi[2], and D. Paulusma[1]

[1] Department of Computer Science, Durham University, Science Labs,
South Road, Durham DH1 3LE, England
{hajo.broersma, daniel.paulusma}@durham.ac.uk
[2] Computer Science, Division of Engineering and Applied Sciences,
California Institute of Technology, U.S.A.
acapponi@cs.caltech.edu

Abstract. We present a new on-line algorithm for coloring bipartite graphs. This yields a new upper bound on the on-line chromatic number of bipartite graphs, improving a bound due to Lovász, Saks and Trotter. The algorithm is on-line competitive on various classes of H-free bipartite graphs, in particular P_6-free bipartite graphs and P_7-free bipartite graphs, i.e., that do not contain an induced path on six, respectively seven vertices. The number of colors used by the on-line algorithm in these particular cases is bounded by roughly twice, respectively roughly eight times the on-line chromatic number. In contrast, it is known that there exists no competitive on-line algorithm to color P_6-free (or P_7-free) bipartite graphs, i.e., for which the number of colors is bounded by any function only depending on the chromatic number.

1 Introduction

In static optimization problems one is often faced with the challenge of determining efficient algorithms that solve a particular problem (nearly) optimally for any given instance of the problem. This task is usually facilitated if the structure of the instances is pretty straightforward. As an example, it is a trivial exercise to determine an algorithm for finding a 2-coloring of a given bipartite graph.

In the area of dynamic optimization the situation gets more complicated. There, one often lacks the knowledge of the complete instances of the problems. As an illustration, compare the previous problem with the slightly changed situation in which the bipartite graph comes in on-line, i.e., vertex by vertex and the algorithm has to assign a color to a vertex as it comes in, i.e., only based on the knowledge of the subgraph that has been revealed so far. This slight change of the problem formulation makes it a lot more difficult: Whereas the static problem was trivial, no algorithm for the dynamic problem can guarantee an optimal solution for every instance. In [9] it has been shown that the worst-case performance ratio between on-line and off-line coloring of a known input graph on n vertices is at least $\Omega(n/log_2 n)$. It is even questionable whether one can expect to determine an on-line algorithm that does reasonably well, in the sense that the number of colors used is bounded in some other reasonable way. In

T. Calamoneri, I. Finocchi, G.F. Italiano (Eds.): CIAC 2006, LNCS 3998, pp. 284–295, 2006.
© Springer-Verlag Berlin Heidelberg 2006

this paper we will focus on particular questions of this type related to coloring bipartite graphs. This type of questions in a more general setting is at the heart of the areas of on-line algorithms and of approximation algorithms.

We first give a short historical excursion starting with a benchmark paper due to Gyárfás and Lehel [6]. They introduced the concept of on-line coloring as a general approach. This was motivated by their translation of a rectangle packing problem related to dynamical storage allocation appearing in [2] into an on-line coloring problem. The latter problem was to decide whether the on-line coloring algorithm known as *First-Fit (FF)* has a constant worst-case performance ratio on the family of interval graphs. We note that since [6] many papers on on-line (coloring) problems have appeared. We refer to [11] for a survey.

In order to have some measure of the performance of on-line algorithms, the notion of competitive algorithms has been introduced in [6]. Intuitively, an on-line coloring algorithm is said to be competitive for a family of graphs $\mathcal{G}$, if for any graph $G \in \mathcal{G}$, the number of colors used by the algorithm on G is bounded from above by a function only depending on the chromatic number of G. In [10] it is shown that *FF* is competitive for interval graphs, with a bounding function that is linear in the chromatic number, and in [3] competitiveness of *FF* for geometric intersection graphs has been proven. It is well-known that *FF* is not competitive for P_6-free bipartite graphs, i.e., bipartite graphs that do not contain an induced path on six vertices: If the vertices of a complete bipartite graph $K_{m,m}$ minus a perfect matching $\{u_1, v_1\}, \{u_2, v_2\}, \ldots, \{u_m, v_m\}$ are presented in the ordering $u_1, v_1, u_2, v_2, \ldots, u_m, v_m$, then *FF* uses m colors. In fact, there are many families of graphs for which no competitive algorithms exist: Two examples given in [6] are the family of trees and the family of P_6-free bipartite graphs. These negative results have led to the definition of a weaker form of competitiveness in [4], although results of this type have been obtained before the term was formally introduced. An on-line coloring algorithm is said to be on-line competitive if the number of colors is bounded from above by a function only depending on the on-line chromatic number of G. It is shown in [7] that *FF* is on-line competitive for trees; it is even optimal for trees, in the sense that if *FF* uses k colors, then the on-line chromatic number of the tree is also k. In [4] it is shown that *FF* is on-line competitive with an exponential bounding function for graphs with girth at least five. There are very few existing results on on-line competitive coloring algorithms.

In the context of algorithmic graph theory it is rather natural to consider forbidden subgraph conditions, as many NP-hard problems turn out to be solvable in polynomial time when restricted to H-free graphs for particular choices of H. Therefore, these graph classes are well-studied throughout a range of NP-hard problems. In the context of coloring, e.g., 3-colorability is polynomially solvable for P_6-free graphs, while 4-colorability remains NP-hard for P_{12}-free graphs, and 5-colorability remains NP-hard for P_8-free graphs. We refer the reader to the survey paper [16] for more details. Note that also well-studied graph classes like chordal graphs can be characterized by forbidden subgraph conditions.

2   Results of This Paper

One of the main open problems concerning on-line competitive coloring algorithms [4] is to decide whether for every k there exists an on-line competitive coloring algorithm for the family of graphs with on-line chromatic number k. Perhaps surprisingly, this is even open for bipartite graphs for $k = 4$, whereas it has been solved for general graphs for $k \leq 3$. (In both [5] and [14] it is proven that for the family of graphs with on-line chromatic number 3 at most 4 colors are needed.) The open problem on bipartite graphs seems to be very hard and emphasizes how much on-line coloring differs from off-line coloring. We are not aware of any recent developments towards settling this problem. Our results are motivated by a number of open problems, but most strongly by the above open problem for bipartite graphs. We solve the problem for several subclasses of bipartite graphs which are defined by forbidding a certain fixed graph H as an induced subgraph. For a relatively small graph H this is an easy exercise, but for larger graphs this gets difficult, in correspondence with the fact that the class of H-free graphs contains the class of H'-free graphs if H' is a subgraph of H. By combining known results and dealing with a few cases ourselves, we show that for every graph H with at most 5 vertices there exists an on-line competitive coloring algorithm for the class of H-free bipartite graphs. Since for P_4-free and P_5-free graphs there even exists a competitive algorithm [6, 8], and since P_6-free bipartite graphs do not admit a competitive algorithm [6], our natural starting point from there is the latter class. The main contribution of this paper is the proof that the on-line coloring algorithm we present for bipartite graphs is on-line competitive for P_6-free bipartite graphs; its bounding function is *linear* in the on-line chromatic number, namely roughly twice the on-line chromatic number. In fact, this gives a 2-approximation algorithm for on-line coloring P_6-free bipartite graphs. We can prove a similar result for the larger class of P_7-free bipartite graphs with a bounding function that is roughly eight times the on-line chromatic number. Due to page limitations we leave its proof for the full paper. Note that the on-line chromatic number for both these graph classes can be arbitrarily high, so these classes are definitely no subclasses of the class of bipartite graphs with on-line chromatic number 4. In this sense, our results have a broader appeal than just solving the aforementioned problem with $k = 4$ for the restricted classes of P_6-free and P_7-free bipartite graphs. It might be possible that our algorithm or variations on it can be used to prove similar results for larger subclasses of bipartite graphs, although we have not been able to do so yet. We will see that our algorithm is competitive for the class of P_5-free bipartite graphs.

The rest of the paper is organized as follows. Section 3 contains the basic notation and definitions. In Section 4 we start our exposition by proving the result on H-free bipartite graphs with $|V(H)| \leq 5$. Next we present the key algorithm of the paper called *BicolorMax*. We prove that it is on-line competitive for P_6-free bipartite graphs, and that the number of colors used by *BicolorMax* on any bipartite graph is bounded from above by the number of mutually remote subgraphs isomorphic to P_5. As a consequence we improve the best known upper bound for the on-line chromatic number of bipartite graphs given in [15] and [11].

3   Preliminaries

Throughout we consider simple graphs $G = (V(G), E(G))$, where $V(G)$ is a set of vertices and $E(G)$ is a set of unordered pairs of vertices, called edges. For graph terminology not defined below we refer to [1]. If $S \subseteq V(G)$, then $G[S]$ denotes the subgraph of G with vertex set S and edge set $\{\{x, y\} \mid x \in S, y \in S\}$. A graph is an *induced subgraph* of G if it is isomorphic to $G[S]$ for some nonempty $S \subseteq V(G)$. A graph is *H-free* if it does not contain the graph H as an induced subgraph. We call two vertex-disjoint graphs *remote* if there are no edges joining them. A maximal connected subgraph of a graph G is called a *component* of G. For any two vertices x, y of a connected graph G we denote by P_{xy} a shortest path between x and y in G, and we define the *distance* $d(x, y, G)$ between x and y in G as $|E(P_{xy})|$. We use K_n, C_n and P_n to denote, respectively, the complete graph, the cycle and the path on n vertices, and we use $K_{m,n}$ to denote the complete bipartite graph with m vertices in one bipartition class and n vertices in the other. A *coloring* of a graph G is a function $c : V(G) \rightarrow \{1, 2, \ldots\}$ such that $c(v) \neq c(w)$ whenever $\{v, w\} \in E(G)$. We denote the set of all colorings of G by $\mathcal{C}(G)$. The smallest number of colors in a coloring of G is the *chromatic number* of G and denoted by $\chi(G)$. We assume that the reader is familiar with the basic concept of an on-line coloring algorithm. For details we refer to [11]. Intuitively, an on-line coloring algorithm properly colors the vertices of a graph one by one, consistently using a fixed strategy, depending only on the subgraph induced by the revealed vertices and their colors, according to an externally determined ordering of the presented vertices. We denote the (finite) set of all on-line coloring algorithms for a graph G by $AOL(G)$. Let $\Pi(G)$ denote the set of all permutations of the vertices of G. If $A \in AOL(G)$ and $\pi \in \Pi(G)$, we denote by $\chi_A(G, \pi)$ the number of colors used by A when the vertices of G are presented according to π. The largest number of colors used by the on-line algorithm A for G is called the *A-chromatic number* of G and denoted by $\chi_A(G)$. Hence $\chi_A(G) = \max_{\pi \in \Pi(G)} \chi_A(G, \pi)$. The smallest number of colors used by an on-line algorithm for G is the *on-line chromatic number* of G, and denoted by $\chi_{OL}(G)$ [6]. Hence $\chi_{OL}(G) = \min_{A \in AOL(G)} \chi_A(G)$. Let $\mathcal{G}$ denote a (possibly infinite) family of graphs. If $A \in AOL(G)$ for every $G \in \mathcal{G}$, we say that A is an on-line coloring algorithm for $\mathcal{G}$ and write $A \in AOL(\mathcal{G})$. An algorithm $A \in AOL(\mathcal{G})$ is said to be *competitive* for $\mathcal{G}$ if there exists a function f such that $\chi_A(G) \leq f(\chi(G))$ for every $G \in \mathcal{G}$; it is *on-line competitive* if $\chi_A(G) \leq f(\chi_{OL}(G))$ for every $G \in \mathcal{G}$.

4   On-Line Competitive Coloring Algorithms

As stated before, there does not exist a competitive on-line coloring algorithm for P_6-free bipartite graphs, but there exists a competitive on-line coloring algorithm for P_5-free bipartite graphs. In fact, combining results from [4, 8, 12, 13], and analyzing a few cases ourselves, we can show there exists an on-line coloring algorithm that is on-line competitive for the class of H-free bipartite graphs for any fixed graph H on at most five vertices.

Proposition 1. *Let H be a (bipartite) graph on at most five vertices. Then there exists an on-line coloring algorithm that is on-line competitive for the class of H-free bipartite graphs.*

Proof. The statement is trivial when H is not bipartite. We may further restrict ourselves to bipartite graphs on exactly five vertices, noting that an F-free bipartite graph with F bipartite on at most four vertices is also H-free for some bipartite graph H on five vertices. We use $H + H'$ to denote the disjoint union of two graphs H and H', and pH to denote the disjoint union of $p \geq 2$ copies of H. Before we make a case distinction we first make the following easy observation:

(1) Let F be a graph and A an on-line coloring algorithm that is on-line competitive for the class of F-free bipartite graphs. Then there exists an on-line coloring algorithm A' that is on-line competitive for the class of $F + K_1$-free bipartite graphs.

This claim can be seen as follows. Initially we use algorithm A to color the vertices of an $F + K_1$-free graph G. If G contains an induced F, then as soon as all vertices of F have been colored all vertices presented afterwards have a neighbor in F. Since G is bipartite, this means that the coloring of G can be finished using only two new colors at most. We now distinguish a number of cases depending on the value of $|E(G)| = m$.

Case I: $m = 0$. Then $H = 5K_1$ and clearly $\chi_{FF} \leq 5$, since FF only uses color 6 on a vertex that has already neighbors with colors 1 to 5. In a bipartite graph these neighbors form an independent set. On-line competitiveness also follows from applying (1) five times.

Case II: $m = 1$. Then $H = K_2 + 3K_1$. It is trivial to see that FF is on-line competitive for the class of K_2-free graphs. After applying (1) three times we get the desired result.

Case III: $m = 2$. Then $H = P_3 + 2K_1$ or $2K_2 + K_1$. For the first subcase we can proceed similarly as in Case II. For the second subcase we use the following result from [8]:

(2) If G is a P_5-free graph without triangles, then $\chi_{FF}(G) \leq 3$.

Noting that $2K_2$-free bipartite graphs are both P_5-free and triangle-free, and combining (1) and (2), yields the result.

Case IV: $m = 3$. Then $H = P_4 + K_1$, $K_{1,3} + K_1$, or $P_3 + K_2$. Noting that P_4-free bipartite graphs are both P_5-free and triangle-free, and combining (1) and (2), yields the desired result for the first subcase. For the second subcase we first observe that $\chi_{FF}(G) \leq 3$ for any $K_{1,3}$-free bipartite graph G (cf. Case I), and then we apply (1) to get the result. Since a $P_3 + K_2$-free bipartite graph is a P_6-free bipartite graph, we can of course immediately apply Theorem 1 (which will be presented later) for the third subcase. It is also not difficult to give a direct proof that our algorithm *BicolorMax* is on-line competitive for this class of graphs.

Case V: $m = 4$. Then $H = K_{1,4}$, $C_4 + K_1$, P_5, or the unique connected graph with degree sequence 3,2,1,1,1 which we denote by $K_{1,3}^+$. For the first subcase we easily get that $\chi_{FF}(G) \le 4$ in a similar way as in Case I. The *girth* of a graph G is the number of edges of a smallest cycle in G. For the second subcase we combine (1) with the following result from [4]:

(3) If G has girth at least five, then $\chi_{FF}(G) \le \binom{2^{\chi_{OL}(G)}}{2}$.

For the third subcase we use (2). The *radius* of a graph G is defined as the minimum of $\max_v d(u, v, G)$ over all vertices u in G. For the fourth subcase we use the following result from [13]:

(4) For every tree T with radius 2, there is an on-line coloring algorithm A that is on-line competitive for the class of T-free graphs.

Case VI: $m = 5$. Then $H = K_{2,3} - e$ for an edge e of $K_{2,3}$. We need a separate proof for this case. We first prove the following claim:

Claim: Let G be bipartite and H-free and let C be a component of G such that C_4 is an induced subgraph of C. Then $C = K_{s,t}$ for some integers $s, t \ge 2$.

We prove this claim as follows. If $C = C_4 = K_{2,2}$, then the claim trivially holds. If not, let $C_4 = uvwxu$, and let $N(p)$ denote the neighbors of vertex p in C. If $N(u) \not\subseteq N(w)$, then G contains H as an induced subgraph. So, by symmetry, $N(u) = N(w)$, and similarly $N(v) = N(x)$. Let $y \in N(u) \cap N(w)$. Then $uvwyu$ is an induced C_4, so as before $N(y) = N(v) = N(x)$. Hence all neighbors of u and w are adjacent to all neighbors of v and x, and vice versa. By repeating the arguments for all induced C_4s, we obtain that $C = K_{s,t}$ for some $s, t \ge 2$.

Since $\chi_{FF}(K_{s,t}) = 2$, the above claim together with (3) implies that $\chi_{FF}(G) \le \max\{\binom{2^{\chi_{OL}(G)}}{2}, 2\}$.

Case VII: $m = 6$. Then $H = K_{2,3}$. Kierstead and Penrice [12] showed that FF is on-line competitive for the class of H-free graphs. $\qquad\qquad\square$

We conclude that the first open question with respect to the (non)existence of on-line competitive coloring algorithms for H-free bipartite graphs concerns bipartite graphs H on 6 vertices, in particular $H = P_6$. In 4.1 we present a new on-line algorithm for coloring general bipartite graphs. We analyze the behavior of this algorithm in 4.3 and 4.4. In 4.3 we present our main results: the algorithm is a linear on-line competitive algorithm for P_6-free bipartite graphs and for P_7-free bipartite graphs. For our proof of the P_6-free case we need a suitable new class of P_6-free bipartite graphs that will be introduced in 4.2. We will not prove the P_7-free case here due to the page limits. In 4.4 we give a new upper bound for the on-line chromatic number of bipartite graphs.

4.1   The Algorithm BicolorMax

Let G be a bipartite graph on n vertices denoted by $1, 2, \ldots, n$. Let $A = \{a_1, a_2, \ldots, a_p\}$ and $B = \{b_1, b_2, \ldots, b_p\}$ be two disjoint ordered sets of colors. For a fixed positive integer $k \leq p$, let $A(k) = \{a_1, a_2, \ldots, a_k\}$ and $B(k) = \{b_1, b_2, \ldots, b_k\}$.

We first give the general idea of our on-line algorithm. Suppose that G is presented to the algorithm. At some stage a new uncolored vertex v of G is revealed, together with its adjacencies to the set S of already colored vertices of G. If v is not adjacent to any previously revealed vertex of G, then v receives color a_1. Otherwise, the choice of the color for v is based on the present colors in the bipartition classes of the subgraph of G induced by v and the vertices of S with colors in $A(k) \cup B(k)$ for some suitable $k \geq 1$. To explain this choice we first need to introduce some additional terminology.

If $F \subseteq V(G)$, then the *hue* of F, denoted by $H(F)$, is the set of all colors used on vertices in F. Let $\pi(G)$ be a permutation of $V(G)$, and assume that $v = \pi(j)$. Let $G_j(k, v)$ denote the subgraph of $G[\{\pi(1), \ldots, \pi(j)\}]$ induced by $v = \pi(j)$ and all the vertices in $\{\pi(1), \ldots, \pi(j-1)\}$ that have been assigned colors from $A(k) \cup B(k)$. We denote by $C_j(k, v)$ the component of $G_j(k, v)$ containing v, and we write $C_j(k, v) := (I_1, I_2)$ to indicate the bipartition of its vertex set. Note that (I_1, I_2) is the unique bipartition of $C_j(k, v)$, because $C_j(k, v)$ is connected. We say that color a_k is *mixed* on $C_j(k, v) = (I_1, I_2)$ if there exist at least two vertices $v \in I_1$ and $w \in I_2$ that have been colored with a_k. We then call (v, w) a *k-mixed pair*.

The algorithm *BicolorMax* is defined inductively. The vertex $\pi(1)$ is colored with a_1. Suppose that vertices $\pi(1), \ldots, \pi(j-1)$ have already been colored and let $v = \pi(j)$ be the next vertex presented to the algorithm.

$BicolorMax(G[\{\pi(1), \ldots, \pi(j-1)\}], v)$

$m := \max(\{0\} \cup \{k : a_k \text{ is mixed on } C_j(k, v)\})$.
if $a_{m+1} \notin H(V(C_j(m+1, v)))$
    $C_j(m+1, v) := (I_1, I_2)$ such that $v \in I_1$
else
    $C_j(m+1, v) := (I_1, I_2)$ such that $a_{m+1} \in H(I_1)$.
if $v \in I_1$
    assign color a_{m+1} to v
else
    assign color b_{m+1} to v.

It is easy to check that *BicolorMax* is a polynomial time on-line coloring algorithm for bipartite graphs. We leave the details to the reader, but we illustrate the algorithm with the following example.

Example 1. Let G be a $K_{4,4}$ without a perfect matching, i.e., with $V(G) = \{1, 2, 3, 4, 5, 6, 7, 8\}$, bipartition in $\{1, 3, 5, 7\}$ and $\{2, 4, 6, 8\}$, and only edges $\{1, 2\}$, $\{3, 4\}$, $\{5, 6\}$, and $\{7, 8\}$ omitted. If the vertices are revealed in the order of increasing numbers, the algorithm assigns colors $a_1, a_1, b_1, b_1, a_2, b_2, a_2, b_2$, respectively. The last color is assigned since a_1 is mixed in the subgraph of G induced by $\{1, 2, 3, 4, 8\}$, while a_2 is assigned to a vertex in the other bipartition class of $C_8(2, 8) = G$ than the vertex 6. Suppose that G is extended and a new vertex 9 is revealed. Then 9 is respectively assigned color a_1 if 9 is only adjacent to 7, color b_1 if 9 is adjacent to 1 and 7, color b_2 if 9 is adjacent to 1, 3 and 7, and color a_2 if 9 is adjacent to 2, 4 and 6. For a $K_{n,n}$ without a perfect matching with $n \geq 5$ the algorithm will continue assigning a_2 and b_2 if the vertices are presented in an order alternating between the two classes of the bipartition, as in the above example for $n = 4$. In contrast, recall that FF uses n colors in this case.

4.2 A Class of P_6-Free Bipartite Graphs

The objective is to show that *BicolorMax* is an on-line competitive algorithm for P_6-free bipartite graphs. As a first step, we inductively define a class of P_6-free bipartite graphs (see Figure 1). Before giving the formal definition, intuitively, the next member is composed of two remote copies of the previous member with complementary adjacencies with respect to the bipartition. The latter property enables us to define a permutation that forces a large number of colors on any on-line coloring algorithm for the large members of this class. It will turn out that a member H_k from this class has on-line chromatic number at least k, and that if *BicolorMax* uses color a_k on a P_6-free bipartite graph G, then H_{k+1} is an induced subgraph of G.

Each graph H_i of the class has a root vertex $r(H_i)$, and:

- H_1 is a graph consisting of a single root vertex.
- H_2 is a graph consisting of an edge, one of whose end vertices is the root vertex.
- H_3 is a path on four vertices, one of whose internal vertices is the root vertex.
- H_k, $k \geq 4$ consists of a root vertex v and two disjoint copies H_{k-1}^1 and H_{k-1}^2 of H_{k-1} and edges joining v to all non-neighbors of $r(H_{k-1}^1)$ (including $r(H_{k-1}^1)$) in H_{k-1}^1 and all neighbors of $r(H_{k-1}^2)$ in H_{k-1}^2.

It is easy to check that for all $k \geq 1$ the graph H_k is bipartite and P_6-free. We note that the above defined class is different from the class of P_6-free bipartite graphs defined in [6]. The graphs H_k have the following useful properties.

Lemma 1. *The two remote copies H_{k-1}^1 and H_{k-1}^2 of H_{k-1} in H_k ($k \geq 4$) each contain:*

- (i) *a set of pairwise remote subgraphs isomorphic to $H_1, \ldots, H_{k-2}$ with all the vertices in the bipartition class containing their root vertex adjacent to $r(H_k)$;*
- (ii) *a set of pairwise remote subgraphs isomorphic to $H_1, \ldots, H_{k-2}$ with all the vertices in the bipartition class not containing their root vertex adjacent to $r(H_k)$.*

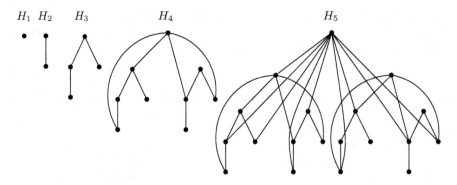

Fig. 1. The graphs H_1, H_2, H_3, H_4, H_5

Proof. By induction on k. This can easily be checked. Note that a subgraph in (i) can use some vertices of H_k that a graph in (ii) also uses. □

The structural properties of H_k imply that its on-line chromatic number is at least k.

Proposition 2. *For any $k \geq 1$, $\chi_{OL}(H_k) \geq k$.*

Proof. By induction on k. It is routine to check this for $k = 1, 2, 3$. Suppose that $k \geq 4$ and that the result holds for H_k with $4 \leq k \leq t$. Consider H_{t+1} and an on-line algorithm A for coloring H_{t+1}. The first time the ith color is used by A we identify it as color i. We choose an ordering on $V(H_{t+1})$ such that the vertices of pairwise remote copies of $H_1, \ldots, H_t$ are presented until color i is used on H_i $(i = 1, \ldots, t)$; then, if $i \leq t - 1$, we immediately start presenting the vertices of H_{i+1}. By the adjacency relations from the definition of H_{t+1} and the properties of Lemma 1, the ordering of the vertices of $H_1, \ldots, H_t$ can be chosen in such a way that $r(H_{t+1})$ is adjacent to the (not necessarily root) vertices that received colors $1, \ldots, t$. Hence a new color $t + 1$ is forced upon A. □

4.3   BicolorMax is On-Line Competitive

Before we present our main result on the on-line competitiveness of *BicolorMax*, we make a number of useful observations in the following three lemmas.

Lemma 2. *Let G be a bipartite graph. Let BicolorMax color vertex $v = \pi(j)$ with a_m or b_m, $m \geq 2$. If (x, y) is a k-mixed pair in $C_j(k, v)$ with $k \leq m - 1$, then any path between x and y in $C_j(k, v)$ must pass through v.*

Proof. Suppose there exists a path in $C_j(k, v)$ between x and y not passing through v. Let $x = \pi(r)$ and let $y = \pi(s)$. We assume without loss of generality that y has been presented to *BicolorMax* after x, i.e., $s > r$. Suppose x belongs to $C_s(k, y)$, implying that $a_k \in H(V(C_s(k, y)))$. Since y is colored with a_k, color a_{k-1} is mixed on $C_s(k-1, y)$. Then *BicolorMax* would have colored y with color b_k. Hence x does not belong to the component $C_s(k, y)$. Suppose there exists an index i with $s < i < j$ such that x and y belong to the component $C_i(k, \pi(i))$. This means that a_k is mixed on $C_i(k, \pi(i))$. Then $BicolorMax$ would never use a color a_h with $h \leq k$ to color $\pi(i)$. This implies that such an index i does not exist. We conclude that every path between x and y in $C_j(k, v)$ must pass through v.   □

Lemma 3. Let G be a P_6-free bipartite graph. Let $BicolorMax$ color vertex $v = \pi(j)$ with a_m, $m \geq 2$. Let z be a vertex in $C_j(m-1, v)$ assigned color a_{m-1}. If z has odd distance from v in $C_j(m-1, v)$, then $d(v, z, C_j(m-1, v)) = 1$. Otherwise $d(v, z, C_j(m-1, v)) = 2$.

Proof. Since $BicolorMax$ uses a_m for v, color a_{m-1} is mixed on $C_j(m-1, v)$. This means that there exists a vertex z^* with color a_{m-1}, such that z and z^* are in different classes of the bipartition of $C_j(m-1, v)$. By Lemma 2, a shortest path P_{zz^*} must be formed by joining shortest paths P_{zv} and P_{vz^*}. Suppose $d(v, z, C_j(m-1, v))$ is odd. Then z^* has even distance from v in $C_j(m-1, v)$ implying that $d(v, z, C_j(m-1, v)) \geq 2$. If $d(v, z, C_j(m-1, v)) \geq 3$, then P_{zz^*} contains an induced P_6. Hence $d(v, z, C_j(m-1, v)) = 1$. Suppose $d(v, z, C_j(m, v))$ is even. If $d(v, z, C_j(m-1, v)) \geq 4$, then P_{zz^*} contains an induced P_6. Hence $d(v, z, C_j(m-1, v)) = 2$.   □

Lemma 4. Let G be a P_6-free bipartite graph. If $BicolorMax$ uses color a_k on vertex $v = \pi(j)$, $k \geq 2$, then $C_j(k-1, v)$ contains H_{k+1} as an induced subgraph with $v = r(H_{k+1})$.

Proof. By induction on k. The case $k = 2$ is trivial. Let $k \geq 3$. Since $BicolorMax$ uses color a_k on vertex v, there exists a $(k-1)$-mixed pair (x, y) in $C_j(k-1, v)$. Assume $x = \pi(r)$ and $y = \pi(s)$. By Lemma 2 the components $C_r(k-2, x)$ and $C_s(k-2, y)$ are remote. By the inductive hypothesis x is the root of an induced copy H_k^1 of H_k in $C_r(k-2, x)$ and y is the root of an induced copy H_k^2 of H_k in $C_s(k-2, y)$. Lemma 3 implies that we may without loss of generality assume that distance $d(x, v, C_j(k-1, v)) = 2$ and distance $d(y, v, C_j(k-1, v)) = 1$. We claim that v is adjacent to all neighbors of x in H_k^1 and to all non-neighbors of y in H_k^2. Suppose x has a neighbor x' in H_k^1 not adjacent to v. Let y' be a neighbor of y in H_k^2, and let z be a common neighbor of x and v in $C_j(k-1, v)$. Then the path $x'xzvyy'$ is an induced P_6 in G, which is a contradiction. By using similar arguments we can prove that v is adjacent to all non-neighbors of y in H_k^2. Hence, we obtain an induced H_{k+1} in $C_j(k-1, v)$ with $v = r(H_{k+1})$.   □

We now present our main theorem showing that $BicolorMax$ is a linear online competitive algorithm for the class of P_6-free bipartite graphs. Denote by $\chi_{Bm}(G)$ the maximum number of colors used by $BicolorMax$ for coloring G.

Theorem 1. If G is a P_6-free bipartite graph, then $\chi_{Bm}(G) \leq 2\chi_{OL}(G) - 1$.

Proof. Let k be the highest index such that $BicolorMax$ uses color a_k on a vertex in the P_6-free bipartite graph G. Since $BicolorMax$ only uses b_i with $i \leq k$ if a_i has been used before, $\chi_{Bm}(G) \leq 2k$. For $k = 1$ the statement of the theorem obviously holds. Suppose $k \geq 2$. Due to Lemma 4 the graph G contains a copy of H_{k+1} as an induced subgraph. Proposition 2 implies that $\chi_{OL}(G) \geq \chi_{OL}(H_{k+1}) \geq k+1$. $\hfill\square$

Using a similar but more involved analysis, we were able to prove the following result, showing that *BicolorMax* is also on-line competitive for the class of P_7-free bipartite graphs. We will postpone the proof to the full paper.

Theorem 2. *If G is a P_7-free bipartite graph, then $\chi_{Bm}(G) \leq 8\chi_{OL}(G) + 8$.*

4.4   A New Upper Bound on χ_{OL} for Bipartite Graphs

In [15], Lovász, Saks and Trotter define an on-line coloring algorithm A for general graphs that has $\chi_A(G) \leq 2\log_2(n)$ when applied to any bipartite graph G on n vertices (See also [11]). Below we give a tighter upper bound for the on-line chromatic number of a bipartite graph in terms of subgraphs isomorphic to P_5. We note that it is not possible to prove an upper bound in terms of induced subgraphs isomorphic to P_6, since it follows from Proposition 2 and also from a result in [6] that no competitive algorithm exists for the class of P_6-free bipartite graphs.

Theorem 3. *Let G be a bipartite graph in which each component has at most s pairwise remote induced subgraphs isomorphic to P_5. If $s = 0$, then $\chi_{Bm}(G) \leq 4$. If $s > 0$, then $\chi_{Bm}(G) \leq 2\log_2(s) + 6$.*

Proof. We prove the theorem by showing that a component C of G contains at least 2^{k-3} pairwise remote induced subgraphs isomorphic to P_5, if *BicolorMax* uses color a_k on C with $k \geq 3$. We use induction on k. It is easy to check that a component C contains an induced P_5, if *BicolorMax* uses color a_3 on a vertex of C. Let $k \geq 4$. Suppose $v = \pi(j)$ is colored by a_k. Then there exists a $(k-1)$-mixed pair (x,y) in $C_j(k-1,v)$. By Lemma 2, x and y belong to two different components in $G_j(k-1,v) - v$ both containing 2^{k-4} pairwise remote induced subgraphs isomorphic to P_5. $\hfill\square$

The above proof shows that if *BicolorMax* uses color a_3 on a bipartite graph G, then G contains an induced P_5. This implies that *BicolorMax* is competitive for the class of P_5-free bipartite graphs.

5   Conclusions and Future Work

We have introduced the new on-line coloring algorithm *BicolorMax* for bipartite graphs. We have shown that the number of colors used by this algorithm on a bipartite graph G is bounded from above by the number of remote induced subgraphs of G isomorphic to P_5. As a consequence we improved the best known upper bound for the on-line chromatic number of bipartite graphs given in [15]. For any P_6-free (respectively, P_7-free) bipartite graph G, *Bicolor-Max* has been shown to use at most twice (respectively, eight times) as many colors as any optimal on-line coloring algorithm for G. In a future continuation of this work, we would like to face the problem of deciding whether for any $n \geq 8$, a linear on-line competitive algorithm can be defined for the class of P_n-free bipartite graphs. We also consider analyzing *BicolorMax* and related algorithms for other classes of H-free bipartite graphs, in particular for graphs H with 6 vertices. A seemingly difficult and interesting open case is the (non)existence of an on-line competitive algorithm for the class of C_6-free bipartite graphs.

References

1. J.A. BONDY AND U.S.R. MURTY, *Graph Theory with Applications,* Macmillan, London and Elsevier, New York (1976).
2. M. CHROBAK AND M. ŚLUSAREK, Problem 84-23. *Journal of Algorithms 5* (1984), 588.
3. T. ERLEBACH AND J. FIALA, On-line coloring of geometric intersection graphs. *Computational Geometry: Theory and Applications 23* (2002), 243–255.
4. A. GYÁRFÁS, Z. KIRÁLY, AND J. LEHEL, On-line competitive coloring algorithms. *Technical report TR-9703-1* (1997), available at http://www.cs.elte.hu/tr97/.
5. A. GYÁRFÁS, Z. KIRÁLY, AND J. LEHEL, On-line 3-chromatic graphs. II. Critical graphs. *Discrete Mathematics 177* (1997), 99–122.
6. A. GYÁRFÁS AND J. LEHEL, On-line and first-fit colorings of graphs. *Journal of Graph Theory 12* (1988), 217–227.
7. A. GYÁRFÁS AND J. LEHEL, First fit and on-line chromatic number of families of graphs. *Ars Combinatorica 29C* (1990), 168-176.
8. A. GYÁRFÁS AND J. LEHEL, Effective on-line coloring of P_5-free graphs. *Combinatorica 11* (1991), 181–184.
9. M.M. HALLDÓRSSON, Online coloring known graphs. *Electronic Journal of Combinatorics 7* (2000), Research Paper 7, 9pp.
10. H.A. KIERSTEAD, The linearity of first-fit coloring of interval graphs. *SIAM Journal on Discrete Mathematics 1* (1988), 526–530.
11. H.A. KIERSTEAD, Coloring graphs on-line. In: Fiat and Woeginger, eds.: *Online algorithms: the state of the art* (1998), no. 1442 in Lecture Notes in Computer Science, Springer Verlag, 281–305.
12. H. KIERSTEAD AND S.G. PENRICE Radius two trees specify χ-bounded classes. *Journal of Graph Theory 18* 1994, 119–129.
13. H. KIERSTEAD, S.G. PENRICE, AND W. TROTTER, On-line graph coloring and recursive graph theory. *SIAM Journal on Discrete Mathematics 7* 1994, 72–89.
14. K. KOLOSSA, On the on-line chromatic number of the family of on-line 3-chromatic graphs. *Discrete Mathematics 150* (1996), 205–230.
15. L. LOVÁSZ, M. SAKS, AND W.T. TROTTER, An on-line graph coloring algorithm with sublinear performance ratio. *Discrete Mathematics 75* (1989), 319–325.
16. B. RANDERATH AND I. SCHIERMEYER, Vertex colouring and forbidden subgraphs – a survey. *Graphs and Combinatorics 20* (2004), 1–40.

Distributed Approximation Algorithms for Planar Graphs

Andrzej Czygrinow[1], Michał Hańćkowiak[2], and Edyta Szymańska[2,*]

[1] Department of Mathematics and Statistics,
Arizona State University, Tempe, AZ 85287-1804, USA
andrzej@math.la.asu.edu
[2] Faculty of Mathematics and Computer Science,
Adam Mickiewicz University, Poznań, Poland
mhanckow@amu.edu.pl, edka@amu.edu.pl

Abstract. In this paper we construct two distributed algorithms for computing approximations of a largest matching and a minimum dominating set in planar graphs on n vertices. The approximation ratio in both cases approaches one with n tending to infinity and the number of synchronous communication rounds is poly-logarithmic in n. Our algorithms are purely deterministic.

1 Introduction

The distributed model of computation has gained a lot of attention after the pioneering work by Awerbuch et. al. [AGLP89] and many others in the mid eighties of the last century. The most fundamental challenge in distributed networks is how the local structure of a network impacts its global properties. This leads to a completely different computational paradigm than the sequential model or the parallel PRAM model. Not surprisingly many problems which admit efficient sequential protocols, as maximum matching or maximal independent set, to name a few, require a completely new algorithmic approach and yield interesting open problems in discrete mathematics.

The model considered in this paper was introduced by Linial in [L92] and named $\mathcal{LOCAL}$ in [P00]. In this model, the network is represented by an undirected graph, each vertex of which corresponds to a processor, and each edge corresponds to a communication channel between two processors. The network is synchronized and computations proceed in discrete rounds. In a single round a vertex can send and receive messages to and from its neighbors, and perform some local computations. Neither the amount of local computations nor the lengths of messages are restricted in any way. In addition, we assume that vertices have unique identifiers. There are several measures of efficiency of distributed protocols but we will concentrate on its time complexity, that is, a maximum number of rounds needed to find a solution. An algorithm is efficient if its time complexity is poly-logarithmic in n.

* The third author thanks the Department of Mathematics and Computer Science at Emory University for providing an office space and computer access.

T. Calamoneri, I. Finocchi, G.F. Italiano (Eds.): CIAC 2006, LNCS 3998, pp. 296–307, 2006.

Very few classical graph-theoretic problems admit efficient distributed algorithms. For example, even the maximal independent set problem, for which an efficient deterministic PRAM algorithm exists [L86], still has an unknown distributed complexity. Another approach towards a better understanding of a computational model, is to study the approximability of problems in that model. This has motivated intensive research on approximation algorithms in the distributed model. For the state of the art of distributed approximation we refer the reader to an excellent survey by Elkin [E04].

In this paper we design distributed algorithms for planar graphs and exploit the fact the planar graphs are minor monotone. In a given graph $G = (V, E)$ every set of pairwise disjoint edges constitutes a matching. Let $\beta(G)$ denote the cardinality of a largest matching in G. The *maximum matching* problem is to find a matching M in graph G of size $\beta(G)$. A *dominating set* in a graph G is a subset D of vertices such that for every vertex $v \notin D$ a neighbor u of v belongs to D. By $\gamma(G)$ we will denote the cardinality of a smallest dominating set in G, also known as the *domination number* . The *minimum dominating set* problem is to find a dominating set D in graph G of size $\gamma(G)$. We propose two purely deterministic distributed algorithms with the poly-logarithmic time complexity. For every planar graph on n vertices, our first algorithm finds a matching M such that $|M| \geq (1 - O(\frac{1}{\log n}))\beta(G)$ (see Theorem 1). The second algorithm works for planar graphs that do not contain $K_{2,\log n}$ as a subgraph. In every such graph on n vertices it finds a dominating set D such that $|D| \leq (1+O(\frac{1}{\log n}))\gamma(G)$ (see Theorem 2). Although this technical assumption certainly restricts the applicability of the method, the subclass of $K_{2,\log n}$-free planar graphs is quite large and contains, for example, outer-planar graphs (they do not contain $K_{2,3}$).

To give an overview of previous research, let us mention that there exists no efficient distributed protocol for finding a maximum matching or a minimum dominating set even when restricted to very particular families of networks. As shown by Linial in [L92], finding a maximum matching in a cycle on n vertices requires $\Omega(n)$ rounds and the same bound holds for minimum dominating set. More recently, it has been shown in [KMW04] that the number of rounds required in order to achieve a constant or even only a poly-logarithmic approximation ratio for constructing an inclusion maximal matching and a minimum dominating set is at least $\Omega(\sqrt{\log n/\log\log n})$ or $\Omega(\log \Delta/\log\log \Delta)$, where Δ denotes the maximum degree of the graph.

A maximal matching problem admits a $O(\log n)$ time randomized distributed algorithm (see, for example [L86]). Later, in [HKP99] a deterministic, poly-logarithmic time algorithm for this problem was given. The techniques from [HKP99] were applied in [CHS04] and [CHSz04] to give a 2/3-approximation for maximum matching in general graphs. Moreover, based on these ideas, in [CH03] a $(1 - \epsilon)$-approximation (for any fixed $\epsilon > 0$) for bipartite graphs was derived.

For the minimum dominating set problem, Kutten and Peleg [KP95] gave an efficient distributed algorithm which finds a dominating set of size at most $n/2$ in general graphs. The first non-trivial approximation ratio, $O(\log \Delta)$ was achieved in [JRS01] by a randomized method. Further, in [KW03] a $O(k\Delta^{2/k} \log \Delta)$-approximation in constant time was obtained using the LP relaxation techniques with randomization. Similar, randomized result for the connected dominating set can be found in [DPRS03]. In contrast, our results are purely deterministic and are among only few examples of distributed protocols where the poly-logarithmic time complexity with a very good approximation ratio is achieved without the use of random bits.

Approximations of weighted versions of the maximum matching and minimum dominating set problems were recently studied in [CH04]. Our proof techniques rely on a clustering procedure introduced in [CH04]. We further develop the method in this work and hope that it might be applied for other problems. By a cluster we mean a subset of the vertex set that induces a connected subgraph. The clustering procedure partitions the nodes of the input graph into clusters. If the diameter of a cluster is poly-logarithmic in n, then, in the $\mathcal{LOCAL}$ model, we can compute every function efficiently. Therefore, having the vertices grouped into clusters we find the maximum matching in every cluster. The union of the matchings yields a matching of size approximating $\beta(G)$. The situation is similar but more complicated in the case of the minimum dominating set. We first make sure that all vertices of large degree are included in the dominating set. Then the clustering is performed and a set of vertices dominating the remaining vertices is constructed within the clusters.

In both problems the number of rounds of our algorithms is a poly-logarithmic function determined by the diameters of the clusters. At the same time, we control the number of edges connecting different clusters and based on that value the approximation ratio is derived. For both algorithms, better approximation ratios can be achieved at the expense of higher running times.

The rest of the paper is organized as follows. In Section 2, we present the clustering algorithm. Sections 3 and 4 contain the description and analysis of the approximation algorithms for the maximum matching and the minimum dominating set, respectively.

2   Clustering Algorithm

In this section, we give a clustering algorithm which will be applied to find matchings as well as dominating sets. We will use the low-degree decomposition of a planar graph from [CH04].

Definition 1. *A low-degree decomposition of a planar graph* $G = (V, E)$ *is a partition of* V *into* K *independent sets* $V_1, \ldots, V_K$ *that satisfies two conditions:*

1. $K = O(\log |V|)$.
2. *For every* $i = 1, \ldots, K - 1$, *if* $v \in V_i$ *then* $deg(v, \bigcup_{l=i+1}^{K} V_l) \leq 6$.

It is not difficult to prove that every planar graph admits a low-degree decomposition. In addition, as shown in [CH04], such a decomposition can be found efficiently by a distributed algorithm.

DECOMPOSITION

Input: Planar graph G, number n such that $|V(G)| \le n$.
Output: Low-degree decomposition $V_1, \ldots, V_{\log_k n}$ of G with $k = 36/35$.

1. Let $U := V(G)$, $i := 1$.
2. Iterate $\log_{36/35} n$ times:
   (a) Let A be the set of vertices in $G[U]$ of degree at most 6.
   (b) Use the Cole-Vishkin algorithm from [CV86] to find a maximal independent set I in the subgraph of $G[U]$ induced by A.
   (c) $V_i := I$, $i := i + 1$, $U := U \setminus I$.

Lemma 1. *[CH04] Let $G = (V, E)$ be a planar graph such that the identifiers of V are in $\{1, \ldots, n\}$. Then the procedure* DECOMPOSITION *finds a low-degree decomposition of G in $O(\log^* n \log n)$ rounds.*

Our approximation algorithms will use a similar clustering strategy as the one in [CH04]. In addition to procedure CLUSTERING we introduce a subprocedure SMALLCLUSTERS. The latter computes clusters of a constant diameter and in each cluster finds a set of vertex disjoint stars with special properties that can be used by CLUSTERING to compute "big clusters". Thanks to this approach we save on the time complexity for constructing the clusters (see Lemma 4).

SMALLCLUSTERS

Input: Planar graph $G = (V, E)$ with weights on edges $\omega : E \mapsto \mathcal{R}^+$ and number n such that $|V| \le n$ and $ID(v) \le n$.
Output: Set of vertex-disjoint stars in G.

1. $H := G$.
2. Iterate $\log 10 / \log \frac{12}{11}$ times:
   (a) Call DECOMPOSITION to find a partition $W_1, \ldots, W_{\log_{36} n}$ of H. In addition, let $Z_i := \bigcup_{l > i} W_l$.
   (b) For every vertex w, in parallel, if $w \in W_i$ and $N(w) \cap Z_i \ne \emptyset$ then:
      – Let $u(w)$ be a vertex in $N(w) \cap Z_i$ such that
      $$\omega(\{w, u(w)\}) = \max_{v \in N(w) \cap Z_i} \omega(\{w, v\}).$$
      – Add $\{w, u(w)\}$ to the auxiliary graph F.
   (c) Each connected component of F is a tree of diameter $O(\log n)$. For each tree T in F, in parallel, find a set of disjoint stars $S_1, S_2, \ldots,$ in T of the maximum weight.
   (d) Modify H as follows:
      – In each star, contract vertices to create a new vertex. Let $V(H)$ consist of new vertices and those vertices which were not contracted.
      – For every $v, w \in V(H)$ set the weight of $\{v, w\}$ to be the sum of weights of edges between vertices contracted to v and vertices contracted to w.

3. If $V(H) = \{v_1, ..., v_M\}$ then for each v_i let V_i be the set of vertices contracted to v_i in all of the above iterations.
4. In each $G[V_i]$, in parallel, compute a set of disjoint stars $Q_1^{(i)}, \ldots, Q_{M(i)}^{(i)}$ of the largest possible weight.
5. Return the set of stars $Q_j^{(i)}$, for $i = 1, \ldots, M; j = 1, \ldots, M(i)$.

Let κ be the supremum of all real numbers r such that every weighted planar graph G contains a set of vertex-disjoint stars with the total weight of at least an r fraction of the weight of G. We need the following lemma.

Lemma 2. $\kappa \geq \frac{1}{5}$.

Proof. A star forest is a forest in which every connected component is a star and the star arboricity of a graph G, $st(G)$, is the minimum number of star forests that partition $E(G)$. Hakimi et al. [HMS96] showed that if G is planar then $st(G) \leq 5$ and so there is a set of vertex-disjoint stars with weight of at least $\omega(G)/5$ where $\omega(G) = \sum_{e \in E} \omega(e)$.

Lemma 3. Let $Q_1, \ldots, Q_L$ be the disjoint stars in G obtained from SMALL-CLUSTERS. Then

$$\omega(\bigcup_i Q_i) \geq \frac{9}{10}\kappa\omega(G).$$

Proof. Let σ_i be the maximum diameter of a subgraph of G which corresponds to a vertex of H in the ith iteration. Then $\sigma_i \leq 3\sigma_{i-1} + 2$ with $\sigma_0 = 0$ which gives $\sigma_i < 2 \cdot 3^i$ and so $\sigma_k < 2 \cdot 3^{27}$ for $k = \log 10 / \log \frac{12}{11}$. Therefore each subgraph $G[V_i]$ in step 4 has a constant diameter and its optimal set of stars can be computed in a constant number of rounds. Let P_i be the sum of weights of edges in H in the ith iteration. In the next iteration $w(F)$ is at least $P_i/6$ and the stars $S_1, S_2, \ldots,$ in each tree of F have the weight of at least $\omega(T)/2$. Consequently, the weight of the graph in the $(i+1)$st iteration, P_{i+1}, is at most $\frac{11}{12}P_i$ and $P_0 = \omega(G)$. This gives $P_k \leq \frac{1}{10}\omega(G)$ for $k = \log 10 / \log \frac{12}{11}$. By Lemma 2, the weight of stars in $G[V_i]$ is larger then $\frac{9}{10}\kappa\omega(G)$.

The procedure SMALLCLUSTERS is now used in CLUSTERING given below.

CLUSTERING
Input: Planar graph $G = (V, E)$ and number n such that $|V| \leq n$ and $ID(v) \leq n$. Number $c \geq 1$.
Output: Partition of V into L sets $V_1, \ldots, V_L$.

1. $H := G$ and let $\omega(e) := 1$ for any $e \in E(H)$.
2. Iterate $c \log \log n / \log \frac{1}{1 - \frac{9}{10}\kappa}$ times:
   (a) Call SMALLCLUSTERS to find set of disjoint stars $S_1, S_2, \ldots$ in H.
   (b) Modify H as in step 2(d) of SMALLCLUSTERS

3. If $V(H) = \{v_1, ..., v_L\}$ then for each v_i let V_i be the set of vertices contracted to v_i in all of the above iterations.
4. Return sets $V_1, \ldots, V_L$.

We summarize the CLUSTERING in the next lemma.

Lemma 4. *Let* $V_1, \ldots, V_L$ *be the clusters in* G *obtained from* CLUSTERING. *Then*

1. *For every* i, $G[V_i]$ *is a subgraph of diameter* $O(\log^d n)$, *where*

$$d = c\log 3/\log \frac{1}{1 - \frac{9}{10}\kappa} < 5.54c.$$

2. *The number of edges connecting different clusters is* $O(|E(G)|/\log^c n)$.
3. CLUSTERING *can be performed in* $O(\log\log n \log^* n \log^{1+d} n)$ *rounds.*

Proof. Analogously to the proof of Lemma 3 we have $\sigma_i < 2 \cdot 3^i$ and so $\sigma_k < 2\log^d n$ for $k = c\log\log n/\log \frac{1}{1-\kappa\frac{9}{10}}$. Then, for the second part, we have $P_{i+1} \leq (1 - \kappa\frac{9}{10})P_i$ and $P_0 = |E(G)|$, and so $P_k = O(|E(G)|/\log^c n)$. Finally, the third part of the lemma follows from the fact that we have $O(\log\log n)$ iterations of step 2 and in each iteration we invoke SMALLCLUSTERS that calls the DECOMPOSITION a constant number of times. DECOMPOSITION, in turn, needs $O(\log^* n \log n)$ rounds. Since the diameter of each cluster (which corresponds to a vertex of H) is $O(\log^d n)$, CLUSTERING needs $O(\log^* n \log^{1+d} n)$ rounds.

3   Maximum Matching

In this section, we will give a distributed algorithm which approximates a maximum matching in a planar graph G. The algorithm consists of two main parts. First we modify the graph G to obtain a new graph $\bar{G}$ and then we invoke the clustering algorithm for $\bar{G}$ and find a maximum matching locally in each cluster. Recall that the total number of edges connecting different clusters is small in comparison with the number of vertices in the graph. However, a maximum matching in a planar graph can be much smaller than the number of vertices and so if clustering is invoked in such a graph its result would not yield a good approximation. The preprocessing phase addresses this issue. It obtains from a graph G a subgraph $\bar{G}$ with the property that $\beta(G) = \beta(\bar{G}) = \Omega(|V(\bar{G})|)$.

The first phase of the algorithm, the preprocessing, eliminates (by deleting some of the vertices) two special subgraphs of G: the stars and the double-stars. We say that G contains a k-*star* if for some $v, v_1, \ldots, v_k \in V(G)$, $\{v, v_i\} \in E(G)$ for every i, and $deg_G(v_i) = 1$ for every i. In a similar way, we say that G contains a k-*double-star* if for some $u, v, v_1, \ldots, v_k \in V(G)$, $\{u, v\} \in E(G)$ and $\{v, v_i\} \in E(G)$ for every i, and $deg_G(v_i) = 2$ for every i. Every such structure contributes at most two edges to any maximum matching in G. In the next two lemmas we shall show that if H contains neither 2-stars nor 3-double-stars then $\beta(H) = \Omega(|V(H)|)$.

Lemma 5. *Let $H = (V, E)$ be a planar graph and let $\tau = |\{v \in V : deg(v) \geq 3\}|$. Then $\beta(H) \geq (\tau + 4)/6$.*

Proof. Let M be a matching in H with $|M| = \beta(H)$. Let V_1 be the set of M-saturated vertices. Then, since M is a maximum matching, $V \setminus V_1$ induces the empty subgraph of H. Let $V_2 := (V \setminus V_1) \cap \{v \in V : deg(v) \geq 3\}$. Consider the bipartite graph $F = H[V_1, V_2]$. As F is planar, $|E(F)| \leq 2(|V_1| + |V_2|) - 4$. On the other hand, $3|V_2| \leq |E(F)|$. Thus $|V_2| + 4 \leq 2|V_1|$. However, $|V_1| = 2\beta(H)$ and $|V_2| \geq \tau - 2\beta(H)$ yields $\beta(H) \geq (\tau + 4)/6$.

Lemma 6. *Let $G = (V, E)$ be a planar graph with $n = |V|$ and no isolated vertices. If G contains neither 2-stars nor 3-double-stars then $\beta(G) = \Omega(n)$.*

Proof (Sketch). By Lemma 5 we may concentrate only on the set $W = \{v \in V(G) : deg(v) = 2\}$ and the case when, say $|W| \geq 14n/15$. Then, for a subset $W' \subseteq W$ such that if $w \in W'$ then $deg(w, W) \geq 1$, using planarity and the assumption about the absence of 2-stars and 3-double-stars, we have $|W'| \geq |W|/2$ and $\beta(G) \geq |W'|/3 = 7n/45$.

PREPROCESS
Input: Planar graph G.
Output: Planar graph $\bar{G}$ with no 2-stars and no 3-double stars.

1. For every vertex v, in parallel, find the largest k-star $v, v_1, \ldots, v_k$ with the center in v. If $k > 1$ then delete $v_2, \ldots, v_k$.
2. For every pair of vertices u, v which are at distance two, in parallel, find the largest k-double-star $u, v, v_1, \ldots, v_k$ with centers in u and v. If $k > 2$ then delete $v_3, \ldots, v_k$.
3. Return the new graph $\bar{G}$.

Clearly $\bar{G}$ contains neither 2-stars nor 3-double-stars, as in the second step we did not create any vertices of degree one. Thus, by Lemma 6, $\beta(\bar{G}) = \Omega(|V(\bar{G})|)$. In addition, it is easy to see that

$$\beta(\bar{G}) = \beta(G). \tag{1}$$

We can now describe our approximation algorithm.

APPROXMAXMATCHING
Input: Planar graph G.
Output: Matching M in G.

1. Call PREPROCESS to obtain $\bar{G}$.
2. Call CLUSTERING with $c = 1$ to partition $V(\bar{G})$ into clusters $V_1, \ldots, V_L$.
3. For every i, in parallel, find a maximum matching M_i in $\bar{G}[V_i]$.
4. Return $M := M_1 \cup M_2 \cup \cdots \cup M_L$.

Theorem 1. APPROXMAXMATCHING *finds in a planar graph G on n vertices a matching M with*

$$|M| \geq (1 - O(1/\log n))\,\beta(G).$$

The algorithm runs in $O(\log\log n \log^* n \log^{1+d} n)$ rounds, where $d = 5.54$.

Proof. Consider a maximum matching M^* in $\bar{G}$ and let M_i^* be the subset of M^* which contains all edges with both endpoints in V_i. In addition, let C be the set of edges that connect different clusters. We have

$$|M^*| \leq |C| + \sum_{i=1}^{L} |M_i^*| \leq |C| + \sum_{i=1}^{L} |M_i| = |C| + |M|.$$

By Lemma 4 (part 2), $|C| \leq |V(\bar{G})|/\log n$ which in view of Lemma 6 gives $|C| \leq O(\beta(\bar{G}))/\log n$. Consequently,

$$|M| \geq \beta(\bar{G}) - |C| = \beta(\bar{G})(1 - O(1/\log n))$$

which by (1) gives $|M| \geq (1 - O(1/\log n))\beta(G)$.

4   Minimum Dominating Set

We will now turn our attention to the minimum dominating set problem. We assume that $G = (V, E)$ is a planar graph on n vertices such that for any two vertices $u, v \in V$ $|N(u) \cap N(v)| \leq \log n$. Again the algorithm has two phases. In the first phase we add to a dominating set vertices with degrees of at least $\log^2 n$. Then we consider two sets of vertices. Let V_{SN} be the set of vertices of degree smaller than $\log^2 n$ which do not have neighbors of degree at least $\log^2 n$, that is

$$V_{SN} = \{v \in V : \forall_{u \in N[v]} deg(u) < \log^2 n\}.$$

Let $V_{BN} \subset V \setminus V_{SN}$ be the set of vertices which have degree smaller than $\log^2 n$ but have a neighbor in V_{SN}, that is

$$V_{BN} = \{v \in V \setminus V_{SN} : deg(v) < \log^2 n, \exists_{u \in V_{SN}} \{u, v\} \in E\}.$$

In the second phase of the algorithm we shall find a clustering using CLUSTERING in the graph induced by $V_{SN} \cup V_{BN}$ and locally, in each cluster V_i, we will find a set of the smallest size which dominates $V_i \cap V_{SN}$.

APPROXMINDS
Input: Planar graph $G = (V, E)$.
Output: A dominating set D in G.

1. Let $D := \emptyset$.
2. For every vertex v, in parallel, if $deg(v) \geq \log^2 n$ then add v to D.

3. Let $G' = G[V_{BN} \cup V_{SN}]$. Call CLUSTERING with constant $c = 5$ to partition $V(G')$ into clusters $V_1, \ldots, V_L$. Let V_i' be the set of vertices v in $V(G') \setminus V_i$ such that for some $u \in V_i$, $\{v, u\} \in E$ and let V_i'' be the set of vertices $v \in V_i$ such that for some $u \in V_i'$, $\{v, u\} \in E$.
4. For every $i = 1, \ldots, L$, in parallel, find a smallest set $D_i' \subseteq V_i$ which dominates $(V_{SN} \cap V_i) \setminus V_i''$. Let $D_i := D_i' \cup V_i'$.
5. For every $i = 1, \ldots, L$, in parallel, add all vertices from D_i to D.
6. Return D.

In the lemma below we analyze the first phase of the algorithm where vertices of degree at least $\log^2 n$ are added to D.

Lemma 7. *Let $G = (V, E)$ be a planar graph such that for any two distinct vertices $u, v \in V$, $|N(u) \cap N(v)| \leq \log n$ and let $B = \{v \in V : deg(v) \geq \log^2 n\}$. If D^* is a dominating set in G then $|B \setminus D^*| = O\left(|D^*|/\log n\right)$.*

Proof. We will show that $|D^*| = \Omega(|B \setminus D^*| \log n)$. For that we first prove that there is a subset $\{w_1, \ldots, w_k\} \subseteq B \setminus D^*$ of at least $k = |B \setminus D^*|/10$ vertices such that each w_i has a set $S_i \subseteq (N(w_i) \setminus \{w_1, \ldots, w_k\})$ of $\frac{\log^2 n}{4}$ neighbors and $S_i \cap S_j = \emptyset$ whenever $i \neq j$. Indeed, note that as $G[B \setminus D^*]$ is planar there is an independent set I in $G[B \setminus D^*]$ of at least $2k$ vertices. Take $w_1 \in I$ arbitrarily and let S_1 be a set of $\frac{\log^2 n}{4}$ neighbors of w_1. Now suppose that $\{w_1, \ldots, w_l\}$ have been selected with $l < k$. Consider the bipartite subgraph of G with bipartition $W = I \setminus \{w_1, \ldots, w_l\}$ and $S = \bigcup_{i=1}^{l} S_i$. Then $G[W, S]$ is a planar graph and so

$$|E(W, S)| \leq 2(|W| + |S|) - 4 = 2\left(|W| + \frac{l \log^2 n}{4}\right) - 4 < 2\left(|W| + \frac{|W| \log^2 n}{4}\right) - 4.$$

Consequently,

$$|E(W, V \setminus S)| > \log^2 n |W| - 2\left(|W| + \frac{|W| \log^2 n}{4}\right) + 4 =$$

$$|W|\left(\frac{\log^2 n}{2} - 2\right) + 4 \geq \frac{|W| \log^2 n}{4}$$

and we can select w_{l+1} from W which is connected with at least $\log^2 n/4$ vertices from $V \setminus S$.

Let $(w_1, S_1), \ldots, (w_k, S_k)$ be as above. Let D be a subset of $V \setminus \{w_1, \ldots, w_k\}$ which dominates $S = \bigcup_{i=1}^{k} S_i$ in G. We claim that $|D| = \Omega(k \log n)$. Consider $D' = D \cap S$. If $|D'| \geq k \log n$ then we are done. Otherwise, let $S_i' = S_i \setminus D$. Since $|S| - |D'| \geq k\left(\frac{\log^2 n}{4} - \log n\right)$ at least $k/2$ of w_i's have $|S_i'| \geq \frac{\log^2 n}{8}$. Otherwise

$$\sum_{i=1}^{k} |S_i'| < \frac{k}{2}\left(\frac{\log^2 n}{8} + \frac{\log^2 n}{4}\right) < |S| - |D'|$$

which is not possible. Without loss of generality, we can assume that for $i = 1 \ldots, k/2$, $|S'_i| = \frac{\log^2 n}{8}$. Note that $\bigcup_{i \le k/2} S'_i \cap D = \emptyset$. Consider the auxiliary bipartite graph $H = (V_1, V_2)$ obtained by setting $V_1 = D$ and contracting each S'_i to one vertex and adding it to V_2. Put an edge between $v \in V_1$ and $S'_i \in V_2$ if v dominates at least one vertex from S'_i in G. First observe that H is planar as all edges correspond to edges in G and so a subdivision of $K_{3,3}$ or K_5 in H will yield the subdivision of $K_{3,3}$ or K_5 in G. Thus $|E(H)| \le 2(|V_1| + |V_2|) - 4 = 2\left(\frac{k}{2} + |D|\right) - 4$. Degree of each S'_i in H is at least $\frac{\log n}{8}$ as if for some i, there are less than $\frac{\log n}{8}$ vertices dominating S'_i then one of them has more than $8|S'_i|/\log n = \log n$ neighbors in S'_i and so more than $\log n$ common neighbors with w_i. Thus,

$$\frac{k \log n}{8} \le |E(H)| \le k + 2|D| - 4$$

and so

$$|D^*| \ge |D| \ge \left(\frac{\log n}{8} - 1\right)\frac{k}{2} + 2 = \Omega(k \log n) = \Omega(|B \setminus D^*| \log n).$$

In the next lemma, we analyze the second phase.

Lemma 8. *Let* $G = (V, E)$ *be a planar graph and let* $D' = \bigcup_{i=1}^{L} D_i$ *be the union of sets obtained by* APPROXMINDS *in step four. Let* D^*_{SN} *be a set of the smallest size which dominates* V_{SN} *in* G . *Then* $|D'| \le (1 + O(1/\log n))|D^*_{SN}|$.

Proof. First note that $D^*_{SN} \subseteq V_{SN} \cup V_{BN}$. Since every vertex in $V_{SN} \cup V_{BN}$ has degree of at most $\log^2 n$ we have

$$|D^*_{SN}| \ge |V_{SN}|/\log^2 n \text{ and } |V_{BN}| \le |V_{SN}|\log^2 n.$$

On the other hand, by Lemma 4, the number of edges connecting different clusters $e_{clusters}$ is at most

$$|E(G[V_{SN} \cup V_{BN}])|/\log^5 n < 3(\log^2 n + 1)|V_{SN}|/\log^5 n \le$$

$$3 \log^2 n(\log^2 n + 1)|D^*_{SN}|/\log^5 n = O(|D^*_{SN}|/\log n).$$

Thus

$$e_{clusters} = O(|D^*_{SN}|/\log n). \tag{2}$$

We claim that

$$|D'| \le |D^*_{SN}| + 2e_{clusters}. \tag{3}$$

Indeed, vertices from $D^*_{SN} \cap V_i$ dominate all vertices from $(V_{SN} \cap V_i) \setminus V''_i$ as any vertex in the latter set has all of its neighbors in V_i. Thus

$$\sum_{i=1}^{L} |D'_i| \le \sum_{i=1}^{L} |D^*_{SN} \cap V_i| = |D^*_{SN}|,$$

and so

$$|D'| \leq \sum_{i=1}^{L}(|D'_i| + |V'_i|) \leq |D^*_{SN}| + \sum_{i=1}^{L}|V'_i| \leq |D^*_{SN}| + 2e_{clusters}$$

which verifies (3). Finally, by (3) and (2),

$$|D'| \leq (1 + O\left(1/\log n\right))|D^*_{SN}|.$$

We can now summarize the performance of APPROXMINDS.

Theorem 2. *Let $G = (V, E)$ be a planar graph on n vertices such that for any two distinct vertices u, v, $|N(u) \cap N(v)| \leq \log n$. Then APPROXMINDS finds a dominating set D in G with*

$$|D| \leq (1 + O(1/\log n))\gamma(G).$$

Procedure APPROXMINDS runs in $O(\log \log n \log^* n \log^{1+d} n)$ rounds, where $d = 27.7$.

Proof. To see that D is a dominating set note that after the second step of AP-PROXMINDS all vertices with degree of at least $\log^2 n$ or which have a neighbor of such a degree are dominated by D. Therefore, only vertices from V_{SN} are not dominated at this moment. However $D_i := D'_i \cup V'_i$ dominates all vertices in $V_{SN} \cap V_i$ and so D is a dominating set in G. Now let D^* be a dominating set in G of the minimum size and, as in Lemma 7, let $B = \{v : deg(v) \geq \log^2 n\}$. We have $D = B \cup D'$ where $D' = \bigcup_{i=1}^{L} D_i$ and so $|D| \leq |B| + |D'|$. By virtue of Lemma 7,

$$|B| = |B \cap D^*| + |B \setminus D^*| = |D^* \cap B| + O\left(\frac{|D^*|}{\log n}\right).$$

In addition, $|D^*_{SN}| \leq |D^* \cap S|$ as every vertex in V_{SN} can be dominated only by vertices of degree less than $\log^2 n$. Consequently, be Lemma 8,

$$|D'| \leq (1 + O\left(1/\log n\right))|D^*_{SN}| \leq (1 + O\left(1/\log n\right))|D^* \cap S|.$$

Thus

$$|D| \leq |D^* \cap B| + O\left(\frac{|D^*|}{\log n}\right) + (1 + O\left(1/\log n\right))|D^* \cap S| = (1 + O\left(1/\log n\right))|D^*|.$$

Finally, by Lemma 4, the number of rounds is $O(\log \log n \log^* n \log^{1+d} n)$.

References

[AGLP89] B. Awerbuch, A. V. Goldberg, M. Luby, S. A. Plotkin, Network Decompo-sition and Locality in Distributed Computation, Proc. 30th IEEE Symp. on Foundations of Computer Science , 1989, pp. 364-369.

[CV86]   R. Cole, U. Vishkin, Deterministic coin tossing with applications to optimal parallel list ranking, Information and Control, 1986, 70, pp. 32-53.

[CH03]   A. Czygrinow, M. Hańćkowiak, Distributed Algorithm for Better Approxi-
           mation of the Maximum Matching, COCOON 2003 LNCS 2697 , 2003, pp.
           242-251.

[CH04]   A. Czygrinow, M. Hańćkowiak, Distributed algorithms for weighted prob-
           lems in sparse graphs, to appear in Journal of Discrete Algorithms, in Press,
           (Available online 7 September 2005)

[CHS04]   A. Czygrinow, M. Hańćkowiak, E. Szymańska, Distributed algorithm for
           approximating the maximum matching, Discrete Applied Mathematics,
           Volume 143, Issues 1-3, (2004), 62–71.

[CHSz04]   A. Czygrinow, M. Hańćkowiak, E. Szymańska, A fast distributed algorithm
           for approximating the maximum matching, Algorithms - ESA 2004 LNCS
           3221, (2004), 252–263.

[DPRS03]   D. Dubhashi, A. Mei, A. Panconesi, J. Radhakrishnan, and A. Srinivasan,
           Fast Distributed Algorithms for (Weakly) Connected Dominating Sets and
           Linear-Size Skeletons, In Proc. of the ACM-SIAM Symposium on Discrete
           Algorithms (SODA), pp. 717-724, 2003.

[E04]   M. Elkin, An Overview of Distributed Approximation, in ACM SIGACT
           News Distributed Computing Column Volume 35, Number 4 (Whole num-
           ber 132), Dec. 2004, pp. 40-57.

[HMS96]   S. Hakimi, J. Mitchem, E. Schmeichel, Star arboricity of graphs, Discrete
           Mathematics, 149, 1-3, (1996), 93–98.

[HKP99]   M. Hańćkowiak, M. Karoński, A. Panconesi, A faster distributed algorithm
           for computing maximal matching deterministically, Proceedings of PODC
           99, the Eighteen Annual ACM SIGACT-SIGOPS Symposium on Principles
           of Distributed Computing, pp. 219-228.

[JRS01]   L. Jia, R. Rajaraman, and R. Suel, An Efficient Distributed Algorithm for
           Constructing Small Dominating Sets, In Proc. of the 20th ACM Sympo-
           sium on Principles of Distributed Computing (PODC), pp. 33-42, 2001.

[KP95]   S. Kutten, D. Peleg, Fast distributed construction of k-dominating sets
           and applications, Proceedings of the fourteenth annual ACM symposium
           on Principles of distributed computing, 1995, pp. 238 - 251.

[KW03]   F. Kuhn, R. Wattenhofer, Constant-Time Distributed Dominating Set Ap-
           proximation, 22nd ACM Symposium on the Principles of Distributed Com-
           puting (PODC), Boston, Massachusetts, USA, July 2003.

[KMW04]   F. Kuhn, T. Moscibroda, and R. Wattenhofer, What Cannot Be Com-
           puted Locally!, Proceedings of 23rd ACM Symposium on the Principles of
           Distributed Computing (PODC), 2004, pp. 300-309.

[L92]   N. Linial, Locality in distributed graph algorithms, SIAM Journal on Com-
           puting, 1992, 21(1), pp. 193-201.

[L86]   M. Luby, A simple parallel algorithm for the maximal independent set
           problem, SIAM J. Comput., vol 15(4), 1986, pp. 1036-1053.

[P00]   D.Peleg, Distributed Computing: A Locality-Sensitive Approach, SIAM,
           2000.

A New NC-Algorithm for Finding a Perfect Matching in d-Regular Bipartite Graphs When d Is Small

Raghav Kulkarni

The Department of Computer Science, University of Chicago, Chicago, USA
raghav@cs.uchicago.edu

Abstract. The perfect matching problem for general graphs reduces to the same for regular graphs. Even finding an NC algorithm for the perfect matching problem in cubic (3-regular) or 4-regular graphs will suffice to solve the general problem (see [DK 92]). For regular bipartite graphs an NC algorithm is already known [LPV 81], while [SW 96] give an NC algorithm for cubic-bipartite graphs.

We present a new and conceptually simple parallel algorithm for finding a perfect matching in d-regular bipartite graphs. When d is small (polylogarithmic) our algorithm in fact runs in NC. In particular for cubic-bipartite graphs, our algorithm as well as its analysis become much simpler than the previously known algorithms for the same. Our techniques are completely different from theirs.

Interestingly, our algorithm is based on a method used by [MV 00] for finding a perfect matching in planar-bipartite graphs. So, it is remarkable that, circumventing the planarity, we could still make the same approach work for a non-planar subclass of biparitite graphs.

1   Introduction

The *perfect matching problem* is of particular interest to a variety of people including combinatorists, algorithmists and complexity-theorists. In parallel settings, the complexity of the problem is still unresolved. In this paper we propose an approach based on an interior point method for the *perfect matching problem*, especially for bipartite graphs.

Given a graph $G = (V, E)$ with n vertices and m edges, a *perfect matching* in G is a subgraph M of G such that every vertex in G has degree exactly 1 in M. The *decision* problem is to determine whether G has a perfect matching. The *search* problem is to construct a perfect matching, if one exists. Both these problems have a *Randomized* NC algorithm ([KUW 86], [MVV 87]) but are not known to be in deterministic NC even for bipartite graphs.

For special classes of graphs, however, there are deterministic NC algorithms. These classes include bipartite-planar graphs [MN 95], regular bipartite graphs [LPV 81], small-genus bipartite graphs, [MV 00] bipartite graphs having polynomially bounded permanent [GK 87]. In [SW 96], another NC algorithm was presented for cubic bipartite graphs.

T. Calamoneri, I. Finocchi, G.F. Italiano (Eds.): CIAC 2006, LNCS 3998, pp. 308–319, 2006.
© Springer-Verlag Berlin Heidelberg 2006

This work is motivated by [MV 00], where a simple and elegant NC algorithm was presented for bipartite planar graphs. There, counting was used to get a point inside the perfect matching polytope and then starting from this point, navigating outwards, a vertex of the prefect matching polytope was reached. However, it is not clear how to use this approach for non-planar graphs as the analysis of the algorithm in [MV 00] crucially uses planarity.

In this paper, we use the same approach for regular bipartite graphs. In particular, we show that for d-regular bipartite graphs, when d is polylogarithmic, one can get a point inside the perfect matching polytope and navigate outwards to seek a vertex of the perfect matching polytope in NC. Our first observation is that getting a point inside the perfect matching polytope of a regular bipartite graph is easy. This observation is simple but crucial. Next we crucially exploit the notion of *2-3 graphs* (developed by [KM 04] in the context of planar graphs) coupled with a delicately chosen potential function. The techniques we use are very elementary.

Our main contribution here is making the approach taken by [MV 00] and [KM 04] for planar graphs, work for a non-planar subclass of bipartite graphs, namely d-regular bipartite graphs having polylogarithmic d. These two subclasses of bipartite graphs are totally different in structure. This suggests that the same approach should work for much more general subclass of bipartite graphs. Moreover, our algorithm is conceptually simple, especially for cubic-bipartite graphs, and we hope that the techniques here will generalize.

The organization of this paper is as follows. Section 2 builds preliminaries used in this paper. Section 2.1 defines some matching polytopes. A special class of graphs, namely 2-3 graphs is defined in Section 2.2. Section 3 talks about the cycle space of 2-3 graphs. Section 3.1 describes the proceudure *find-big-even* and section 3.2 has the procedure *manip* which will be used in the algorithms of next sections. Section 4 describes a new NC algorithm to find a perfect matching in bipartite cubic graphs. In Section 5, we generalize the algorithm given in Section 3 for d-regular bipartite graphs. This gives an NC algorithm when d is small (polylogarithmic). We discuss the possibility of generalizing our algorithm for regular non-bipartite gaphs and conclude with some remarks and immediate open questions in Section 6.

2   Preliminaries

In this paper, we allow graphs to have self loops and multiple edges between two vertices, i.e., the term multigraph is abused as graphs unless specified otherwise.

The *weighted graph* $\mathcal{G} = (G, g)$ is a graph $G = (V, E)$, together with the weight function $g : E \to \mathbb{Q}$. We denote by g_e the weight of the edge e, i.e., $(g_e := g(e))(\forall e \in E)$. When we talk about the weighted graph $\mathcal{G}$ then it is assumed that G is the underlying unweighted graph and g is the weight function on the edges of G.

2.1   Some Polytopes Related to Matching

Given a graph G on n vertices and m edges, following are the definitions of the some of the polytopes in $\mathbb{R}^m$ associated with it. Every matching M in G

corresponds to a $0-1$ vector v_M in $\mathbb{R}^m$ namely $v_M(e) = 1$ if $e \in M = 0$ otherwise. We will use this corresponds throughout this paper.

Matching Polytope. The matching polytope, $\mathcal{M}(G)$, is the convex hull of all the matchings in G.

If G is a bipartite graph then it turns out that $\mathcal{M}(G)$ is given by the following contraints: $(g_e \geq 0)(\forall e \in E(G))$ and $(\sum_{e \perp v} g_e \leq 1)(\forall v \in V)$ ($e \perp v$ means e incident on v). However, for general graphs we need additional constraints.

Perfect Matching Polytope. The perfect matching polytope $\mathcal{PM}(G)$, is the convex hull of all the perfect matchings in G. For bipartite graphs, $\mathcal{PM}(G)$ is described by the following inequalities: $(g_e \geq 0)(\forall e \in E(G))$ and $(\sum_{e \perp v} g_e = 1)(\forall v \in V(G))$. However, this is not the case for general graphs. We need some additional contraints called "odd-cut" constraints (see [LP 86]). For a weighted graph, the *minimum weight perfect matching polytope* is the convex hull of all the perfect matchings of the minimum weight.

Fractional Perfect Matching Polytope (or 2-matching Polytope). The fractional matching polytope, $\mathcal{FPM}(G)$ is defined by the following inequalities: $(g_e \geq 0)(\forall e \in E(G))$ and $(\sum_{e \perp v} g_e = 1)(\forall v \in V(G))$.

It turns out that for bipartite graphs $\mathcal{FPM} = \mathcal{PM}$ but this is not true in general. Note that any interior point (of any matching polytope) corresponds to a weighted graph . We will treat an interior point as a weighted graph.

2.2   The Notion of 2-3 Graphs

A graph G is said to be a *2-3 graph* if
(a) the degree of every vertex of G is either 2 or 3
(b) both the neighbours of a degree 2 vertex in G have degree 3 in G.

A *3-bounded path* in a 2-3 graph G is a path in which both the end points have degree 3 in G but rest of the vertices on the path have degree 2 in G.

The definition of 2-3 graphs is motivated by the following theorem.

Theorem 1 (KM 04). *Given (i) graphs $\overline{G}$, G and interior points $\overline{\mathcal{G}} = (\overline{G}, \overline{g})$ and $\mathcal{G} = (G, g)$ of $\mathcal{FPM}(\overline{G})$ and $\mathcal{FPM}(G)$ respectively and (ii) a "backtracker" $\pi: E(\overline{G}) \to \{0,1\} \times E(G) \bigcup \{0,1\}$ there is an NC procedure which outputs*

(i) a 2-3 graph G' and an interior point $\mathcal{G}' = (G', g')$ of $\mathcal{FPM}(G')$ such that $(g'_{e'} > 0)(\forall e' \in E(G'))$ and
(ii) a "backtracker" $\pi': E(\overline{G}) \to \{0,1\} \times E(G') \bigcup \{0,1\}$, where a "backtracker" $\pi: E(\overline{G}) \to \{0,1\} \times E(G) \bigcup \{0,1\}$ is a function such that
if $\pi: \overline{e} \mapsto \mathbf{0}$ then $\overline{g}_{\overline{e}} = 0$,
if $\pi: \overline{e} \mapsto \mathbf{1}$ then $\overline{g}_{\overline{e}} = 1$,
if $\pi: \overline{e} \mapsto (i, e)$ then $\overline{g}_{\overline{e}} = g_e$ if $i = 0$ and $\overline{g}_{\overline{e}} = 1 - g_e$ if $i = 1$.

We call such a procedure as *make2-3* and we assume that it takes as input weighted graphs $\overline{\mathcal{G}}$, $\mathcal{G}$ and a "backtracker" $\pi : E(\overline{G}) \rightarrow \{0,1\} \times E(G) \bigcup \{\mathbf{0},\mathbf{1}\}$ and outputs a weighted 2-3 graph $\mathcal{G}'$ and a "backtracker" which is given by the function $\pi' : E(\overline{G}) \rightarrow \{0,1\} \times E(G') \bigcup \{\mathbf{0},\mathbf{1}\}$. In our algorithm, we will write this shortly as the following: $(\mathcal{G}', \pi') = make2 - 3(\overline{\mathcal{G}}, \pi, \mathcal{G})$. The procedure is quite simple and straightforward (see [KM 04]). Now we briefly describe why these "backtrackers" come into the picture.

The "backtracker" $\pi : E(\overline{G}) \rightarrow \{0,1\} \times E(G) \bigcup \{\mathbf{0},\mathbf{1}\}$, is useful in the following way. If we could find a perfect matching in G then we can translate this into a perfect matching of $\overline{G}$ by using the "backtracker." The translation is obvious: Let M be the perfect matching in G then $M(e) = 1$ if e is in M and 0 otherwise. The perfect matching $\overline{M}$ is given as follows: if $\pi : \overline{e} \mapsto \mathbf{0}$ then $\overline{M}(\overline{e}) = 0$,

if $\pi : \overline{e} \mapsto \mathbf{1}$ then $\overline{M}(\overline{e}) = 1$,

if $\pi : \overline{e} \mapsto (i,e)$ then $\overline{M}(\overline{e}) = M(e)$ if $i = 0$ and $\overline{M}(\overline{e}) = 1 - M(e)$ if $i = 1$. $\overline{M}$ is just the set of edges in $\overline{G}$ with weight 1.

Thus "backtrackers" are useful and the 2-3 graphs are quite general with respect to these "backtrackers" and we will use the *make2-3* procedure at every step of our algorithm to convert the graph into a 2-3 graph and will continue searching for a perfect matching in this new graph as using the "backtracker" one can backtrack a perfect matching of the original graph in NC.

3 The Cycle Space of a 2-3 Graph

In this section we will prove some results which have significance in the further sections. We also describe a procedure *find-big-even* which will be used as a subroutine in our algorithm.

Given a graph $G = (V, E)$, consider the vector space $\mathbb{F}_2^{|E|}$. Any subgraph H of G corresponds to a vector v_H in $\mathbb{F}_2^{|E|}$ and vice versa. The correspondence can be given as $v_H(e) = 1$ iff $e \in E(H)$ and $v_H(e) = 0$ otherwise. The *cycle space* of G is the subspace spanned by the vectors corresponding to cycles in G.

An element of the cycle space of G is called a *cycle vector* in G. Every cycle vector corresponds to a disjoint union of cycles. A cycle vector is called an *even cycle vector* if all the cycles in the cycle vector are of even length.

Given a spanning tree T of a graph G, every non-tree edge is in a unique cycle in G, called a *fundamental cycle* in G with respect to T. The non-tree edge is called a *fundamental edge* in G with respect to T. The set of all such fundamental cycles corresponding to the nontree edges, is called the set of *fundamental cycles* in G with respect to T. The set of fundamental cycles in G with respect to a spanning tree T of G forms a basis for the cycle space of G.

We call a vertex a 3-vertex if its degree is exactly 3. We denote by $V^{(3)}(G)$ the set of all 3-vertices in G.

Let $k := |V^{(3)}(G)|$ be the number of 3-vertices in G. We say that an even cycle vector in G is a *big even cycle vector* if it contains $\Omega(k)$ 3-vertices of G.

3.1   Finding a *Big Even Cycle Vector*

We shall describe the procedure *find-big-even* which will find a *big even cycle vector* in a bipartite 2-3 graph.

find-big-even(G)
1. find a spanning tree T of G
2. find S = the set of fundamental cycles in G with respect to T
3. return $\mathcal{C} = \oplus_{C \in S} C$.

Lemma 1. *The procedure* find-big-even *runs in NC.*

Proof. Easy to check.   □

To show that it outputs a *big even cycle vector*, we prove the following lemma.

Lemma 2. *If G is a bipartite 2-3 (multi)graph having k 3-vertices then the dimension of its cycle space is $\Omega(k)$. (at least $k/2$)*

Proof. Consider any spanning tree T of G. We claim that the number of nontree edges is $\Omega(k)$. Let l be the number of vertices in G having degree 2. Hence, total number of edges in G is $l + \frac{3}{2}k$. The number of tree edges is $l + k - 1$. Hence, the number of nontree edges is greater than $\frac{k}{2}$.   □

Lemma 3. find-big-even (G) *finds a* big even cycle vector *in G.*

Proof. Consider a spanning tree T of G. S is the set of fundamental cycles in G with respect to T. $\mathcal{C} = \oplus_{C \in S} C$. $|\mathcal{C}| = \Omega(k)$. Now, when we add all the fundamental cycles in G with respect to T, all the fundamental edges are still preserved because each fundamental edge is in a unique fundamental cycle. In a 2-3 graph, every edge has at least one endpoint of degree 3. Hence we still have $\Omega(k)$ 3-vertices of G in $\mathcal{C}$. This proves that $\mathcal{C}$ is the required *big even cycle vector*.   □

3.2   Manipulating the *Big Even Cycle Vector* in *Right* Direction

Here we describe procedures *simple-manip* and *manip* to manipulate even cycles so that some of the edges get destroyed. It will be easy to check that both these procedure run in NC.

Given an interior point $\mathcal{G}$ of $\mathcal{FPM}(G)$ and an even cycle C in G one can move to another interior point $\overline{\mathcal{G}}$ of $\mathcal{FPM}(G)$ using the procedures *simple-manip* or *manip*.

simple-manip $(\mathcal{C}, \mathcal{G})$
1. do parallely for every cycle $C \in \mathcal{C}$
2. pick a minimum weight edge (say e) in C of weight w (say)
3. add w to the weights of edges in C at odd distance from e
4. subtract w from the weights of edges in C at even distance from e.

Lemma 4 (MV 00). *The procedure* simple-manip *runs in NC, preserves bipartiteness, and we are still inside $\mathcal{FPM}$ at the end of* simple-manip.

Proof. Note that *simple-manip* preserves the contraint at every vertex that the sum of the weights of edges incident on that vertex is exactly 1. So, we are still inside the perfect matching polytope.   □

This procedure will be used as a subroutine for the algorithm in the next section for finding a perfect matching in cubic bipartite graphs. Somewhat more sophisticated version of *simple-manip* is required for d-regular bipartite graphs (in Section 6). That procedure is described here as *manip*.

The even cycle manipulation with C. Suppose we have an interior point of $\mathcal{FPM}(G)$ and C is an even cycle in G.

1. Fix an edge f in C.
2. Let C_{odd} be the edges in C at odd distance from f.
3. Similarly, let C_{even} be the edges in C at even distance from f.
4. Choose one of the sets C_{odd} or C_{even}.
5. Let e be the minimum weight edge in that set and w be the weight of e.
6. Add w to the weight of all edges in C at odd distance from e.
7. Subtract w from the weight of all edges at even distance from e.
8. We get a new interior point of $\mathcal{FPM}(G)$ in which weight of e is 0.

This procedure is called as *even cycle manipulation*. This can be done for any closed walk of even length. If we have an even cycle vector then we can manipulate each of its even cycle.

Lemma 5. *The even cycle manipulation can be done in NC and it doesn't leave the* $\mathcal{FPM}$ *of the graph.*

Proof. Trivial to check.   □

Note that when we manipulate an even cycle C in G, we had choice between C_{odd} and C_{even} in Step 4 of the *even cycle manipulation*. We now define what is the *right direction* which we have to choose during the algorithm.

Making the right choice. Suppose C contains the following 3-vertices in G: $v_1, v_2, \ldots, v_\ell$. Let $e_i{}^{odd}$ be the edge in C incident on v_i and belonging to C_{odd}. Let $e_i{}^{even}$ be the edge in C incident on v_i and belonging to C_{even}. Let $x_i = w_{e_i{}^{odd}}$. Let $y_i = w_{e_i{}^{even}}$. If $\sum_{i=1}^{\ell}(x_i - y_i) \le 0$ then the *right direction* is choosing C_{odd} otherwise the *right direction* is choosing C_{even}. In our algorithm, we always choose the *right direction*. So, the procedure manip is as described below.

manip $(\mathcal{C}, \mathcal{G})$
1. for every even cycle C in $\mathcal{C}$
2. (* choose the right direction *)
if $\sum_{i=1}^{\ell}(x_i - y_i) \le 0$ then choose C_{odd}
else choose C_{even}
3. manipulate C as described above in the chosen direction.

Lemma 6. *The procedure* manip *runs in NC, preserves bipartiteness and we are still inside* $\mathcal{FPM}$ *at the end of* manip.

Proof. Easy to check.   □

We will use the procedures *make2-3*, *find-big-even*, *simple-manip* and *manip* as subroutines in the algorithms in next sections.

4   Finding a Perfect Matching in Bipartite Cubic Graphs in NC

Now we describe an NC algorithm to find a perfect matching in bipartite cubic graphs. [SW 96] have already given an NC algorithm for the same but the approach here is totally different as compared to theirs. They maintain a subgraph called pseudo perfect matching (degree of every vertex is odd) at every step and try decreasing the number of 3-vertices in the pseudo perfect matching by a constant fraction. Here, we start from an interior point of the perfect matching polytope and move towards a vertex withoght leaving the polytope.

4.1   Algorithm

Given a cubic bipartite graph $\overline{G}$
1. (a) get an interior point $\mathcal{G}$ of $\mathcal{PM}(\overline{G})$ by assigning $(\overline{g}_{\overline{e}} = \frac{1}{3})(\forall \overline{e} \in E(\overline{G}))$.
(b) let $\mathcal{G} = \overline{\mathcal{G}}$ and let $\pi : E(\overline{G}) \rightarrow \{0,1\} \times E(G) \bigcup \{0,1\}$ be a "backtracker" such that $(\pi : \overline{e} \mapsto (0,e))(\forall \overline{e} \in E(\overline{G}))$.   (* Initialize the "backtracker" π. *)
2. while (G is nonempty)
{
(a) $\mathcal{C} = $ *find-big-even* $(\mathcal{G})$
(* Take $\mathcal{C}$ to be the XOR of all fundamental cycles of a spanning tree of G. *)
(b) $\mathcal{G} = $ *simple-manip* $(\mathcal{C}, \mathcal{G})$
(* Delete alternate edges of each even cycle in $\mathcal{C}$. *)
(c) $(\mathcal{G}, \pi) = $ make2-3 $(\overline{\mathcal{G}}, \pi, \mathcal{G})$
(* Replace each 3-bounded path (section 2.2) in G by single edge, update π. *)
}
3. backtrack a perfect matching in $\overline{G}$ using π.
(* When G is empty then π maps each edge of $\overline{G}$ to **0** or **1**, the edges mapped to **1** form a perfect matching in $\overline{G}$. *)

4.2   Analysis

The following sequence of lemmas will show that the above algorithm runs correctly in NC.

Lemma 7. *During the course of the algorithm, the weights of the edges of G will be 0, 1/3 or 2/3.*

Proof. The procedures *simple-manip* as well as *make2-3* don't change the denominators of the weights.   □

Lemma 8. *At the beginning of each iteration of the* while *loop, G is cubic-bipartite graph with weight 1/3 on each edge.*

Proof. Note that G is a 2-3 graph at the beginning of each iteration of *while* loop. So, any edge of G has one end point of degree 3. By the previous lemma, the weights of the edges of G are multiples of $1/3$ and since G is a 2-3 graph output by *make2-3* there are no zero weight edges. So, the weight of any edge is at least $1/3$. Now for any vertex of degree 3, the sum of the weights of edges incident on it should add up to 1. Hence, the weight of each of the edges incident on it should be $1/3$. But every edge in a 2-3 graph is incident on a 3-vertex. So, the graph is again cubic bipartite with weight $1/3$ on each edge.   □

Lemma 9. *At the end of each iteration of the* while *loop, the size (number of edges) of G decreases by a contant fraction.*

Proof. Note that during an iteration, *simple-manip* will destroy half the edges of C by making their weight 0 because the weight of any edge is $1/3$. Due to *simple-manip*, the size of the graph has reduced by a constant fraction.

Theorem 2. *The above algorithm finds a perfect matching in cubic-bipartite graphs in NC.*

Proof. The previous lemma implies that, the while loop terminates in $O(logn)$, proving that the algorithm runs in NC.   □

We need more sophisticated version of *simple-manip* (the way it is described in Section 3.2) for extending the same approach for d-regular bipartite graphs. The next section describes the extension of the result in this section. As it was the case with cubic-bipartite graphs here, the notion of 2-3 graphs will play crucial role in the analysis in the next section.

5 Finding a Perfect Matching in d-Regular Bipartite Graphs for Polylogarithmic d in NC

There is already an NC algorithm for d-regular bipartite graphs by [LPV 81]. Our algorithm is totally different from theirs and conceptually simple though it works in NC only when d is polylogarithmic.

5.1 Algorithm

Given a d-regular bipartite graph $\overline{G}$,
1. (a) get an interior point of $\mathcal{PM}(\overline{G})$ by assigning $(\overline{g}_{\overline{e}} = 1/d)(\forall \overline{e} \in E(\overline{G}))$
(b) let $\overline{\pi} : E(\overline{G}) \rightarrow \{0,1\} \times E(\overline{G}) \bigcup \{0,1\}$ be a "backtracker" such that $(\overline{\pi} : \overline{e} \mapsto (0, \overline{e}))(\forall \overline{e} \in E(\overline{G}))$.
2. $(\mathcal{G}, \pi) = make2\text{-}3 \ (\overline{G}, \overline{\pi}, \overline{G})$
3. while (G has no vertex of degree 3)
{
(a) $\mathcal{C} = find\text{-}big\text{-}even \ (\mathcal{G})$
(b) $\mathcal{G} = manip \ (\mathcal{C}, \mathcal{G})$
(c) $(\mathcal{G}, \pi) = make2\text{-}3 \ (\overline{G}, \pi, \mathcal{G})$
}

4. (a) Now all the vertices in G have degree 2. So, a perfect matching in G can be found very easily by taking alternate edges of each even cycle in G.
(b) Using the "backtracker" π one can get back a perfect matching in $\overline{G}$ from the perfect matching in G.

5.2   Analysis

Now we will show that the above algorithm runs in NC. To show this we consider certain integer potential function for weighted graphs. A similar potential function is used in [S 98]. We show that the potential of $\mathcal{G}$ decreases by large amount after each iteration of the while loop. Finally, when the potential becomes zero, we get a perfect matching easily.

Lemma 10. *During the course of the algorithm, the weights of the edges of G are multiples of $1/d$.*

Proof. Both *manip* and *make2-3* don't change the denominators of the weights. $\square$

The integer potential function Φ

By the previous lemma we can assume that $g_e = \frac{w_e}{d}$ where w_e is an integer.
The potential of an edge is defined as $\Phi(e) := w_e(d - w_e)$.
The potential of a vertex is $\Phi(v) := \sum_{e \perp v} \Phi(e)$.
The potential of the graph G is $\Phi(G) := \sum_{v \in V^{(3)}(G)} \Phi(v)$. Recall that $V^{(3)}(G)$ is the set of vertices of degree 3 in G.

Lemma 11. *If even cycle C contains ℓ 3-vertices of G then after manipulating C in the* right *direction, the potential of G decreases at least by 2ℓ.*

Proof. Without loss of generality, say the right direction was to choose C_{odd}. Say the min weight edge in C_{odd} has weight w. Let $x_i = w_{e_i \, odd}$. Let $y_i = w_{e_i \, even}$. (As in Section 3.2.) Let $\Delta(\Phi)$ denote the change in potential due to manipulating C. Let G' be the graph obtained by manipulating C in G.

Then, $\Delta(\Phi) = \Phi(G') - \Phi(G)$.

Now, $\Phi(G') = \sum_{v \in C \cap V^{(3)}(G')} \Phi'(v)$, where Φ' is the new potential of a vertex after the manipulation.

Similarly, $\Phi(G) = \sum_{v \in C \cap V^{(3)}(G)} \Phi(v)$.

But, $C \cap V^{(3)}(G') \subseteq C \cap V^{(3)}(G)$.

Therefore, $\Phi(G') \leq \sum_{v \in C \cap V^{(3)}(G)} \Phi'(v)$.

Now, $\Phi'(v_i) - \Phi(v_i) = (x_i - w)(d - x_i + w) + (y_i + w)(d - y_i - w) - (x_i)(d - x_i) - (y_i)(d - y_i)$
$= x_i w - wd + wx_i - w^2 - y_i w + wd - wy_i - w^2$
$= 2(x_i - y_i)w - 2w^2$.

Therefore, $\Delta(\Phi) \leq \sum_{i=1}^{\ell} 2(x_i - y_i)w - 2w^2$.

Since we chose the *right direction* to manipulate C, $\sum_{i=1}^{\ell}(x_i - y_i) \leq 0$. Hence, the change in potential of G, $\Delta(\Phi) \leq -2w^2\ell$. Hence, the potential of G decreases at least by 2ℓ as w^2 is at least 1. $\square$

Lemma 12. *If the number of 3-vertices in G is k at the beginning of a while loop, then at the end of the while loop, the potential of G decrease by at least $\frac{k}{2}$.*

Proof. The *big even cycle vector* contains at least $\frac{k}{4}$ 3-vertices in G. For each cycle C in the *big even cycle vector*, after manipulating, the potential decreases at least by 2 times #3−*vertices in C*. Hence, when we manipulate a *big even cycle vector* in G containing at least $\frac{k}{4}$ 3-vertices in G, the potential of G decreases at least by $\frac{k}{2}$.   □

Lemma 13. *In $O(d^2)$ iterations of while loop, the potential of G becomes 0.*

Proof. Initially, the potential of G at the start of the first iteration of *while* loop is $O(d^2 k)$. By previous lemma, the potential decrease by $\Omega(k)$ in each iteration of while loop. Hence, in $O(d^2)$ steps the potential becomes 0 and *while* loop terminates.

Lemma 14. *When the* while *loop ends, we get a perfect matching in the original graph.*

Proof. When the *while* loop terminates, G is a disjoint union of even cycles and a perfect matching in G can be found easily and could be backtracked to the perfect matching in $\overline{G}$ using π.   □

Theorem 3. *For polylogarithmic d, the above algorithm runs correctly in NC.*

Proof. The number of iterations of while loop is $O(d^2)$. In particular, for poly-logarithmic d, the algorithm runs in NC.   □

Note that the algorithm presented above works in NC if you start with any small *magnitude* interior point of a bipartite graph, i.e., if the least common multiple of the denominators of the weights is polylogarithmic.

6   Discussion

The starting point of our algorithm - to get an interior point of the perfect matching polytope - was simple but crucial. First of all, a non-bipartite regular graph need not have a perfect matching and even if it has a perfect matching, it is not clear how to find an interior point of the $\mathcal{PM}$ of such a graph to provide the start for the algorithm. In certain cases, however we can get a start.

6.1   Getting an Interior Point for Regular Expander Graphs

A graph G is said to be an α-expander if for every $S \subset V$ such that $|S| \leq \frac{|V|}{2}$, the number of edges (size of the cut $(S, \overline{S})$) from S to its complement $(\overline{S} = V \backslash S)$ is at least α times the cardinality of S. The α is called the expansion factor of G.

Lemma 15. *If G is a simple (no multiple edges) d-regular expander graph on even number of vertices with the expansion factor $\alpha > \frac{d-1}{d}$, then $(g_e = 1/d)(\forall e \in E(G))$ gives an interior point of the perfect matching polytope of G.*

Proof. It suffices to check "odd-cut" contraints (sum of the weights of the edges in a cut $(S, \overline{S})$ at least one whenever S is odd) [LP 86]. For S of size at least

d, expanstion property guarantees that there are at least d edges from S to $\overline{S}$. Hence, the size of cut $(S, \overline{S})$ is at least 1. If size of S is less than d, assuming that G is simple, one can show that at least d edges should go out of S by counting the total of degrees. □

Corollary 1. *Such an expander graph always has a perfect matching. In fact it has at least d perfect matchings.*

Corollary 2. *One can check in NC whether $(g_e = 1/d)(\forall e \in E)$ gives an interior point of $\mathcal{PM}(G)$ for a d-regular expander graph with expansion α, if d and α are constants.*

Another main difficulty is in maintaining the "odd-cut" constraints in the perfect matching polytope. The procedure *manip* just maintains the constraints at a vertex but the odd-cut contraints might get violated. By just maintaining the constraints at each node of a regular graph, we can get a perfect 2-matching: a vertex of the 2-matching polytope (Section 2.1).

Lemma 16. *A perfect 2-matching in non-bipartite regular graphs can be found in NC.*

Proof. The problem of finding a perfect 2-matching in regular graphs reduces to the problem of finding a perfect matching in regular bipartite graphs (see for instance [KR 98]). □

6.2   Conclusion

We have presented a different and conceptually simple parallel algorithm for finding a perfect matching in d-regular bipartite graphs. In particular, when d is small (polylog), our algorithm runs in NC. The connection between our algorithm and the algorithm of [MV 00] for bipartite-planar graphs is notable. [MV 00] uses planarity crucially and we could still use their approach to get an NC algorithm in a non-planar subclass of bipartite graphs. It is also remarkable that the notion of 2-3 graphs, developed in the context of planar graphs in [KM 04] plays a crucial role in our algorithm. This suggests that these techniqes seem more general. As in case of [MV 00], the "odd-cut" constraints are difficult to maintain. The problem here is more basic, even it is difficult to get a starting point inside $\mathcal{PM}$. We have observed that in certain cases it is possible to get a starting point and without maintaining the "odd-cut" contraints, one can still get a perfect 2-matching in regular graphs, yet, as of now, the quest for the perfect matching continues.

Acknowledgements

I sincerely thank Prof. Janos Simon for helpful discussions and Prof. Meena Mahajan for introducing the problem and for numerous useful suggestions.

References

[DK 92]   E. Dahlhouse and M. Karpinski. Perfect matching for regular graphs is AC^0-hard for the general matching problem. *J. Comput. Syst. Sci., 44, p. 94-102,* 1992.

[GK 87]   D. Grigoriev and M. Karpinski. The matching problem for bipartite graphs with polynomially bounded permanent is in NC. In *Proceedings of 28th IEEE Conference on Foundations of Computer Science, pages 166-172.* IEEE Computer Society Press, 1987.

[KM 04]   Raghav Kulkarni, Meena Mahajan. Seeking a vertex of the planar matching polytope in NC. *In the Proceedings of the 12th European Symposium on Algorithms ESA, LNCS vol. 3221, pages 472-483.* Springer, 2004.

[KR 98]   M. Karpinski and W. Rytter. Fast parallel algorithms for graph matching problems. *Oxford Science Publications,* 1998.

[KUW 86]   R Karp, E Upful, A Wigderson. Constructing a perfect matching is in random NC. *Combinatorica, 6:35-48,* 1986.

[LP 86]   Lovasz and Plummer. Matching theory. *Mathematical Studies, Annals of Discrete Maths,*Vol. 25, North-Holland, Amsterdam, 1986.

[LPV 81]   G. Lev, M. Pippenger, L. Valiant. A fast parallel algorithm for routing in permutation networks, *IEEE Transactions on Computers, C-30:93-100,* 1981.

[MN 95]   G Milller and J Naor. Flow in planar graphs with multiple sources and sinks. *SIAM Journal of Computing,* 24:1002-1017, 1995.

[MV 00]   M Mahajan, K Varadarajan. A new NC algorithm to find a perfect matching in planar and bounded genus graphs. In *Proceedings of the Thirty-Second Annual ACM Symposium on Theory of Computing (STOC),* pages 351-357, 2000.

[MVV 87]   K Mulmuley, U Vazirani, V Vazirani. Matching is as easy as matrix inversion. *Combinatorica,* 7(1): 105-131, 1987.

[S 98]   Alexander Schrijver: Bipartite Edge Coloring in O(Delta m) Time. SIAM J. Comput. 28(3): 841-846 (1998)

[SW 96]   R Sharan, A Wigderson. A new NC algorithm for perfect matching in cubic bipartite graphs. *Proc. of ISTCS 96,* pp. 56-65, 1996.

Fixed-Parameter Tractability Results for Feedback Set Problems in Tournaments

Michael Dom, Jiong Guo*, Falk Hüffner*, Rolf Niedermeier, and Anke Truß

Institut für Informatik, Friedrich-Schiller-Universität Jena,
Ernst-Abbe-Platz 2, D-07743 Jena, Germany
{dom, guo, hueffner, niedermr, tanke}@minet.uni-jena.de

Abstract. Complementing recent progress on classical complexity and polynomial-time approximability of feedback set problems in (bipartite) tournaments, we extend and partially improve fixed-parameter tractability results for these problems. We show that FEEDBACK VERTEX SET in tournaments is amenable to the novel iterative compression technique. Moreover, we provide data reductions and problem kernels for FEEDBACK VERTEX SET and FEEDBACK ARC SET in tournaments, and a depth-bounded search tree for FEEDBACK ARC SET in bipartite tournaments based on a new forbidden subgraph characterization.

1 Introduction

Feedback set problems deal with destroying cycles in graphs using a minimum number of vertex or edge removals [10]. Although feedback set problems usually are NP-hard for undirected as well as for directed graphs, the algorithmic treatment by means of approximation, exact, or parameterized algorithms seems to be significantly easier in the undirected case where more and better results are known. In particular, in the case of directed graphs the research so far mainly focused on a special class of graphs, so-called tournaments, since they appear in applications such as voting systems, rankings, and graph drawing.

A tournament is a directed graph where there is exactly one arc between each pair of vertices. Also due to important applications, feedback set problems in tournaments recently received considerable interest, e.g., [1, 2, 3, 4, 5, 6, 16, 20]. For instance, the NP-hardness of FEEDBACK ARC SET in tournaments has recently been addressed by at least four independent groups of researchers [1, 2, 5, 6]. Here, we contribute new results concerning the algorithmic tractability of FEEDBACK ARC SET (FAS) and FEEDBACK VERTEX SET (FVS) in tournaments and bipartite tournaments.

Table 1 surveys known and new complexity results for feedback set problems in (bipartite) tournaments. Concerning polynomial-time approximability, the following results are known. For FVS in tournaments (FVST), the trivial factor 3 has been improved to 2.5 [3] whereas for FVS in bipartite tournaments

---

* Supported by the Deutsche Forschungsgemeinschaft, Emmy Noether research group PIAF (fixed-parameter algorithms), NI 369/4.

T. Calamoneri, I. Finocchi, G.F. Italiano (Eds.): CIAC 2006, LNCS 3998, pp. 320–331, 2006.

Table 1. Complexity results for feedback set problems in tournaments. Herein, n denotes the number of vertices and k denotes the size of the desired feedback solution set.

	Complexity	Approximation		Fixed-parameter tractability	
		factor	runtime	runtime	kernel
FVST	NP-c [18]	2.5 [3]	$O(n^3)$	$O(2^k \cdot n^2(\log n + k))$ [§3]	$O(k^3)$ [§4.1]
FVSBT	NP-c [4]	3.5 [4]	$O(n^3)$	$O(3.12^k \cdot n^4)$ [19]	?
FAST	NP-c [2,5,6]	2 [20]	—	$O(2.42^k \cdot n^{2.38})$ [16]	$O(k^2)$ [§4.2]
FASBT	?	?	?	$O(3.38^k \cdot n^6)$ [§5]	?

(FVSBT) the trivial factor 4 has been improved to 3.5 [4]. For FAS in tournaments (FAST) a factor-2 approximation is known [20] whereas we are not aware of any approximation results for FAS in bipartite tournaments (FASBT).

Alternatively, it is reasonable to study feedback set problems from a parameterized point of view [8,14]. For instance, in undirected graphs, there has been recent progress showing that a feedback vertex set of size at most k can be found in $c^k \cdot n^{O(1)}$ time for some constant c [7,13], where n is the number of graph vertices. The corresponding question for directed graphs is famously open. Restricting the consideration to the class of tournaments, Raman and Saurabh [16] have given the first positive result by giving fixed-parameter algorithms for weighted FVST and weighted FAST running in $O(2.42^k \cdot n^{O(1)})$ time. For the unweighted case of FVST, the previously fastest algorithm is obtained by a reduction to 3-HITTING SET and runs in $O(2.18^k \cdot n^{O(1)})$ time [9]. The algorithm for FVSBT with a running time of $O(3.12^k \cdot n^4)$ is derived in a similar way [19].

We improve the time bound of exactly solving unweighted FVST to $O(2^k \cdot n^{O(1)})$, demonstrating the applicability of an elegant technique—so-called iterative compression—in contrast to the more standard depth-bounded search tree methodology employed by Raman and Saurabh [16] and Fernau [9]. Moreover, we present a data reduction providing a size-$O(k^3)$ problem kernel for FVST. As we show, this is only one instance of a problem kernel for a larger class of vertex deletion problems. Furthermore, complementing the $O(2.42^k \cdot n^{O(1)})$-time fixed-parameter algorithm for FAST, we develop an $O(3.38^k \cdot n^{O(1)})$-time algorithm for FASBT which is based on a novel characterization by forbidden subgraphs. Finally, we also demonstrate a size-$O(k^2)$ problem kernel for FAST, complementing the search tree result of Raman and Saurabh [16]. Table 1 summarizes all results.

We feel that an important contribution of this paper—besides improving known upper bounds—is to show the applicability of innovative and practically relevant techniques such as data reduction and iterative compression to feedback set problems in tournaments. In particular, to the best of our knowledge, here we demonstrate for the first time the applicability of iterative compression to directed feedback set problems—previous applications only addressed the undirected case [7,13,17].

2   Preliminaries

In this paper we deal with fixed-parameter algorithms that emerge from the field of parameterized complexity analysis [8, 11, 14]. An instance of a parameterized problem consists of a problem instance I and a parameter k. A parameterized problem is *fixed-parameter tractable* if it can be solved in $f(k) \cdot |I|^{O(1)}$ time, where f is a computable function solely depending on the parameter k, not on the input size $|I|$.

A directed graph or digraph D consists of a vertex set V and an arc set E with $n := |V|$ and $m := |E|$. Each arc is an ordered pair of vertices. We consider only digraphs without loops, that is, $(v, v) \notin E$ for all $v \in V$. We call a digraph $D' = (V', E')$ an *induced subgraph* of $D = (V, E)$ if $V' \subseteq V$ and $E' = \{(u, v) \mid u, v \in V' \text{ and } (u, v) \in E\}$. The subgraph of D induced by a vertex subset V' is denoted by $D[V']$. With *reversing* an arc (u, v) we mean that we delete the arc (u, v) from E and insert (v, u) into E. A *tournament* $T = (V, E)$ is a digraph where there is exactly one arc between each pair of vertices. A digraph is a *bipartite tournament* if its vertex set is the union of two disjoint sets V_1 and V_2 such that each arc consists of one vertex from each of V_1 and V_2 and between each vertex from V_1 and each vertex from V_2 there is exactly one arc. A *cycle* is a sequence of distinct vertices $v_1, \ldots, v_s$ with $(v_i, v_{i+1}) \in E$ for all $1 \le i < s$ and $(v_s, v_1) \in E$. A *triangle* is a cycle of length 3. A *topological sort* of a digraph $D = (V, E)$ is a sequence $v_1, v_2, \ldots, v_n$ of the vertices in V in which each vertex appears exactly once and $i < j$ for each arc $(v_i, v_j) \in E$. Clearly, a digraph has a topological sort iff it is acyclic, that is, it does not contain a cycle.

The FEEDBACK VERTEX (ARC) SET in tournaments (FV(A)ST) problem is defined as follows:

Input: A tournament T and a nonnegative integer k.
Task: Find a set F of at most k vertices (arcs) whose removal results in an acyclic digraph.

The set F is called a *feedback vertex (arc) set*. When the input digraph is restricted to bipartite tournaments, we have the FEEDBACK VERTEX (ARC) SET in bipartite tournaments (FV(A)SBT) problem.

The following property with respect to acyclicity of tournaments is well-known.

Lemma 1. *A tournament is acyclic iff it contains no triangles.*

For the purpose of showing a problem kernel for FVST in Sect. 4.1, we reduce FVST to the 3-HITTING SET (3HS) problem defined as follows:

Input: A finite set S, a collection C of size-3 subsets of S, and a nonnegative integer k.
Task: Find a subset S' of S with $|S'| \le k$ such that S' contains at least one element from each subset in C.

Due to the following lemma shown by Raman and Saurabh [16], we can reverse arcs instead of deleting them when dealing with FAST and FASBT. This is useful because it allows us to apply feedback arc sets without leaving the class of (bipartite) tournaments.

Lemma 2. *Let F be a minimal feedback arc set of a digraph D. Then the graph formed from D by reversing the arcs in F is acyclic.*

3   Iterative Compression for Feedback Vertex Set in Tournaments

In this section we present a fixed-parameter algorithm solving FEEDBACK VER-TEX SET in tournaments in $O(2^k \cdot n^2 (\log n + k))$ time. This algorithm is based on the concept of *iterative compression*, which was introduced by Reed et al. [17]. The heart of our algorithm is a *compression routine*, which computes from a tournament and a feedback vertex set of size $k + 1$ a new feedback vertex set of size k, or proves that no smaller feedback vertex set exists.

Using such a compression routine, FEEDBACK VERTEX SET for a tourna-ment T can be solved by successively considering induced subgraphs of T with increasing sizes. Let $\{v_1, \ldots, v_n\}$ be the vertex set V of T. Then the induced subgraphs $T_i := T[\{v_1, \ldots, v_i\}]$ are considered iteratively for $i = 1$ to $i = n$. The optimal feedback vertex set X_1 for the tournament T_1 is empty. For $i > 1$, assume that an optimal feedback vertex set X_{i-1} for T_{i-1} is known. Obviously, $X_{i-1} \cup \{v_i\}$ is a feedback vertex set for T_i. Using the compression routine, we can either determine that $X_{i-1} \cup \{v_i\}$ is optimal, or otherwise compute an op-timal feedback vertex set for T_i. For $i = n$, we thus have computed an optimal feedback vertex set for T. It remains to describe the compression routine.

Compression Routine. To make the task of looking for a smaller feedback vertex set for a tournament $T = (V, E)$ easier, we would like to restrict our search to feedback vertex sets that are disjoint from a given one. This can be achieved by a brute-force enumeration of all $O(2^k)$ possibilities to partition the given feedback vertex set X into two vertex sets S and $X \setminus S$. For each partition, we then look only for solutions that contain all of $X \setminus S$ (they can immediately be deleted from the tournament), but none of S.

Up to this point, the algorithm is analogous to the iterative compression al-gorithm for undirected FEEDBACK VERTEX SET [7, 13]. The core part of the compression routine, however, is completely different; in particular, we will be able to solve the remaining task of finding a smaller feedback vertex set that is disjoint from the given one S in polynomial time, whereas in [7, 13] still expo-nential time is required.

The central observation is that both $T[S]$ and $T[V \setminus S]$ are acyclic ($T[S]$ because otherwise there is no feedback vertex set without vertices from S, and $T[V \setminus S]$ because S is a feedback vertex set). Then, the topological sort of a maximum acyclic subtournament of T containing all of S can be thought of as resulting from inserting a subset of $V \setminus S$ into the topological sort of S. On the one hand, the order of the inserted subset must not violate the topo-logical sort of $T[V \setminus S]$. On the other hand, we can achieve by a data reduc-tion rule that for every $v \in V \setminus S$, the subtournament $T[S \cup \{v\}]$ is acyclic

Input: Tournament $T = (V, E)$ and a feedback vertex set S for T.
Output: A minimum feedback vertex set F for T with $F \cap S = \emptyset$.
1   **if** $T[S]$ contains a cycle: **return nil**
2    $s_1, \ldots, s_{|S|} \leftarrow$ topological sort of $T[S]$
3    $R \leftarrow \emptyset$
4   **while** there is a triangle u, v, w with $u, v \in S$ and $w \in V \setminus S$:
5        $R \leftarrow R \cup \{w\}$
6        $T \leftarrow T$ with w deleted
7   **for each** $v \in V \setminus S$:
8        $p[v] \leftarrow \min(\{i \mid (v, s_i) \in E\} \cup \{|S| + 1\})$
9    $L \leftarrow$ topological sort of $T[V \setminus S]$
10   $P \leftarrow V \setminus S$ sorted by p, with position in L as tie-breaker
11   $Y \leftarrow$ vertices in a longest common subsequence of L and P
12   **return** $R \cup ((V \setminus S) \setminus Y)$

Fig. 1. A subroutine for the compression step

and therefore v has a "natural" position within the topological sort of S. We then obtain the maximum acyclic subtournament as the longest common subsequence of the topological sort of $T[V \setminus S]$ and $V \setminus S$ sorted by natural position within S.

We describe this in more detail using the subroutine displayed in Fig. 1. First we check whether S induces a cycle in T: if so, no feedback vertex set for T disjoint from S can be found, and we abort (line 1). Then we apply data reduction to the instance: whenever there is a triangle with two vertices in S, we can only get rid of this triangle by deleting the third vertex (lines 4–6). After applying this reduction rule exhaustively, for any $v \in V \setminus S$ the subtournament $T[S \cup \{v\}]$ clearly does not contain triangles anymore and therefore is acyclic by Lemma 1. This means that we can insert v at some point in the topological sort $s_1, \ldots, s_{|S|}$ of S without introducing cycles. Since T is a tournament, there is thus some integer $p[v]$ such that for $i < p[v]$, there is an arc from s_i to v, and for $i \geq p[v]$, there is an arc from v to s_i (Fig. 2):

$$(v, s_i) \in E \iff i \geq p[v]. \tag{1}$$

We calculate p in lines 7–8: when we encounter the first s_i in the topological sort of S where $(v, s_i) \in E$, we can insert v before s_i; if there is no such s_i, we set $p[v]$ to $|S| + 1$, and (1) still holds.

We now construct a sequence P from p (line 10), where vertices from $V \setminus S$ that are positioned by p between the same two vertices of S are ordered according to their relative position in the topological sort of $T[V \setminus S]$. Clearly, any acyclic subtournament of T containing all of S must have a topological sort where the vertices from $V \setminus S$ occur in the same order as in P. The same holds for the topological sort L of $T[V \setminus S]$, which is calculated in line 9. This leads to the following lemma.

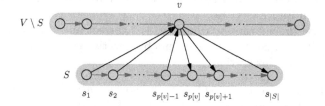

Fig. 2. Illustration of equivalence (1). For clarity, only some of the arcs are shown.

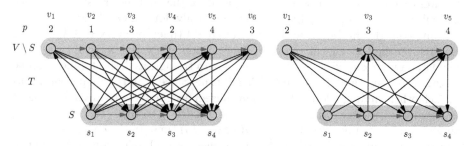

Fig. 3. Example for the subroutine in Fig 1. For clarity, only some of the arcs within the acyclic subtournaments $T[S]$ and $T[V \setminus S]$ are shown. Left: Tournament T after data reduction with $L = v_1, v_2, v_3, v_4, v_5, v_6$ and $P = v_2, v_1, v_4, v_3, v_6, v_5$. A longest common subsequence is v_1, v_3, v_5, yielding the acyclic graph shown on the right.

Lemma 3. *After line 10 of the algorithm in Fig. 1, T is acyclic iff the sequences L and P are equal.*

Proof. "$\Rightarrow$": If L and P are not equal, then there are $v, w \in V \setminus S$ with $(v, w) \in E$ but $p[v] > p[w]$. Then by (1) we have $(w, s_{p[w]}) \in E$ and $(v, s_{p[w]}) \notin E \Rightarrow (s_{p[w]}, v) \in E$, and T is not acyclic.

"$\Leftarrow$": By Lemma 1, it suffices to look for triangles to decide whether T is acyclic. Since $T[S]$ and $T[V \setminus S]$ are acyclic and we destroyed all triangles with two vertices in S, there can only be triangles with exactly two vertices in $V \setminus S$. If L and P are equal, then for all $v, w \in V \setminus S$ with $(v, w) \in E$ we have $p[v] \leq p[w]$. Then by (1) there cannot be any s_i with $(w, s_i) \in E$ and $(s_i, v) \in E$, and there can be no triangle in T. □

With the same justification, Lemma 3 holds for induced subgraphs of T and the corresponding sequences L and P. Clearly, deleting a vertex $v \in V \setminus S$ from T affects L and P only insofar as v disappears from L and P. Therefore, the cheapest way to make T acyclic by vertex deletions can be found by finding the cheapest way to make L and P equal by vertex deletions; this is exactly the complement of the longest common subsequence of L and P. We then obtain the desired feedback vertex set for T by adding the vertices of this complement to those of R, which were determined to be in any feedback vertex set in the reduction step (lines 11–12). Figure 3 shows an example for the execution of the subroutine from Fig. 1.

In summary, the subroutine from Fig. 1 is correct and can be used to solve FEEDBACK VERTEX SET in tournaments by iterative compression as described at the beginning of this section.

Theorem 1. FEEDBACK VERTEX SET *in tournaments of n vertices with k vertex deletions can be solved in $O(2^k \cdot n^2(\log n + k))$ time.*

Proof. We have shown how to solve FEEDBACK VERTEX SET in tournaments using iterative compression. It remains to analyze the runtime. First we examine the subroutine from Fig. 1. Lines 1–2 can be easily done in $O(|S|) = O(k)$ time. Finding triangles in line 4 can be done in $O(nk)$ time: for every $v \in V \setminus S$, we iterate over the topological sort of S; if we encounter a vertex s_i with $(v, s_i) \in E$ and later a vertex s_j with $(s_j, v) \in E$, we have a triangle as desired. Line 9 can be done in $O(n)$ time and line 10 in $O(n \log n)$ time. Since L and P are permutations of each other, finding a longest common subsequence reduces to finding a longest increasing subsequence, which can be done in $O(n \log n)$ time [12]. In summary, the subroutine can be executed in $O(n(\log n + k))$ time. In the compression routine, the subroutine is called $O(2^k)$ times, once for each partition of X into two subsets. The compression routine itself is called n times when inductively building up the graph structure. In total, we have a runtime of $O(2^k \cdot n^2(\log n + k))$. □

4   Problem Kernels by Data Reduction

Developing good kernelizations is among the most important contributions of fixed-parameter algorithmics for hard problems [8, 14]. A *data reduction rule* replaces, in polynomial time, a given problem instance (I, k) by a "simpler" instance (I', k') such that (I, k) is a yes-instance iff (I', k') is a yes-instance. An instance to which none of a given set of reduction rules applies is called *reduced* with respect to these rules. A parameterized problem is said to have a *problem kernel* if, after the application of the reduction rules, the resulting reduced instance has size $f(k)$ for a function f depending only on k.

4.1   Feedback Vertex Set in Tournaments

With Lemma 1, it is easy to observe that FEEDBACK VERTEX SET in tournaments (FVST) is a special case of 3-HITTING SET (3HS). Based on the kernelization method for 3HS [15], we show that FVST admits a kernel.

Theorem 2. FEEDBACK VERTEX SET *in tournaments admits a problem kernel with an $O(k^3)$-vertex tournament, and it can be found in $O(n^3)$ time.*

Proof. The basic idea of the kernelization process is to do a trivial transformation from a given FVST instance to a 3HS instance and to perform the known kernelization process [15] on this constructed 3HS instance. The kernel of the FVST instance is then constructed from the reduced 3HS instance—this is the core contribution. In the following, we first describe these three steps and give an estimation of the runtime. Then, we prove the size bound of the kernel and the correctness of the kernelization process.

The transformation from a given FVST instance $(T = (V, E), k)$ to a 3HS instance (S, C, k) with $S := V$ is easy: By Lemma 1, it suffices to enumerate all triangles in T and, for each triangle, add its three vertices as a three-element subset into the subset collection C. This transformation can be done in $O(n^3)$ time. Note that $|C| \leq n^3$.

Then, we apply the data reduction rules for 3HS given in [15] to the generated 3HS instance. Herein, the second rule removes some elements from S, which have to be contained in every size-k solution of the 3HS instance. We use a set H to store these elements; H is initialized as an empty set.

Rule 1. If there is a pair of elements x and y appearing together in more than k three-element subsets, then delete all these subsets from C and add a two-element subset $\{x, y\}$ to C.

Rule 2. If there is an element x appearing in more than k^2 three-element subsets or in more than k two-element subsets, then delete all subsets containing x from C, add x to H, and decrease the parameter k by one.

A 3HS instance can be transformed in $O(\max\{|S|, |C|\}) = O(n^3)$ time into a reduced instance [15].

Finally, from the reduced 3HS instance (S', C', k'), we construct an FVST instance (T', k') with $k' = k - |H|$. First, we replace the two-element subsets in C' by some three-element subsets. Note that any two-element subset $\{x, y\}$ was added to C' by an application of Rule 1; this application did remove a set A of three-element subsets from C with $|A| > k$. We partially "reverse" this application, that is, we delete $\{x, y\}$ from C', choose exactly $k' + 1$ three-element subsets from A and add them to C'. We choose the $k' + 1$ subsets such that they do not contain any element from H; because $k = k' + |H|$, this is always possible. After replacing all two-element subsets in C', we define S'' as the set containing all elements of S appearing in at least one subset in C'. Then the tournament $T' = (V', E')$ is constructed by setting $T' := T[S'']$. Due to Rule 2, the subset collection of the reduced 3HS instance contains $O(k^2)$ two-element subsets; otherwise, there is no solution. We can construct T' from C' in $O(k^3)$ time.

Summarizing the runtimes of the three steps, the runtime of the kernelization process for FVST is $O(n^3)$.

In the construction of T', we add for each two-element subset exactly $k' + 1$ three-element subsets. There are at most $(k')^2$ two-element subsets in the subset collection of the reduced 3HS instance. Together with the problem kernel of 3HS shown in [15] with $|C'| = O(k^3)$, we have $O((k')^3)$ elements in S''. Therefore, $|V'| = O(k^3)$.

It remains to show the correctness of the kernelization process: tournament T has a feedback vertex set of size at most k iff T' has a feedback vertex set of size at most k'.

Given a feedback vertex set F for T with $|F| \leq k$, $F' := V' \cap F$ is a feedback vertex set for T': with the transformation from the FVST instance to the 3HS instance and the kernelization process for 3HS, the elements in H generated by Rule 2 correspond to vertices v in T that are in more than k triangles that, except for v, are vertex-disjoint. Thus, the vertices corresponding to the elements in H are clearly in every feedback vertex set of T, and $H \subseteq F$. Moreover, since T' is an induced subgraph of T, F' is a feedback vertex set of T'. From $H \cap V' = \emptyset$, we have $|F'| \leq |F| - |H| = k'$, that is, F' is a feedback vertex set of T' with at most k' vertices.

Given a feedback vertex set F' of T' with at most k' vertices, $F' \cup H$ is a feedback vertex set of T: Every triangle in T corresponds to a three-element subset in C. If such a three-element subset contains no element from H, then either it is not changed during the kernelization process or it is removed since it contains two elements x and y which appear together in more than k three-element subsets in C. For the former case we have a triangle in T' due to the construction of T' and, thus, at least one vertex of this triangle is in F'. Considering the latter case, after the kernelization process of 3HS, there is a two-element subset $\{x, y\} \in C'$. While constructing T', we have added $k' + 1$ three-elements subsets containing x and y to C'; this results in at least $k' + 1$ triangles in T' containing x and y. Thus, $\{x, y\} \cap F' \neq \emptyset$. Summarizing both cases, $F' \cup H$ is a feedback vertex set of T with at most k vertices. $\square$

The basic idea for the kernelization of FVST can be generalized to any vertex deletion problem whose goal graph can be characterized by a finite set of forbidden subgraphs consisting of three vertices; this results in the following theorem.

Theorem 3. *If a vertex deletion problem on directed or undirected graphs has a goal graph that can be characterized by a finite set of forbidden subgraphs consisting of three vertices, then this problem admits a problem kernel consisting of a graph with $O(k^3)$ vertices, where k denotes the number of allowed vertex deletions.*

4.2   Feedback Arc Set in Tournaments

We present a simple data reduction rule for FEEDBACK ARC SET in tournaments (FAST), which leads to a kernel for this problem consisting of a tournament with $O(k^2)$ vertices. Without loss of generality, we assume that each vertex of the input tournament $(T = (V, E), k)$ is in at least one triangle.

Data reduction rule. If there is an arc in more than k triangles, then reverse this arc, add this arc to the solution, and decrease the parameter k by one.

Theorem 4. FEEDBACK ARC SET *in tournaments admits a problem kernel consisting of an $O(k^2)$-vertex tournament that can be found in $O(kn^3)$ time.*

Proof (sketch). Suppose that we have a reduced FAST instance (T, k) where T has a feedback arc set F with at most k arcs. Then each triangle contains at least one arc from F. Due to the data reduction rule, each arc in F can be in at most k triangles. $\square$

5   Search Tree for Feedback Arc Set in Bipartite Tournaments

Raman and Saurabh [16] have shown that if a tournament T does not contain a particular four-vertex tournament denoted by G, then the cycles in T are pairwise vertex-disjoint. Using this, their $O(2.42^k \cdot n^{2.38})$-time algorithm solves FAST in a two-phase manner: First, it uses a depth-bounded search tree approach to get rid of all cycles contained in subtournaments G appearing in T by reversing at most k arcs; this also destroys all subtournaments G in T. In the second phase, in each tournament output by the search tree it destroys all remaining, pairwise disjoint triangles by reversing an arbitrary arc in each triangle. If after these two phases there is an acyclic tournament with at most k arcs reversed, then T has a feedback arc set with size at most k.

   Following the same approach, we derive a fixed-parameter algorithm for FEED-BACK ARC SET in bipartite tournaments (FASBT). We use the following lemma, which is easy to prove.

Lemma 4. *A bipartite tournament is acyclic iff it contains no cycle of length four.*

By Lemma 4, in order to derive a forbidden subgraph characterization for bipartite tournaments where all cycles of length four are disjoint, we consider two length-four cycles in a bipartite tournament. If they are not vertex-disjoint, then they have one, two, or three common vertices. These three possibilities lead to bipartite tournaments which contain G_1 or G_2 shown in Fig. 4 as induced subgraph. The following lemma strengthens this finding.

Lemma 5. *If a bipartite tournament B contains neither G_1 nor G_2 (shown in Fig. 4) as an induced subgraph, then the cycles in B are pairwise disjoint.*

Proof. With Lemma 4, we first consider length-four cycles. By distinguishing three cases, namely two length-four cycles sharing one, two, and three vertices, respectively, one can easily show that a $\{G_1, G_2\}$-free bipartite tournament contains no two length-four cycles having a common vertex. Moreover, observe that in a bipartite tournament B, a subgraph of B induced by the vertices lying on

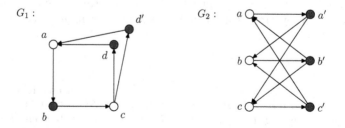

Fig. 4. Forbidden subgraphs for bipartite tournaments where all cycles of length four are disjoint. The color of the vertices describes the bipartition.

a cycle with length greater than four contains several length-four cycles which are not vertex-disjoint. Thus, a $\{G_1, G_2\}$-free bipartite tournament contains no cycle with a length greater than four. This completes the proof.   □

Based on Lemma 5 our algorithm solving FASBT has the same two phases as the algorithm by Raman and Saurabh [16], namely a search tree algorithm destroying all cycles contained in the induced subgraphs G_1 and G_2 from Fig. 4 and a polynomial-time second phase getting rid of the remaining, vertex-disjoint cycles. For destroying the cycles in G_1, the search tree algorithm makes a branching into six subcases, namely, reversing (a, b), reversing (b, c), reversing (c, d) and (c, d'), reversing (c, d) and (d', a), reversing (d, a) and (c, d'), and reversing (d, a) and (d', a). For each reversed arc, the parameter k is decreased by one. The size of depth-bounded search trees can be estimated using *branching vectors* [14]. The branching vector here is $(1, 1, 2, 2, 2, 2)$, corresponding to a search tree size of $O(3.24^k)$. Dealing with G_2, we make a branching into 17 subcases and, in each subcase, reverse two or three arcs. We omit the details of this branching. The worst-case runtime is determined by the branching for G_2, with a search tree size of $O(3.38^k)$. Note that finding one of G_1 and G_2 in an n-vertex bipartite tournament needs $O(n^6)$ time. When destroying vertex-disjoint cycles in the second phase, reversing arcs on cycles does not generate new cycles and, thus, we need only $O(n)$ time. The following theorem then follows.

Theorem 5. FEEDBACK ARC SET *in bipartite tournaments of n vertices with k arc deletions can be solved in $O(3.38^k \cdot n^6)$ time.*

6   Outlook

Table 1 surveys and compares complexity results on feedback set problems in tournaments. As can be seen there, the class of bipartite tournaments is not yet well explored. From a parameterized view, the grand challenge is to answer the question whether FVS in general directed graphs is fixed-parameter tractable or not, a long-standing open problem. On the route to this, further studying generalizations of tournaments might be fruitful. Besides attacking problems left open in Table 1, clearly further improvements concerning the efficiency of the described algorithms are very desirable. Due to the considerable practical relevance of the considered problems in applications such as voting systems, rankings, and graph drawing, they are natural candidates for algorithm engineering.

References

1. N. Ailon, M. Charikar, and A. Newman. Aggregating inconsistent information: ranking and clustering. In *Proc. 37th STOC*, pages 684–693. ACM, 2005.
2. N. Alon. Ranking tournaments. *SIAM Journal on Discrete Mathematics*, 20(1):137–142, 2006.
3. M.-C. Cai, X. Deng, and W. Zang. An approximation algorithm for feedback vertex sets in tournaments. *SIAM Journal on Computing*, 30(6):1993–2007, 2001.

4. M.-C. Cai, X. Deng, and W. Zang. A min-max theorem on feedback vertex sets. *Mathematics of Operations Research*, 27(2):361–371, 2002.
5. P. Charbit, S. Thomassé, and A. Yeo. The minimum feedback arc set problem is NP-hard for tournaments. *Combinatorics, Probability and Computing*, 2005. To appear.
6. V. Conitzer. Computing Slater rankings using similarities among candidates. Technical Report RC23748, IBM Thomas J. Watson Research Center, Yorktown Heights, NY, 2005.
7. F. K. H. A. Dehne, M. R. Fellows, M. A. Langston, F. A. Rosamond, and K. Stevens. An $O(2^{O(k)}n^3)$ FPT algorithm for the undirected feedback vertex set problem. In *Proc. 11th COCOON*, volume 3595 of *LNCS*, pages 859–869. Springer, 2005. To appear in *Theory of Computing Systems*.
8. R. G. Downey and M. R. Fellows. *Parameterized Complexity*. Springer, 1999.
9. H. Fernau. A top-down approach to search-trees: Improved algorithmics for 3-hitting set. Technical Report TR04-073, Electronic Colloquium on Computational Complexity, 2004.
10. P. Festa, P. M. Pardalos, and M. G. C. Resende. Feedback set problems. In D. Z. Du and P. M. Pardalos, editors, *Handbook of Combinatorial Optimization, Vol. A*, pages 209–258. Kluwer, 1999.
11. J. Flum and M. Grohe. *Parameterized Complexity Theory*. Springer, 2006.
12. M. L. Fredman. On computing the length of longest increasing subsequences. *Discrete Mathematics*, 11(1):29–35, 1975.
13. J. Guo, J. Gramm, F. Hüffner, R. Niedermeier, and S. Wernicke. Improved fixed-parameter algorithms for two feedback set problems. In *Proc. 9th WADS*, volume 3608 of *LNCS*, pages 158–168. Springer, 2005. To appear in *Journal of Computer and System Sciences*.
14. R. Niedermeier. *Invitation to Fixed-Parameter Algorithms*. Oxford University Press, 2006.
15. R. Niedermeier and P. Rossmanith. An efficient fixed parameter algorithm for 3-Hitting Set. *Journal of Discrete Algorithms*, 1(1):89–102, 2003.
16. V. Raman and S. Saurabh. Parameterized algorithms for feedback set problems and their duals in tournaments. *Theoretical Computer Science*, 351(3):446–458, 2006.
17. B. Reed, K. Smith, and A. Vetta. Finding odd cycle transversals. *Operations Research Letters*, 32(4):299–301, 2004.
18. E. Speckenmeyer. On feedback problems in digraphs. In *Proc. 15th WG*, volume 411 of *LNCS*, pages 218–231. Springer, 1989.
19. A. Truß. Parameterized algorithms for feedback set problems in tournaments (in German). Diplomarbeit, Institut für Informatik, Friedrich-Schiller-Universität Jena, Dec. 2005.
20. A. van Zuylen. Deterministic approximation algorithms for ranking and clustering problems. Technical Report 1431, School of Operations Research and Industrial Engineering, Cornell University, Ithaca, NY, Sept. 2005.

Parameterized Algorithms for HITTING SET: The Weighted Case

Henning Fernau[1,2,3,4]

[1] Univ. Trier, FB 4—Abteilung Informatik, 54286 Trier, Germany
[2] Univ. Hertfordshire, Comp. Sci., College Lane, Hatfield, Herts AL10 9AB, UK
[3] Univ. Tübingen, WSI für Informatik, Sand 13, 72076 Tübingen, Germany
fernau@informatik.uni-tuebingen.de
[4] Univ. Newcastle, Comp. Sci., University Drive, Callaghan, NSW 2308, AUS

Abstract. We are going to analyze simple search tree algorithms for WEIGHTED d-HITTING SET. Although the algorithms are simple, their analysis is technically rather involved. However, this approach allows us to even improve on elsewhere published algorithm running time estimates for the more restricted case of (unweighted) d-HITTING SET.

1 Introduction

Our approach—in general. We exhibit how to systematically design and analyze search tree algorithms within the framework of parameterized algorithmics [2]. Here, we advocate a top-down approach as opposed to a rather bottom-up design, because the resulting algorithms tend to be simpler than via the opposite approach, and they sometimes pretty much resemble heuristic pruning techniques as used in branch-and-cut algorithms for solving hard problems. Moreover, this approach is quite modular in the sense that it produces algorithms whose *search tree backbone*, i.e., the branching pattern of the algorithm as such, is not affected by the optimization techniques reflected in what we will call *heuristic priorities* (according to which the branching is performed) and the employed *reduction rules*. This not only modularizes correctness proofs for such algorithms, but also favors rapid prototyping of implementations. We will exemplify this approach by developing and analyzing simple algorithms for WEIGHTED d-HITTING SET (d-WHS) problems. No prior research on parameterized algorithms has been reported for these problems.

Problem statement. WEIGHTED d-HITTING SET (d-WHS) can be viewed as a "weighted vertex cover problem" on hypergraphs. More formally, this problem can be stated as follows:

Given: A weighted hypergraph $G = (V, E, w)$ with *edge size* bounded by d, i.e., $\forall e \in E(|e| \leq d)$, and a *weight function* $w : V \rightarrow [1, \infty)$
Parameter: a non-negative integer k
Question: Is there a *(weighted) hitting set* C of total weight of at most k, i.e., $\exists C \subseteq V \forall e \in E(C \cap e \neq \emptyset)$ and $w(C) := \sum_{x \in C} w(x) \leq k$?

T. Calamoneri, I. Finocchi, G.F. Italiano (Eds.): CIAC 2006, LNCS 3998, pp. 332–343, 2006.
© Springer-Verlag Berlin Heidelberg 2006

Why HITTING SET*?* HITTING SET problems show up in many places; e.g., Reiter's ground-breaking research on *model-based diagnosis* [10,14] relates the automatic diagnosis of systems to HITTING SET. The thrive for minimum hitting sets is in that context motivated by the parsimony principle in two ways: (a) the simplest diagnosis tends to find the actual cause, and (b) when the diagnosis implies exchanging (possibly) faulty components (as a consequence of a self-diagnosis of an autonomous system, e.g., in space), then a minimum hitting set might also be the cheapest repair solution; in that particular scenario, however, the weighted case seems to be even more interesting than the unweighted one. As a further application, in [8], connections between a two-tree drawing problem that is important in bioinformatics and 4-WHS are shown, where the weights reflect further natural restrictions from biological background knowledge. The algorithmics of this paper can be immediately transferred to both applications.

Previous work. For the unweighted case (which is a special case of the weighted setting if all weights are equal to one), there is one published paper presenting a search tree algorithm for UNWEIGHTED d-HITTING SET (d-HS), $d > 2$, from a parameterized perspective [11]. The exponential base of the running time estimate for these algorithms tends to $d-1$ with growing d, although in the simplest case $d = 3$, it is still relatively far off from that bound: that basis is $1 + \sqrt{2}$. By an intricate case analysis of a comparatively complicated algorithm, they were able to arrive at an $\mathcal{O}^*(2.270^k)$ algorithm for the (unweighted) 3-HS problem (i.e., all weights equal one). This was improved in [5] to about $\mathcal{O}^*(2.179^k)$ by using a similar methodology as explained here for the weighted case.

Notice that we are dealing with search tree algorithms and apply a parameterized analysis of the search tree size. If we then say that the algorithm has $\mathcal{O}^*(f(k))$ running time, where k is the parameter, this means that the search tree has size (number of leaves) $\mathcal{O}(f(k))$, since the work in each search tree node will be at worst polynomial in n. In actual fact, all analysis that follows will be a clever estimate on the size of the search tree.

For the special case of 2-HS, likewise known as VERTEX COVER, in a kind of race (using more and more intricate case analysis) an $\mathcal{O}(1.285^k + kn)$-algorithm [1] has been obtained. For 2-WHS, likewise known as WEIGHTED VERTEX COVER, the best that was obtained is on $\mathcal{O}^*(1.396^k)$, see [12]. Our approach seems not to be suitable to tackle the case $d = 2$.

The results of this paper. As in the unweighted case [5], our analysis is based on the introduction of a second *auxiliary parameter* that allows us to account for "gains" obtained by using appropriate reduction rules and heuristic priorities. This technique can be useful in other areas of parameterized algorithms, as we believe. We get the following table for the bases c_d of an $\mathcal{O}^*(c_d^k)$ algorithm for d-WHS; the bases are better than those for the *unweighted case* published in [11]:

d	3	4	5	6	7	8	9	10	100	
$c_d \leq$	2.2470	3.1479	4.1017	5.0640	6.0439	7.0320	8.0243	9.0191	99.0002	(1)

General notions and definitions. We introduce some terminology on hypergraphs as needed for HITTING SET. A *hypergraph* $G = (V, E)$ is given by its finite set of *vertices* V and its set of *(hyper)-edges* E, where a hyperedge is a subset of V. The cardinality $|e|$ of a hyperedge e is also called its *size*. The cardinality of the set of edges which contain the vertex v is called the *degree* of v, written $\delta(v)$.

2   Heuristics and Reductions for WEIGHTED *d*-HITTING SET

2.1   A Simple Branching Algorithm

Since each hyperedge must be covered and the weights are all at least one, there exists a trivial $\mathcal{O}^*(d^k)$-algorithm for d-WHS.

```
simple-WHS(G = (V, E, w), k, S):
  IF k > 0 AND G has some edges THEN
    choose some edge e; // to be refined
    S' = ∅; // solution to be constructed
    FOREACH x ∈ e DO // recursively branch
      G' = (V \ {x}, {e ∈ E | x ∉ e});
      S' = simple-WHS(G', k − w(x), S ∪ {x})
      IF S' ≠ failure THEN break
    return S'
  ELSIF E = ∅ THEN return S ELSE return failure
```

Obviously, the base of the exponential running time of this algorithm heavily depends on the necessary amount of branching. Observe that according to the problem specification, in a d-WHS instance, there might be edges of size *up to d* already in the very beginning. "Small edges" may also be introduced later during the run of the algorithm. A natural heuristic would first branch on small edges. We would therefore refine:

```
simple-WHS(G, k, S):
  IF k > 0  AND G has some edges THEN
    choose some edge e of smallest size;
    ... // as before
```

Can we make use of this "heuristic priority" in our analysis? We therefore now define reduction rules which we will always exhaustively apply at the beginning of each recursive call. Moreover, we switch towards a "binary branching" at vertices (instead of branching on edges), as can be seen in Alg. WHS-ST below.

2.2   Reduction Rules

First reduction rule: vertex domination. The *vertex domination rule* that was used in [5, 11] for the unweighted case is invalid in full generality in the weighted case, but has to be replaced by the following *weighted vertex domination rule*: If, for all edges e, $x \in e$ implies $y \in e$ and if $w(y) \leq w(x)$, then delete x.

This reduction rule implies the following one (reduction rule for degree-one-vertices): If $x, y \in e$ with $\delta(x) = 1$ and $w(y) \leq w(x)$, then remove x. The soundness of this rule is easily seen: the only reason for taking a vertex x into the hitting set, in a situation as described by the reduction rule, is that it might be *cheap*. Conserving expensive vertices makes no sense. This reduction rule immediately implies:

Lemma 1. *In a reduced instance, there is no edge with more than one vertex of degree one.*

The next lemma is again an easy consequence from the weighted vertex domination rule and is of particular importance when $d > 3$.

Lemma 2. *In a reduced instance, for any two edges e_1 and e_2, there is at most one $x \in e_1 \cap e_2$ with $\delta(x) = 2$.*

Other rules stated in [5] literally transfer to the weighted case:

Second reduction rule: edge domination. An edge e is dominated by another edge f if $f \subset e$. Then, we delete e, since covering f will automatically also cover e.

Third reduction rule: small edges. Delete all edges of size one and place the corresponding vertices into the hitting set.

The small edge rule, together with the vertex domination rule, proves the non-existence of *isolated edges* in the following precise sense:

Lemma 3. *In a reduced instance, there is no edge e such that all vertices $x \in e$ have degree one.*

Fourth reduction rule: edge cover rule. If G contains a component C that is of maximum vertex degree two, then resolve C in polynomial time.

This (last) rule is justified by the following lemma:

Lemma 4. *If G is a weighted hypergraph of maximum vertex degree of two, then a minimum weighted hitting set can be found in polynomial time.*

Proof. To G, there corresponds an edge-weighted graph G' whose vertices are the edges of G and whose vertex-adjacency relation is the edge-adjacency relation of G. A minimum weighted hitting set of G corresponds to a minimum weighted edge cover of G' that can be computed in polynomial time (via matching). ∎

2.3   Branching Rules and Their Analysis

The idea of making favorable branches first has also another bearing, this time on the way we are going to analyze the search tree algorithm, based on an *auxiliary parameter* ℓ. Let $T^\ell(k)$, $\ell \geq 0$ denote the size (more precisely, the number of leaves) of the search tree when assuming that at least ℓ edges in the given instance (with parameter k) have a size of (at most) $d - 1$. The intuition is that $T^3(k)$ would describe a situation which is "more like" $(d-1)$-WHS than $T^2(k)$. The underlying idea is that search trees with many small edges are smaller than search trees with only a few; hence:

$$\forall k : T^\ell(k) \geq T^{\ell+1}(k). \tag{2}$$

Regarding an upper bound on the size $T(k)$ of the search tree of the whole problem, we can equate $T(k) = T^0(k)$ by following the same intuition. Eq. (2) also shows that, upon analyzing a T^ℓ-situation, we can always assume that there are exactly ℓ edges that have a size of at most $d - 1$, and these small edges do have a size of exactly $d - 1$.

Our algorithm will make choices with the bias of what we will call *heuristic priorities*. They can be refined if necessary along the analysis of the algorithm. The simplest list to start with might contain a single rule that should be intuitively clear: Choose a vertex of highest degree within an edge of smallest size. We will update the list of priorities whenever necessary.

```
WHS-ST(G = (V, E, w), k, S):
  exhaustively apply reduction rules;
  IF k > 0 THEN
    IF E = ∅ THEN return S;
    choose some vertex x according to the heuristic priorities
    S' = ∅; // solution to be constructed
    E' = { e ∈ E | x ∉ e };
    S' = WHS-ST((V \ {x}, E'), k − w(x), S ∪ {x});
    IF S' ==failure THEN
      E'' = { e \ {x} | e ∈ E };
      S' = WHS-ST((V \ {x}, E''), k, S);
    return S'
  ELSIF G contains some edges or k < 0
    return failure
  ELSE // G contains no edges and k is zero
    return S
```

In the very beginning, given the instance (G, k), we call WHS-ST$(G, k, \emptyset)$. We assume that reduction rules may also change the parameter value k and the solution S. The algorithm is quite generic: the list of reduction rules may grow and we might also change the heuristic priorities. The simple *binary branching* structure of WHS-ST enables a straight-forward inductive proof of its correctness:

Theorem 1. *If the reduction rules are correct, then* WHS-ST$(G, k, \emptyset)$ *either returns a correct hitting set to the d-WHS instance (G, k) or it returns* failure, *if there is no solution of size at most k.*

3   A Simple Branching Analysis

We will now undertake a *simple* analysis, only considering T^0, T^1 and (partially) T^2 and T^3.

Lemma 5. $T^0(k) \le T^0(k − 1) + T^3(k)$.

Proof. Whenever we select an edge of size d to branch on (according to the heuristic priorities), we can find an edge that contains a vertex x of degree three or larger due to the edge cover rule. One branch is that x is put into the hitting set. This reduces the admissible weight by at least one. If x is not put into the hitting set, then at least three new edges of size two are created.   ∎

T^1-*branching.* In the next lemma, we show a first step into a strategy which will finally give us better branching behaviors. Namely, we try to exploit the effect of reduction rules triggered in different sub-cases. This already necessitates a refinement in the choice of heuristic priorities: within a smallest edge e of size $j < d$, we prefer branching at $x \in e$ that maximizes the number of incident edges of size $j + 1$.

Lemma 6. $T^1(k) \leq \max\{T^0(k-1) + T^1(k-1) + T^2(k-1) + (d-4)T^3(k-1), T^0(k-1) + T^1(k-1) + (j-2)T^2(k-1) + (d^2 - (2j+1)d + (j^2+j))T^0(k-2) : j = 2, 3, \ldots, d-2\}$ *if* $d \geq 4$; *if* $T^0(k) \geq (d-1)^k$ *or if* $d = 3$, *this may be simplified:* $T^1(k) \leq T^0(k-1) + (d-2)T^1(k-1)$.

Proof. The instance G has an edge $e = \{x_1, x_2, \ldots, x_{d-1}\}$ of size $(d-1)$.

<u>Case 1.</u> If there is an edge f of size d such that $2 \leq j := |e \cap f| \leq d - 2$ (this can only happen if $d \geq 4$), then we would first branch at the vertices in $e \cap f$; due to weighted vertex domination, at least one of the j branches that take one of the vertices of $e \cap f$ into the hitting set is an $T^1(k-1)$-branch and $j-2$ are even $T^2(k-1)$-branches (or better). If none of the vertices from $e \cap f$ goes into the hitting set, then, in order to cover e, there are $d-1-j$ many possibilities left, and in order to cover f, there are $d-j$ remaining possibilities. This explains the other $((d-j)-1)(d-j)$ many $T^0(k-2)$-branches. It can be shown that we may neglect these cases in our time analysis when assuming $T^0(k) \geq (d-1)^k$.

<u>Case 2.</u> If the previous case does not occur, then for all edges $f \neq e$, $|e \cap f| \leq 1$. Assume that x_1 is the vertex of maximum degree in e, so that we branch at x_1. If $\delta(x_1) = 1$, we can deterministically resolve the case with the reduction rules (apply $d-1$ times the weighted vertex domination rule and then the small edge rule) and get *one* $T^0(k-1)$-branch. This is obviously better than the inequality claimed in the lemma. Therefore, we can now assume that $\delta(x_1) \geq 2$. If we take x_1 into the hitting set, then we get a $T^0(k-1)$-branch. If we do not take x_1 into the hitting set, we create one new edge e_1 of size $(d-1)$ and we get the edge $e' = e \setminus \{x_1\}$ of size $(d-2)$. In the next recursive call, e' is the edge of smallest size. There is no other edge of that size, since Case 1 did not apply. We therefore continue branching at the vertex (say x_2) of maximum degree in e'. Again, $\delta(x_2) = 1$ is better than the case we are going to pursue next. If $\delta(x_2) \geq 2$, then we again have two cases: either we take x_2 into the hitting set or not. If x_2 goes into the hitting set, then this is a $T^1(k-1)$-branch; namely, since Case 1 did not apply, $x_2 \notin e_1$, so that the small edge e_1 will be preserved. If x_2 does not go into the hitting set, then there will be a new edge e_2 of size $(d-1)$ ("new" due to edge domination). In the next recursive call, $e'' = e \setminus \{x_1, x_2\}$ is the edge of smallest size. The argument continues and shows that branches of type $T^j(k-1)$ will show up, for $j = 2, 3, \ldots, d-2$. This shows the claim, taking into account that $T^j(k-1) \leq T^3(k-1)$ for $j \geq 3$ due to Eq. (2).   ∎

Estimating branching numbers. By using the inequality $T^3(k) \leq T^2(k) \leq T^1(k)$, Lemmas 5 and 6 yield:

$$T^0(k) \leq T^0(k-1) + T^1(k) \tag{3}$$
$$T^1(k) \leq T^0(k-1) + (d-2)T^1(k-1)$$

With c_d being the largest positive real root of the characteristic polynomial $x^2 - dx + d - 2$, i.e.,

$$c_d = \frac{d + \sqrt{d^2 - 4d + 8}}{2} = \frac{d + \sqrt{(d-2)^2 + 4}}{2} \geq d - 1 \tag{4}$$

we can see that by setting $T^0(k) = c_d^k$ and $T^1(k) = (c_d - 1)c_d^{k-1}$, the inequalities system (3) can be solved. The larger d, the closer c_d gets to $d - 1$. Hence:

d	3	4	5	6	10	100
$T(k) \leq$	2.62^k	3.42^k	4.31^k	5.24^k	9.13^k	99.0103^k

Obviously, this is worse than what Niedermeier and Rossmanith got in [11] for the (general) unweighted case (due to the lack of the vertex domination rule in full generality), but shows the same "limit behavior" (when d is large). Can we do better? Let us give a simple trial to incorporate T^2 and T^3 into the analysis in the special case of WEIGHTED 3-HITTING SET.

4   WEIGHTED 3-HITTING SET

We will use subscripts in the functions that describe the search tree sizes to indicate this special case. We branch according to the following heuristic priorities.
Let s be the size of the smallest edge in the instance $G = (V, E, w)$.
Let E_s be the collection of smallest size edges.
$(P_3 1)$ Let the set of (first) branching candidates B be $\bigcup_{e \in E_s} e$.
$(P_3 2)$ If e is a smallest edge that is disjoint with all other $e' \in E_s$, refine $B = e$.
$(P_3 3)$ If no such isolated smallest edge exists, then update B to collect the vertices of maximum degree in the hypergraph $(\bigcup_{e \in E_s} e, E_s)$.
$(P_3 4)$ Select $x \in B$ to be a vertex of maximum degree in G.
It is easy to check that the analyses of Lemmas 5 and 6 are still valid under these heuristic priorities.

Lemma 7. $T_3^2(k) \leq \max\{T_3^1(k-1) + T_3^2(k-1), T_3^0(k-1) + T_3^0(k-2)\}$

Proof. We consider first the situation that the two edges e_1 and e_2 of size two are disjoint (see $(P_3 2)$). Then, basically the analysis of Lemma 6 applies, showing the claim. More precisely, we have $T_3^2(k) \leq T_3^1(k-1) + T_3^2(k-1)$.

Otherwise, $e_1 \cap e_2 \neq \emptyset$, i.e., $e_1 = \{x, y\}$ and $e_2 = \{x, z\}$. According to the heuristic priority $(P_3 3)$, we branch at x. If we take x into the hitting set, we get a $T_3^0(k-1)$-branch. Not taking x into the hitting set enforces y and z into the hitting set, which is a $T_3^0(k-2)$-branch. ∎

Lemma 8. $T_3^3(k) \leq \max \begin{cases} T_3^1(k-1) + T_3^0(k-2), \\ T_3^0(k-1) + T_3^0(k-3), \\ T_3^2(k-1) + T_3^3(k-1) \end{cases}$

Proof. If there is a edge e of size two that has non-empty intersection with any other edge of size two, due to (P_32) we branch on e without destroying the at least two other edges of size two. The reasoning given in Lemma 6 therefore yields the upper bound $T_3^2(k-1) + T_3^3(k-1)$ in this case.

If the first case does not apply, the all edges of size two are connected. Let e_1, e_2, e_3 be three connected edges of size two. If $x \in e_1 \cap e_2 \cap e_3$ exists, then we branch at x due to (P_33). This gives the (trivial) upper bound of $T_3^0(k-1) + T_3^0(k-3)$. Otherwise, we branch at some x contained in two small edges due to (P_33); w.l.o.g.: $x \in e_1 \cap e_2$. Since $x \notin e_3$, the case that we take x into the hitting set is indeed a $T_3^1(k-1)$-branch. This explains the upper bound $T_3^1(k-1) + T_3^0(k-2)$. ∎

Theorem 2. WEIGHTED 3-HITTING SET *can be solved in time* $\mathcal{O}^*(2.2470^k)$.

The algebra justifying this claim can be found in the long version of the paper. We only mention that the exact solution of the inequalities system can be described by the largest positive root c_3 of the polynomial $x^3 - 2x^2 - x + 1$, which then gives $T_3^0(k) = c_3^k$, $T_3^1 = c_3^k/(c_3 - 1)$, $T_3^2(k) = c_3^k/(c_3 - 1)^2$, and $T_3^3(k) = c_3^{k-1}(c_3 - 1)$. This worst case is realized when all T^3-branches are according to the $T_3^1(k-1) + T_3^0(k-2)$-estimate. Improving on that particular case would not help too much, however, since the other extreme cases show also branching behaviors worse than 2.2^k. Observe that this also means that a search tree in the $T^3(k)$-case is only about half the size of a search tree in the $T^0(k)$-case.

5   WEIGHTED d-HITTING SET with $d \geq 4$

How well do our considerations transfer to the more general case? We analyze possible T_d^2-branches in what follows. Since the obtained bases are quite satisfactory, we refrain from analyzing the T_d^3-branches. In our analysis, we apply the following heuristic priorities to a given (reduced) instance $G = (V, E, w)$:

   Let s be the size of the smallest edge in the instance $G = (V, E, w)$.

   Let E_s be the collection of smallest size edges.

$(P1)$ Let the set of (first) branching candidates B be $\bigcup_{e \in E_s} e$.

$(P2)$ Define $G_B = (B, E_s)$ and update B to be the set of vertices in G_B of maximum degree.

$(P3)$ Choose a vertex $x \in B$ of maximum degree in G.

   One can check that Lemmas 5 and 6 are still valid when assuming these priorities.

Analyzing T^2. We will distinguish several cases in what follows:

Lemma 9. *Let e_1 and e_2 be two edges of size $d - 1$. If $e_1 \cap e_2 = \emptyset$, then we can estimate $T_d^2(k) \leq T_d^1(k-1) + (d-2)T_d^2(k-1)$.*

This can be basically inherited from Lemma 6 due to edge domination. As we will see, this is the second worst case branching. Being the simplest case, we give some details. As justified in long version, we have to solve the next set of equations:

$$T_d^0(k) = T^0(k-1) + T^2(k) \tag{5}$$
$$T_d^1(k) = T^0(k-1) + T^1(k-1) + (d-3)T^2(k-1)$$
$$T_d^2(k) = T_d^1(k-1) + (d-2)T_d^2(k-1)$$

This yields, after some algebra:

$$0 = T_d^0(k+1) - dT_d^0(k) + (d-1)T_d^0(k-1) - T_d^0(k-2). \tag{6}$$

Theorem 3. *Let c_d denote the largest positive real root of the polynomial $x^3 - dx^2 + (d-1)x - 1$. Then $T_d^0(k) = c_d^k$, $T_d^1(k) = \alpha_{d,1}c_d^k$ with $\alpha_{d,1} = (c_d - d + 2)(c_d - 1)/c_d$ and $T_d^2(k) = \alpha_{d,2}c_d^k$ with $\alpha_{d,2} = (c_d - 1)/c_d$ solve the system (5).*

The following table lists some of the exponential bases c_d for (5):

d	3	4	5	6	7	8	9	10	100	
$c_d \leq$	2.3248	3.1479	4.0780	5.0490	6.0330	7.0237	8.0178	9.0139	99.0002	(7)

Lemma 10. *Let e_1 and e_2 be two edges of size $d-1$. If $|e_1 \cap e_2| = j \in \{1, 2, \ldots, d-2\}$, then we can estimate*

$$T_d^2(k) \leq T_d^0(k-1) + T_d^1(k-1) + (j-2)T_d^2(k-1) + (d-1-j)^2 T_d^0(k-2),$$

thereby assuming that $T_d^0(k) \geq (d-1)^k$, i.e., $c_d \geq d-1$.

Proof. The priorities $(P1)$ and $(P2)$ let us branch at a vertex $x \in e_1 \cap e_2$. If $j > 1$, the weighted vertex domination rule moreover guarantees that there is a vertex of degree at least three in $e_1 \cap e_2$, and $(P3)$ will select one such vertex x for branching. Hence, when x is not taken into the hitting set, then we gain at least one edge of size $d-1$ if $j > 1$ due to vertex domination, see Lemma 2, since we will continue selecting vertices within $e_1 \cap e_2$ according to $(P2)$. The case that edges that intersect with $e_1 \cap e_2$ might contain more than one vertex in this intersection turns out not to be the worst case (assuming $(d-1)^k$ as a lower bound of our approach) along the lines of Lemma 6. If $e_1 \cap e_2$ is "exhausted", then in the case that we take none of the vertices from $e_1 \cap e_2$ into the hitting set, we are left with two very small edges $e_1' = e_1 \setminus e_2$ and $e_2' = e_2 \setminus e_1$. $P1$ lets us continue branching at say e_1'. Having selected $x \in e_1'$ to go into the hitting set, e_2' will be the smallest edge (of size $(d-1-j)$), and hence $P1$ continues to branch on e_2' in the next recursion step. This explains that we get (very grossly estimated) $(d-1-j)^2$ many $T_d^0(k-2)$-branches. ∎

In order to prove Theorem 4, the following technical lemma is important:

Lemma 11. *If $j > 1$ and $d > 3$, then*

$$T_d^1(k-1) + (d-2)T_d^2(k-1)$$
$$\geq T_d^0(k-1) + T_d^1(k-1) + (j-2)T_d^2(k-1) + (d-1-j)^2 T_d^0(k-2)$$

for $T_d^0(k) = c^k$ and $T_d^2(k) = c^k - c^{k-1}$ with $d - 1 \leq c$, independent of T_d^1.

We need a somewhat stronger result (compared to Lemma 10) in the case $j = 1$ that describes our worst case (for $d > 4$):

Lemma 12. *Let e_1 and e_2 be two edges of size $d - 1$. If $|e_1 \cap e_2| = 1$, then we can estimate*

$$T_d^2(k) \leq T_d^0(k-1) + (d-3)T_d^1(k-2) + [(d-2)(d-3) + 1]T_d^0(k-2).$$

Moreover,

$$T_d^1(k-1) + (d-2)T_d^2(k-1) \geq T_d^0(k-1) + (d-3)T_d^1(k-2) + [(d-2)(d-3)+1]T_d^0(k-2)$$

for T_d^ℓ as defined in Theorem 4 below.

Proof. We only explain the branching in what follows (for the algebra, see the long version). Assume that $\{x\} = e_1 \cap e_2$. x is selected for branching according to $(P1)$. If x does not go into the hitting set, then we may continue branching on e_1. The claim is that, for any $y \in e_1 \setminus \{x\}$ (with one possible exception, if $\delta(y) = 1$ for some $y \in e_1$; but due to Lemma 1, there is at most one vertex of degree one in e_1 and $(P3)$ avoids branching at that vertex), there is an edge $e_y \neq e_1$ with $y \in e_y$ such that there is a vertex $z_y \in e_2 \setminus e_1$ with $z_y \notin e_y$. For, if $(e_2 \setminus \{x\}) \subseteq e_y$, then the edge domination rule would have triggered. The branch that takes y and z_y into the hitting set is a $T_d^1(k-2)$-branch (possibly better). ∎

Theorem 4. *d-WHS can be solved in time $\mathcal{O}^*(c_d^k)$, where c_d is the largest positive root of the characteristic polynomial $x^4 - 3x^3 - (d^2 - 5d + 5)x^2 + x + (d^2 - 6d + 9)$. Some values of c_d are listed below:*

d	4	5	6	7	8	9	10	100
$c_d \leq$	3.1845	4.1017	5.0640	6.0439	7.0320	8.0243	9.0191	99.0002

(8)

With a more sophisticated analysis (and again varied heuristic priorities), we could let Theorem 3 describe the worst case for 4-WHS.

Corollary 1. *4-WHS can be solved in time $\mathcal{O}^*(3.1479^k)$.*

Is it worthwhile trying to further improve on the exponential bases as derived in this section? In principle, yes of course; however, one would need a different approach for substantial improvements: (a) the second-worst case is only slightly better than the worst case that we analyzed, and (b) with growing d, the lower bound $(d-1)$ assumed in (some) estimates is already quite well approximated.

6   Conclusions

We are currently developing and analyzing a new, top-down methodology for parameterized search tree algorithms. Up to now, we have appliedthis methodology to d-HITTING SET [5], biplanarization problems [7] (thereby improving on the constants derived in [3]), linear arrangement problems (in the long version of [6]) and to WEIGHTED d-HITTING SET (this paper). In order to apply this method, we need a kind of second *auxiliary parameter* in the problem which we try to improve on in case the main parameter cannot be improved upon *binary branching*. In the case of (WEIGHTED) HITTING SET, the number of edges of small size is such an auxiliary parameter. Our results show that this methodology is a quite powerful tool of algorithm analysis. For example, while the gap between the running times of the (very sophisticated) best search tree algorithms for WEIGHTED VERTEX COVER and for VERTEX COVER [1, 12] do differ significantly (both algorithms being approximately of the same complexity), this paper shows that with our analysis method of a comparatively simple algorithm for 3-WHS, we can even (slightly) improve on the previous analysis of a much more sophisticated algorithm for UNWEIGHTED 3-HS [11].

It may be interesting to compare the way the analysis of the recurrences guided by the auxiliary parameter is undertaken in this paper with the analysis method of Wahlström [15] or with Eppstein's quasiconvex method [4]. It would be also interesting to see this approach applied to other, different problems with accordingly different auxiliary parameters.

More generally speaking, there seems to be a recent thrive in Exact Algorithmics towards "simple" algorithms. The recent MINIMUM DOMINATING SET algorithm of Fomin, Grandoni and Kratsch is only one more example (see [9]) that incidentally also uses a (special) HITTING SET algorithm. This direction of research certainly brings practical and theoretical research on attacking hard problems closer together, since one could also envisage a kind of interplay between algorithm analysis and algorithm testing in the near future. Can an appropriate analysis then "explain" certain observed phenomena of the implementation? The modular decomposition of such an algorithm into the actual recursive "search tree backbone" and the reduction rules and (in particular) the heuristic priorities also opens up a whole area of experimental algorithmics: under which circumstances (or, in a more theoretical formulation: for which classes of hypergraphs) is a certain set of rules the most successful? Can this be proved? Due to the simple overall structure of the algorithms, also an analysis of expected running times (possibly adding coin tossing into the heuristic priorities) might be possible.

Incidentally, improvements in parameterized algorithms for d-HITTING SET also entail improvements in exact algorithms for MINIMUM d-HITTING SET, measured in terms of number of vertices: in the case of 3-HITTING SET, the use of the algorithm exhibited in [5] improves Wahlström's algorithm [15] from $\mathcal{O}^*(1.6538^n)$ down to $\mathcal{O}^*(1.6483^n)$ (personal communication by Wahlström). The results of this paper will immediately entail new running time bounds for exact algorithms for MINIMUM WEIGHTED HITTING SET. For example, along the lines sketched by Raman, Saurabh and Sikdar, in [13], we get an exact algorithm for MINIMUM WEIGHTED 4-HITTING SET that runs in time $\mathcal{O}^*(1.97^n)$, using our parameterized WEIGHTED 4-HITTING SET algorithm.

References

1. J. Chen, I. A. Kanj, and W. Jia. Vertex cover: further observations and further improvements. *Journal of Algorithms*, 41:280–301, 2001.
2. R. G. Downey and M. R. Fellows. *Parameterized Complexity*. Springer, 1999.
3. V. Dujmović, M. R. Fellows, M. Hallett, M. Kitching, G. Liotta, C. McCartin, N. Nishimura, P. Ragde, F. A. Rosamond, M. Suderman, S. Whitesides, and D. R. Wood. A fixed-parameter approach to two-layer planarization. In P. Mutzel, M. Jünger, and S. Leipert, editors, *9th International Symp. on Graph Drawing GD'01*, volume 2265 of *LNCS*, pages 1–15. Springer, 2002.
4. D. Eppstein. Quasiconvex analysis of backtracking algorithms. In *Proc. 15th Symp. Discrete Algorithms SODA*, pages 781–790. ACM and SIAM, January 2004.
5. H. Fernau. A top-down approach to search-trees: Improved algorithmics for 3-Hitting Set. Technical Report TR04-073, Electronic Colloquium on Computational Complexity ECCC, 2004.
6. H. Fernau. Parameterized algorithmics for linear arrangement problems. In U. Faigle, editor, *CTW 2005: Workshop on Graphs and Combinatorial Optimization*, pages 27–31. University of Cologne, Germany, 2005. Long version submitted to a special issue of *Discrete Applied Mathematics*.
7. H. Fernau. Two-layer planarization: Improving on parameterized algorithmics. *Journal of Graph Algorithms and Applications*, 9:205–238, 2005.
8. H. Fernau, M. Kaufmann, and M. Poths. Comparing trees via crossing minimization. In R. Ramanujam and Sandeep Sen (editors): *Foundations of Software Technology and Theoretical Computer Science FSTTCS 2005*, vol. 3821 of *LNCS*, pp. 457–469. Berlin: Springer, 2005.
9. F. V. Fomin, F. Grandoni, and D. Kratsch. Measure and conquer: domination – a case study. In L. Caires, G. F. Italiano, L. Monteiro, C. Palamidessi, and M. Yung, editors, *Automata, Languages and Programming, 32nd International Colloquium, ICALP*, volume 3580 of *LNCS*, pages 191–203. Springer, 2005.
10. J. de Kleer, A. K. Mackworth, and R. Reiter. Characterizing diagnoses and systems. *Artificial Intelligence*, 56:197–222, 1992.
11. R. Niedermeier and P. Rossmanith. An efficient fixed-parameter algorithm for 3-Hitting Set. *Journal of Discrete Algorithms*, 1:89–102, 2003.
12. R. Niedermeier and P. Rossmanith. On efficient fixed parameter algorithms for weighted vertex cover. *Journal of Algorithms*, 47:63–77, 2003.
13. V. Raman, S. Saurabh, and S. Sikdar. Improved exact exponential algorithms for vertex bipartization and other problems. In M. Coppo et al., editors, *Italian Conference on Theoretical Computer Science ICTCS*, volume 3701 of *LNCS*, pages 375–389. Springer, 2005.
14. R. Reiter. A theory of diagnosis from first principles. *Artificial Intelligence*, 32:57–95, 1987.
15. M. Wahlström. Exact algorithms for finding minimum transversals in rank-3 hypergraphs. *Journal of Algorithms*, 51:107–121, 2004.

Fixed-Parameter Tractable Generalizations of Cluster Editing

Peter Damaschke

School of Computer Science and Engineering,
Chalmers University, 41296 Göteborg, Sweden
ptr@cs.chalmers.se

Abstract. In the CLUSTER EDITING problem, a graph has to be changed to a disjoint union of cliques by at most k edge insertions or deletions. Several reasons suggest a generalized problem where the target graph can have some overlapping cliques. We show that the problem remains fixed-parameter tractable (FPT) in the combination of both parameters: k and a second parameter t describing somehow the complexity of overlap structure. For this result we need a structural property of twins in graphs enabling a certain elimination scheme that finally leads to a small enough subgraph we can branch on. We also give a nontrivial algorithm for problem minimizing the number of disjoint clusters, based on a concise enumeration of all solutions to the original CLUSTER EDITING problem. This generic scheme may become interesting also for other multicriteria FPT problems.

1 Introduction

CLUSTER EDITING requires to transform a graph $G = (V, E)$ with n vertices and m edges by at most k edge changes into a *cluster graph*, that is, a disjoint union of complete graphs. G and k are given. A *change* is an edge insertion or deletion. This problem from [2, 1, 13] has applications in computational biology, such as phylogeny reconstruction [2], and classification of gene expression data [14, 15], where vertices represent genes, and edges join co-regulated genes belonging to the same functional group. CLUSTER EDITING is also a purely graph-theoretic approach to clustering in general, where G reflects similarities of items, and an underlying clustering is sought that explains the data in the sense that only a few of the binary relations differ from the empirical data. CLUSTER EDITING is NP-hard even for a prescribed number of clusters [13], but easily seen to be fixed-parameter tractable (FPT): In order to reach a cluster graph one has to destroy all induced P_3 (chordless paths with three vertices). This gives a trivial branching rule and $O^*(3^k)$ algorithm. (We refer to [5] for an introduction to fixed-parameter tractability and all the basic techniques of the field.) A nontrivial bound $O(2.27^k + n^3)$ was shown in [8]. In [9] the bases have been further reduced considerably, as a demonstration example for a computer program for search tree construction. As discussed in [6], it is not always clear whether algorithms with provably smaller base perform better on the computer, as they exploit more

T. Calamoneri, I. Finocchi, G.F. Italiano (Eds.): CIAC 2006, LNCS 3998, pp. 344–355, 2006.

complicated branching rules which may create too much overhead for realistic input sizes. However these questions are beyond our scope.

In several applications, the same item may be involved in different clusters. In the gene expression context for example, some groups of genes can play a role in several functional groups. Clusters are still cliques, but they may overlap. The model of disjoint clusters would then be inappropriate and give artificial and meaningless classifications. This issue has also been raised in text document clustering. To quote from [7]: "Documents in a collection can rarely be described as members of a single/exclusive category. In fact most documents will tend to straddle in their subject between two or more different subjects." Labeling each document into a single class "can drastically affect retrieval abilities once a classification model is built". Another source of interest is the identification of modules in interaction networks of genes or proteins. Groups of biologic molecules are often involved in several cellular processes. In the evolution model for gene interaction networks of [16], gene duplications produce true twins in the graph, i.e., cliques of vertices with identical neighborhood. Thus, solutions of the TWIN GRAPH EDITING problem we define below may be used to estimate the rate of creation or loss of single interactions, compared to the duplication rate. In summary, there are good reasons to consider more general target graphs than just cluster graphs. They may be composed of a cliques that have a limited number of intersections. The "complexity" of overlapping cliques in the target graph may be controlled by a second parameter that indicates how far a graph is away from being a cluster graph, cf. also the general distance-from-triviality paradigm in [10]. Note that we do not "create" a completely new problem, but extend a problem that has already received attention, driven by new motivations, and we will show that fixed-parameter tractability is preserved. We mention that [12] studied generalizations of another FPT graph problem, VERTEX COVER.

The parameterization proposed below seems natural, even though the definition is slightly technical. First we need some more notation. A chordless path and cycle of n vertices is denoted P_n and C_n, respectively. As usual, $N(u)$ denotes the open neighborhood of vertex u, that is, the set of vertices adjacent to u, and $N[u] = N(u) \cup \{u\}$ is the closed neighborhood. If $N(u) = \emptyset$ then u is an *isolated* vertex. We may use these notions also with respect to an induced subgraph, which will be clear from context. Vertices u, v are called *true twins* if $N[u] = N[v]$. This relation is symmetric and transitive, thus it gives rise to an equivalence relation. We define the *twin graph* $T(G)$ of G as follows. Each equivalence class in G becomes a vertex of $T(G)$, and two vertices of $T(G)$ are adjacent iff edges exist between the corresponding equivalence classes in G. In other words, $T(G)$ is isomorphic to any induced subgraph of G obtained by choosing one representative vertex from each equivalence class. Twin graphs appeared under different names also in other contexts, e.g., as critical clique graphs in [11].

The TWIN GRAPH EDITING problem is: Given a graph G and parameters k and t, can we obtain by at most k edge changes a graph whose twin graph has at most t edges?

Note that G is a cluster graph iff $T(G)$ has no edges. Hence CLUSTER EDITING is the special case with $t = 0$. Alternative parameters would be the number of non-isolated vertices in $T(G)$, which is polynomially equivalent to t, or the number c of maximal cliques that are involved in overlaps, i.e., not disjoint to all others. But since c overlapping cliques cannot generate more than 2^c equivalence classes of true twins becoming non-isolated vertices in $T(G)$, this is also captured by our smoother parameter t. TWIN GRAPH EDITING is still tractable, by the following result that will be shown in this paper.

Theorem 1. TWIN GRAPH EDITING *is FPT in combined parameters k and t.*

Overview of our contributions. In Section 2 we give an FPT algorithm proving Theorem 1. Already membership in FPT is not trivial. It crucially depends on a graph-theoretic lemma, preoved in Section 3, that might be interesting in itself. In Section 4 we study problem variants where also the number c of clusters, overlapping or not, reached by at most k changes shall be minimized. Sometimes a coarser clustering gives a better overview of the data. Clearly, c is monotone decreasing in k, and c can properly decrease. For instance, a P_4 can be split in two clusters by one deletion or, alternatively, completed to one clique by three insertions. CLUSTER EDITING with at most k changes minimizing the number of clusters remains FPT. We reduce the trivial base 3 in the time bound to 2.48. A noticeable feature of our algorithm is that it utilizes concise descriptions of enumerations of all solutions to the original CLUSTER EDITING problem with inclusion-minimal edit sequences, a notion we have recently introduced in [3, 4]. We describe certain simple-structured parts of the solution space only implicitly. For optimizing an additional objective (in our case, the number of clusters), we only need one optimal solution from every such part. Then, compactness of the description and simplicity of the solution sets result in the improved base. We emphasize that this approach would work for *any* multicriteria problem where a solution with a particular property is sought, provided that the problem in question is FPT in at least one parameter, and a concise enumeration of all its solutions is available. It seems that our result in Section 4 is the first of this type. For our particular problem we finally observe that just minimizing the number of clusters can force unmotivated mergings, only for cardinality reasons. To overcome this undesirable effect we propose in Section 5 so-called natural clusterings that satisfy two modest requirements. It turns out that they are equivalent to minimal edit sequences, and hence also a natural clustering with a minimum number of clusters can be found within the same time bounds as in Section 4, and in a simpler way: A bizarre subproblem from Section 4, a bin packing problem with a least-sum-of-squares objective, becomes superfluous in this setting. We summarize the results:

Problem version	Complexity for k changes
twin graph with t edges	$O^*((3t(3t + 1)/2)^k)$
twin graph with t vertices	$O^*((t(t + 1)/2)^k)$
cluster graph, minimum number of clusters	$O^*(2.48^k)$
natural clustering, minimum number of clusters	$O^*(2.48^k)$

2 Twin Graph Editing Is FPT

We call a graph G *twin-free* if it has no true twins, equivalently $T(G) = G$. Since $T(G)$ is always twin-free, we have $T(T(G)) = T(G)$. A vertex x is called a *discriminator* of edge yz if exactly one of the edges xy, xz exists. Clearly, two adjacent vertices are no true twins iff they have a discriminator. Predicate $D(x, y, z)$ means that x is the *only* discriminator of edge yz, and xy is an edge. Note that $D(x, y, z)$ also excludes the existence of a further discriminator v with $D(v, z, y)$.

Let G, k, t be an instance of TWIN GRAPH EDITING. To establish an FPT algorithm we seek a subgraph H whose size is limited by a function of the parameters and where at least one change is forced. Then we can branch on the possible changes in H, and repeated application gives a search tree as usual. We call H the branching graph.

Lemma 1. *If H is a twin-free induced subgraph of G with more than t edges, any solution* TWIN GRAPH EDITING *must insert or delete an edge in H.*

Proof. Since H is twin-free, every edge in H has a discriminator in H. If we fail to change anything in H, this remains true in the edited graph G', thus we get no true twins in H. It follows that $T(G')$ still contains $T(H)$ as an induced subgraph and has therefore more than t edges, a contradiction. □

While this choice of a branching graph is fairly obvious, the difficulty is twofold. (1) H might be too large, i.e., not bounded by any function of the parameters, and then we cannot simply branch on H. (2) Even if a small enough branching graph is guaranteed to exist, we have to identify one without exhaustive search, as this would bring parameter t into the exponent of n in the runtime. The problem is that we cannot take any subgraph of a too large H. Neither the twin-free nor the non-twin-free property is preserved in induced subgraphs: deleting a vertex can make a graph twin-free, and on the opposite, some vertices may become twins in the subgraph when some edge loses all its discriminators. What we need is the following:

Lemma 2. *In a twin-free graph H there is always a vertex r such that $D(r, s, t)$ does not hold for any edge st in H. We call r a* redundant discriminator.

We defer the proof to the next section. As a consequence of Lemma 2 we can remove r and all incident edges from H, and no remaining edge will lose all its discriminators, so that $H - r$ is twin-free again. Now we are able to prove Theorem 1.

Given G, we first compute for every edge of G the list of discriminators. This is trivially done in $O(mn)$ time. All P_3 and hence all discriminators may also be found via matrix multiplication, which is faster if G is dense. Now we easily obtain a (twin-free!) subgraph H of G isomorphic to $T(G)$ in $O(m)$ time: Edges with empty discriminator lists build disjoint cliques, and from every such clique we keep one vertex and remove the others and all their incident edges.

If H has at most t edges, we are done. Otherwise we remove a redundant discriminator r from H (existence of r is given by Lemma 2) and then all vertices that became isolated in $H - r$. Since $H - r$ is still twin-free, so is the smaller subgraph. Thus we can continue the process and never get stuck, and in every step we can select an *arbitrary* vertex r which does not occur alone in any discriminator list. Redundant discriminators are held in a separate set RD. In every step we update the data structure by removing the edges incident with r, their discriminator lists, all occurrences of r in the lists of other edges, and the isolated vertices. From RD we remove r and the vertices that grew lonely in a list. All this can be done within $O(mn)$ time in total.

We stop as soon as a subgraph H' contains at most t edges, whereas the previous H had more than t edges. This H we use as a branching graph. Since H' has at most t edges, trivially H' has at most $2t$ vertices. The vertices of H' that became isolated in this step are all adjacent to r, since they have not been isolated in H. Together with r they induce a star graph which is twin-free. If more than t such vertices exist, we obviously get a twin-free subgraph with only $t + 2$ vertices. Otherwise H has at most $3t + 1$ vertices. We have to change the status of one of the at most $3t(3t + 1)/2$ vertex pairs in H, which implies a time complexity of $O^*((3t(3t + 1)/2)^k)$ in the worst case. This completes the proof of Theorem 1. There seems to be room for improving this naive time bound, however we conjecture that some $t^{O(k)}$ term is inevitable.

3   Existence of a Redundant Discriminator

Lemma 3. *Let be $D(x, v, u)$. Then:*
(i) $D(v, t, y)$ implies $t = x$, and (y, x, v, u) is an induced P_4.
(ii) $D(u, t, y)$ implies $y = x$, and t, v are identical or adjacent.

Proof. First assume that $\{x, v, u\} \cap \{t, y\} = \emptyset$. Since v and u are adjacent to exactly the same vertices except x, each of $D(v, t, y)$ and $D(u, t, y)$ would imply also the other relation, contradicting the definition of predicate D. Thus, in (i) it must be $t = x$ or $t = u$. But if $D(v, u, y)$ then y would be another discriminator of vu. There remains $t = x$, hence $D(v, x, y)$. If yu were an edge then y would again be another discriminator of vu. In (ii) it must be $t = v$ or $y = x$. But if $t = v$ then y would be another discriminator of vu, unless $y = x$. Finally, if t, v were distinct but not adjacent then t would be another discriminator of vu. $\square$

Using this building block we shall prove Lemma 2 now. First we give the idea. Consider any vertex x. If x is not already a redundant discriminator then $D(x, v, u)$ for some edge vu. If none of v, u is already a redundant discriminator, there must exist edges they are unique discriminators of, and so on. The difficulty is that such edges can in fact be established, and they can involve new vertices. (Hence the construction shows by itself that the argument cannot be simplified.) On the other hand, Lemma 3 imposes strong enough restrictions that enforce a certain repeated pattern. Due to finiteness of the graph we must abort the construction at some point, and then we are left with some redundant discriminator. Now the detailed exposition follows.

Consider a graph H where every vertex is the unique discriminator of some edge. We show by induction that H must contain pairwise distinct vertices u_j, v_j, x_j, y_j for all $j \geq 1$ that have certain properties listed below. It follows that such a finite H cannot exist. Let W_j be the set of all u_i, v_i, x_i, y_i with $i \leq j$, in particular $W_0 = \emptyset$. The properties are the following.

(1) (u_j, v_j, x_j, y_j) in this ordering form an induced P_4.
(2) Vertices v_j, x_j are adjacent to all vertices in W_{j-1}.
(3) Vertex u_j is adjacent to all u_i, v_i in W_{j-1}, but not to the x_i, y_i. Similarly, y_j is adjacent to all x_i, y_i in W_{j-1}, but not to the u_i, v_i.
(4) We have $D(v_j, s, y_j)$ for some $s \in W_j$. Similarly, $D(x_j, s, u_j)$ for another $s \in W_j$.
(5) All u_i, v_i with $i \leq j$ have exactly the same neighbors outside W_j. Similarly, all x_i, y_i with $i \leq j$ have exactly the same neighbors outside W_j.

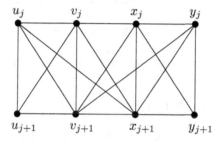

Fig. 1. Two consecutive layers of the graph used in the proof

We establish induction base $j = 1$. Let x_1 be any vertex. By assumption on H there exist u_1, v_1 with $D(x_1, v_1, u_1)$. Also by assumption, v_1 is the only discriminator of some edge. Lemma 3 (i) allows only $D(v_1, x_1, y_1)$ for some new vertex y_1 and also implies (1) and (4) for $j = 1$. Conditions (2),(3) are vacuously true, and (5) follows from $D(x_1, v_1, u_1)$ and $D(v_1, x_1, y_1)$.

For the induction step, suppose that (1)-(5) are fulfilled for some j. By assumption on H, u_j is the only discriminator of some edge, and (4) for j says that $D(x_j, s, u_j)$ for some $s \in W_j$. Lemma 3 (ii) yields $D(u_j, t, x_j)$ for some t being identical or adjacent to s. Case $t = v_j$ is impossible, since by (1), edge $v_j x_j$ has already another discriminator y_j, contradicting $D(u_j, v_j, x_j)$. If $t \in W_{j-1}$ then, due to (3), vertex t must be one of the u_i or v_i. But then, due to (1),(2), vertex y_i with the same i is another discriminator of $t x_j$, contradicting $D(u_j, t, x_j)$. This shows $t \notin W_j$. Define $x_{j+1} := t$. Since x_{j+1} has neighbors u_j and x_j, property (5) yields that x_{j+1} is adjacent to all of W_j. Thus we have established (2) for x_{j+1}. By symmetry there also exists v_{j+1} that satisfies (2). Remember that $D(u_j, x_{j+1}, x_j)$ and, symmetrically, $D(y_j, v_{j+1}, v_j)$. Assume $v_{j+1} = x_{j+1}$. Then we had also $D(y_j, x_{j+1}, v_j)$. By Lemma 3 (i), any edge x_{j+1} is the unique discriminator of must contain both u_j and y_j, which is obviously not possible. Thus

$v_{j+1} \neq x_{j+1}$. Furthermore these two vertices are adjacent, since otherwise v_{j+1} would be another discriminator of $x_j x_{j+1}$, besides u_j. (Note that $v_{j+1} x_j$ is an edge by (2).)

Applying Lemma 3 (i) again to $D(y_j, v_{j+1}, v_j)$, we see that $D(v_{j+1}, y_j, z)$ for some z being not adjacent to v_{j+1}, hence $z \notin W_j$ due to (2). Since (5) holds for j, and v_j, v_{j+1} have the same neighbors (except y_j), vertex z is adjacent to all the x_i, y_i in W_j, but to none of the u_i, v_i in W_j. Defining $y_{j+1} := z$ we get $D(v_{j+1}, y_j, y_{j+1})$. Altogether, y_{j+1} satisfies (3), and also (4) holds, with $j + 1$ in the role of j, and y_j is the role of s. By symmetry we also get a vertex u_{j+1} which satisfies the other half of (3) and (4), respectively. In particular we have $D(x_{j+1}, u_j, u_{j+1})$. Since the new vertices u_{j+1} and y_{j+1} have distinct neighborhoods in W_j, they are distinct. Moreover they are not adjacent, otherwise u_{j+1} would be another discriminator of $y_j y_{j+1}$. Edges $x_{j+1} y_{j+1}$ and $u_{j+1} v_{j+1}$ do exist, since without them, v_{j+1} and x_{j+1} would be another discriminator of $u_j u_{j+1}$ and $y_j y_{j+1}$), respectively. This establishes (1) for $j + 1$.

Finally we recover (5) for $j + 1$. By induction hypothesis and relations $D(x_{j+1}, u_j, u_{j+1})$, $D(y_j, v_{j+1}, v_j)$, both u_{j+1} and v_{j+1} have outside W_{j+1} the same neighborhoods as all other u_i, v_i have. Once more, the argument holds symmetrically for x_{j+1} and y_{j+1}. This completes the induction step and the proof.

Now we have shown Lemma 2, and hence also Theorem 1.

4   Minimizing the Number of Clusters

In TWIN GRAPH EDITING we measured the complexity of the target graph by the number of *edges* of its twin graph, in view of the fact that the number of clusters was arbitrary in the basic CLUSTER EDITING problem. Nevertheless one may alternatively aim at a minimum number of classes of true twins. We define the TWIN GRAPH EDITING (V) problem variant as follows: Given a graph G and parameters k and t, can we obtain by at most k edge changes a graph whose twin graph has at most t vertices?

Theorem 2. TWIN GRAPH EDITING (V) *is FPT in combined parameters* k *and* t.

Proof. The proof is very much the same as for Theorem 1, replacing "t edges" with "t vertices". It becomes even slightly simpler: Our branching graph H has exactly $t + 1$ vertices, we do not need an additional argument to bound the size of H. This finally implies a time complexity of $O^*((t(t + 1)/2)^k)$ in the worst case.   □

Minimizing the number of vertices in the twin graph also provokes the question of finding solutions to CLUSTER EDITING with the additional demand to minimize the number of clusters. That is, we pose the following CLUSTER EDITING (MIN) problem: Given a graph G and parameter k, can we obtain by at most k edge changes a cluster graph, and if so, find one with the smallest number of clusters?

CLUSTER EDITING (MIN) is FPT by a trivial argument: In order to reach a cluster graph, we must destroy all induced P_3 by edge changes. This gives branching number 3. We can even enumerate all solutions with at most k changes in $O^*(3^k)$ time and then simply pick a solution with minimum number of clusters, in the same time bound. The number of solutions to an instance of CLUSTER EDITING can be up to 3^k, thus one might think that the time for CLUSTER EDITING (MIN) cannot be reduced below $O^*(3^k)$ in this way. However, in [4] we introduced concise representations of all minimal solutions. We outline the necessary facts for CLUSTER EDITING needed below, these things can be discussed similarly for any modification problem. Given an input graph G with n vertices and parameter k, *minimal solutions* are those reachable from G by an inclusion-minimal set of at most k changes, i.e., sequences of changes that lead to a cluster graph, without reaching the same or another cluster graph before. We may imagine a "state graph" whose vertices are all graphs with n vertices, and where two graphs are "adjacent" if they differ in exactly one edge. A search tree can be seen as a subtree of this state graph, rooted at G. Informally, the idea of *concise* representations of all minimal solutions is to abort paths of a search tree already before a target graph (here: a cluster graph) is reached, if the current graph is so simple that all minimal solutions reachable from it are described in some simple way, e.g., by a set-theoretic formula. This has to be made precise by an ad-hoc definition for any concrete problem. Thus, the description can be significantly smaller than the number of minimal solutions. In particular we proved the following in [4], the formulation below is adapted to the current needs.

Theorem 3. *A description of all minimal solutions of an instance of* CLUSTER EDITING *can be computed by a search tree algorithm that uses only branching rules with branching numbers no larger than 2.562. All leaves of the resulting search tree represent graphs which may contain, besides cliques, also P_3 as connected components. All minimal solutions are reached by further, independent changes in these P_3 components, namely, either one edge insertion (to get a cluster of size 3) or one edge deletion (to get two clusters of size 1 and 2).*

The general idea behind our algorithm for CLUSTER EDITING (MIN) is to compute an concise enumeration, as in Theorem 3, and then to expand the paths of this aborted search tree further, but this time only in order to obtain just *one* optimal solution starting from the graph represented by every leaf of the search tree. In the following we denote any such graph L. Finally, the global optimum from all tree paths is taken. We only have to bound the branching numbers in the search trees we attach to the tree from Theorem 3.

First we deal with the case that L itself is already a cluster graph, albeit with too many clusters. The following lemma restricts the transformations we still have to take into account. Note that k is here the number of remaining changes still allowed in L, rather than the given parameter value.

Lemma 4. *Let L and H be two cluster graphs, such that H has at most c clusters and can be reached from L by at most k edge changes. Then, there exists such a transformation where clusters of L are only merged but never split. Moreover, each cluster C of L either appears as a cluster in H, or C is attached to a cluster in $H - C$ of minimum size.*

Proof. Consider a transformation τ of L into H, and any cluster C in L. The transformation induced by τ on $L - C$ yields an induced subgraph $H - C$ of H which is, clearly, also a cluster graph. Since the order of changes is immaterial for the result, we may rearrange τ so that all deletions or insertions of edges incident to vertices in C are performed last. Now, either let C be a cluster in H, or attach C to a smallest cluster in $H - C$. We claim that all other solutions are only worse: Let B be the cluster of $H - C$ merged with C in the latter case. Then we have to insert $|B|$ edges adjacent to every vertex in C. If we, instead, attached vertices from C to any other cluster of $H - C$, this number would only increase. If we split C in smaller cliques and attached them to several clusters of $H - C$, we would also need edge deletions in C. In any case, we would need more changes to obtain the same number of clusters in H.   □

What we learn from Lemma 4 is that an optimal solution to CLUSTER EDITING (MIN), when L is already a cluster graph, can be obtained by successive merging of clusters, until the allowed number of changes is exceeded. Moreover, in every step we can take a currently *minimum size* cluster C and merge it with one of the other clusters. This is correct by the following arguments. If C participates in further merge operations at all, we can merge C *right now* with another cluster. The other alternative is that C becomes a cluster in the target graph H. But then it is better to replace in this process a larger cluster with C, since this only reduces the number of edge insertions. Hence, it is never a mistake to merge smallest cluster with some other cluster. However, it is not clear which cluster should be the second merging partner. A greedy rule would not work, small counterexamples are easy to find. Some more side remarks are in order here. Instead of transforming a graph L into a cluster graph using at most a prescribed number k of changes, thereby minimizing the number c of clusters, we may also ask how many changes are necessary to reach a cluster graph with a prescribed (maximum) number c of clusters. Clearly, the two problems are polynomially equivalent. We conjecture that CLUSTER EDITING (MIN) is NP-hard already for cluster graphs L as input, since it is a balancing problem of similar nature as BIN PACKING. Note that the version with prescribed c can be rephrased as follows, since only cardinalities of clusters in L matter: Given integers $x_1, \ldots, x_d$ and $c < d$, divide the x_i into c groups such that the sum of products $x_i x_j$, taken over all index pairs i, j in the same group, is minimized. Another equivalent formulation is: Divide the x_i into c groups such that the sum of squares of group sums is minimized.

Luckily, in our context we need not worry too much about the complexity of this subproblem, since we are only interested in the branching number. Lemma 4 and the subsequent paragraph gives us a branching rule: Merge a smallest cluster C with another cluster, and branch for all possible merge partners, however, consider only one representative cluster for each appearing cardinality. If $|C| = x$, and the different sizes of clusters other than C are $x_1 < x_2 < \ldots < x_d$, the numbers of edge insertions in these d cases are $xx_1, xx_2, \ldots, xx_d$. Thus, the characteristic equation for the inverse branching number y is $\sum_{i=1}^{d} y^{xx_i} = 1$. For any d, the worst case (y minimal) appears if $x = 1$ and $x_i = i$ for all i, that is, $\sum_{i=1}^{d} y^i = 1$. Now we see $y > 1/2$, hence the branching number for this cluster merging phase is always smaller than 2.

It remains to consider the P_3 components in the graphs L at the leaves of a search tree obtained by Theorem 3. In any transformation of L into a cluster graph, we first reach a minimal solution (when we encounter a cluster graph for the first time), and then we can, in general, transform it further into a cluster graph with fewer clusters. But all minimal solutions are already given by Theorem 3: Any P_3 is transformed into a cluster graph in two different ways, each with one change. (Again, since only cardinalities of clusters are relevant, symmetric cases need not be considered.) This gives immediately the branching number 2 for treating the P_3 components. Once we are left with cluster graphs only, we perform the merging phase. Altogether it follows:

Theorem 4. CLUSTER EDITING (MIN) *can be solved in* $O^*(2.562^k)$ *time.*

In [4] we also achieved an improved version of Theorem 3 with base 2.4, however at cost of more complicated non-clique components in the graphs L: There can appear P_l and C_l of arbitrary lengths l, and seven further 6-vertex graphs. Based on this result we reduce the base in Theorem 4 to 2.48. Due to space limitations we only sketch the proof. As above, we have to turn all types of non-clique components into cluster graphs, for all possible combinations of cardinalities of clusters. For the exceptional 6-vertex graphs mentioned above, simple but tedious case distinctions show that the branching numbers are always below 2. From a P_l component in L we may successively split off small paths, by deleting one edge, and complete them to cliques by edge insertions. Whenever a P_i has been cut off, we must add $\binom{i}{2} - i + 1$ edges. The root of $2y + y^2 + y^4 + y^7 + y^{11} + y^{16} + \ldots = 1$ is a lower bound for the inverse branching number of this path splitting. We get $y \approx 0.403$ and $1/y \approx 2.48$. For C_l components in L the argument is the same, but for cutting out the first path we even need two deletions, so that the branching number cannot be worse.

Corollary 1. CLUSTER EDITING (MIN) *can be solved in* $O^*(2.48^k)$ *time.*

5 Natural Clusterings

Clusterings with a forced small number of clusters tend to merge small clusters just because this is cheap, i.e., it requires only few edge insertions. But this does not mean that vertices put in such garbage collecting clusters are really related from the viewpoint of the application at hand. Thus one may argue that CLUSTER EDITING (MIN) is somewhat ill-posed in this respect. To overcome this effect, we propose to restrict the permitted clusterings in the following way.

Definition 1. *Given a graph $G = (V, E)$, we call a cluster graph H on vertex set V a natural clustering of G if:*
(1) The vertex set of every cluster in H induces a connected subgraph of G.
(2) For any two clusters C and C' in H, not all possible edges between C and C' exist in edge set E of G.

Condition (1) is intended to suppress unmotivated mergings, while (2) shall avoid unmotivated splits. We define the problem CLUSTER EDITING (MIN,NAT): Given a graph G and parameter k, find a natural clustering of G with the smallest number of clusters, by at most k edge changes (if existing).

Recall that a minimal solution to the CLUSTER EDITING problem for graph G is defined as a cluster graph H obtained from G by an inclusion-minimal set of changes (that transforms G into any cluster graph). Neatly, the following equivalence holds:

Lemma 5. *The natural clusterings of a graph are exactly the minimal solutions to CLUSTER EDITING.*

Proof. Let H be a natural clustering of G, and τ the set of changes turning G into H. We claim that a proper subset σ of τ cannot lead to another cluster graph, hence H is a minimal solution. If σ omits some of the edge deletions in τ, then at least two clusters of H, say C and C', fall into the same connected component of G. Since σ does not make more edge deletions than τ, it follows that C and C' remain connected after σ. But then σ has to insert all missing edges between C and C' (due to (2) there are some), whereas τ does not insert any such edge. Thus, σ is not a subset of τ, a contradiction. We conclude that all edges deleted by τ are deleted by σ, too. After the deletions, σ has to complete the resulting connected components to cliques. Hence σ must insert all missing edges in the clusters of H, unless some cluster of H is not a connected subgraph of G. But condition (1) excludes the latter case. This implies $\sigma = \tau$, a contradiction. The claim is proved.

Conversely, let H be a minimal solution to CLUSTER EDITING, and τ defined as above. By minimality, a proper subset σ of τ cannot produce another cluster graph. Now assume that (1) is violated, i.e., some cluster C of H is a disconnected subgraph of G. If we do not insert the edges between the connected components of C, we get a proper subset of τ generating another cluster graph (with C divided in several clusters), a contradiction. Assume that (2) is violated, so that E contains all possible edges between two of the clusters, say C and C'. Since C and C' are clusters in H, τ has deleted all these edges. If we choose not to delete them, we get a proper subset of τ producing another cluster graph (with cluster $C \cup C'$ rather than C and C'), a contradiction.   □

Theorem 5. CLUSTER EDITING (MIN,NAT) *can be solved in $O^*(2.562^k)$ time.*

This follows directly from Lemma 5 and Theorem 3. Only the non-clique components must be turned into cluster graphs, while the entire cluster merging phase is superfluous when we are interested in natural clusterings only. Hence, also Corollary 1 carries over to natural clusterings.

References

1. N. Bansal, A. Blum, S. Chawla. Correlation clustering, *Machine Learning* 56 (2004), 89-113
2. Z.Z. Chen, T. Jiang, G. Lin. Computing phylogenetic roots with bounded degrees and errors, *SIAM J. Comp.* 32 (2003), 864-879
3. P. Damaschke. Parameterized enumeration, transversals, and imperfect phylogeny reconstruction, *Theoretical Computer Science* 351 (2006), 337-350, special issue: *1st International Workshop on Parameterized and Exact Computation IWPEC 2004*
4. P. Damaschke. On the fixed-parameter enumerability of cluster editing, *31st International Workshop on Graph-Theoretic Concepts in Computer Science WG 2005, LNCS* 3787, 283-294
5. R.G. Downey, M.R. Fellows. *Parameterized Complexity*, Springer, 1999
6. H. Fernau. A top-down approach to search trees: improved algorithmics for 3-hitting set, ECCC Report 73 (2004)
7. H. Frigui, O. Nasraoui. Simultaneous clustering and dynamic keyword weighting for text documents, in: M. Berry (ed.), *Survey of Text Mining*, Springer 2004, 45-70
8. J. Gramm, J. Guo, F. Hüffner, R. Niedermeier. Graph-modeled data clustering: Fixed-parameter algorithms for clique generation, *Theory of Computing Systems* 38 (2005), 373-392, preliminary version in *5th CIAC 2003, LNCS* 2653, 108-119
9. J. Gramm, J. Guo, F. Hüffner, R. Niedermeier. Automated generation of search tree algorithms for hard graph-modification problems, *Algorithmica* 39 (2004), 321-347
10. J. Guo, F. Hüffner, R. Niedermeier. A structural view on parameterizing problems: Distance from triviality, *Parameterized and Exact Computation*, 1st Int. Workshop IWPEC'2004, Proceedings, *LNCS* 3162, 162-173
11. G.H. Lin, T. Jiang, P.E. Kearney. Phylogenetic k-root and Steiner k-root. *11th ISAAC 2000, LNCS* 1969, 539-551
12. N. Nishimura, P. Ragde, D.M. Thilikos. Fast fixed-parameter tractable algorithms for nontrivial generalizations of vertex cover, *7th WADS 2001, LNCS* 2125, 75-86, journal version to appear in *Discrete Applied Math.*
13. R. Shamir, R. Sharan, D. Tsur. Cluster graph modification problems, *Discrete Applied Math.* 144 (2004), 173-182, preliminary version in: *28th WG 2002, LNCS* 2573, 379-390
14. R. Sharan, A. Maron-Katz, R. Shamir. CLICK and EXPANDER: A system for clustering and visualizing gene expression data, *Bioinformatics* 19 (2003), 1787-1799
15. R. Sharan, R. Shamir. Algorithmic approaches to clustering gene expression data, in: *Current Topics in Computational Molecular Biology*, MIT Press, 2002, 269-300
16. S. Wu, X. Gu. Gene network: Model, dynamics and simulation, *11th COCOON 2005, LNCS* 3595, 12-21

The Linear Arrangement Problem
Parameterized Above Guaranteed Value

Gregory Gutin[1,*], Arash Rafiey[1], Stefan Szeider[2,**], and Anders Yeo[1,*]

[1] Department of Computer Science, Royal Holloway University of London,
Egham, Surrey TW20 OEX, England, United Kingdom
`gutin, arash, anders@cs.rhul.ac.uk`
[2] Department of Computer Science, Durham University,
Durham DH1 3LE, England, United Kingdom
`stefan.szeider@durham.ac.uk`

Abstract. A linear arrangement (LA) is an assignment of distinct integers to the vertices of a graph. The cost of an LA is the sum of lengths of the edges of the graph, where the length of an edge is defined as the absolute value of the difference of the integers assigned to its ends. For many application one hopes to find an LA with small cost. However, it is a classical NP-complete problem to decide whether a given graph G admits an LA of cost bounded by a given integer. Since every edge of G contributes at least one to the cost of any LA, the problem becomes trivially fixed-parameter tractable (FPT) if parameterized by the upper bound of the cost. Fernau asked whether the problem remains FPT if parameterized by the upper bound of the cost minus the number of edges of the given graph; thus whether the problem is FPT "parameterized above guaranteed value." We answer this question positively by deriving an algorithm which decides in time $O(m + n + 5.88^k)$ whether a given graph with m edges and n vertices admits an LA of cost at most $m + k$ (the algorithm computes such an LA if it exists). Our algorithm is based on a procedure which generates a problem kernel of linear size in linear time for a connected graph G. We also prove that more general parameterized LA problems stated by Serna and Thilikos are not FPT, unless P = NP.

1 Introduction

All graphs considered in this paper do not have loops or parallel edges. A *linear arrangement (LA)* of a graph $G = (V, E)$ is a one-to-one mapping $\alpha : V \to \{1, \ldots, |V|\}$. The *length* of an edge $uv \in E$ relative to α is defined as

$$\lambda_\alpha(uv) = |\alpha(u) - \alpha(v)|.$$

The *cost* $c(\alpha, G)$ of an LA α is the sum of lengths of all edges of G relative to α. LAs of minimal cost are *optimal*; $ola(G)$ denotes the cost of an optimal LA of G.

---

 * Research supported in part by the IST Programme of the European Community, under the PASCAL Network of Excellence, IST-2002-506778.
** Research supported in part by the Nuffield Foundation, NAL/01012/G.

T. Calamoneri, I. Finocchi, G.F. Italiano (Eds.): CIAC 2006, LNCS 3998, pp. 356–367, 2006.
© Springer-Verlag Berlin Heidelberg 2006

The **Linear Arrangement Problem (LAP)** is the problem of deciding whether, given a graph G and an integer k, G admits a linear arrangement of cost at most k. The problem has numerous application; in particular, the first published work on the subject appears to be the 1964 paper of Harper [14], where a polynomial-time algorithm for finding optimal linear arrangement for n-cubes is developed, which has applications in error-correcting codes. Goldberg and Klipker [13] were first to obtain a polynomial-time algorithm for computing optimal linear arrangements of trees. Faster algorithms for trees were obtained by Shiloach [17] and Chung [2]. However, we cannot hope to find optimal linear arrangements for the class of all graphs in polynomial time since LAP is a classical NP-complete problem [11, 12].

Recently, LAP was studied under the framework of parameterized complexity [6, 18]. We recall some basic notions of parameterized complexity here, for a more in-depth treatment of the topic we refer the reader to [4, 5, 6, 10, 16]. A parameterized problem Π can be considered as a set of pairs (I, k) where I is the *problem instance* and k (usually an integer) is the *parameter*. Π is called *fixed-parameter tractable (FPT)* if membership of (I, k) in Π can be decided in time $O(f(k)|I|^c)$, where $|I|$ is the size of I, $f(k)$ is a computable function, and c is a constant independent from k and I. Let Π and Π' be parameterized problems with parameters k and k', respectively. An *fpt-reduction R from Π to Π'* is a many-to-one transformation from Π to Π', such that (i) $(I, k) \in \Pi$ if and only if $(I', k') \in \Pi'$ with $k' \leq g(k)$ for a fixed computable function g and (ii) R is of complexity $O(f(k)|I|^c)$. A *reduction to problem kernel* (or *kernelization*) is an fpt-reduction R from a parameterized problem Π to itself. In kernelization, an instance (I, k) is reduced to another instance (I', k'), which is called the *problem kernel*. It is easy to see that a decidable parameterized problem is FPT if and only if it admits a kernelization (see, e.g., [5, 16]); however, the problem kernels obtained by this general result have impractically large size. Therefore, one tries to develop kernelizations that yield problem kernels of smaller size, if possible of size linear in the parameter.

The following is a straightforward way to parameterize LAP [6, 18]:

Parameterized LAP
Instance: A graph G. *Parameter:* A positive integer k.
Question: Does G have an LA of cost at most k?

An edge has length at least 1 in any LA. Thus, for a graph G with m edges we have $\mathrm{ola}(G) \geq m$; in other words, m is a guaranteed value for $\mathrm{ola}(G)$. Consequently, parameterized LAP is FPT by trivial reasons (we reject a graph with more than k edges and solve LAP by brute force if the graph has at most k edges). Hence it makes sense to consider the *net cost* $\mathrm{nc}(\alpha, G)$ of an LA α defined as follows:

$$\mathrm{nc}(\alpha, G) = \sum_{e \in E} (\lambda_\alpha(e) - 1) = \mathrm{c}(\alpha, G) - m.$$

We denote the net cost of an optimal LA of G by $\mathrm{ola}^+(G)$. Indeed, the following non-trivial parameterization of LAP is considered by Fernau [6, 7]:

LA parameterized above guaranteed value (LAPAGV)
Instance: A graph G. *Parameter:* A positive integer k.
Question: Does G have an LA of net cost at most k?

Parameterizations above a guaranteed value were first considered by Mahajan and Raman [15] for the problems Max-SAT and Max-Cut; such parameterizations have lately gained much attention [6, 16]. However, apparently only a few nontrivial problems parameterized above guaranteed value are known to be FPT.

Fernau [6, 7] raises the question of whether LAPAGV is FPT (the status of this problem is reported open in Cesati's compendium [1]). We answer this question positively by deriving a kernelization procedure for LAPAGV that yields problem kernels of linear size in linear time for connected graphs G. Moreover, using the method of bounded search trees, we develop an algorithm that solves LAPAGV for the obtained kernel more efficiently than by brute force. In summary, we obtain an algorithm that decides in time $O(m+n+5.88^k)$ whether a given graph with m edges and n vertices admits an LA of cost at most $m+k$. Our algorithm also produces an optimal LA if $\text{ola}^+(G) \leq k$. A key concept of our kernelization is the suppression of vertices of degree 2, a standard technique used in the design of parameterized algorithms (e.g., for finding small feedback vertex sets in graphs [4]). For LAPAGV, however, we need a more sophisticated approach where we suppress only vertices of degree 2 that satisfy a certain condition depending on the parameter k.

Fernau [8] proposes a bounded search tree approach to prove that LAPAGV is FPT. The description of the approach is incomplete (for example, it is unclear how to deal with vertices of degree 2 without rejecting any yes-instances) and an inequality, which is required by Fernau's approach to show that LAPAGV is FPT, is not proved. These conclusions are confirmed in our private communication with Fernau (February, 2006) and it remains to be seen whether a bounded search tree approach can be used to prove that LAPAGV is FPT.

Serna and Thilikos [18] formulate more general parameterized LA problems (see Section 4) and ask whether their problems are FPT. We prove that the problems are not FPT (unless P = NP) by demonstrating that for almost all fixed values of the parameter, the corresponding decision problems are NP-complete. This implies that the problems are para-NP-complete [10]. We conclude the paper by Theorem 8, which indicates that our FPT result cannot be extended much further, in a sense.

For a graph G and a set X of its vertices, $V(G)$, $E(G)$ and $G[X]$ denote the vertex set of G, the edge set of G, and the subgraph of G induced by X, respectively. An edge e in a graph G is a *bridge* if $G - e$ has more components than G has. A connected graph with at least two vertices and without bridges is called *2-edge-connected*. A *bridgeless component* of a graph G is a maximal induced subgraph of G with no bridges. Observe that the bridgeless components of G are the connected components that we get after removing all bridges from G. A bridgeless component is either a 2-edge-connected graph or is isomorphic to K_1; in the latter case we call it *trivial*. Further graph-theoretic terminology can be found in Diestel's book [3].

2   Kernelization

In the next section, we use the following simple lemma to solve LAPAGV for the general case of an arbitrary graph input G. The lemma allows us to confine our attention to connected graphs in the rest of this section.

Lemma 1. Let $G_1, \ldots, G_p$ be the connected components of a graph G. Then $\mathrm{ola}^+(G) = \sum_{i=1}^{p} \mathrm{ola}^+(G_i)$.

Let α be an LA of a graph G. It is convenient to use for subgraphs G' of G the notation $\mathrm{nc}(\alpha, G') = \sum_{uv \in E(G')} (\lambda_\alpha(uv) - 1)$.

Lemma 2. Let G be a graph, let $X \subseteq V(G)$, and let u, v be two distinct vertices of G that belong to the same connected component of $G - X$. Let α be an LA of G with $\alpha(u) < \alpha(x) < \alpha(v)$ for every $x \in X$. Then $\mathrm{nc}(\alpha, G - X) \geq |X|$.

Proof. We proceed by induction on $|X|$. If $|X| = 0$ then the lemma holds vacuously. Hence we assume $|X| \geq 1$ and pick $x \in X$. We define $G' = G - x$, $X' = X \setminus \{x\}$, and we let α' be the LA of G' obtained from α by setting, for $y \in V(G')$, $\alpha'(y) = \alpha(y)$ if $\alpha(y) < \alpha(x)$, and $\alpha'(y) = \alpha(y) - 1$ otherwise. By induction hypothesis, $\mathrm{nc}(\alpha', G' - X') \geq |X'|$. By assumption, $G - X$ contains a path P from u to v; hence P contains at least one edge $w_1 w_2$ with $\alpha(w_1) < \alpha(x) < \alpha(w_2)$ (and $w_1, w_2 \notin X$). By definition of α', we have $\lambda_\alpha(w_1 w_2) = \lambda_{\alpha'}(w_1 w_2) + 1$. Since for all other edges $e \in E(G' - X')$ we have $\lambda_\alpha(e) \geq \lambda_{\alpha'}(e)$, $\mathrm{nc}(\alpha, G - X) \geq \mathrm{nc}(\alpha', G' - X') + 1$ follows. $\square$

Let G be a connected graph and let α be an LA of G. We say that two subgraphs A, B of G are α-*comparable* if either $\alpha(a) < \alpha(b)$ holds for all $a \in V(A), b \in V(B)$, or $\alpha(a) > \alpha(b)$ holds for all $a \in V(A), b \in V(B)$. Moreover, let e be a bridge of G and let G_1, G_2 be the two connected components of $G - e$. For a positive integer k, we say that e is k-*separating* if both $|V(G_1)|, |V(G_2)| > k$.

Lemma 3. Let G be a connected graph and let k be a positive integer such that $k \geq \mathrm{ola}^+(G)$. Then for every optimal LA α of G and every k-separating bridge e of G, the two connected components of $G - e$ are α-comparable.

Proof. Let α be an optimal LA. Let e be a k-separating bridge of G and let G_1, G_2 be the two connected components of $G - e$. Since e is a k-separating bridge, we have $|V(G_1)|, |V(G_2)| > k$. We denote the extremal values of the vertices of G_1 and G_2 with respect to α by $l_i = \min_{v \in V(G_i)} \alpha(v)$ and $r_i = \max_{v \in V(G_i)} \alpha(v)$, $i = 1, 2$. We may assume that $l_1 < l_2$. First we show that $r_1 < r_2$. Assume to the contrary that $r_1 > r_2$. Now $\alpha^{-1}(l_1)$ and $\alpha^{-1}(r_1)$ belong to the same connected component of $G - V(G_2)$, and Lemma 2 implies $\mathrm{nc}(\alpha, G) \geq |V(G_2)| > k$, contradicting the assumption $\mathrm{nc}(\alpha, G) \leq k$. Hence indeed $l_1 < l_2$ and $r_1 < r_2$.

Next we show that $r_1 < l_2$. Assume to the contrary that $l_2 < r_1$. From α we obtain a new LA α' of G, changing the order of vertices in $X = \{\, x \in V(G) : l_2 \leq \alpha(x) \leq r_1 \,\}$ such that G_1 and G_2 become α'-comparable, without changing the relative order of vertices within G_1 or changing the relative order of vertices within G_2. Since e is a bridge, we have

$$\mathrm{nc}(\alpha, G) = \mathrm{nc}(\alpha, G - e) + \lambda_\alpha(e) - 1 \text{ and } \mathrm{nc}(\alpha', G) = \mathrm{nc}(\alpha', G - e) + \lambda_{\alpha'}(e) - 1. \quad (1)$$

Although $\lambda_{\alpha'}(e)$ can be greater than $\lambda_\alpha(e)$, we will show that an increase of the length of e is more than compensated by the reduced cost of $G-e$ under α'. Again using Lemma 2 we conclude that $\mathrm{nc}(\alpha', G_i) \leq \mathrm{nc}(\alpha, G_i) - |X \cap V(G_{3-i})|$ holds for $i = 1, 2$ (observe that the vertices $\alpha^{-1}(l_i), \alpha^{-1}(r_i)$ are in the same component of $G - V(G_{3-i})$, and for each vertex x in $X \cap V(G_i)$ we have $\alpha(l_i) < \alpha(x) < \alpha(r_i)$). In summary, we have

$$\mathrm{nc}(\alpha', G - e) \leq \mathrm{nc}(\alpha, G - e) - |X|. \tag{2}$$

Using the fact that $|\alpha(x) - \alpha'(x)| \leq |X| - 1$ holds for all vertices $x \in V(G)$, it is easy to see that

$$\lambda_{\alpha'}(e) \leq \lambda_\alpha(e) + |X| - 1. \tag{3}$$

Indeed, if at least one of the ends of e is in $V(G) \setminus X$, then clearly $\lambda_{\alpha'}(e) \leq \lambda_\alpha(e) + |X| - 1$; otherwise, if both ends of e are in X, then $\lambda'_\alpha(e) \leq |X| - 1$, and since $\lambda_\alpha(e) \geq 1$, we have even $\lambda_{\alpha'}(e) \leq \lambda_\alpha(e) + |X| - 2$.

By (1), (2) and (3), we obtain $\mathrm{nc}(\alpha', G) \leq \mathrm{nc}(\alpha, G) - 1$. This contradicts the assumption that α is an optimal LA. Hence $l_1 < r_1 < l_2 < r_2$, and so G_1 and G_2 are α-comparable as claimed. □

Lemma 4. *If G is a connected bridgeless graph of order $n \geq 1$, then $\mathrm{ola}^+(G) \geq (n-1)/2$.*

Proof. If $n \leq 2$, then the inequality trivially holds. Thus, we may assume that $n \geq 3$ and G is 2-edge-connected. Let α be an optimal LA of G and put $u = \alpha^{-1}(1)$ and $w = \alpha^{-1}(n)$. Since G is 2-edge-connected, Menger's Theorem (see, e.g., [3]) implies that there are two paths P, P' between u to w such that $E(P) \cap E(P') = \{u, w\}$. Observe that the subgraph G' of G induced by $E(P) \cup E(P')$ is a collection of $t \geq 1$ edge-disjoint cycles. Let n' be the number of vertices in G'. Since G' has $t - 1$ vertices of degree 4 and $n' - t + 1$ vertices of degree 2, $|E(G')| = (n' - t + 1) + 2(t - 1) = n' + t - 1$. Since $n' \leq n$ and $t \leq \frac{n-1}{2}$, we conclude that $|E(G')| \leq \frac{3}{2}(n - 1)$. Observe that $\mathrm{nc}(\alpha, P) \geq n - 1 - |E(P)|$ and $\mathrm{nc}(\alpha, P') \geq n - 1 - |E(P')|$. Hence, $\mathrm{ola}^+(G) = \mathrm{nc}(\alpha, G) \geq \mathrm{nc}(\alpha, G') \geq 2(n - 1) - |E(G')| \geq (n - 1)/2$. □

Let α be an optimal LA of G. We call a vertex $u \in V(G)$ α-*special* if $G - u$ is connected and $\alpha(u) \notin \{1, n\}$.

Lemma 5. *Let G be a connected graph. Let X be a vertex set of G such that $G[X]$ is connected and let $G - X$ have connected components $G_1, G_2, \ldots, G_r$ with $n_1, n_2, \ldots, n_r$ vertices, respectively, such that $n_1 \leq n_2 \leq \ldots \leq n_r$. Then $\mathrm{ola}^+(G) \geq \mathrm{ola}^+(G[X]) + \sum_{i=1}^{r-2} n_i$.*

Proof. Let α be an optimal LA of G. If $r \leq 2$, then $\sum_{i=1}^{r-2} n_i = 0$ and, thus, this lemma holds. Now assume that $r \geq 3$. Observe that each nontrivial G_i has a pair u_i, v_i of distinct vertices such that $G_i - u_i$ and $G_i - v_i$ are connected. If G_i is trivial, i.e., it has just one vertex x, then set $u_i = v_i = x$. Since $r \geq 3$, for some $j \in \{1, 2, \ldots, r\}$, we have $\alpha(u_j) \notin \{1, n\}$ and $\alpha(v_j) \notin \{1, n\}$. Now we claim that there is a vertex $u \in V(G_j)$ such that $G - u$ is connected. Indeed, we set $u = u_j$ if there are edges between v_j and $G[X]$, we set $u = v_j$, otherwise. We have proved that G has an α-special vertex u not in X. Let α_u be an LA of $G - u$ defined as follows: $\alpha_u(x) = \alpha(x)$ for all $x \in V(G)$ with $\alpha(x) < \alpha(u)$, and $\alpha_u(x) = \alpha(x) - 1$ for all $x \in V(G)$ with $\alpha(x) > \alpha(u)$. Since G is connected, it has an edge yz such that $\alpha(y) < \alpha(u) < \alpha(z)$. Observe that $\lambda_\alpha(yz) = \lambda_{\alpha_u}(yz) + 1$. Hence, we have $\text{ola}^+(G) = \text{nc}(\alpha, G) \geq \text{nc}(\alpha_u, G - u) + 1 \geq \text{ola}^+(G - u) + 1$. Thus,

$$\text{ola}^+(G) \geq \text{ola}^+(G - u) + 1 \text{ for an } \alpha\text{-special vertex } u \text{ of } G \tag{4}$$

Run the following procedure: while $G - X$ has a least three components, choose a β-special vertex $u \notin X$ of G for an optimal LA β of G and replace G with $G - u$. By the end of this procedure, we have deleted some t vertices from G obtaining a subgraph H of G. By (4), we have $\text{ola}^+(G) \geq \text{ola}^+(G[X]) + t$. Observe that $H - X$ has at most two components, if all vertices of at least $r - 2$ components $G_1, G_2, \ldots, G_r$ are deleted from G during the procedure. Thus, $t \leq \sum_{i=1}^{r-2} n_i$ and $\text{ola}^+(G) \geq \text{ola}^+(G[X]) + \sum_{i=1}^{r-2} n_i$. □

Lemma 6. *Let k be a positive integer and let G be a connected graph with n vertices with $\text{ola}^+(G) \leq k$. Then either G has a k-separating bridge or $n \leq 4k+1$.*

Proof. Assume that G does not have a k-separating bridge. If G is a bridgeless graph, then by Lemma 4 we know that $n \leq 2k + 1$. So, we may assume that G has a bridge. Choose a bridge e_1 with maximal $\min\{|V(F_1)|, |V(F_0)|\}$, where F_1, F_0 are the components of $G - e_1$. Assume, w.l.o.g., that $|V(F_1)| \leq |V(F_0)|$. Since e_1 is not a k-separating bridge, $|V(F_1)| \leq k$ follows of necessity. Let F_0^* denote the bridgeless component of F_0 that contains a vertex incident to e_1. If $F_0 = F_0^*$ then $|V(F_0)| \leq 2k + 1$ follows by Lemma 4 and we are done; hence we assume that $F_0 \neq F_0^*$.

Let $e_2, \ldots, e_r$ denote the bridges of F_0 that are incident to vertices in F_0^*. Moreover, let $F_2, \ldots, F_r$ denote the connected components of $F_0 - V(F_0^*)$ such that each e_i is incident with a vertex of F_i, $i = 2, \ldots, r$. Assume that $|V(F_2)| \geq |V(F_3)| \geq \ldots \geq |V(F_r)|$. Suppose that $|V(F_2)| > |V(F_1)|$. Then the component of $G - e_2$ different from F_2 has more vertices than F_1, which is impossible by the choice of e_1 and the assumption that G has no k-separating bridges. We conclude that $|V(F_1)| \geq |V(F_2)|$. By Lemma 5, $\text{ola}^+(G) \geq \text{ola}^+(F_0^*) + \sum_{i=3}^{r} |V(F_i)|$. Thus, $\sum_{i=3}^{r} |V(F_i)| \leq k - \text{ola}^+(F_0^*)$. Since $|V(F_2)| \leq |V(F_1)| \leq k$ and, by Lemma 4, $|V(F_0^*)| \leq 2 \cdot \text{ola}^+(F_0^*) + 1$, we obtain that $n = |V(F_0^*)| + \sum_{i=1}^{r} |V(F_i)| \leq (2 \cdot \text{ola}^+(F_0^*) + 1) + (3k - \text{ola}^+(F_0^*)) = 3k + \text{ola}^+(F_0^*) + 1 \leq 4k + 1$. □

Lemma 7. *Let k be a positive integer and let G be a connected graph with the following structure:*

1. *G has bridgeless components $C_1, C_2, \ldots, C_t$, $t \geq 2$, such that every two consecutive components C_i and C_{i+1} are linked by a single edge e_i, which is a k-separating bridge in G, $i = 1, 2, \ldots, t - 1$.*
2. *Let $L = G[\bigcup_{i=1}^{t} V(C_i)]$. The graph $G' = G - V(L)$ has connected components $G_1, G_2, \ldots, G_r$ such that each G_j has edges only to one subgraph $C_{\pi(j)}$, $\pi(j) \in \{1, 2, \ldots, t\}$.*

Let J_p be the indices of all G_j such that $\pi(j) = p$, $p = 1, 2, \ldots, t$. Let $n_i = \max\{|V(G_j)| : j \in J_i\}$, $i = 1, 2, \ldots, t$. Then $\mathrm{ola}^+(G) \geq \mathrm{ola}^+(L) + |V(G')| - n_1 - n_t$.

Proof. Let α be an optimal LA of G. Let

$$A_p = \left(\bigcup_{j \in J_1 \cup J_2 \cup \cdots \cup J_p} V(G_j) \right) \cup \left(\bigcup_{j=1}^{p} V(C_j) \right)$$

for $p = 1, 2, \ldots, t$. By Lemma 3, the two components of $G - e_1$ are α-comparable. We may assume, w.l.o.g., that $\alpha(x) < \alpha(y)$ for each $x \in A_1$, $y \notin A_1$. Because of the assumption and since the two components of $G - e_2$ are α-comparable, we have $\alpha(x) < \alpha(y) < \alpha(z)$ for each $x \in A_1$, $y \in A_2 - A_1$ and $z \notin A_2$. Continuing this argument, we can prove that $\alpha(x_i) < \alpha(x_{i+1})$ for each $x_i \in A_i$ and $x_{i+1} \in A_{i+1} \setminus \bigcup_{j=1}^{i} A_j$.

By the above conclusion and the arguments similar to those used in the proof of Lemma 5, we can prove that each G_j, apart from at most one graph G_p with $p \in J_1$ and at most one graph G_q with $q \in J_t$, has an α-special vertex u. As in Lemma 5, it follows that $\mathrm{ola}^+(G - u) \leq \mathrm{ola}^+(G) - 1$. Now we apply a procedure similar to that used in the proof of Lemma 5: until $|J_1| \leq 1$, $|J_t| \leq 1$ and $J_2 = \cdots = J_{t-1} = \emptyset$, choose a β-special vertex $u \in V(G')$ for an optimal LA β of G and replace G with $G - u$ and G' with $G' - u$. The procedure will have at most $|V(G')| - n_1 - n_t$ steps each decreasing $\mathrm{ola}^+(G)$ by at least 1. Hence $\mathrm{ola}^+(G) \geq \mathrm{ola}^+(L) + |V(G')| - n_1 - n_t$. □

Let G be a graph and let v be a vertex of degree 2 of G. Let vu_1, vu_2 denote be the edges incident with v. Assume that $u_1u_2 \notin E(G)$. We obtain a graph G' from G by removing v (and the edges vu_1, vu_2) from G and adding instead the edge u_1u_2. We say that G' is obtained from G by *suppressing* vertex v. Furthermore, if the two edges incident with v are k-separating bridges for some positive integer k, then we say that v is *k-suppressible*. The last definition is justified by the following lemma.

Lemma 8. *Let G be a connected graph and let v be an $\mathrm{ola}^+(G)$-suppressible vertex of G. Then $\mathrm{ola}^+(G) = \mathrm{ola}^+(G')$ holds for the graph G' obtained from G by suppressing v.*

Proof. Let u_1, u_2 denote the neighbors of v and let G_1, G_2 denote the connected components of $G - v$ with $u_i \in V(G_i)$, $i = 1, 2$. Consider an optimal LA α of G. As above we use the notation $l_i = \min_{w \in V(G_i)} \alpha(w)$ and $r_i = \max_{w \in V(G_i)} \alpha(w)$, $i = 1, 2$, and we assume, w.l.o.g., that $l_1 < l_2$. Since vu_1, vu_2 are $\mathrm{ola}^+(G)$-separating bridges, Lemma 3 implies that α assigns to the vertices of G_i an interval of consecutive integers. Thus, we conclude that $l_1 < r_1 < \alpha(v) < l_2 < r_2$. We define an LA α' of G' by setting $\alpha'(w) = \alpha(w)$ for $w \in V(G_1)$ and $\alpha'(w) = \alpha(w) - 1$ for $w \in V(G_2)$. Evidently $\mathrm{ola}^+(G') \leq \mathrm{nc}(\alpha', G') = \mathrm{nc}(\alpha, G) = \mathrm{ola}^+(G)$.

Conversely, assume that α' is an optimal LA of G'. We proceed symmetrically to the first part of this proof. Let $l_i = \min_{w \in V(G_i)} \alpha'(w)$ and

$r_i = \max_{w \in V(G_i)} \alpha'(w)$, $i = 1, 2$, and assume, w.l.o.g., that $l_1 < l_2$. Observe that $u_1 u_2$ is an $\mathrm{ola}^+(G')$-separating bridge of G', hence Lemma 3 applies. Thus $l_1 < r_1 < l_2 < r_2$. We define an LA α of G by setting $\alpha(w) = \alpha'(w)$ for $w \in V(G_1)$, $\alpha(v) = r_1 + 1$, and $\alpha'(w) = \alpha(w) + 1$ for $w \in V(G_2)$. Evidently $\mathrm{ola}^+(G) \leq \mathrm{nc}(\alpha', G) = \mathrm{nc}(\alpha', G') = \mathrm{ola}^+(G')$. Hence $\mathrm{ola}^+(G) = \mathrm{ola}^+(G')$ as claimed. □

Theorem 1. *Let k be a positive integer, and let G be a connected graph without k-suppressible vertices. If $\mathrm{ola}^+(G) \leq k$, then G has at most $5k + 2$ vertices and at most $6k + 1$ edges.*

Proof. Let $n = |V(G)| > 1$, and let $\mathrm{ola}^+(G) \leq k$. Any LA of G can have at most $n - 1$ edges of length 1, and each additional edge contributes at least 1 to the net cost. Thus, $m \leq n - 1 + k$ and it suffices to show that $n \leq 5k + 3$. If G does not have a k-separating bridge, then by Lemma 6 we have $n \leq 5k + 1$. Assume now that G has a k-separating bridge. Let $e = uv$ be such a bridge, and let H_1, H_2 be two connected component of $G - e$, where H_1 contains u. Let C^u (C^v) be the bridgeless components containing u (v). Let $C_1^u, C_2^u, \ldots, C_p^u$ ($C_1^v, C_2^v, \ldots, C_q^v$) be all connected components of $H_1 - V(C^u)$ ($H_2 - V(C^v)$). Observe that each of the components C_i^u (C_i^v) is linked to C^u (C^v) by a bridge. Assume that $|V(C_i^x)| \leq |V(C_j^x)|$ for $i < j$, where $x \in \{u, v\}$. By Lemma 5, we have $\sum_{i=1}^{i=p-1} |V(C_i^u)| \leq k$ and $\sum_{i=1}^{i=q-1} |V(C_i^v)| \leq k$. If the bridge between C_p^u and C^u (C_q^v and C^v) is k-separating, we consider the bridgeless component of C_p^u (C_q^v) containing an endvertex of the bridge and the connected components obtained from C_p^u (C_q^v) by deleting the vertices of the bridgeless component. Continuation of the procedure above as long as possible will bring us the following decomposition of G:

1. G has bridgeless components $C_1, C_2, \ldots, C_t$, $t \geq 2$, such that every two consecutive components C_i and C_{i+1} are linked by a single edge e_i, which is a k-separating bridge in G, $i = 1, 2, \ldots, t - 1$.

2. Let $L = G[\bigcup_{i=1}^t V(C_i)]$. The graph $G' = G - V(L)$ has connected components $G_1, G_2, \ldots, G_r$ such that each G_j has edges only to one subgraph $C_{\pi(j)}$, $\pi(j) \in \{1, 2, \ldots, t\}$.

Since we have carried out the above procedure as long as possible, all bridges between G' and L are not k-separating. Thus, $|V(G_j)| \leq k$ for each $j = 1, 2, \ldots, t$. Recall that J_p is the set of indices of all G_j such that $\pi(j) = p$, $p = 1, 2, \ldots, t$, and $n_i = \max\{|V(G_j)| : j \in J_i\}$, $p = 1, 2, \ldots, t$. By Lemma 7, $\mathrm{ola}^+(G) \geq \mathrm{ola}^+(L) + |V(G')| - n_1 - n_t$. Since $n_1 \leq k$, $n_t \leq k$ and $\mathrm{ola}^+(G) \leq k$, we obtain

$$|V(G')| \leq 3k - \mathrm{ola}^+(L). \tag{5}$$

Since G has no k-suppressible vertices, the bridgeless components $C_2, C_3, \ldots, C_{t-1}$ are not trivial. Observe that $\sum_{i=2}^{t-1} \mathrm{ola}^+(C_i) \leq \mathrm{ola}^+(L)$. By Lemma 4, every component $\mathrm{ola}^+(C_i) \geq 1$, $2 \leq i \leq t - 1$, and thus $t - 2 \leq \mathrm{ola}^+(L)$. By Lemma 4, $|V(C_i)| \leq 2 \cdot \mathrm{ola}^+(C_i) + 1$ for each $i = 1, 2, \ldots, t$. Hence,

$$|V(L)| = \sum_{i=1}^t |V(C_i)| \leq 2 \left(\sum_{i=1}^t \mathrm{ola}^+(C_i) \right) + t \leq 3 \cdot \mathrm{ola}^+(L) + 2. \tag{6}$$

Combining (5) and (6), we obtain $|V(G)| = |V(G')| + |V(L)| \le (3k - \mathrm{ola}^+(L)) + (3 \cdot \mathrm{ola}^+(L) + 2) \le 3k + 2 \cdot \mathrm{ola}^+(L) + 2 \le 5k + 2.$ □

Theorem 2. *Let $f(n, m)$ be the time sufficient for checking whether $\mathrm{ola}^+(G) \le k$ for a connected graph G with n vertices and m edges. Then $f(n, m) = O(m + n + f(5k + 2, 6k + 1))$*

Proof. (Sketch.) Using a depth-first-search (DFS) algorithm, we can determine the cut vertices of G in time $O(n+m)$ [19]. Let T be a spanning rooted tree of G (say, as obtained by the DFS algorithm). By a bottom-up traversal of T we find the set S of all k-suppressible vertices of G in time $O(n + m)$. Note that if H is the graph obtained by suppressing some $v \in S$, some vertices of $S \setminus \{v\}$ may not be k-suppressible in H; however, any k-suppressible vertex of H belongs to $S \setminus \{v\}$.

We compute a set $S' \subseteq S$ starting with the empty set and successively adding some of the vertices of S to S'. We visit the vertices of G according to a bottom-up traversal of T (i.e., if v is a descendant of v' then we visit v before v'). During this traversal we assign to each vertex v an integer t'_v which is the number of vertices in S' that belong to the subtree of T rooted at v. Performing a further bottom-up traversal of T we suppress the vertices in S' one after the other, and we are left with a graph G' which has no k-suppressible vertices. If $|V(G')| > 5k + 2$ or $|E(G')| > 6k + 1$, then we know from Theorem 1 that $\mathrm{ola}^+(G') > k$. It follows from Lemma 8 that $\mathrm{ola}^+(G) > k$ as well, and we can reject G. On the other hand, if $|V(G')| \le 5k + 2$ and $|E(G')| \le 6k + 1$, then we can find an optimal LA α' for G' in time $f(5k + 2, 6k + 1)$. By means of the construction in the proof of Lemma 8 we can transform in time $O(n + m)$ the arrangement α' into an optimal LA α of G.   □

3   Computing Optimal Linear Arrangements

Let n and k be nonnegative integers. Let $\mathcal{T}_n$ be the set of trees with n vertices. Let $T \in \mathcal{T}_n$ and let $X \subseteq V(T)$ be arbitrary. Let $OLA_T^+(n, k, X)$ be the set of LAs α of T with net cost at most k and such that $\alpha(x) \in \{1, n\}$ for all $x \in X$. Note that $OLA_T^+(n, k, X) = \emptyset$ if $|X| \ge 3$. Now define $t(n, k, i)$ as follows:

$$t(n, k, i) = \max\{\, |OLA_T^+(n, k, X)| : T \in \mathcal{T}_n, |X| = i \,\}.$$

In other words, no tree T of order n has more than $t(n, k, i)$ LAs such that the net cost is at most k and i prescribed vertices have to be mapped to either 1 or n (and $t(n, k, i)$ is the minimum such value). For a connected graph G, let T_G be a spanning tree of G. Since $\mathrm{ola}^+(T_G) \le \mathrm{ola}^+(G)$ we only have to check all LAs in $OLA_{T_G}^+(n, k, \emptyset)$ (but still considering all edges in G and not just T_G) to decide whether $\mathrm{ola}^+(G) \le k$. Since $|OLA_{T_G}^+(n, k, \emptyset)| \le t(n, k, 0)$ the values of $t(n, k, i)$ are of interest (especially when $i = 0$). We will give an upper bound for $t(n, k, i)$. Proofs are omitted due to space limitations. The proofs show how to generate all LAs in $OLA_{T_G}^+(n, k, \emptyset)$. Note that $t(n, k, 3) = 0$.

Theorem 3. *For all $n \geq 2$, $k \geq 0$ and $0 \leq i \leq 3$, we have the following:*

$$t(n, k, i) \leq 2^{0.119n + 1.96k - 1.4625i + 4}.$$

Using this theorem, we can prove the following result.

Theorem 4. *Let n be the number of vertices in a connected graph G and let k be a nonnegative integer. If $n \leq 5k + 2$, then we can check whether $\mathrm{ola}^+(G) \leq k$ and compute $\mathrm{ola}^+(G)$ provided $\mathrm{ola}^+(G) \leq k$ in time $O(2^{2.5551k})$.*

Now we are ready to prove the main result of this paper.

Theorem 5. *Let $G = (V, E)$ be a graph and let k be a nonnegative integer. We can check whether $\mathrm{ola}^+(G) \leq k$ and compute $\mathrm{ola}^+(G)$ provided $\mathrm{ola}^+(G) \leq k$ in time $O(|V| + |E| + 5.88^k)$.*

Proof. Let $G_1, G_2, \ldots, G_p$ be the connected components of G. We can check, in time $O(|V(G_i)|)$, whether $\mathrm{ola}^+(G_i) = 0$ since $\mathrm{ola}^+(G_i) = 0$ if and only if G_i is a path. Thus, in time $O(|V|)$, we can detect all components of G of net cost zero. By Lemma 1, we do not need to take these components into consideration when computing $\mathrm{ola}^+(G)$. Thus, we may assume that for all components G_i, $i = 1, 2, \ldots, p$, we have $\mathrm{ola}^+(G_i) \geq 1$. Thus, if $\mathrm{ola}^+(G) \leq k$, then $\mathrm{ola}^+(G_i) \leq k - p + 1$. By Lemma 1, Theorems 2 and 4, and the fact that $\mathrm{ola}^+(G_i) \leq k - p + 1$ if $\mathrm{ola}^+(G) \leq k$, we can check whether $\mathrm{ola}^+(G) \leq k$ and compute $\mathrm{ola}^+(G)$ provided $\mathrm{ola}^+(G) \leq k$ in time $O(\sum_{i=1}^{p}(|V(G_i)| + |E(G_i)|) + p2^{2.5551(k-p+1)}) = O(|V| + |E| + 5.88^k)$. □

4   More General Parameterizations of LAP

Serna and Thilikos [18] introduce the following related problems. They ask whether either problem is FPT.

Vertex Average Min Linear Arrangement (VAMLA)
Instance: A graph G. *Parameter:* A positive integer k.
Question: Does G have a linear arrangement of cost at most $k|V(G)|$?

Edge Average Min Linear Arrangement (EAMLA)
Instance: A graph G. *Parameter:* A positive integer k.
Question: Does G have a linear arrangement of cost at most $k|E(G)|$?

Both problems are not FPT (unless P = NP), which follows from the next two theorems.

Theorem 6. *For any fixed integer $k \geq 2$, it is NP-complete to decide whether $\mathrm{ola}(H) \leq k|V(H)|$ for a given graph H.*

Proof. Let G be a graph and let r be an integer. We know that it is NP-complete to decide whether $\mathrm{ola}(G) \leq r$ (LAP). Let $n = |V(G)|$. Let k be a fixed integer, $k \geq 2$. Define G' as follows: G' contains k copies of G, j isolated vertices and a clique with i vertices (all of these subgraphs of G' are vertex disjoint). We have $n' = |V(G')| = kn + i + j$.

By the definition of G' and the fact that $\mathrm{ola}(K_i) = \binom{i+1}{3}$, we have $k \cdot \mathrm{ola}(G) = \mathrm{ola}(G') - \mathrm{ola}(K_i) = \mathrm{ola}(G') - \binom{i+1}{3}$. Therefore, $\mathrm{ola}(G) \le r$ if and only if $\mathrm{ola}(G') \le kr + \binom{i+1}{3}$. If there is a positive integer i such that $kr + \binom{i+1}{3} = kn'$ and the number of vertices in G' is bounded from above by a polynomial in n, then G' provides a reduction from LAP to VAMLA with the fixed k. Observe that $kr + \binom{i+1}{3} \ge k(kn + i)$ for $i = 6kn$. Thus, by setting $i = 6kn$ and $j = r + \frac{1}{k}\binom{i+1}{3} - kn - i$, we ensure that G' exists and the number of vertices in G' is bounded from above by a polynomial in n.   □

The proof of the following theorem is similar, but G' is defined differently: G' contains k copies of G, a path with j edges and a clique with i vertices (all of these subgraphs of G' are vertex disjoint).

Theorem 7. *For any fixed integer $k \ge 2$, it is NP-complete to decide whether $\mathrm{ola}(H) \le k|E(H)|$ for a given graph H.*

The *profile* of a linear arrangement α of a graph G is

$$\mathrm{prf}(\alpha, G) = \sum_{z \in V} (\alpha(z) - \min\{\, \alpha(w) : w \in N[z] \,\});$$

here $N[z]$ denotes the closed neighborhood of vertex z in G. Serna and Thilikos [18] introduce also the following problem and ask whether it is FPT.

Vertex Average Profile (VAP)
Instance: A graph $G = (V, E)$. *Parameter:* A positive integer k.
Question: Does G have a linear arrangement of profile $\le k|V|$?

Similarly to Theorem 6 we can prove that the problem is NP-complete for every fixed $k \ge 2$. Observe that VAMLA, EAMLA and VAP are in para-NP. Moreover, it follows directly form our results that the three problems are para-NP-complete (see Corollary 2.16 in Flum and Grohe's book [10]).

Similarly to Theorem 7 we can prove the following:

Theorem 8. *For each fixed $0 < \epsilon \le 1$, it is NP-complete to decide whether $\mathrm{ola}^+(H) \le |E(H)|^\epsilon$ for a given graph H.*

Notice that Theorem 5 implies that we can decide, in polynomial time, whether $\mathrm{ola}(H) \le |E(H)| + \log|E(H)|$ for a graph H. Theorem 8 indicates that the possibility to strengthen the last result is rather limited. It would be interesting to determine the complexity of the problem of verifying whether $\mathrm{ola}(H) \le |E(H)| + \log^2 |E(H)|$ for a graph H.

References

1. M. Cesati, Compendium of parameterized problems, Sept. 2005. http://bravo.ce.uniroma2.it/home/cesati/research/compendium.pdf
2. F.R.K. Chung, On optimal linear arrangements of trees. *Comp. & Maths. with Appls.* 10 (1984), 43–60.

3. R. Diestel, *Graph Theory*, Springer–Verlag, New York, 2nd ed., 2000.

4. R.G. Downey and M.R. Fellows, *Parameterized Complexity*, Springer–Verlag, New York, 1999.

5. V. Estivill-Castro, M.R. Fellows, M.A. Langston, and F.A. Rosamond, FPT is P-Time extremal structure I. In H. Broersma, M. Johnson, and S. Szeider, editors, *Algorithms and Complexity in Durham 2005, Proceedings of the first ACiD Workshop*, volume 4 of *Texts in Algorithmics*, pages 1–41. King's College Publications, 2005.

6. H. Fernau, *Parameterized Algorithmics: A Graph-theoretic Approach*. Habilitation thesis, U. Tübingen, 2005.

7. H. Fernau, *Parameterized Algorithmics for Linear Arrangement Problems*. Talk at Dagstuhl, July 2005, slides at `http://www.dagstuhl.de/files/Materials/05/05301/05301.FernauHenning.Slides.pdf`

8. H. Fernau, *Parameterized Algorithmics for Linear Arrangement Problems*. Manscript, July 2005, `http://homepages.feis.herts.ac.uk/~comrhf/papers/ola.pdf`

9. J. Flum and M. Grohe, Describing parameterized complexity classes. *Information and Computation* 187 (2003), 291–319.

10. J. Flum and M. Grohe, *Parameterized Complexity Theory*, Springer, 2006.

11. M. R. Garey and D. R. Johnson, *Computers and Intractability*, W.H. Freeman & Comp., New York, 1979.

12. M. R. Garey, D. S. Johnson, and L. Stockmeyer, Some simplified NP-complete graph problems. *Theoret. Comput. Sci.* 1 (1976), 237–267.

13. M.K. Goldberg and I.A. Klipker, *Minimal placing pf trees on a line*. Tech. Report, Physico-Technical Institute of Low Temperatures, Ukranian SSR Acad. of Sciences, USSR, 1976. [In Russian]

14. L.H. Harper, Optimal assignments of numbers to vertices. *J. Soc. Indust. Appl. Math.* 12 (1964) 131–135.

15. M. Mahajan and V. Raman, Parameterizing above guaranteed values: MaxSat and MaxCut. *J. Algorithms* 31 (1999), 335–354.

16. R. Niedermeier. *Invitation to Fixed-Parameter Algorithms*. Oxford Lecture Series in Mathematics and Its Applications. Oxford University Press, 2006. Forthcoming.

17. Y. Shiloach, A minimum linear arrangement algorithm for undirected trees. *SIAM J. Comp.* 8 (1979), 15–32.

18. M. Serna and D.M. Thilikos, Parameterized complexity for graph layout problems. EATCS Bulletin 86 (2005), 41–65.

19. R.E. Tarjan, Depth first search and linear graph algorithms. *SIAM J. Comput.* 1 (1972), 146–160.

Universal Relations and #P-Completeness

Hervé Fournier[1],[*] and Guillaume Malod[2],[*]

[1] Laboratoire PRiSM, Université de Versailles St-Quentin en Yvelines, France
herve.fournier@prism.uvsq.fr
[2] Laboratory of Prof. Masahiko SATO, Graduate School of Informatics,
Kyoto University, Japan
malod@kuis.kyoto-u.ac.jp

Abstract. This paper follows the methodology introduced by Agrawal and Biswas in [AB92], based on a notion of universality for the relations associated with NP-complete problems. The purpose was to study NP-complete problems by examining the effects of reductions on the solution sets of the associated witnessing relations. This provided a useful criterion for NP-completeness while suggesting structural similarities between natural NP-complete problems. We extend these ideas to the class #P. The notion we find also yields a practical criterion for #P-completeness, as illustrated by a varied set of examples, and strengthens the argument for structural homogeneity of natural complete problems.

1 Introduction

Complexity classes such as P, NP or #P are great theoretical notions to further our knowledge of the resources needed to solve computational problems. Their usefulness however goes beyond the theoretical setting, because knowing the right class for a given problem is a precious hint as to the kind of algorithms one should look for.

Agrawal and Biswas [AB92] study the structure of NP-complete sets. In a decision problem one must determine for an instance x whether there exists a y such that $R(x, y)$ holds, where R is the solution checking relation. Agrawal and Biswas focus on the relations $R(x, y)$ to which any other relation can be reduced in a way which preserves solutions, roughly meaning that there is an application between solutions sets, and call them *universal*. In the setting of polynomial time checkable relations, they show that the decision problem corresponding to a universal relation is NP-complete. By giving a simple criterion for NP-completeness based on this definition and applying it to a varied set of examples, they argue that their result underlines a structural similarity between natural NP-complete sets, in the spirit of the work of Berman and Hartmanis [BH77]. Indeed Agrawal and Biswas show that their notion of universality is related to structural properties such as paddability and self-reducibility.

The notion of universality was subsequently used by Buhrman et. al. [BKT94] to provide sufficient conditions for NP optimization problems that admit efficient

---

[*] This work was partially funded as CEFIPRA Project 2602-1.

T. Calamoneri, I. Finocchi, G.F. Italiano (Eds.): CIAC 2006, LNCS 3998, pp. 368–379, 2006.

approximation algorithms. It was subsequently extended by Portier [Por98] to problems defined on an arbitrary structure, in the framework of Poizat [Poi95], and recently Choudhary, Sinha and Biswas [CSB04] defined it for non-deterministic logspace. Our aim in this paper is to adapt this notion to the class #P, which is the class of functions counting the number of solutions for relations checkable in polynomial time, a class at least as hard as the polynomial hierarchy, as shown by Toda [Tod91]. Showing that a function is in #P is thus a convincing argument for its intractability, and a criteria for #P-completeness would be a useful tool. Universality for #P has been studied in [CK03], where an elaborate definition of #P universality is given, based on Valiant's technique for showing #P-completeness (cf. [Val79]) by recovering the coefficients of a polynomial from its value at suitable points. We give here a definition of universality suited to #P which is both simpler and closer to the definition used for NP by Agrawal and Biswas. The relative simplicity translates into a *usable* criterion for #P-completeness.

Section 2 introduces the background notions. Section 3 defines universality for #P and shows that it implies completeness (proposition 1). Section 4 provides the main point of the article, with the #P-universality criterion and the proof that it implies universality (theorem 1). Proposition 1 and theorem 1 together become a practical criterion for #P-completeness, which we apply to examples from different backgrounds in section 5.

2 Definitions and Notations

We present here the framework with which we will work. For each instance x of a problem there is a set C_x of candidate solutions and the set S_x of actual solutions. This is another way of saying that we focus on the relation between an instance x and a candidate solution y which holds iff y belongs to S_x. One can for instance consider the set of 3CNF formulas over the variables $x_1, x_2, \ldots$; for a formula F, the set of candidate solutions is the set of truth-value assignments for the variables in F. The set of solutions is the set of satisfying assignments for F. Another example is the problem of finding a maximal independent set in a graph, i.e. a maximal subset of the vertices such that any two nodes are never connected by an edge. The set of instances is the set of graphs, candidate solutions are subsets of the vertices and solutions are maximal independent sub-graphs.

The complexity of a computational question is the growth of some computational resource when the size of the instance increases, where the size usually means the length of the encoding. In this paper however we will need a slightly more general definition of size, which we will call a measure.

Definition 1. *A measure for a problem is an application m from the set of instances into $\mathbb{N}^k$ for a given k, such that there is a polynomial $p(n_1, \ldots, n_k)$, with $p(m(x))$ bounding the length of the encodings of x and of any possible solution for x. We will write $m(x) + n$ for the tuple $(m_1 + n, \ldots, m_k + n)$ if n is an integer, and $m(x) + m(y)$ for $(m_1 + n_1, \ldots, m_k + n_k)$ if $m(y) = (n_1, \ldots, n_k)$.*

For instance, the number of clauses is a valid measure for a 3CNF formula F, because it gives us a bound on the length of the formula and its number of variables. Encoding a graph of size n by giving its adjacency matrix yields a word of length $O(n^2)$, and therefore taking the number n of vertices is a valid measure for graphs. We could also have chosen to measure a graph with two integers, one being the number of vertices and the other being the number of edges. The idea behind measures is that sometimes the "size" of an instance depends on several independent parameters.

Definition 2. *A relation is called a P-relation if the associated solution checking problem can be checked in deterministic polynomial time, if the number of candidate solutions is simply exponential in the measure of an instance, and if it satisfies the two conditions on names and renaming, as described below.*

The first condition is that the encoding chosen include "names" in the following sense. There exists an application which, given the encoding of an instance x, returns a set V_x of integers such that the set of candidate solutions for x can be identified with the powerset of V_x. We will thus not distinguish between candidate solutions and subsets of V_x. Let us return to our examples to clarify this. Consider a 3CNF formula ϕ, encoded in such a way that the variables are numbered. Then the set V_ϕ described above is simply the set of the numbers of the variables appearing in the formula. An assignment gives the value 0 or 1 to a variable, and can be described by listing only the variables which are given value 1, thus identifying the set of candidate solutions with the powerset of V_ϕ. For the maximal independent set problem, we can choose an encoding which labels the vertices of a graph G with integers; V_G is the set of integers labeling a vertex in G; candidate solutions are subsets of the vertices, i.e. they are subsets of V_G. This formalizes an intuition of Agrawal and Biswas taken from [AB92]: natural problems often consist of "atomic" units joined together and such that a solution is a subset of the atomic units satisfying certain properties.

The second condition is purely technical condition but will be natural on specific examples. We call it the *renaming property*. We suppose that there is a polynomial time computable function which, given an instance x of X, an integer $a \in V_x$ and an integer $b \notin V_x$, computes an instance y of X such that $V_y = (V_x \backslash \{a\}) \cup \{b\}$ and x is isomorphic to y by a bijection which maps a unto b. In effect, the renaming property states that we can arbitrarily rename a variable in a formula or a vertex in a graph, providing it does not clash with an existing name.

3   Universality for #P

Let us now give a definition of universality adapted to #P. In the definition, $|A|$ denotes the cardinality of the set A.

Definition 3. *Let X be a P-relation. It is said to be #P-universal if for any P-relation Y there exists a polynomial time computable function which, given an instance y of Y, computes an instance x of X, two integers $k, M > 0$ such that*

$M \cdot 2^{|V_y|} < 2^k$, *and a map* α *from* V_y *to* V_x *such that for all* $t \subseteq V_y$, *if it is a solution for* y, *then*

$$|\{s \in S_x \mid \alpha(t) = s \cap \alpha(V_y)\}| \equiv M \pmod{2^k},$$

and otherwise (if t *is not a solution for* y*), then*

$$|\{s \in S_x \mid \alpha(t) = s \cap \alpha(V_y)\}| \equiv 0 \pmod{2^k}.$$

For example, suppose that we have a reduction from the maximal independent problem to 3SAT. This means that given a graph G we can compute a 3SAT formula ϕ and a map α from the vertices of G to the variables of ϕ such that: any maximal independent set of G yields exactly M satisfying assignments for ϕ via the map α, and the number of remaining satisfying assignments for ϕ is 0, all these numbers being modulo an adequate power of 2. Now if we know the number of solutions of ϕ we can compute the number of maximal independent sets of G. In the general case this gives us the following proposition.

Proposition 1. *The counting problem associated to a #P-universal relation is #P-complete for Cook[1]-reductions.*

Proof. Let X be a #P-universal relation and g the associated counting function: $g(x) = |S_x|$. The function g is obviously in #P. Let Y be a P-relation and h the associated counting function. By definition of universality, there is a computable time function which given an instance y of Y computes two integers $k, M > 0$ and an application α with the above properties. Thus,

$$|S_x| = \sum_{t \subseteq V_y} |\{s \in S_x \mid \alpha(t) = s \cap \alpha(V_y)\}|$$

$$= \sum_{t \in S_y} |\{s \in S_x \mid \alpha(t) = s \cap \alpha(V_y)\}| + \sum_{t \notin S_y} |\{s \in S_x \mid \alpha(t) = s \cap \alpha(V_y)\}|$$

$$\equiv M \cdot |S_y| \pmod{2^k}.$$

As $M \cdot 2^{|V_y|} < 2^k$ by definition, we have $M \cdot |S_y| < 2^k$. Thus $h(y) = (g(x) \bmod 2^k)/M$, and this computation can be done in polynomial time.

4 Sufficient Conditions for Universality

Suppose that X is a P-relation and that there exists an integer $k_0 \in \mathbb{N} \setminus \{0\}$ such that the following three properties hold:

Block. There exist $M_b \in \mathbb{N} \setminus \{0\}$ and a polynomial time computable function which, given an integer $k \geqslant k_0$ in unary encoding, computes an instance b of X, elements $d_1, d_2, d_3 \in V_b$ and a subset t of $\{d_1, d_2, d_3\}$ such that:

- $|\{s \in S_b \mid s \cap \{d_1, d_2, d_3\} = t\}| \equiv 0 \pmod{2^k}$.
- for any subset u of $\{d_1, d_2, d_3\}$ different from t, $\big|\{s \in S_b \mid s \cap \{d_1, d_2, d_3\} = u\}\big| \equiv M_b \pmod{2^k}$.

Join. There exist $M_j \in \mathbb{N} \setminus \{0\}$ and a polynomial time computable function which, given two instances x_1, x_2 of X, such that V_{x_1} and V_{x_2} are disjoint, and $k \geqslant k_0$ in unary encoding, computes an instance x of X such that:

- $V_{x_1} \cup V_{x_2} \subseteq V_x$.
- if $s_1 \in S_{x_1}$ and $s_2 \in S_{x_2}$, then

$$|\{s \in S_x \mid s \cap V_{x_1} = s_1 \text{ and } s \cap V_{x_2} = s_2\}| \equiv M_j \pmod{2^k}.$$

- $|\{s \in S_x \mid s \cap V_{x_1} \notin S_{x_1} \text{ or } s \cap V_{x_2} \notin S_{x_2}\}| \equiv 0 \pmod{2^k}$.
- $m(x) \leqslant m(x_1) + m(x_2) + k^{O(1)}$.

Couple. There exist $M_c \in \mathbb{N} \setminus \{0\}$ and a polynomial time computable function which, given an integer $k \geqslant k_0$ in unary encoding, an instance x of X, and $a, b \in V_x$ with $a \neq b$, computes an instance y of X such that:

- $V_x \subseteq V_y$.
- for all $s \in S_x$, if exactly one of a or b belongs to s, then

$$|\{t \in S_y \mid t \cap V_x = s\}| \equiv M_c \pmod{2^k},$$

otherwise:

$$|\{t \in S_y \mid t \cap V_x = s\}| \equiv 0 \pmod{2^k}.$$

- $m(y) \leqslant m(x) + k^{O(1)}$.

The example of 3SAT should help understand these conditions and show that they can be very intuitive in the case of specific examples.

Block. Consider the clause $\phi = d_1 \vee d_2 \vee d_3$, then V_ϕ is $\{d_1, d_2, d_3\}$ and $t = \emptyset$ is the only subset which does not yield a solution for ϕ; all the other subsets of $\{d_1, d_2, d_3\}$ yield exactly one solution. The general case extends this in the following ways: computations must hold only modulo a given power of 2; the special subset t may be different from $\emptyset$; the other subsets are not constrained to yielding exactly one solution but a constant number M_b instead.

Join. Given two 3SAT formulas ϕ_1 and ϕ_2 with distinct variables, the conjunction $\psi = \phi_1 \wedge \phi_2$ is a 3SAT formula such that: a solution for ϕ_1 coupled with a solution for ϕ_2 yields exactly one solution for ψ; the number of other solutions for ψ is 0; the measure of ψ is bounded by the sum of the measures of ϕ_1 and ϕ_2. Differences in the general case: computing modulo a given 2^k, getting a constant number of solutions M_j for each couple, allowing the measure to increase polynomially in k.

Couple. Given a 3SAT formula ϕ and two variables a and b in ϕ, the formula $\psi = \phi \wedge (a \vee b \vee b) \wedge (\neg a \vee \neg b \vee \neg b)$ is such that any solution of ϕ which satisfies $(a \text{ XOR } b)$ yields exactly one solution for ψ and the number of other solutions for ψ is 0. Differences in the general case: computations modulo a given 2^k, getting a constant number of solutions M_j for each solution of ϕ, allowing the measure of the new instance to increase polynomially in k.

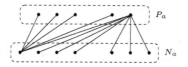

Fig. 1. Coupling the variables in F_2

Theorem 1. *If a P-relation has the above properties, it is #P-universal.*

Proof. Let X be a P-relation which has the properties detailed above. We wish to show that it is #P-universal. Consider another P-relation Y. We follow the steps given in the introduction in greater detail.

Because 3SAT is #P-complete for parsimonious reductions, from any instance y of Y we can compute in polynomial time a 3CNF-formula F_0 such that there is a bijection between the satisfying assignments of F_0 and the solutions for y. Following the construction in [BDG88], one can build an application from V_y to V_{F_0} which yields this bijection between the solution sets. The measure of F_0 is polynomial in the measure of y.

For i in $\{1, 2, 3\}$, define ν_i as $\emptyset$ if $d_i \in t$ and $\neg$ otherwise. We add clauses $\nu_1 a \vee \neg \nu_2 a \vee \epsilon a$ for each variable a in F_0, where ϵ is $\emptyset$ if ν_1 is $\neg$ and $\neg$ otherwise. The measure of the formula F_1 thus obtained is still polynomial in the measure of y and it has the same set of solutions as F_0, because these additional clauses are always satisfied.

We then replace each clause in the following manner. Call n the measure of F_1, i.e. F_1 has n clauses $C_1, \ldots, C_n$, and suppose each clause C_i is of the form $\mu_{i,1} a_{i,1} \vee \mu_{i,2} a_{i,2} \vee \mu_{i,3} a_{i,3}$, where the $a_{i,j}$ are variables and the $\mu_{i,j}$ are either $\neg$ or $\emptyset$. We introduce three new variables $c_{i,1}, c_{i,2}, c_{i,3}$ and replace C_i by the clause $\nu_1 c_{i,1} \vee \nu_2 c_{i,2} \vee \nu_3 c_{i,3}$. If ν_j and $\mu_{i,j}$ have the same value, we say that $c_{i,j}$ represents $a_{i,j}$; otherwise we say that $c_{i,j}$ represents $\neg a_{i,j}$. Call the resulting formula F_2. It has the same measure as F_1.

For each variable a in F_1 let P_a be the set of variables of F_2 which represent a and N_a the set of variables which represent $\neg a$. Both sets are non-empty because of the clauses we added when we built F_1. We shall consider the solutions which are coupled according to the graph in figure 1, where the top vertices are the variables in P_a, the bottom vertices are the variables in N_a and the edges means that the variables corresponding to the vertices are coupled. All the variables in P_a are therefore coupled to the first variable in N_a and all the variables in N_a are coupled to the last variable in P_a. There is a bijection between the set of solutions of F_1 and the set of solutions of F_2 which are coupled in the previous manner for all variables in F_1. Indeed any coupled assignment which is a solution of F_2 gives the same value to all the variables representing a given variable a, and the same opposite value to all those representing $\neg a$. The construction ensures that if we give this first value to a, and build an assignment for F_1 by treating all the variables in a similar way, all the clauses of F_1 are satisfied because all the clauses of F_2 are satisfied. The reciprocal construction of a unique satisfying assignment of F_2 from a satisfying assignment of F_1 is easy.

Let us now choose the smallest integer k such that $M_b^n M_j^{n-1} M_c^{3n} \cdot 2^{|V_v|} < 2^k$. We build n instances of block b, called $b_1, \ldots, b_n$ and rename them with the renaming property so that the associated sets $V_{b_1}, \ldots, V_{b_n}$ are pairwise disjoint and the specified elements in each block b_i are called $d_{i,1} d_{i,2}, d_{i,3}$. We bring these instances together with $n-1$ join operations to get an instance x_1. There is an application from V_{F_2} to V_{x_1}, defined by mapping $c_{i,j}$ to $d_{i,j}$, which is such that any solution of F_2 yields a number of solutions for x_1 which is congruent to $M_b^n M_j^{n-1}$ modulo 2^k. We perform all the couplings (suppose there are l) and get an instance x_2. For each variable a in F_1 there is at least one variable in F_2 which represents a. This variable can be mapped to an element of V_{x_2} via the previous mapping. We thus get an application from V_{F_1} to V_{x_2} such that a solution for V_{F_1} yields a number of solutions for x_2 which is congruent to $M_b^n M_j^{n-1} M_c^l$ modulo 2^k. Using the application from V_y to V_{F_0} (and thus to V_{F_1}) we obtain the application from V_y to V_{x_1} required in the definition.

We must however check that this computation can be done in polynomial time. Remark that at most $3n$ coupling operations are necessary. We first check that the size of k is polynomially bounded in the measure of y. We know that Y is a P-problem, so that $|V_y|$ is bounded by $p(m(y))$ for some polynomial p. The number n, which is the number of clauses of F_2, is also polynomially bounded in $m(y)$. The inequation $M_b^n M_j^{n-1} M_c^{3n} \cdot 2^{|V_v|} < 2^k$ can thus be satisfied by an integer k whose value, and therefore whose unary encoding, is polynomial in $m(y)$. We should finally check the size of the resulting instance x. Thanks to the growth conditions in the three properties, it is bounded by $n \cdot m(b) + (n-1)k^{O(1)} + 3nk^{O(1)}$. The measure of the instance b, which is computed from k (in unary encoding) in polynomial time, is also polynomially bounded in $m(y)$.

5   Examples

The first obvious example would be 3SAT: the necessary arguments have been given just after the criterion for universality. Further examples are given in this section. Monotone 2SAT is a logical problem and an example of a relation whose decision problem is in P but whose counting problem is #P-complete. Maximal independent set and Hamiltonian cycles are graph problems; the proofs that they satisfy the criterion are very short and boil down to finding the right graph gadgets. Knapsack is an application of the criterion in yet another setting.

5.1   Monotone 2SAT

Monotone 2SAT is similar to 3SAT, but it demands a little more work. Monotone 2SAT instances are 2CNF formulas without negative literals. Let $k_0 = 1$.

Block. For $k \geqslant k_0$, the formula corresponding to the block (modulo 2^k) is the following one: $B = \bigwedge_{i=1}^{k} (d_1 \vee u_i) \wedge (d_2 \vee u_i) \wedge (d_3 \vee u_i)$. If at least one of the variables d_1, d_2 or d_3 is false, then all the u_i must be true for an assignment to satisfy B. On the other hand, if all three d_i are true, then there are 2^k ways to satisfy B. This corresponds to the definition of Block with the subset $t = \{d_1, d_2, d_3\}$.

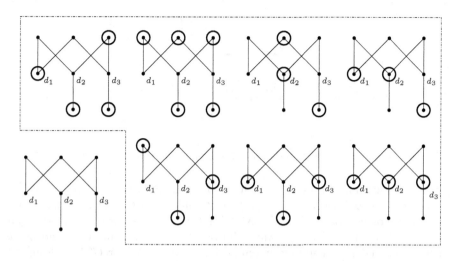

Fig. 2. Block for maximal independent set and its solutions

Join. The join of F and G is the formula $F \wedge G$.

Couple. Coupling x_a and x_b in the formula $F(\bar{x})$ (modulo 2^k) is done by the formula $F(\bar{x}) \wedge (x_a \vee x_b) \wedge \bigwedge_{i=1}^{k}(x_a \vee u_i) \wedge (x_b \vee u_i)$. This ensures first that x_a and x_b cannot both be false. If both are true there are 2^k possible assignments for the variables u_i.

5.2 Maximal Independent Set

We focus now on a graph problem. The definition, measure and name set have been given in the introduction. Let $k_0 = 2$.

Block. $M_b = 1$. For any $k \geqslant k_0$, consider the graph G of figure 2. This figure also shows all maximal independent subsets of G: none of them contain d_2 and d_3 but not d_1, but all other subsets of $\{d_1, d_2, d_3\}$ correspond to exactly one solution.

Join. Suppose we have an integer $k \geqslant k_0$ and two disjoint graphs G_1 and G_2. Consider the graph G which is the union of G_1 and G_2. There is a bijection between the solution set of G and the Cartesian product of the solution sets of G_1 and G_2, so that the required cardinality condition holds for $M_j = 1$. The growth condition is obvious.

Couple. Suppose we have an integer $k \geqslant k_0$, a graph G, and two vertices a and b appearing in G. Consider the graph G' obtained from G in the following manner. We first add the edge (a, b) if it is not already present. Then we add $2k$ vertices $\{u_1, \ldots, u_k\}$ and $\{v_1, \ldots, v_k\}$. At last, we add the $3k$ edges (a, u_i), (u_i, v_i) and (v_i, b) for $1 \leqslant i \leqslant k$, as shown on figure 3.

The set of solutions for G' can be partitioned in the following way. Any solution for G which contains neither a nor b yields exactly 2^k solutions of G',

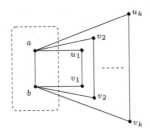

Fig. 3. Couple for maximal independent set

because any maximal subset of G' which contains neither a nor b must contain exactly one vertex in each pair (u_i, v_i). Any solution for G which contains both a and b cannot be extended to a maximal independent set for G' because a and b are linked by an edge in G'. Any solution for G which contains a and not b can be extended into a unique maximal independent set for G' by adding all the vertices v_i. The symmetrical situation for a and b is similar. Therefore if a solution for G is not coupled with regard to vertices a and b, then it yields either 0 or a multiple 2^k solutions of G'. A coupled solution for G yields exactly one solution for G'. Therefore $M_c = 1$. As for the growth condition, we have added $2k$ vertices to G, so that the measure of G' is bounded by the measure of G plus k^2 for $k \geqslant k_0 = 2$.

5.3 Hamiltonian Cycles

An instance for this problem is a graph and a solution is a subset of the edges which is a Hamiltonian cycle. Figure 4 gives the Block, the Join construction and a XOR gadget which we will use to couple.

Block. One should check that there is exactly one Hamiltonian cycle for each non-empty subset of $\{d_1, d_2, d_3\}$, and no Hamiltonian cycle avoiding these three edges.

Join. Suppose we now wish to join two instances G and G'. We choose a vertex s in G and split into two vertices s_1 and s_2. All the outgoing edges of s become outgoing edges of s_1 and all the incoming edges of s become incoming edges of s_2, so that there is now a one-to-one correspondence between Hamiltonian cycles of G and Hamiltonian paths from s_1 to s_2 in the new graph. We modify G' in

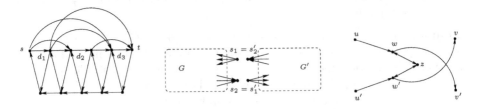

Fig. 4. Block, Join and XOR for Hamiltonian cycles

the same way, splitting a node s' into s'_1 and s'_2. We then identify s_2 and s'_1, and s'_2 and s_1. Any Hamiltonian cycle of this graph is made of a Hamiltonian path of G from s_1 to s_2 and a Hamiltonian path of G' from s'_1 to s'_2.

Couple. If we wish to couple edges (u, v) and (u', v') in a graph G, we start by deleting these edges and connect the vertices u, v, u' and v' with the XOR gadget. This gadget is such that the Hamiltonian cycles of the resulting graph must contain one of the edges (u, w) or (u', w') but not both.

5.4 Knapsack

Here is an example from a different setting. An instance of Knapsack is given by a set of integer weights $c_1, \ldots, c_n$ and an integer b called the sum, where the c_i and b are all strictly positive. A solution is a subset s of $\{1, \ldots, n\}$ such that $\sum_{i \in s} c_i = b$. The measure of an instance will be given by two integers: n (the number of integers c_i) and the bitsize of $b + \sum_{i \in \{1, \ldots, n\}} c_i$.

Block. Consider the instance with weights $1, 1, 1, 1, 2$ and sum 4. Solutions for this instance cannot omit the first three weights. Moreover, for any subset of these three weights, there is only one way to complement the sum to 4.

Join. Let us take two instances $a_1, \ldots, a_j, a$ and $b_1, \ldots, b_k, b$. Call S and T respectively the sums $a + \sum_{i=1}^{j} a_i$ and $b + \sum_{i=1}^{k} b_i$. Let our new instance be composed of the weights $a_1, \ldots, a_j, Sb_1, \ldots, Sb_k$ and sum $a + Sb$. The first integer in the measure of this new instance is $j + k$. The second is the bitsize of the following integer:

$$b + \sum_{i=1}^{j} a_i + S \cdot \left(\sum_{i=1}^{k} b_i \right) + S \cdot b = S(T + 1).$$

The measure can thus be bounded by the sum of the measures of the two initial instances plus a constant. It is easy to see that there is a bijection between couples of solutions for the initial instances and solutions for the new one. Indeed, suppose we have a subset J of $\{1, \ldots, j\}$ and a subset K of $\{1, \ldots, k\}$ such that $\sum_{i \in J} a_i + S \cdot \left(\sum_{i \in K} b_i \right) = a + Sb$, then we have the following equation: $\sum_{i \in J} a_i - a = S \cdot \left(b - \sum_{i \in K} b_i \right)$. The absolute value of the left-hand side is strictly smaller than S, while on the right-hand side it is either 0 or greater than S. Therefore it must be 0 on both sides, and J and K yield solutions for the initial instances.

Couple. Suppose we have an instance $a_1, \ldots, a_k, a$ and we wish to couple a_i and a_j. Let S be the sum $a + \sum_{l=1}^{k} a_l$. We consider the new instance obtained by replacing the weight a_i is with $a_i + S$, the weight a_j with $a_j + S$ and the sum a with $a + S$. Any solution for this new instance cannot omit or include both $a_i + S$ and $a_j + S$. Now suppose we have a solution which includes only one of them, for instance we have a subset K of $\{1, \ldots, k\} \setminus \{i\}$ such that $a_i + S + \sum_{l \in K} a_l = a + S$. Then $a_i + \sum_{l \in K} a_l = a$ and we get a solution for the initial instance. The measure of this new instance is composed of k and the bitsize of $\sum_{l=1}^{k} a_l + a + 3S = 4S$, and therefore can be bounded by the initial measure plus a constant.

6   Conclusion

There are two ways to see the work done in this paper. On a theoretical level, it argues for the existence of structural similarities between difficult problems, an idea which has been studied for NP and which is here applied to #P. As such it is an attempt to better understand why some problems are easy and some are difficult. Our work shows basically two ingredients for a relation to yield a #P-complete problem. One is a kind of building block/inductive structure, already noticed by Agrawal and Biswas in the case of NP, with stricter conditions on solution sets in order to adapt it to #P. The other is the possibility to compute modulo a big integer, as is crucial for instance in the proof of the completeness of the Permanent. If we consider the first ingredient, finding the closest possible (w.r.t. NP) structural criteria for #P-universality is a good way to study the famous question of whether all relations which yield NP-complete decision problems also yield #P-complete counting problems. The second ingredients is a rough explanation of why some easy decision problems have #P-complete counting equivalents. The interplay between these two ingredients makes the class #P rich and interesting.

The other focus of this work, which is perhaps more apparent in this short version, is to give a useful criterion for #P-completeness. When one wishes to prove that a problem is #P-hard, one often tries to find a known #P-hard problem which seems "near" enough, so that the reduction will be easier to exhibit. Our criterion takes advantage of the argument from the previous paragraph, namely the existence of common structure, to eliminate the search for a suitable known #P-hard problem. In other words the universality criterion plays the role of a generic #P-hard problem, but one which should be "close" enough in most cases, because the distance is bridged by theorem 1. There may well be a #P-hard problem more suitable for a given example, i.e. yielding a simpler reduction, but we believe that the universality criterion corresponds to a large class of natural problems, as hinted at by the variety of examples, for which proofs can be built in a systematic way.

References

[AB92]   Manindra Agrawal and Somenath Biswas. Universal relations. In *Structure in Complexity Theory Conference*, pages 207–220, 1992.

[BDG88]   J. L. Balcázar, J. Díaz, and J. Gabarró. *Structural complexity 1*. Springer-Verlag New York, Inc., New York, NY, USA, 1988.

[BH77]   Leonard Berman and Juris Hartmanis. On isomorphisms and density of NP and other complete sets. *SIAM J. Comput.*, 6(2):305–322, 1977.

[BKT94]   Harry Buhrman, Jim Kadin, and Thomas Thierauf. On functions computable with nonadaptive queries to NP. In *Structure in Complexity Theory Conference*, pages 43–52, 1994.

[CK03]   G. Chakravorty and R. Kumar. #P universality. Technical report, Indian Institute of Technology, 2003.

[CSB04]   V. Chaudhary, A. K. Sinha, and S. Biswas. Universality for nondeterministic logspace. presented at Indo-German Workshop on Algorithms, Bangalore, 2004.

[Pap94]   Christos H. Papadimitriou. *Computational complexity*. Addison-Wesley Publishing Company, Reading, MA, 1994.

[Poi95]   Bruno Poizat. *Les Petits Caillloux*, volume 3 of *Nur Al-Mantiq Wal-Ma'rifah*. Aléas, Lyon, 1995.

[Por98]   Natacha Portier. Résolutions universelles pour des problèmes NP-complets. *Theor. Comput. Sci.*, 201(1-2):137–150, 1998.

[Tod91]   Seinosuke Toda. PP is as hard as the polynomial-time hierarchy. *SIAM J. Comput.*, 20(5):865–877, 1991.

[Val79]   Leslie G. Valiant. The complexity of enumeration and reliability problems. *SIAM J. Comput.*, 8(3):410–421, 1979.

Locally 2-Dimensional Sperner Problems Complete for the Polynomial Parity Argument Classes*

Katalin Friedl[1], Gábor Ivanyos[2], Miklos Santha[3], and Yves F. Verhoeven[3,4]

[1] BME, H-1521 Budapest, P.O.Box 91., Hungary
friedl@cs.bme.hu
[2] MTA SZTAKI, H-1518 Budapest, P.O. Box 63., Hungary
Gabor.Ivanyos@sztaki.hu
[3] CNRS–LRI, UMR 8623, bâtiment 490, Université Paris XI, 91405 Orsay, France
santha@lri.fr
[4] ENST, 46 rue Barrault, 75013 Paris, France
yves.verhoeven@normalesup.org

Abstract. In this paper, we define three Sperner problems on specific surfaces and prove that they are complete respectively for the classes PPAD, PPADS and PPA. This is the first time that locally 2-dimensional Sperner problems are proved to be complete for any of the polynomial parity argument classes.

1 Introduction

The complexity class TFNP, the family of all total NP-search problems, was introduced by Megiddo and Papadimitriou [9]. It contains several important, computationally probably hard problems for which no classical polynomial time algorithms are known. On the other hand, these problems are also somewhat easy in the sense that they can not be NP-hard unless NP = co-NP. The class TFNP is a semantic complexity class and thus doesn't seem to have complete problems. It is therefore natural to look for syntactically definable subclasses of TFNP. Indeed, several such subclasses have been identified along the lines of the mathematical proofs establishing the existence of a solution. The important subclasses Polynomial Pigeonhole Principle (PPP) and Polynomial Local Search (PLS) were defined respectively in [12] and [7]. The elements of PPP are problems which by their combinatorial nature obey the pigeonhole principle and therefore have a solution. In a PLS problem, one is looking for a local optimum for a particular objective function, in some easily computable neighborhood structure.

The parity argument subclasses PPA, PPAD, and PSK of TFNP were defined by Papadimitriou in [11, 12]. The class PSK was renamed PPADS in [1]. These

---

* Research supported by the EU 5th framework programs RESQ IST-2001-37559, Centre of Excellence ICAI-CT-2000-70025, the EU 6th framework program QAP, the OTKA grants T42559, T46234, and by the ACI CR 2002-40, ACI SI 2003-24, ANR Blanc AlgoQP grants of the French Research Ministry.

T. Calamoneri, I. Finocchi, G.F. Italiano (Eds.): CIAC 2006, LNCS 3998, pp. 380–391, 2006.

classes can be characterized by some simple graph theoretical principles. The class Polynomial Parity Argument (PPA) is the class of NP search problems, where the existence of the solution is guaranteed by the fact that in every finite graph whose vertices are of degree at most two, the number of leaves is even. The class PPAD is the directed version of PPA, and its basic search problem is the following: in a directed graph, where the in-degree and the out-degree of every vertex is at most one, given a source, find another source or a sink. In the class PPADS the basic search problem is more restricted than in PPAD: given a source, find a sink.

Another point that makes the parity argument classes interesting is that there are several natural problems from different branches of mathematics that belong to them. For example, in a graph with odd degrees, when a Hamiltonian path is given, a theorem of Smith [15] ensures that there is another Hamiltonian path. It turns out that finding this second path belongs to the class PPA [12]. A search problem coming from a modulo 2 version of Chevalley's theorem [12] from number theory is also in PPA. Complete problems in PPAD are the search versions of Brouwer's fixed point theorem, Kakutani's fixed point theorem, Borsuk-Ulam theorem, and Nash equilibrium (see [12]).

The classical Sperner's Lemma [14] states that in a triangle with a regular triangulation whose vertices are labeled with three colors, there is always a trichromatic triangle. This lemma is of special interest since some customary proofs for the above topological fixed point theorems rely on its combinatorial content. However, it is unknown whether the corresponding search problem, that Papadimitriou [12] calls **2D-SPERNER**, is complete in PPAD. Variants of Sperner's Lemma also give rise to other problems in the parity argument classes. Papadimitriou [12] has proved that a 3-dimensional analogue of **2D-SPERNER** is in fact complete in PPAD. In [6], Grigni described a non-oriented version of 3-dimensional Sperner's Lemma that is complete for the class PPA. In this paper we show that appropriately chosen locally 2-dimensional versions of the problem are already complete for PPAD, for PPADS, and for PPA, respectively.

This work was completed early 2005 [5]. Recently it has been announced by Chen and Deng that they have proven the PPAD completeness of **2D-SPERNER** in reference 2 in [2].

2   Results

An *NP-search problem* is specified by a polynomial time relation $\mathcal{R}(x, y)$, such that for some polynomial $p(n)$, for every x and y such that $\mathcal{R}(x, y)$, we have $|y| \leq p(|x|)$. Given an input x to the problem, the task is to find a y such that $\mathcal{R}(x, y)$ if there is one, and else report failure. We call an NP-search problem *total* if for every x there exists a solution y. The class of total NP-search problems is called TFNP by Megiddo and Papadimitriou [9].

For two problems $\mathcal{R}_1$, $\mathcal{R}_2$ in TFNP, we say that $\mathcal{R}_1$ is *reducible to* $\mathcal{R}_2$ if there exist two functions f and g computable in polynomial time such that $f(x)$

is a legal input to $\mathcal{R}_2$ whenever x is an input to $\mathcal{R}_1$, and $\mathcal{R}_2(f(x), y)$ implies $\mathcal{R}_1(x, g(x, y))$.

The parity argument classes are defined via concrete problems, by closure under reduction. The **LEAF** problem is defined as follows. The input is a pair $(M, 0^k)$ where M is the description of a polynomial time Turing machine that on every input outputs a set of size at most 2, and k is a positive integer. Moreover, M is such that $M(0^k) = \{1^k\}$, and $0^k \in M(1^k)$. Such an input specifies an undirected graph $G_k = (V, E)$, where $V = \{0, 1\}^k$, and $\{u, v\}$ is in E if $u \in M(v)$, and $v \in M(u)$. The output of the problem is a leaf of G_k different from 0^k. The class PPA is the set of total search problems reducible to **LEAF**. In the search problems defining the classes PPADS and PPAD, the Turing machine defines a directed graph, where the in-degree and the out-degree of every vertex is at most one, and where 0^k is always a source. The output in the case of PPADS is a sink, and in the case of PPAD a sink or source different from 0^k.

After some preliminaries in Section 3 the definitions of the three Sperner problems of interest for us will be given in Section 4: **OSPS** and **SOSPS** for the oriented cases, and **SPS** for the non-oriented case. Our main results are proven in Section 5: **OSPS** is complete for PPAD (**Theorem 2**) and **SOSPS** is complete for PPADS (**Theorem 3**). The proof of the completeness of **SPS** for PPA is left for the full paper.

The results of this paper are motivated by an open problem of Papadimitriou in [12], asking whether **2D-SPERNER** is PPAD-complete. The main reason why the 3-dimensional Sperner problem could be proved complete in PPAD is that there exists an embedding of the complete graph of any size in the 3-dimensional Euclidean space without any two edges crossing. Of course, such an embedding is impossible in the plane, and it is not clear how to circumvent this difficulty when one tries to extend Papadimitriou's proof in 2 dimensions. Our approach consists in exhibiting such an embedding in compact 2-dimensional manifolds, i.e. surfaces, of non-zero genus, and proving the completeness of Sperner problems on these surfaces for the classes PPAD, PPADS and PPA. Therefore, our results show that the difficulty of the Sperner problems is independent of the local dimension of the instance, if it is at least 2.

3   Preliminaries

Unless otherwise stated, the graphs considered in the paper will be undirected. If S is any set, $\equiv$ is an equivalence relation over S, and a is an element of S, then $[a]_\equiv$ denotes the equivalence class of a in S for the relation $\equiv$.

3.1   Surfaces

Definition 1 (triangles). *Let $\mathfrak{R}$ be the equivalence relation over triples of distinct elements such that we have $(a, b, c) \, \mathfrak{R} \, (a', b', c')$ if (a', b', c') is obtained from (a, b, c) by cyclic permutation. An equivalence class T of $\mathfrak{R}$ is called a triangle. If T is the equivalence class of (a, b, c), then $\overline{T}$ denotes the equivalence class of (a, c, b).*

For a pair (a,b), let $\overline{(a,b)}$ denote the pair (b,a). For every triangle T and elements a and b, $(a,b) \prec T$ indicates that there exists an element c such that $(a,b,c)\,\mathfrak{R}\,T$, and $\{a,b\} \prec T$ indicates that either $(a,b) \prec T$ or $\overline{(a,b)} \prec T$.

A finite set of triangles $\mathcal{T}$ is called a triangle arrangement. If $\mathcal{T}$ is a triangle arrangement, its skeleton graph $G_{\mathcal{T}}$ is the graph $G_{\mathcal{T}} = (V,E)$, where $V = \bigcup_{T \in \mathcal{T}} T$, and $\{a,b\}$ is an edge if there is a triangle $T \in \mathcal{T}$ such that $\{a,b\} \prec T$. A vertex (resp. edge) of $\mathcal{T}$ is a vertex (resp. edge) of the skeleton graph of $\mathcal{T}$.

We will often specify a triangle T, which is an equivalence class of $\mathfrak{R}$, by a an element of T.

Definition 2 (pseudosurfaces). A pseudo-surface $\mathcal{T}$ is a triangle arrangement $\mathcal{T}$ such that for every edge (a,b) of E there are at most two different triangles $T \in \mathcal{T}$ such that $\{a,b\} \prec T$. The pseudo-surface $\mathcal{T}$ is oriented if for every two triangles T and T' in $\mathcal{T}$ and every edge $\{a,b\} \in E$, when $(a,b) \prec T$ and $(a,b) \prec T'$ we have $T = T'$. The boundary of $\mathcal{T}$, denoted by $\partial \mathcal{T}$, is the set of all edges $e \in E$ for which there exists exactly one triangle $T \in \mathcal{T}$ with $e \prec T$. The dual graph $H_{\mathcal{T}}$ of $\mathcal{T}$ is the graph $H_{\mathcal{T}} = (\mathcal{T}, E')$ such that there is an edge between two triangles $T \neq T'$ in $H_{\mathcal{T}}$ if there are two vertices a and b in $\mathcal{T}$ such that $\{a,b\} \prec T$ and $\{a,b\} \prec T'$.

Definition 3. A surface $\mathcal{S}$ is a pseudo-surface such that $H_{\mathcal{S}}$ is connected and $\partial \mathcal{S}$ is a union of disjoint cycles of $G_{\mathcal{S}}$.

Notice that our definition of surface coincides with the usual definition of triangulated surface.

3.2 Flow Graphs

Definition 4. Let $\mathcal{S}$ be a surface, V be set of vertices of $\mathcal{S}$, $H_{\mathcal{S}} = (V',E')$ be its dual graph. A function $\ell : V \to \{0,1,2\}$ is called a labeling of $\mathcal{S}$. A triangle $T \in \mathcal{S}$ is said to be fully labeled if it is equivalent to a triple (a,b,c) such that $\{\ell(a),\ell(b),\ell(c)\} = \{0,1,2\}$. A fully labeled triangle T has direct orientation if there exists (a,b,c) in its equivalence class such that $(\ell(a),\ell(b),\ell(c)) = (0,1,2)$. Otherwise, it has indirect orientation.

The undirected flow graph $U_{\mathcal{S}} = (V',E'')$ of $\mathcal{S}$ (relatively to ℓ) is a subgraph of $H_{\mathcal{S}}$, such that there is an edge between two triangles T and T' of $\mathcal{S}$ if there are two vertices a and b of $\mathcal{S}$ such that $\{a,b\} \prec T$, $\{a,b\} \prec T'$, and $\{\ell(a),\ell(b)\} = \{0,1\}$.

If $\mathcal{S}$ is oriented, then we define the directed flow graph $D_{\mathcal{S}} = (V',E''')$ of $\mathcal{S}$ (relatively to ℓ) as a a directed graph, such that there is an edge between two triangles T and T' of $\mathcal{S}$ if there are two vertices a and b of $\mathcal{S}$ such that $(a,b) \prec T$, $\overline{(a,b)} \prec T'$, and $(\ell(a),\ell(b)) = (0,1)$.

The proof of the following theorem is straightforward.

Theorem 1. Let $\mathcal{S}$ be a surface, and ℓ be a labeling of $\mathcal{S}$. Then,

(i) the degree of every vertex of the undirected flow graph $U_{\mathcal{S}}$ is at most 2,
(ii) if $\mathcal{S}$ is oriented, then the in-degree and out-degree of every vertex of the directed flow graph $D_{\mathcal{S}}$ are at most 1.

Corollary 1 (Sperner's lemma for surfaces with empty boundary). *Let S be a surface with empty boundary, and ℓ be a labeling of S. Then,*

(i) the number of fully labeled triangles in the undirected flow graph U_S is even,

(ii) if S is oriented, then there are as many fully labeled triangles with direct orientation as fully labeled triangles with indirect orientation in the directed flow graph D_S.

Proof. First, observe that the fully labeled triangles in S are exactly the nodes of degree 1 in U_S, and that the fully labeled triangles having direct (*resp.* indirect) orientation in S are exactly the nodes of out-degree (*resp.* in-degree) 1 in D_S. Since by Theorem 1 (*i*) in U_S the maximal degree is at most two, the number vertices having degree 1 is even. By Theorem 1 (*ii*) in D_S the in- and outdegrees are at most 1, therefore there has to be the same number of sources as sinks.

3.3   Rotation Systems

Definition 5. *Let $G = (V, E)$ be a graph. For every vertex $v \in V$, a local rotation of G at v is a cyclic permutation π_v of the neighbors of v in G. A rotation system for G is a set $\Pi = \{\pi_v \mid v \in V\}$ of local rotations. Let T be a triangle arrangement, and v be a vertex of T. A local rotation π_v of G_T at v is a local orientation of T at v if, for every neighbor v' of v in G_T, $(v', v, \pi_v(v'))$ is a triangle of T.*

Fact 1. *Let S be an oriented surface with empty boundary, and let v be a vertex of S. There exists a unique local orientation π_v of S at v such that, for every neighbor v' of v in G_S, $(v', v, \pi_v(v'))$ is a triangle of S.*

Definition 6. *Let S be an oriented surface with empty boundary. The rotation system defined in Fact 1 is called the rotation system of S.*

Definition 7. *Let $(G_n)_{n \in \mathbb{N}} = (V_n, E_n)_{n \in \mathbb{N}}$ be a family of undirected graphs where $|V_n| = n$, and $\Pi_n = \{\pi_v \mid v \in V_n\}$ be a rotation system for G_n. The rotation system Π_n is said to be efficiently computable if there exists a Turing machine M such that*

(i) on input n and pair (v, v'), with $\{v, v'\} \in E_n$, computes the vertices v'' and v''' such that $\pi_v(v') = v''$ and $\pi_v^{-1}(v') = v'''$ using time polynomial in $\log n$,

(ii) on input n and triple (v, v', v''), with $\{v, v'\}$ and $\{v, v''\}$ in E_n, computes the smallest non-negative integer i such that $\pi_v^i(v') = v''$ using time polynomial in $\log n$. Later, we will refer to the integer i by $\log_{v'}^{\pi_v}(v'')$.

Lemma 1. *If m is an integer that is equal to 7 modulo 12, then the complete graph K_m is the skeleton graph of an oriented surface S_m with empty boundary. Moreover, the rotation system of S_m can be efficiently computed.*

The surface S_m is completely specified by giving an appropriate rotation system for K_m. There are actually several such rotation systems [3, 8]. The proof of the efficient computability of the rotation system is straightforward. It is based on the constructions in [10, 3]. We omit the details.

3.4   Regular Subdivisions

In the following definition, we will formalize the notion of "a regular subdivision" of a surface, which consists in substituting every triangle of the surface with a "regular subdivision" of it, as shown on Figure 1, such that the small triangles of the subdivision have the same orientation as the large triangle that is subdivided.

We will make use of the free Abelian monoid $\mathbb{N}[V]$ over the set of vertices V of a surface $\mathcal{S}$: the elements are those of the form $\sum_{v \in V} c_v \cdot v$, where c_v is a non-negative integer, and v is a vertex of $\mathcal{S}$. For any subset $V' \subseteq V$ and positive integer r let $\mathbb{N}_r[V']$ denote those elements $\sum_{v \in V'} c_v \cdot v$ of $\mathbb{N}[V]$ such that $\sum_{v \in V'} c_v = r$. If $s = \sum_{v \in V} s_v \cdot v$ and $t = \sum_{v \in V} t_v \cdot v$ are two elements of $\mathbb{N}[V]$, we denote by $d(s, t)$ the distance $1/2 \sum_{v \in V} |s_v - t_v|$.

Definition 8. *Let $\mathcal{S}$ be a surface, and r be a positive integer. Let $\mathcal{S}^{(r)}$ be a triangle arrangement whose triangles are of the form (s_1, s_2, s_3) with $\{s_1, s_2, s_3\} \subseteq \mathbb{N}_r[\{a, b, c\}]$, for some triangle (a, b, c) in $\mathcal{S}$, such that there exists $\varepsilon \in \{-1, 1\}$ with $s_2 = s_1 + \varepsilon(a - b)$ and $s_3 = s_1 + \varepsilon(a - c)$. We call $\mathcal{S}^{(r)}$ the regular r-subdivision of $\mathcal{S}$.*

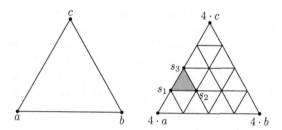

Fig. 1. A triangle (a, b, c) and its regular 4-subdivision

Notice that two vertices of $\mathcal{S}^{(r)}$ are neighbors if and only if they are at distance 1. It implies that the distance between two vertices in the skeleton graph of $\mathcal{S}^{(r)}$ is equal to their distance according to d.

4   Sperner Problems

The NP-search problems for which we prove completeness in Section 5 are the following. The surface $\mathcal{S}_m$ is the one given by Lemma 1. Its skeleton graph is K_m. The surface $\mathcal{S}_m^{(4)}$ is the regular 4-subdivision of $\mathcal{S}_m$.

Oriented Sperner Problem for the Surface $\mathcal{S}_m^{(4)}$ (OSPS)

Input: an integer m equal to 7 modulo 12, the description of a Turing machine M that on input vertex v of $\mathcal{S}_m^{(4)}$ outputs a label $\ell(v)$ in $\{0, 1, 2\}$ using time polynomial in $\log m$, and also a fully labeled triangle T of $\mathcal{S}_m^{(4)}$, which has indirect orientation.

Output: a fully labeled triangle $T' \neq T$ of $\mathcal{S}_m^{(4)}$.

Strict Oriented Sperner Problem for the Surface $\mathcal{S}_m^{(4)}$ (SOSPS)

Input: an integer m equal to 7 modulo 12, the description of a Turing machine M that on input vertex v of $\mathcal{S}_m^{(4)}$ outputs a label $\ell(v)$ in $\{0, 1, 2\}$ using time polynomial in $\log m$, and also a fully labeled triangle T of $\mathcal{S}_m^{(4)}$, which has indirect orientation.

Output: a fully labeled triangle T' of $\mathcal{S}_m^{(4)}$, which has direct orientation.

To prove completeness for a non-oriented Sperner problem, we will use the non-oriented surface $\mathcal{N}_m$, derived from the regular 12-subdivision $\mathcal{S}_m^{(12)}$ of $\mathcal{S}_m$ by adding some cross-caps. Its precise definition will not be given in this extended abstract.

Sperner Problem for the Surface $\mathcal{N}_m^{(12)}$ (SPS)

Input: an integer m equal to 7 modulo 12, the description of a Turing machine M that on input vertex v of $\mathcal{N}_m^{(12)}$ outputs a label $\ell(v)$ in $\{0, 1, 2\}$ using time polynomial in $\log m$, and also a fully labeled triangle T of $\mathcal{N}_m^{(12)}$.

Output: a fully labeled triangle $T' \neq T$ of $\mathcal{N}_m^{(12)}$.

We would like to emphasize that these Sperner problems are in fact not promise problems, since the input requirements can be syntactically enforced. Let us describe this in details for the case of **OSPS**. We can easily provide a syntactical way to force the Turing machine to always give a correct output. For instance, one can assume that every output value not in $\{0, 1, 2\}$ is interpreted as 0. We can also ensure syntactically that T is a fully labeled triangle which has indirect orientation with the help of an arbitrary polynomial time computable total order $<$ on the vertices of $\mathcal{S}_m^{(4)}$. Let $s_1 < s_2 < s_3$ be the vertices of T. The label of s_3 is fixed to 2. The vertex s_1 will get label 0 and s_2 label 1 if (s_1, s_2, s_3) is in the equivalence class T, and the labels are exchanged in the opposite case.

In fact, the membership of each of these problems in the class TFNP follows immediately from Corollary 1.

5   Completeness Results for Oriented Sperner Problems

Let m be a positive integer equal to 7 modulo 12. We will work with the regular 4-subdivision $\mathcal{S}_m^{(4)}$ of $\mathcal{S}_m$.

Theorem 2. *The problem* **OSPS** *is PPAD-complete.*

Proof. To see membership in PPAD, we reduce **OSPS** to the natural complete problem for PPAD. First, notice that from Theorem 1, we know that the directed flow graph $D_{\mathcal{S}_m^{(4)}}$ has in- and out-degree at most 1 at every vertex. Notice also that, given a polynomial Turing machine that outputs the label of vertices of $\mathcal{S}_m^{(4)}$, it is easy to design a polynomial time Turing machine that, given a vertex T in the directed flow graph $D_{\mathcal{S}_m^{(4)}}$ outputs its predecessor and its successor, if they exist: the Turing machine only has to calculate the labels of the vertices in T, and to calculate which are the neighbors of T in $H_{\mathcal{S}_m^{(4)}}$. Finally, observe that, as we previously mentioned in the proof of Corollary 1, the fully labeled triangles having direct (*resp.* indirect) orientation in S are exactly the nodes of out-degree (*resp.* in-degree) 1 in $D_{S_m^{(12)}}$. These three arguments show that there is a reduction (in the sense of total problems) from **OSPS** to the natural complete problem for **PPAD**.

We turn to the proof of completeness. Let k be any positive integer. Let $G = (V, E)$ be a graph which is specified by an instance of the natural complete problem for PPAD (see Section 2). It is an undirected graph over $V = \{0, 1\}^k$, such that each vertex has in-degree at most one, and out-degree at most one. Moreover, 0^k is a source in G. Let us denote by M the polynomial time Turing machine that, given a vertex $v \in V$, outputs its predecessor and its successor, if they exist. From G we make an instance of **OSPS** such that a solution can be efficiently turned into a source or a sink of the graph G different from 0^k.

Let m be the smallest integer greater than 2^k that is equal to 7 modulo 12. We assume that V is included in the set of vertices of S_m. We denote by $\Pi = \{\pi_v \mid v \text{ vertex of } S_m\}$ the rotation system for S_m.

Informally, we give a labeling such that the directed flow graph $D_{S_m^{(4)}}$ imitates the graph G as follows: if (a, b) is an edge of G, then there will be a path in $D_{S_m^{(4)}}$ along the edges near the (a, b) side of the triangle "above" (a, b) (that is the triangle $\{a, b, \pi_a^{-1}(b)\}$). If moreover (b, c) is an edge in G then there will be a path around b in the direction given by the rotation system, leading to the triangle above (b, c). To manage the latter, we need a tool for deciding whether, for a vertex $d \notin \{a, b, c\}$, the edge $\{b, d\}$ is "between" $\{a, b\}$ and $\{b, c\}$ according to the rotation π_b. This tool is provided by the function $\log_a^{\pi_b}$ defined in Definition 7: the edge $\{b, d\}$ is between $\{a, b\}$ and $\{b, c\}$ if $0 < \log_a^{\pi_b}(d) < \log_a^{\pi_b}(c)$. The function $\log_a^{\pi_b}$ is efficiently computable by Lemma 1.

We design a Turing machine M' that for every vertex v in $S_m^{(4)}$ outputs a label $\ell(v)$ in $\{0, 1, 2\}$, using M as a subroutine. Let (a, b, c) be a triangle in S_m, $S = \{a, b, c\}$, and let i_a, i_b and i_c be three non-negative integers such that $i_a + i_b + i_c = 4$. Denote by σ the permutation $\binom{a,b,c}{b,c,a}$. Observe that the definition of the rotation system implies that for every $v \in \{a, b, c\}$ the equality $\pi_v(\sigma^{-1}(v)) = \sigma(v)$ holds. On input $z = i_a \cdot a + i_b \cdot b + i_c \cdot c$ the Turing machine M' outputs

$$
\ell(z) = \begin{cases}
0 & \text{if } \exists v, v' \in S, \, i_v + i_{v'} = 4, \, (v, v') \in E, & (1) \\
0 & \text{if } \exists v \in S, \, i_v = 4, \, \exists w \notin S, \, (v, w) \in E \text{ or } (w, v) \in E, & (2) \\
1 & \text{if } \exists v \in S, \, (i_v, i_{\sigma(v)}) \in \{(2,1), (1,2)\}, \, (v, \sigma(v)) \in E, & (3) \\
1 & \text{if } \exists v \in S, \, \exists v' \in \{\sigma^{-1}(v), \sigma(v)\}, \, (i_v, i_{v'}) = (3, 1), & \\
& \quad \exists w, w' \in V, \, (w, v), (v, w') \in E \text{ and } \log_w^{\pi_v}(v') < \log_w^{\pi_v}(w'), & (4) \\
2 & \text{otherwise.} & (5)
\end{cases}
$$

Notice that conditions 1 and 2 can be matched simultaneously, but the value of ℓ is the same. Notice also that, although less obvious, it is impossible for conditions 1 and 4 to be matched simultaneously. The other pairs of conditions can not be matched simultaneously.

Finding the case in which z falls can be done in time polynomial in k, as the Turing machine M, on input $v \in \{a, b, c\}$, outputs the neighbors of v, and the rotation system Π can be efficiently computed.

Using these rules, we describe (see Figure 2) the possible cases for a triangle (a, b, c) in S_m (we assume that the rotation system is clockwise, and hence the orientation is counter-clockwise):

Case 1: $(a, b), (b, c), (c, a) \in E$.

Case 2: $(a, b), (b, c) \in E$, but $(c, a) \notin E$. The value of $\ell(3 \cdot a + c)$ is 2 if a is a source in G, and 1 otherwise. Similarly, the value of $\ell(3 \cdot c + a)$ is 2 if c is a sink in G, and 1 otherwise.

Case 3: $(a, b) \in E$, but (b, c) and (c, a) are not in E. The value of $\ell(4 \cdot c)$ is 2 if c is isolated in G, and otherwise 0. The value of $\ell(a + 3 \cdot c) = \ell(b + 3 \cdot c)$ is 1 if $\log_w^{\pi_c}(b) < \log_w^{\pi_c}(a) < \log_w^{\pi_c}(w')$, and otherwise 2. The value of $\ell(3 \cdot a + c)$ is 2 if a is a source in G, and 1 otherwise. The value of $\ell(3 \cdot b + c)$ is 2 if b is a sink in G, and 1 otherwise.

Case 4: $(a, b), (b, c)$ and (c, a) are not in E. Let v be in $\{a, b, c\}$. We do not enumerate all the possible sub-cases, but only state the essential relations between the labels:

  (i) $\ell(3 \cdot v + \sigma(v)) = 1 \iff \ell(3 \cdot v + \sigma^{-1}(v)) = 1$, as both $3 \cdot v + \sigma^{-1}(v)$ and $3 \cdot v + \sigma(v)$ simultaneously fall in one of the cases (1), (4) and (5) in the definition of ℓ.

  (ii) $\ell(3 \cdot v + \sigma(v)) = 0 \iff \ell(2 \cdot v + 2 \cdot \sigma(v)) = 0$, as if $\ell(3 \cdot v + \sigma(v)) = 0$ or $\ell(2 \cdot v + 2 \cdot \sigma(v)) = 0$ then case (1) in the definition of ℓ must apply,

  (iii) $\ell(3 \cdot v + \sigma^{-1}(v)) = 0 \iff \ell(2 \cdot v + 2 \cdot \sigma^{-1}(v)) = 0$, for the same reasons as in *(ii)*.

These are the only possible cases, up to renaming the vertices a, b and c of the triangle (a, b, c).

We have to prove that this labeling scheme ℓ is correctly defined among different triangles. It is easy to check that it is correctly defined on $4 \cdot v$, where v is a vertex of V: if v is an isolated vertex in G, then in every face to which it belongs only the case (5) in the definition of ℓ applies, and therefore $\ell(4 \cdot v) = 2$. If v is not isolated, then case (2) in the definition of ℓ applies, and therefore $\ell(4 \cdot v) = 0$.

So, finally, proving that the labeling has been correctly defined amounts to proving that the label $\ell(z)$ of a vertex $z = i_a \cdot a + i_b \cdot b$, $0 < i_a, i_b < 4$ with $i_a + i_b = 4$, that we have defined is the same for the two triangles $(a, b, \pi_a^{-1}(b))$ and $(a, \pi_a(b), b)$. We study the different cases:

  − $(i_a, i_b) = (3, 1)$ or $(1, 3)$: if (a, b) or (b, a) is in E, then case (1) in the definition of ℓ applies to z, and $\ell(z) = 0$. Otherwise, either case (4) applies and therefore $\ell(z) = 1$, or case (5) applies and therefore $\ell(z) = 2$.

  − $(i_a, i_b) = (2, 2)$: if (a, b) or (b, a) is in E, then case (1) in the definition of ℓ applies to z, and $\ell(z) = 0$. Otherwise, case (5) applies.

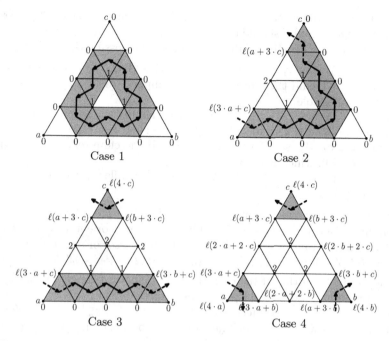

Fig. 2. The different possible cases in the labeling of a triangle (a, b, c) of $\mathcal{S}_m^{(4)}$

Let (a', b', c') be a fully labeled triangle in the subdivision of a triangle (a, b, c) in $\mathcal{S}_m$. We prove that there exists a unique $v = v(a', b', c') \in \{a, b, c\}$ such that $(a', b', c') = (3 \cdot v + \sigma(v), 2 \cdot v + \sigma^{-1}(v) + \sigma(v), 3 \cdot v + \sigma^{-1}(v))$, and v is a source in G if (a', b', c') is a fully labeled triangle having indirect orientation, and a sink if (a', b', c') is a fully labeled triangle having direct orientation. Also, given (a', b', c'), one can efficiently retrieve $v(a', b', c')$. The proof is done for the different cases of Figure 2.

In Case 1 there is no such triangle (a', b', c').

Let us examine Case 2. The possible values for $\ell(3 \cdot a + c)$ and $\ell(a + 3 \cdot c)$ are 1 and 2. Therefore, the only possibilities for (a', b', c') are $(b + 3 \cdot c, a + b + 2 \cdot c, a + 3 \cdot c)$ when $\ell(a + 3 \cdot c) = 2$, and $(3 \cdot a + c, 2 \cdot a + b + c, 3 \cdot a + b)$ when $\ell(3 \cdot a + c) = 2$. These values correspond respectively to the case when c is a sink and (a', b', c') is a fully labeled triangle having direct orientation, and to the case when a is a source and (a', b', c') is a fully labeled triangle having indirect orientation.

Let us turn to Case 3. In this case, we always have $\ell(a + 3 \cdot c) = \ell(b + 3 \cdot c)$. So, the only possibilities for (a', b', c') are $(3 \cdot b + a, a + 2 \cdot b + c, 3 \cdot b + c)$ when $\ell(3 \cdot b + c) = 2$, and $(3 \cdot a + c, 2 \cdot a + b + c, 3 \cdot a + b)$ when $\ell(3 \cdot a + c) = 2$. These values correspond respectively to the case when b is a sink and (a', b', c') is a fully labeled triangle having direct orientation, and to the case when a is a source and (a', b', c') is a fully labeled triangle having indirect orientation.

We finish the case study by proving that in Case 4, there can be no fully labeled triangle (a', b', c'). All the triangles that have twice the label 2 can immediately be discarded. By symmetry between a, b and c, we can assume without loss of generality that a', b' and c' should be in $\{i_a \cdot a + i_b \cdot b + i_c \cdot c \in \mathbb{N}_4[\{a,b,c\}] \mid i_a \geq 2\}$. Assume that (a',b',c') is a fully labeled triangle. The possibilities are:

- $(a',b',c') = (3 \cdot a + c, 4 \cdot a, 3 \cdot a + b)$: $\ell(4 \cdot a) \in \{0,2\}$, so $\ell(3 \cdot a + c) = 1$ or $\ell(3 \cdot a + b) = 1$, and therefore relation *(i)* implies $\ell(3 \cdot a + c) = \ell(3 \cdot a + b)$, which is impossible,
- $(a',b',c') = (3 \cdot a + c, 3 \cdot a + b, 2 \cdot a + b + c)$: similar to the previous case,
- $(a',b',c') = (2 \cdot a + 2 \cdot c, 3 \cdot a + c, 2 \cdot a + b + c)$: $\ell(2 \cdot a + 2 \cdot c) \in \{0,2\}$ and $\ell(2 \cdot a + b + c) = 2$, so $\ell(2 \cdot a + 2 \cdot c) = 0$ and therefore relation *(ii)* implies $\ell(3 \cdot a + c) = 0$, which is impossible,
- $(a',b',c') = (3 \cdot a + b, 2 \cdot a + 2 \cdot b, 2 \cdot a + b + c)$: similar to the previous case, using relation *(iii)*.

Our next step is showing that the map $(a',b',c') \mapsto v(a',b',c')$ is a bijection between fully labeled triangles (a',b',c') having indirect orientation and sources of G. It is onto, as if v is a source in G, v' is the successor of v and $v'' = \pi_v^{-1}(v')$ then $v = v(a',b',c')$, where $(a',b',c') = (3 \cdot v + v'', 2 \cdot v + v' + v'', 3 \cdot v + v')$. The case study also shows that if (a',b',c') is a fully labeled triangle of $\mathcal{S}_m^{(4)}$ having indirect orientation and $v = v(a',b',c')$ then $(a',b',c') = (3 \cdot v + v'', 2 \cdot v + v' + v'', 3 \cdot v + v')$, where v' is the successor of v in G and $v'' = \pi_v^{-1}(v')$. Therefore the map is injective as well. A similar bijection exists between fully labeled triangles (a',b',c') having direct orientation and sinks of G.

Let $(a_0, b_0, c_0) = (0^k, 1^k, \pi_{0^k}^{-1}(1^k))$. The triangle T, which is part of the input for **OSPS**, is $(3 \cdot a_0 + c, 2 \cdot a_0 + b_0 + c_0, 3 \cdot a_0 + b_0)$. We conclude that if we can find a fully labeled triangle $(a',b',c') \neq T$ then we can efficiently retrieve a source or sink v of G with $v = v(a',b',c')$ different from 0^k.

The problem **SOPS** is the hand-made analogue of **OSPS** for PPADS, and therefore it is naturally complete in the class. Indeed, both for the easiness and the hardness results the proofs used for the completeness of **OSPS** in PPAD can be applied. The only substantial remark to be made is that the bijections defined between fully labeled triangles and nodes of degree one bijectively map fully labeled triangles having direct orientations onto sinks. Therefore, we obtain the following result.

Theorem 3. *The problem* **SOSPS** *is PPADS-complete.*

Acknowledgments

The last two authors acknowledge the hospitality of the Isaac Newton Institute, Cambridge, where their work was completed.

References

1. P. Beame, S. Cook, J. Edmonds, R. Impagliazzo and T. Pitassi. The relative complexity of NP search problems. *J. Comput. System Sci.*, 57(1):3–19, 1998.
2. X. Chen and X. Deng. Settling the Complexity of 2-Player Nash-Equilibrium. *ECCC Report* 140, 2005.

3. L. Goddyn, R. Bruce Richter and Jozef Širáň. Triangular embeddings of complete graphs from graceful labellings of paths. Preprint, 2004.

4. K. Fan. Simplicial maps from an orientable n-pseudomanifold into S^m with the octahedral triangulation. *J. Combinatorial Theory*, 2:588–602, 1967.

5. K. Friedl, G. Ivanyos, M. Santha and Y. F. Verhoeven. On the complexity of Sperner's Lemma. Isaac Newton Institute Preprint Series NI05002, 2005.

6. M. Grigni. A Sperner lemma complete for PPA. *Inform. Process. Lett.*, 77(5-6):255–259, 2001.

7. D. Johnson, C. Papadimitriou and M. Yannakakis. How easy is local search? *J. Comput. System Sci.*, 37(1):79–100, 1988.

8. V. P. Korzhik and H.-J. Voss, On the Number of Nonisomorphic Orientable Regular Embeddings of Complete Graphs, *J. Combinatorial Theory Series B* **81**:*58-76 (2001)*.

9. N. Megiddo and C. Papadimitriou. On total functions, existence theorems and computational complexity. *Theoret. Comput. Sci.*, 81:317–324, 1991.

10. B. Mohar and C. Thomassen. *Graphs on surfaces*. Johns Hopkins Studies in the Mathematical Sciences. Johns Hopkins University Press, Baltimore, MD, 2001.

11. C. Papadimitriou. On graph-theoretic lemmata and complexity classes. *In Proc. of 31st FOCS*, pp. 794–801, 1990.

12. C. Papadimitriou. On the complexity of the parity argument and other inefficient proofs of existence. *J. Comput. System Sci.*, 48(3):498–532, 1994.

13. G. Ringel, *Map Color Theorem*, Springer-Verlag, New York, 1974.

14. E. Sperner. Neuer Beweis für die Invarianz der Dimensionzahl und des Gebietes. *Abh. Math. Sem. Hamburg Univ.* 6:265–272, 1928.

15. A. Thomason. Hamilton cycles and uniquely edge colourable graphs. *Ann. Discrete Math.* 3: 259–268, 1978.

Author Index

Lecture Notes in Computer Science

For information about Vols. 1–3899

please contact your bookseller or Springer